FOREST RESOURCE ECONOMICS AND FINANCE

McGraw-Hill Series in Forest Resources

Paul V. Ellefson, University of Minnesota, Consulting Editor

Avery and Burkhart: Forest Measurements
Dana and Fairfax: Forest and Range Policy
Daniel, Helms, and Baker: Principles of Silviculture
Davis and Johnson: Forest Management
Duerr: Introduction to Forest Resource Economics
Ellefson: Forest Resources Policy: Process, Participants, and Programs
Harlow, Harrar, Hardin, and White: Textbook of Dendrology
Klemperer: Forest Resource Economics and Finance
Laarman and Sedjo: Global Forests: Issues for Six Billion People
Nyland: Silviculture: Concepts and Applications
Piirto: Study Guide to Accompany Sharpe, Hendee, Sharpe, and Hendee: Introduction to Forest and Renewable Resources
Sharpe, Hendee, Sharpe, and Hendee: Introduction to Forest and Renewable Resources
Sinclair: Forest Products Marketing
Stoddard, Smith, and Box: Range Management

FOREST RESOURCE ECONOMICS AND FINANCE

W. David Klemperer
Virginia Polytechnic Institute and State University
College of Forestry and Wildlife Resources

McGraw-Hill, Inc.
New York St. Louis San Francisco Auckland Bogotá
Caracas Lisbon London Madrid Mexico City Milan Montreal
New Delhi San Juan Singapore Sydney Tokyo Toronto

This book was set in Times Roman by Publication Services, Inc.
The editors were Anne C. Duffy and David Dunham;
the production supervisor was Paula Keller.
Project supervision was done by Publication Services, Inc.
R. R. Donnelley & Sons Company was printer and binder.

FOREST RESOURCE ECONOMICS AND FINANCE

This book is printed on acid-free paper.

1 2 3 4 5 6 7 8 9 0 DOC DOC 9 0 9 8 7 6 5

ISBN 0-07-035122-8

Library of Congress Cataloging-in-Publication Data

Klemperer, W. David
Forest resource economics and finance / W. David Klemperer.
p. cm.
Includes bibliographical references (p.).
ISBN 0-07-035122-8
1. Forests and forestry—Economic aspects. 2. Forests and forestry—Finance. I. Title.
SD393.K65 1996
634.9′068′1—dc20 95-13668

ABOUT THE AUTHOR

W. DAVID KLEMPERER is professor of forest economics in the Forestry Department at Virginia Polytechnic Institute and State University. His 33 years of professional forestry experience also include work as a county extension forester in Washington state, a forest economist for Associated Oregon Industries, and a self-employed consulting forest economist in Oregon. He has lectured extensively throughout the United States and abroad, has published widely in forestry, economics, and appraisal journals, and has often been a consultant to government and industry. Klemperer has served on numerous boards and committees with the Society of American Foresters, the International Union of Forestry Research Organizations, and other groups and is a member of the American Economic Association. He received a B.S. from the New York State College of Forestry at Syracuse University, and a master's and Ph.D. in forest economics from Oregon State University.

CONTENTS IN BRIEF

CONTENTS

PREFACE

I've written this book for undergraduate forestry students and as a reference for practicing foresters. I assume readers have taken high school algebra, elementary statistics, and at least a one-semester introductory economics course emphasizing microeconomics, although the text reviews some basics. For those whose elementary economics is rusty, I'd encourage a review, especially of microeconomics, in an introductory economics book, examples of which are listed at the end of Chapter 2. I've emphasized solving basic economic problems forestry analysts encounter and learning enough economic theory to understand the solutions. This book is not a substitute for intermediate theory courses in microeconomics and macroeconomics, which should be taken by students aiming for graduate study in forest economics. Since only a small fraction of forestry students choose the latter path, and since students taking forest economics are usually not economics majors, I've stressed practical applications at the expense of certain finer theoretical points. Purists will miss things like indifference curves, price ratios, Edgeworth boxes, contract curves, social welfare functions, continuous time finance, and calculus.

Footnotes and appendixes cover details which aren't vital to understanding the basics but might be interesting to more advanced students. For smoothness in reading, I've minimized citations. A limited number of references and suggested readings are at each chapter's end. The glossary defines specialized terms, which are in bold print the first time they appear in each chapter.

I've included some finance topics often covered in forest management courses because the basics in these areas are applied in most of the later chapters. Also, in forest management courses at Virginia Tech and elsewhere, so much time is spent on computer-based mathematical programming, simulation, and forest-wide planning that some of the investment topics previously stressed in management courses are now covered more fully in the economics class. But the book is organized so that many sections can be skipped without losing continuity.

To list all who have helped and influenced me in writing this book would be a herculean task. I owe an enormous intellectual debt to my colleagues in forestry, fisheries, wildlife, economics, agricultural economics, and business, to people in industry, government, and landowner associations, to my former teachers, and to my undergraduate and graduate students who I hope have learned from me as much as I have learned from them. I am grateful to the following, who reviewed sections of earlier drafts and gave helpful advice. McGraw-Hill and I would like to thank: Robert Abt, North Carolina State University; J. Douglas Brodie, Oregon State University; Hugh Canham, SUNY, College of Environmental Science & Forestry, Syracuse; Paul Ellefson, University of Minnesota; D. Lester Holley, North Carolina State University; William Hoover, Purdue University; John Houghton, University of Wisconsin, Stevens Point; Theodore Howard, University of New Hampshire; Charles McKetta, University of Idaho; Douglas Rideout, Colorado State University; Jeffrey Stier, University of Wisconsin–Madison; and Richard Thompson, California Polytechnic, San Luis Obispo. I would also like to thank Otis Hall, Harry Haney, Carol Hyldahl, Kristen Klemperer, Jack Muench, Jay Sullivan, and Harold Wisdom for their valuable suggestions.

TO THE INSTRUCTOR

This book includes more material than a 3 or 4 semester-credit undergraduate course usually covers, so instructors will need to be selective. Also, certain sections may be more advanced than some instructors might wish for an undergraduate course. Examples of sections to omit are given below. Selected chapters as well as appendixes and footnotes could be suitable as a review with extensions for beginning master's degree students in some programs. Several chapters stand on their own and can be omitted, for example, Chapter 10 on risk analysis, 11 on valuation and appraisal, 13 on the wood processing industry, 16 on regional economics, and 17 on world forestry. Or you can drop sections, as outlined below.

In Chapter 1 you could skip the sections "Forest Outputs and Users" and "Forest Resources in the United States." Although I feel Chapter 3's topics of market failures, environmental economics, optimal damage, and liability rules are vital facets of modern forestry, parts of the chapter can be skipped without loss of continuity (for example, "Damage Fines," "Transactions Costs," and "Income Effects").

If you want less emphasis on finance, skip portions of Chapter 4, but you have to be careful, since many later chapters use several of the payment series formulas. In Chapter 5 on inflation, some later sections are independent and can be skipped, (e.g., "Comparing Forests with Fixed Income Investments," or "Inflation and Housing Demand"). In Chapter 6 on capital budgeting, you could drop "Capital Rationing," "A Summary of Capital Budgeting," and "Applications in Forest Damage Control." In Chapter 11, the initial material on the difference between market value and an individual investor's valuation could be used without the rest of the chapter. Chapter 9, on taxes could be covered in another course, or certain sections could be skipped, for example, the last five sections. You can use Chapter 16 without "Input-Output Analysis" and "Applications." The foregoing are just examples; throughout the book are sections which can be cut, depending on course objectives.

Sequencing is always a challenge. The chapters on finance separate Chapter 2's microeconomics review from timber demand and supply in Chapter 12. The rationale is that you need some knowledge of capital theory to understand timber supply dynamics. Yet basic knowledge about microeconomics and price determination is important for understanding finance and capital allocation. Some may wish to postpone the Chapter 7 and 8 finance applications to forestry until after Chapters 9 and 10 on taxes and risk, but students are usually eager to start talking about trees sooner.

I encourage and welcome all comments and suggestions.

W. David Klemperer

FOREST RESOURCE ECONOMICS AND FINANCE

1

INTRODUCTION

FOREST ECONOMICS

If you're like most forestry students, you probably went into forestry because you loved the outdoors, enjoyed forested areas, and felt some commitment to the wise use of natural resources. Usually the biological and physical aspects of forestry attract students. I was no exception. As an undergraduate, I liked botany, dendrology, and silviculture and wasn't especially interested in economics. In fact, it bored me! It wasn't until my first job as a forestry extension agent in the Pacific Northwest that I began to realize how important economics could be in forestry. Landowners would often ask, "If I invest in reforestation, what kind of a return can I make, compared with other investments?" "Would I be financially ahead to convert some forestland to grazing, or vice versa?" Garden club members asked, "What are we giving up when we set aside more U.S. Forest Service lands for wilderness? How much less profitable is partial cutting than clear-cutting? Are we running out of timber?" I was asked about the pros and cons of log exporting, a major controversy in that region. I complained about air pollution from a local pulp mill and was told, "That's the smell of money." People called me about trees dying of air pollution from a nearby smelter. It became painfully apparent that economics, which I hadn't learned very well, had lots to say about such issues.

I'm assuming that you've successfully completed at least a one-semester college course in principles of economics emphasizing microeconomics. Although this book covers some elementary material, you might wish to review certain basic concepts shown in *italics* or **bold** in Chapters 1 and 2. A list of introductory and intermediate economics texts is at the end of Chapter 2. Specialized terms are defined where first used and also in the glossary. Most glossary terms are in bold the first time they appear in each chapter. Chapter 1 introduces forest economics, gives examples of economics as a way of thinking that doesn't necessarily involve money, and reviews the importance of United States forest resources, outputs, and user groups.

Is Forest Economics Unique?

Cases have been made for forest economics as a special field and as simply the application of ordinary economics to forestry. Most economists would probably agree that forestry calls for economics that's no different from standard economic theory. But several features of forestry, when considered together, create challenging problems, although they aren't unique to forestry, as noted at each item's end:

1. Forests have simultaneous outputs, many of which aren't easily sold in the market. Along with revenue from wood, other plant products, and grazing or hunting, forests also produce nonmonetary outputs valued by society. Examples are scenic beauty, open space, water, flood control, and certain types of recreation. Can you think of analogous unmarketed benefits from other enterprises like agriculture, golf courses, parks, and certain building projects?
2. Harvesting can cause unpriced negative side effects. These are negative outcomes for which the producer needn't pay—for instance, logging-related damage to water quality, fisheries, and scenery; and polluted air from slash burning and wood processing. Obvious examples of environmental damage exist in many other enterprises.
3. From forests, society seems to be pressing for more nonmonetary goods (item 1) and fewer negative side effects (item 2). This brings clashes between (a) public rights to certain natural amenities and (b) private property rights, individual freedom of landowners, and economic development. These conflicts surround the issues of wilderness, clear-cutting, endangered species, and old-growth timber preservation. Analogous struggles occur in mining, agriculture, construction, and many other industries.
4. Some maintain that, for wood production, the standing tree is both the factory and the final product. Harvest the product, and much of the wood-producing mechanism is gone, although the soil and environment to start a new forest remain. Similar arguments apply to most assets held for expected increases in value: aging wine, stocks, art, savings accounts, land, etc. Standard techniques of financial analysis apply.
5. Forestry involves long production periods and uncertainty. While some Christmas trees, fuelwood, and wood crops can be produced in less than 10 years, most tree crops require several decades in the United States: 20 years or more for most pulpwood, 30 to 40 years or more for sawtimber and veneer, and effectively forever for wilderness production. Over these long periods, there's tremendous uncertainty about outcomes. Again, long

time horizons and uncertainty aren't unique; they occur with ventures like education, flood control, or certain speculative investments.

So, while none of the above are unique to forestry alone, together they form a special challenge in the profession. Some of the above issues lead to lively arguments about the proper role of markets in forestry. Where the market generates attractive prices for outputs like wood and good hunting, private landowners provide ample supplies. Thorny questions arise over nonmonetary outputs and environmental damage caused in harvesting timber.

ECONOMICS AS A WAY OF THINKING

Economics helps us choose the best way to use resources. Broadly, economics is the study of how to allocate scarce resources to produce *satisfaction for people.* The last three words are crucially important. Forest managers shouldn't necessarily aim to produce wood or recreation or any single forest output: as we shall see, producing satisfaction may or may not mean producing a given output on a forest. In fact, society's well-being is sometimes improved if we convert some forestlands to other uses and if some nonforestlands are afforested.

Economics is often associated with money, profits, banking, gross domestic product, taxes, rates of return, finance, economic development, and related notions. A young lady once showed surprise when I mentioned that I was a forest economist. She asked, "Isn't that a contradiction in terms? How can you have concern for the resource when you're interested in making money from it?" I explained that economics as a way of thinking needn't necessarily involve dollars, although the field is often interpreted that way. Examples are in the following passages from Leopold (1949):

> One basic weakness in a conservation system based wholly on economic motives is that most members of the land community have no economic value.... a system of conservation based solely on economic self-interest is hopelessly lopsided. [1]

Such a narrow interpretation gives economics a black eye. Yet, elsewhere in the same book we find what I would call economic thinking:

> Like winds and sunsets, wild things were taken for granted until progress began to do away with them. Now we face the question of whether a still higher "standard of living" is worth its cost in things natural, wild, and free.[2]

Read that last sentence again. It shows one form of economic thinking: comparing added benefits (a "higher standard of living") with added costs (the loss in "things natural, wild, and free"). Benefits and costs needn't be measured in dollars; the important thing is that the right questions are asked. In this case, "Do the extra benefits exceed the extra costs?"

---

[1]Leopold (1949), pp. 210, 214.
[2]Leopold (1949), p. vii.

People constantly make incremental decisions in business and in their personal lives where they don't necessarily maximize one desirable output or action. In Leopold's example above, we wouldn't maximize development or natural areas, nationwide. We adjust individual outputs in a way that maximizes total satisfaction. This process is called **optimization,** which means to maximize some objective, in this case, satisfaction. When we deal with nondollar benefits and costs, which are common in forestry, units are hard to define. And everyone's measure of costs and benefits will differ. But these problems don't negate the usefulness of incremental thinking and optimization to maximize satisfaction. We'll come back to these notions repeatedly.

The Equi-Marginal Principle

As another example of economic thinking without considering dollars, look at Figure 1-1, which outlines a student's time allocation problem, assuming the goal is to maximize satisfaction (or **utility**). If you were the student, how many hours per week would you allocate to each activity? The **equi-marginal principle** tells us that utility is maximized if time is allocated so that the last hour spent per week in each activity brings the same added utility. Then no time reallocation can yield a net gain. This principle is illustrated in Figure 1-2, which shows an individual's declining added (or **marginal**) utility from each added hour spent in activities A, B, and C. Note that the vertical axes in the figure are not totals as you'd find in most graphs. See the box on page 6 if you need to review the difference between total and marginal quantities.

Figure 1-2 represents an economic problem because we have an objective (maximize satisfaction) and a constrained input of 15 hours per week (scarce resource). You get the most satisfaction by adding hours to activities that yield the greatest extra utility. Satisfaction is maximized by spending 4 hours per week on activity A, 5 on B, and 6 on C. At that point, the last hour spent in each activity yields an added 4 units of utility—the equi-marginal principle has been met—and

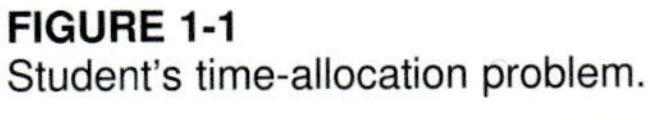

FIGURE 1-1
Student's time-allocation problem.

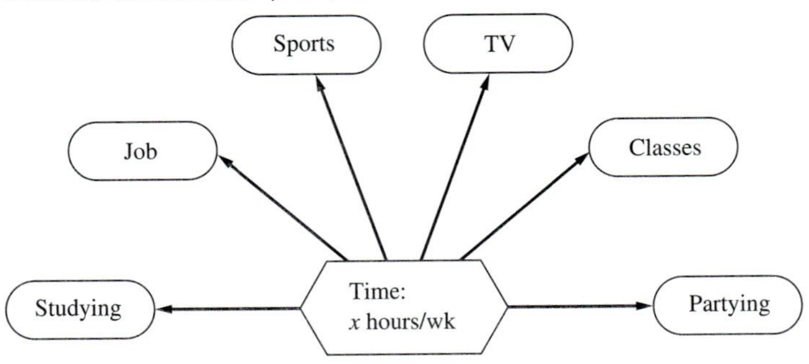

Given limited time, how many hours should the student spend on each activity in order to maximize satisfaction?

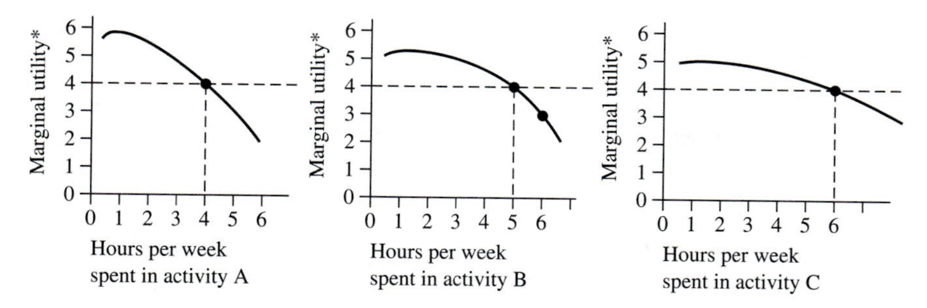

Constraint: Total available time = 15 hours per week.
Objective: Maximize total utility (satisfaction).

To maximize total utility, apply the *equi–marginal principle*: The optimal time allocation is where the last hour spent on each activity yields the same added utility (4 units).

No net gain is possible, given the 15-hour constraint. Try to increase total utility by shifting an hour from one activity to another. The marginal loss will exceed the marginal gain.

*Added utility from the last hour spent, not total utility.

FIGURE 1-2
Individual's optimal allocation of 15 hours per week among different activities.

all 15 available hours have been used. Note that the 15-hour limit is arbitrary; the equi-marginal principle applies under any input constraint, as long as the marginal curves are downward-sloping (more on this shortly).

Try to prove that the time allocation in Figure 1-2 is **optimal,** in other words, that the objective of utility maximization has been met. You can do this by showing that any reallocation of time will reduce total utility. For example, try taking one hour away from activity A and adding one hour to B. The fourth hour in A brought 4 units of utility, so removing that hour would mean a loss of 4 units. Adding this hour to B would bring an added utility of 3 units: see the lone dot on the middle panel's curve in Figure 1-2. Thus, reallocating the total 15 hours created a net loss of 1 unit of utility. In Figure 1-2, try other combinations of taking an hour away from one activity and adding that hour to another activity; you'll see that any reallocation of the 15 hours per week will bring a utility loss that exceeds the gain. Therefore, meeting the equi-marginal principle must be optimal. The term "optimal" is used often in economics; it simply means that some objective has been met (here, to maximize utility).[3]

In Figure 1-2 you could imagine a series of marginal utility graphs for many more activities; the same principle of optimality would hold. It's important that units of output on the vertical axes must all be the same. Here the units are utility:

[3]Note that Figure 1-2 assumes marginal utilities from different activities are independent of one another. Sometimes goods or activities are related in a way that changes your view of one as you add inputs of another. That would not negate the equi-marginal principle but would require more sophisticated graphics: *indifference curve analysis,* which is covered in most introductory or intermediate economics courses. But the latter approach is likely to be confusing to decision makers and is difficult to apply, especially when dealing with more than one good or activity.

THE RELATIONSHIP BETWEEN TOTAL AND MARGINAL QUANTITIES

Figure 1-3a and b reviews the relationship between total and marginal quantities and explains the nature of the marginal utility curves in Figure 1-2. The axes are labeled input and output in order to fit any number of situations; for example, input could be tons of fertilizer per acre per year on a farm, and output could be tons of crop yield. Or inputs could be laborers added to one factory, with output being revenue. Or input could be hours of television watched per week for a person, and output could be units of satisfaction or "utility," paralleling the case of Figure 1-1. On each total output bar in panel (a), the added or "marginal" output with each successive unit of input is shaded. Below the total output, on the same input scale, Figure 1-3b plots only the shaded portions from panel (a); these are the "marginals" and thus are labeled *marginal output*, which means the extra output from the last unit of input. For example, the third unit of input brought 2 more units of total output, so the marginal curve shows 2 units of output at the third unit of input. If the total output curve were total utility, the marginal curve would be "marginal utility," *the same type of curve in Figure 1-2*. Notice how the sixth unit of input in panel (a) of Figure 1-3 actually reduces total output, so the marginal output in panel (b) is negative for the sixth unit of input.

they could also have been, for example, dollars of revenue from income-producing activities or, in general, benefits of some kind. Likewise, on the horizontal axes, units of input must also be the same. Instead of hours, inputs might have been dollars spent on different goods, and we might have sought an optimal expenditure pattern that maximized utility.

In general terms, the equi-marginal principle says that you'll maximize the total benefit from using any limited input if you allocate it so that the last unit of input in each activity brings the same added benefit. Also, a key point in applying the equi-marginal principle is that marginal benefit curves must be downward-sloping: for the utility example, we must be in the range of **diminishing marginal utility**, where added units of input bring a diminishing amount of extra utility. Make sure you review why **diminishing returns** (diminishing marginal outputs) eventually occur: whenever you have a fixed production unit—in this case it's you, the individual, producing utility—and you add units of a given input (here, hours in one activity), the extra output (here, marginal utility) will ultimately decline. Why? Because eventually you overload a fixed system. For the utility case, you finally get fed up with more of anything over a fixed period of time, even pizza. When feeding livestock, ultimately, more feed per day does no good. For trees, eventually, more fertilizer kills them. Examples are almost endless.

One more clarification about Figure 1-2: You can't measure utility in an absolute sense; you can only say that a result yielding more utility is preferred to a result yielding less utility. For this reason we say that utility is measured in an "ordinal" (ranking) sense rather than "cardinal" (absolute).

Different Time Horizons

What about the time element? In forestry, time is extremely important, and we deal very specifically with it. For now, let's briefly consider how important your

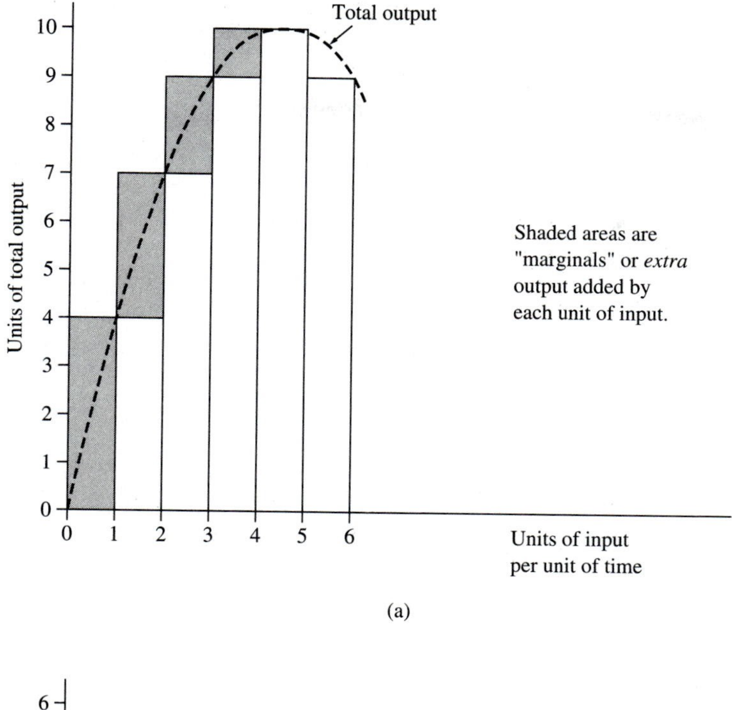

(a)

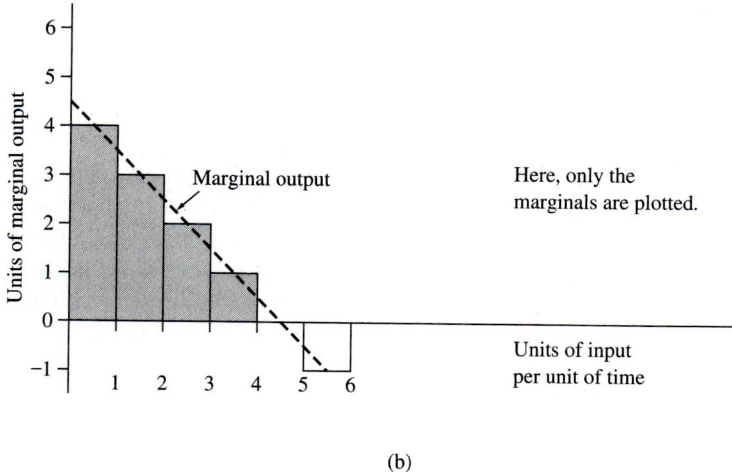

(b)

FIGURE 1-3
Relationship between total and marginal quantities (See Box).

time horizon is. Suppose, in Figure 1-1, you're expecting to live a long productive life. As you consider your optimal allocation of time, you'll probably think about long-term benefits of studying and working. But suppose your physician tells you that you have only one week to live. How might your optimal time allocation change? The way you view future benefits obviously affects the solution. The time frame is important, especially in forestry, and we'll consider it in detail later.

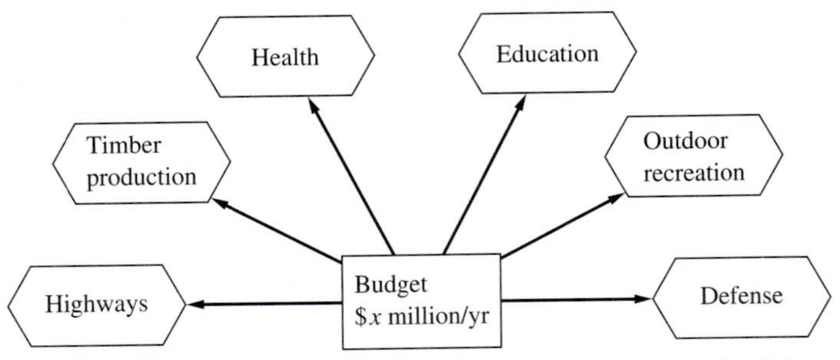

Given limited budget, to maximize satisfaction, how much money should be allocated to each area?

FIGURE 1-4
Legislative Committee budget-allocation problem.

A Legislative Example

Figure 1-4 diagrams possible areas in which a legislative ways and means committee might wish to allocate a year's budget. How much should it spend in each area? This is another optimization problem. Suppose the committee wishes to maximize people's satisfaction or "social welfare" in some sense, subject to a constrained budget. Again the equi-marginal principle is useful. Think of a series of diminishing marginal utility curves for additional millions of dollars spent in each area: eventually as we spend more and more in each area, added satisfaction to society per million spent begins to decline. Theoretically, the optimal spending pattern is where the last million dollars spent in each area brings the same satisfaction. Then no reallocation could increase total satisfaction, following the same logic used in Figure 1-2.

You can probably see several problems in accurately applying the above optimization procedure. In many cases it may be extremely difficult to measure the added benefits of spending a million dollars more in each area, and often outputs couldn't all be measured in the same units. Moreover, each committee member would probably have a different view of social benefits from alternative budget allocations. But none of these doubts negate the usefulness of the equi-marginal principle. The important thing is to think in terms of added costs and benefits and to ask relevant questions; for example, "Are the extra benefits worth the extra costs?" and "If so, how do the extra benefits from an added million dollars spent in area A compare with those from an added million spent in B?" These are examples of *marginal analysis,* the foundation for many decisions in economics. As you'll see throughout this book, decision models will often be much more precise in theory than in actual practice. The trick is to see the opportunity for stepwise improvements in benefits.

Models

We'll often use models like the above figures to explain or predict. Economists are sometimes teased about their endless assumptions in model building. But models simplify complex situations and thus require assumptions, some of them unrealistic. For example, in modeling your checking account with a check register, you make many assumptions. Aeronautical engineers make highly simplifying and unrealistic assumptions when predicting airplane flight behavior based on small stationary models in wind tunnels. They'll make the model small and assume it's large. They keep the model stationary and assume it's moving by only moving the air. The key point is how well a model predicts the real world: how effectively the engineers' models predict flight behavior of large airplanes, or how well your check register explains what's going on in your account. It's OK to make unrealistic assumptions, as long as the model predicts well. But you want to avoid assumptions that destroy predictive ability. For example, if you tested model airplanes in moving water and said, "Assume it's air," you couldn't predict flight behavior very well. In forest economics, we'll be interested in how well our models explain things like landowner and consumer behavior or predict variables such as people's willingness to pay for forest properties or nonmonetary outputs.

Decision Processes

Since much economic analysis in forestry aims to help decision makers, we should briefly consider goal-oriented decisions. The process, which also involves the analyst (who may or may not be the decision maker), is as follows:

1 Define the problem. (The analyst must clarify who the decision maker is—a forest owner, manager, investor, legislative committee, government agency, conservation group, the general public, etc.— and what needs are not being met.)
2 Determine goals. (While this may seem simple enough, decision makers are often not sure of goals—for example, is it to maximize wood output or satisfaction?—and they often have conflicting goals that cannot be met simultaneously.)
3 Isolate alternatives. (What are different ways in which goals could be approached?)
4 Set criteria for evaluating alternatives. (Examples of financial criteria are rate of return and net present value.)
5 Evaluate alternatives in terms of goals. [Which actions bring us closest to the goal(s)?]
6 Feedback. (After implementing an action, sometimes years later, one should learn how effective it was in meeting goals. This step is often omitted.)

The problem of conflicting goals is common. Consider, for example, the often-voiced public forestry goal of providing "the greatest good for the greatest number in the long run." If you wish to create the greatest good per person from a forest, you might have high benefits for a fairly small number. If you want the greatest number of beneficiaries, you could give, say, one toothpick to every person in the world. If you want the greatest good in the long run, you'd save everything for future generations, supplying nothing today. You can't maximize everything at once.

While economic thinking is a useful decision tool in forestry, it certainly can't answer all questions. Economics *cannot* tell us things like the correct distribution of income or of forest outputs between conflicting user groups, the best pattern of forest outputs over time and over space, and the best multiple-use forest plan. In such cases we need political processes, negotiated consensus building, mediation, and compromise. Economic thinking can help in these procedures by clarifying **trade-offs**—the gains and losses in monetary and nonmonetary terms as we change forestry practices. These issues appear throughout the book, especially in the chapter on multiple-use.

Two things that make forestry decisions so fascinating are the large number of forest outputs and the variety of user groups.

FOREST OUTPUTS AND USERS

Many of my analyses will deal with wood products because they're the easiest values to measure, and that's where much forestry decision theory was first developed. But I'll show that much of this theory is just as applicable to nontimber outputs. Also, the interaction between timber and other yields is important. Often nontimber values are greatest, as with wilderness or park areas valued so highly that no timber is cut. Forest recreation is increasing, for example, hiking, bicycling, camping, rock and mountain climbing, photography, boating, river rafting, bird watching, hunting, fishing, nature study, swimming, and picnicking. Forests also provide benefits of open space, scenic beauty, oxygen production, and pollution filtering. They reduce wind erosion and drying of adjacent farmlands, add organic matter to the soil, consume carbon dioxide, prevent erosion, and help watersheds to store and to provide cleaner and more even flows of water. Forested areas are homes for wildlife, can enhance fisheries, and sometimes produce farm crops and forage for livestock. Some of these products can yield income for owners; other outputs aren't so easily valued. But economic reasoning can be applied in decisions about all these forest benefits.

When thinking about forest outputs, at first glance, the user groups seem obvious. For example, the forest industry uses logs, farmers gain some of the soil protection and water benefits, and recreationists and related groups use most of the other nontimber outputs. Right? Well, that's an oversimplification. In most cases the users are consumers, you and I. We're willing to pay for and consume paper, dwellings, furniture, and other products from which the demand for logs is derived. Likewise, we, the consumers, demand food produced by farmers who benefit from forests. And we're also the recreationists and conservationists who demand nontimber benefits and wilderness. Separating forest users into, say, conservationists and the forest industry is unnecessarily divisive. We'll examine in detail how demand for all forest outputs can be derived from consumer behavior. This notion of derived demand is not unique to forestry; however, because the same consumers often demand outputs that compete at the forest level, we're faced with interesting conflicts in forestry.

As you begin your study of forest economics, it's important to know the extent of United States forest resources and the demands being made on them—the next

TABLE 1-1
UNITED STATES LAND AREA, 1992*

Category	Million acres			
	Total U.S.	North	South	West
Total forestland	736.7	168.4	211.8	356.3
Timberland	489.5	157.8	199.3	132.4
Reserved timberland†	35.6	7.5	3.0	25.0
Unproductive forest‡	211.6	3.1	9.5	198.9
Nonforest land	1,526.6	245.0	322.7	959.0
Total land area	2,263.3	413.4	534.5	1,315.3

*USDA Forest Service (1993).
†Not available for commercial timber production.
‡Not capable of producing 20 cubic feet/acre/year.

topic. More detailed resource assessments are in the U.S. Forest Service publications listed at this chapter's end.

FOREST RESOURCES IN THE UNITED STATES[4]

Forestlands

Of the total land area in the United States, 33 percent—about 737 million acres—is forested. Two-thirds of this, or 490 million acres, is "timberland" (defined by the U.S. Forest Service as land capable of producing at least 20 cubic feet of wood per acre per year in natural stands and not reserved for other uses like parks and wilderness areas). Table 1-1 shows that in the North and South,[5] about 95 percent of the forestland is productive timberland, with relatively few acres reserved for nontimber uses or classified as unproductive (not capable of producing 20 cubic feet/acre/year). The table shows that in the West, over half the forest area is in the unproductive category (mainly because of dryness, poor soil, and high elevation). But the West also has some of the most productive forestlands, especially in parts of the Northwest. Overall, 4.8 percent of the United States forest area is reserved for nontimber uses like National Parks and wilderness.

Total U.S. forest area declined 2.5 percent between 1952 and 1992 and actually *increased* by 0.1 percent between 1987 and 1992. The timberland area available for wood production decreased about 4 percent between 1952 and 1992, dropping faster than total forest area because more public forests were reserved for wilderness.

The West has far more timber volume than the other two regions, but not much hardwood (see Table 1-2). Both the North and the South have more hardwood

[4]Most of the statistics in this section come from U.S.F.S. (1989, 1990, 1993, and 1994).

[5]The North includes all states north of Kentucky and Virginia and east of the Mississippi River; the South comprises all states south of the North and as far west as (and including) Texas; the West covers the remaining states, including Alaska and Hawaii.

TABLE 1-2
VOLUME OF GROWING STOCK* ON TIMBERLAND IN THE UNITED STATES, BY SPECIES GROUP AND REGION, 1992

	Species group (million cubic feet)		
Region	**All species**	**Softwoods**	**Hardwoods**
North	207,119 (+27.1%)†	50,976 (+16.3%)	156,143 (+31.0%)
South	250,594 (+12.2%)	102,927 (+1.7%)	147,667 (+20.9%)
West	327,904 (−5.4%)	295,992 (−8.1%)	31,912 (+28.8%)
U.S.	785,617 (+7.2%)	449,895 (−3.7%)	335,722 (+26.2%)

*Live trees of commercial species 5 inches dbh (diameter breast high) and greater (excludes volumes on areas reserved for nontimber uses). *Source:* USFS (1993).
†Figures in parentheses are percentage volume changes between 1977 and 1992.

volume than softwood. The numbers in parentheses in Table 1-2 give the percentage of change in commercial timber volume between 1977 and 1992. You can see that, in spite of steady harvesting, timber inventories have increased in all areas but the West; there, ancient forests grow very slowly, and growth in young stands isn't enough to offset harvests. Overall, between 1977 and 1992, United States **commercial timber** inventories (available for wood production) in softwoods decreased by 3.7 percent and in hardwoods increased by 26.2 percent; summing all species, inventories increased 7.2 percent.

We'll look more closely at harvest and growth relationships in the chapter on timber demand and supply. For now, remember that, despite steadily increasing harvests, nationwide timber inventories have been increasing. It's helpful to realize this as we start our study of forest economics because we often approach timber issues with a mistaken notion of impending wood famine.

Who owns the nation's commercial timberlands? Table 1-3 shows that about 14 percent is held by the forest industry (firms owning wood-processing plants). Nonindustrial private forests (NIPFs) are 59 percent of the area. The remaining

TABLE 1-3
TIMBERLAND AREA OF THE UNITED STATES BY OWNERSHIP AND REGION, 1992*

	Ownership (million acres)			
Region	**Public**	**Forest industry**	**Nonindustrial private**	**Total**
North	30.3	16.2	111.3	157.8
South	20.5	39.0	139.8	199.3
West	80.7	15.2	36.5	132.4
Total	131.5	70.4	287.6	489.5

*Forest area available for commercial timber production (USFS, 1993).

27 percent is in public ownership: of this, 64 percent is national forest; the rest is controlled by other federal agencies and state and local governments. Note in Table 1-3 that well over half the western timberland is publicly owned, while in the North and South combined, less than 15 percent is public. "Timberland" excludes areas reserved for nontimber uses.

In today's global economy, especially with the current emphasis on trade, we shouldn't view United States timber resources in a vacuum. Large volumes of forest products flow in and out of the country, and these interactions, as well as world forest resources, will be covered in the last chapter.

Nontimber Resources

Water All of the above forestlands are part of some watershed and thus produce water. But these lands would produce water, with or without forest cover. Therefore, the major water-related benefit of forests is to improve water storage, flood control, water quality, and distribution of water flow throughout the year. Beneficiaries are users of fish and wildlife resources, recreationists, farmers and other commercial users of water, and domestic users.

Recent studies forecast no nationwide water shortages: by the year 2040, projected United States water consumption is only 10 percent of precipitation. But regional shortages are expected in areas like the Southwest and portions of California. And problems with water quality and flooding, although improving in some areas, still occur. Thus, forests should continue to play a major role in enhancing water resources.

Range Forage Although most livestock grazing occurs on unforested rangeland, some open-grown forests are grazed, especially in the West. This activity often competes with use of forests for watershed protection, wildlife, and recreation. One report projects a 54 percent increase in consumption of domestic range forage by the year 2040, given constant real forage prices. The same forecast suggests that domestic production can keep pace with this increase. But we should expect increasing competition for the limited forest areas that can produce forage for livestock.

Outdoor Recreation Table 1-4 shows outdoor recreation trips away from home in the United States in 1987, excluding hunting and fishing. Also shown, for the year 2040, are projected numbers of trips as a percentage of 1987 levels. These are increases that range from 27 percent for motor boating to over a tripling for sailing and downhill skiing, assuming unchanged costs per trip. Most of these visits take place on or near forested areas. Although nearly all the nation's 737 million forested acres could be used for some type of recreation, a small percentage of this area is managed for intensive recreational use. Such areas include portions of National Parks, state and local government parks, privately managed ski areas (mostly on public lands), and private parks. Most dispersed recreation occurs on public land, especially on U.S. Forest Service and Bureau of Land Management

TABLE 1-4
OUTDOOR RECREATION TRIPS AWAY FROM HOME—1987 AND INCREASE FOR 2040*

Resource category and activity	Trips in 1987 (millions)	Projected number of trips in 2040 as a percentage of 1987 trips (%)
Land		
Wildlife observation	69.5	174
Camping in primitive campgrounds	38.1	164
Backpacking	25.9	255
Nature study	70.7	138
Horseback riding	63.2	190
Day hiking	91.1	293
Photography	42.1	205
Visiting prehistoric sites	16.7	278
Collecting berries	19.0	192
Collecting firewood	30.3	183
Walking for pleasure	266.4	177
Running/jogging	83.7	262
Bicycle riding	114.5	222
Driving vehicles or motorcycles off-road	80.2	130
Visiting museums	9.7	188
Attending special events	73.7	168
Visiting historic sites	73.0	241
Driving for pleasure	421.4	167
Family gatherings	74.4	182
Sightseeing	292.1	212
Picnicking	262.1	144
Camping in developed campgrounds	60.5	186
Water		
Canoeing/kayaking	39.7	169
Stream/lake swimming	238.7	129
Rafting/tubing	9.0	255
Rowing/paddling, etc.	61.8	159
Motor boating	219.4	127
Water skiing	107.4	148
Sailing	35.0	335
Snow and ice		
Cross-country skiing	9.7	195
Downhill skiing	64.3	333
Snowmobiling	17.7	130

*English et al. (1993).

areas. Wilderness preserves include about 89 million acres of federal lands and 2.6 million acres of state forests. Local governments and private groups hold 1.3 million acres of protected natural areas.

Partly because of problems with liability, property damage, and fire danger, most private land isn't open to the public for recreation. About 23 percent is available for public use, either free or for a fee. Over 70 percent of the nation's campsites are on private land, but they are generally not in remote areas.

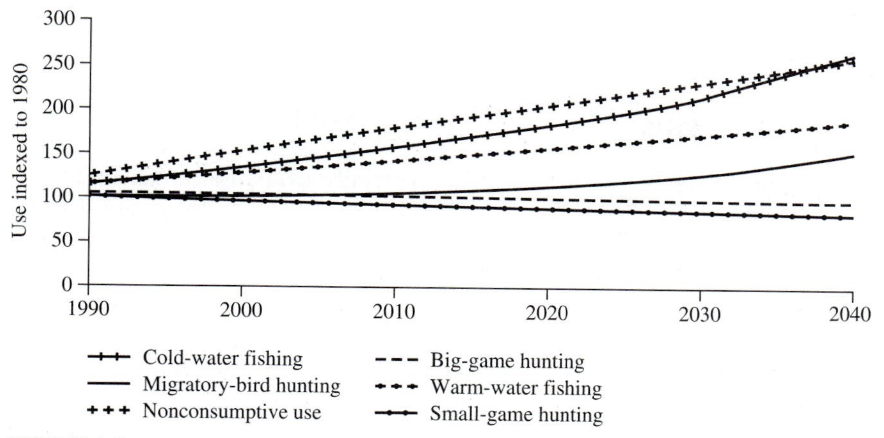

FIGURE 1-5
Projected fish and wildlife related activities in the United States (USDA Forest Service, 1989).

Fishing and hunting are often associated with or affected by forestlands. As shown in Figure 1-5, hunting of large and small game (excluding migratory birds) is projected to decline slightly. The number of small game animals such as pheasants, quail, and rabbits has declined in recent years, but grouse, squirrels, and most big game species are increasing in abundance. Figure 1-5 shows that migratory bird hunting, fishing, and nonconsumptive uses of wildlife—viewing and photographing animals—are expected to increase. Over half the nation's big game habitat is on U.S Forest Service land. However, considering all game, about 68 percent of hunting is done on private land. Many private forest owners lease their lands for hunting and are interested in ways to enhance such revenue.

Projections in Table 1-4 and Figure 1-5 show increasing interest in outdoor recreation. But from an economic efficiency view, this doesn't tell us much because most of this recreation is made available free or below cost. If people had to pay a greater share of the cost, visits would no doubt be fewer. Prices are important and are stressed throughout the book.

Minerals We don't usually think of mining as having much to do with forestry. But most of the country's eastern coal reserves are under Appalachian forests, and numerous metallic minerals lie beneath forested parts of the Rocky Mountains. Oil and gas reserves are under many U.S. Forest Service lands.

Open pit and strip mining can obviously compete with forestry, but even drilling for minerals can cause conflicts if done on recreation areas. Once, after hiking for over a day in what seemed like a pristine wilderness in Washington's Cascade Mountains, I was dismayed to suddenly find an area with exploratory drilling equipment, sheds, roads, fuel barrels, and debris. Although I reminded myself that I was also a user of minerals, the shock was no less real. As competition for the many forest outputs increases, land management will become more and more complex. Although economic thinking can't resolve all these conflicts, it can provide useful inputs.

Synthesis

The U.S. Forest Service projects the commercially available timber inventory of roughly 740 billion cubic feet for 1987 to rise to about 790 billion by the year 2020 and to remain fairly stable through 2040. Over the same period, from 1987 to 2040, the country's timber harvests are projected to increase by roughly 40 percent. In Table 1-4, you'll see that most of the nontimber uses of forests are destined to increase well over 40 percent for the same period. (To get the percentage increase, just subtract 100 from the numbers in the last column of Table 1-4.)

The nuggets to glean from this overview of U.S. forest resources: (1) We aren't running out of timber or forestland. (2) From our forests, consumption of nontimber outputs is rising faster than wood consumption. Keep these trends in mind as you continue through this book.

AN OVERVIEW OF THE TEXT

Chapter 2 reviews elementary microeconomics: price determination and equilibrium in free markets. Chapter 3 covers strengths and weaknesses of free markets in forestry and the role of governments. Several chapters on capital theory and investment analysis—or "finance"—intervene before material on forest output supply and demand equilibria. This break occurs because it's easier to fully grasp such equilibrium dynamics and the material that follows if you first understand capital theory. This is especially important because forestry needs large inputs of capital and time. Chapter 4 shows ways to measure performance of financial assets, giving forestry examples, and Chapter 5 incorporates inflation. Chapter 6 gives more theory and applications in deciding how much to invest where—the "capital budgeting" decision, considering timber and nontimber values. Again, weighing timber and other values, Chapters 7 and 8 look at the decision when, if ever, to cut trees and the question of stand density. Following taxation in Chapter 9, Chapter 10 shows how risk affects forestry investment decisions. Chapter 11 stresses the difference between finding willingness to pay for timberlands and gauging their market values. Short- and long-run timber demand and supply analysis is in Chapter 12. Chapter 13 gives an overview of the United States wood-processing industry. Chapters 14 and 15 cover ways to estimate nontimber values and include them in multiple-use forestry decisions. These chapters are a vital part of modern forestry and appear toward the book's end only because they rely on analyses developed earlier. Chapter 16's focus is measuring impacts of forestry on regional economies, and Chapter 17 considers forestry in a global context. As you go through the book, analyses will sometimes concentrate on specific outputs, but we'll keep coming back to the idea that forests are a system of interrelated outputs and inputs.

The text is designed to flow smoothly without the footnotes and appendixes. These are included for those who want to probe certain issues more deeply.

KEY POINTS

- Forestry has a number of features that, together, create interesting challenges: many outputs, several of which are nonmonetary; unpriced negative side effects; clashes between public amenity rights and private property rights; long production periods; and wood production in which the tree is both factory and product. But none of these features are unique to forestry, and standard economic theory applies.
- Economics considers how to use scarce resources to produce satisfaction for people. Economic thinking doesn't necessarily have to deal with notions of money, banking, profits, economic development, or gross domestic product.
- An example of economic thinking is the equi-marginal principle: In using a limited input like time or money to produce a single output (e.g., satisfaction or profit) with several different processes, output will be maximized when the last unit of input in each process brings the same marginal (extra) output.
- We need simplifying assumptions to build economic models of complex systems. The important thing is how well a model explains or predicts.
- Goal-oriented decision processes are complex in forestry due to the large numbers of outputs and users. Often the same consumers demand wood and nonwood outputs from forests.
- Forestlands cover 33 percent of the United States. About two-thirds of this is commercial timberland, which, in total, shows increasing timber inventories. Thus, no timber famine is on the horizon. Our timberlands have continued to provide wood, water, range forage, fish and wildlife, outdoor recreation, scenic values, minerals, and protection and improvement of soil, water, air, and climate.
- Fifty-nine percent of the nation's forestlands are held by nonindustrial owners. Most of the western forest area is public, while private forests dominate in the South. Demands for most nontimber uses of forests are expected to increase sharply over the next few decades.

QUESTIONS AND PROBLEMS

1-1

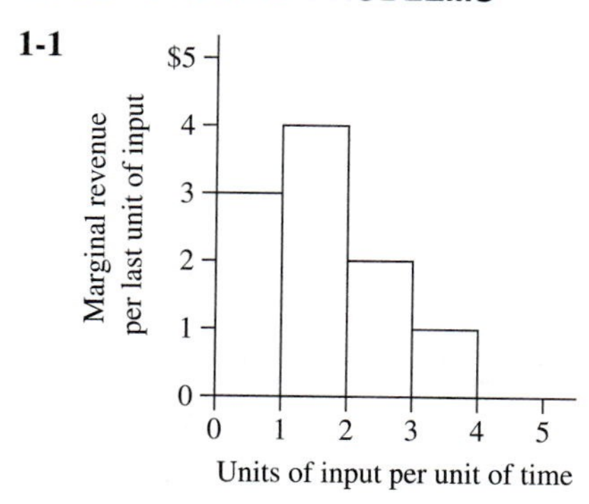

In the preceding bar graph of marginal revenue per added unit of input, what is the total revenue at 5 units of input?

1-2 In the form of a bar graph, draw the total revenue curve from which the preceding marginal revenue graph was drawn.

1-3 Consumer's Choice of Goods A and B

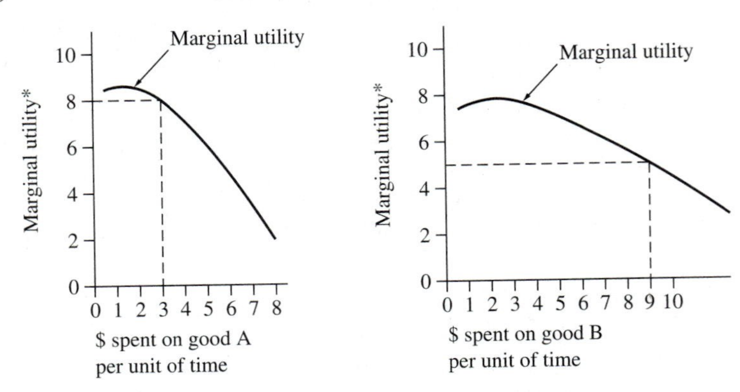

*Marginal utility (*added* utility or satisfaction from last dollar spent).

In the above graph, $3 spent on A and $9 on B is a suboptimal spending pattern. Why? Assume a $12 budget.

1-4 In the previous question, keeping the same total budget ($12), show the optimal spending pattern. Why is it optimal? In terms of using resources, what does optimal mean?

1-5 If the price of B is $0.50/unit, how many units of B would be purchased at the optimum?

1-6 In general terms, define the equi-marginal principle. Specifically, how could a National Forest apply this principle to allocate a year's budget optimally?

1-7 Between 1977 and 1992, was the total United States commercial timber inventory increasing or decreasing?

1-8 What kinds of nontimber uses are expected to increase most rapidly on forestlands?

1-9 Besides recreationists, what other groups use forestlands?

REFERENCES

Duerr, W. A. 1960. *Fundamentals of Forestry Economics.* McGraw-Hill, New York. 579 pp.

Duerr, W. A. 1993. *Introduction to Forest Resource Economics.* McGraw-Hill, New York. 485 pp.

Ellefson, P. V. 1992. *Forest Resources Policy.* McGraw-Hill, New York. 504 pp.

English, D. B. K., C. J. Betz, J. M. Young, J. C. Bergstrom, H. K. Cordell. 1993. *Regional Demand and Supply Projections for Outdoor Recreation.* U.S. Forest Service GTR RM-230. Rocky Mt. Forest and Range Experiment Station. Fort Collins, CO. 39 pp.

Gregory, G. R. 1987. *Resource Economics for Foresters.* John Wiley, New York. 477 pp.

Leopold, A. 1949. *A Sand County Almanac.* Oxford University Press, New York. 226 pp. (1986 paperback edition).

U.S. Forest Service. 1989. *RPA Assessment of the Forest and Rangeland Situation in the United States, 1989.* Forest Resource Report No. 26, Washington, DC. 72 pp.

U.S. Forest Service. 1990. *An Analysis of the Timber Situation in the United States: 1989-2020.* General Technical Report RM-199. Rocky Mountain Forest and Range Experiment Station, Ft. Collins, CO. 268 pp.

U.S. Forest Service. 1993. *Forest Resources of the United States, 1992.* Gen. Tech. Report RM-234. Rocky Mt. Forest and Range Experiment Station, Ft. Collins, CO. 132 pp.

U.S. Forest Service. 1994. *RPA Assessment of the Forest and Rangeland Situation in the United States—1993 Update.* Forest Resource Report No. 27, Washington, DC. 75 pp.

2

MICROECONOMICS REVIEW

The review of introductory microeconomics starts with a market economy where goods and services are sold at a price. Later you'll see how nonmonetary values and costs can be woven into this framework. This review covers demand, development of supply theory through a simple fish farm example, the interaction of supply and demand, and market equilibrium. This is by no means a complete review but highlights important concepts applied throughout the book. Examples will be simplified to show basic principles and ways of economic thinking. More detail and other forestry examples will follow. Taxes and inflation, although briefly mentioned here, will be covered in later chapters. Notation for Chapter 2 is in Table 2-1.

TABLE 2-1
NOTATION FOR CHAPTER 2

E_a =	arc elasticity of demand or supply
E_p =	point elasticity of demand or supply
P =	price per unit
P_e =	equilibrium price
Q =	quantity of a good or service (units per unit of time, for example, thousand of board feet per year)
Q_d =	quantity demanded
Q_e =	equilibrium quantity
Q_s =	quantity supplied
S_n =	shifters of demand or supply curves
Δ =	change in

DEMAND

Economics deals with how an economy meets certain desires of people. In this framework, a desire is expressed as **demand,** which has a special meaning in economics: *demand for a good (or service) shows the quantities of that good consumed at different prices by a given group over some time period.*

Demand Curves from the Consumer's View

Each consumer has a demand curve for separate goods and services, as shown in the hypothetical curves for Nathan (a) and Kristen (b) in Figure 2-1 for paper products. You can imagine your own reactions to different prices of paper or any other good: with high prices you would tend to consume less per unit of time than with low prices. One reason is that paying higher prices will reduce your income for purchases. Curves (a) and (b) show this, as well as the fact that people may respond in different ways to changing prices. The demand curve also slopes downward because when you consume more of a good per unit of time, eventually your willingness to pay (reflecting utility or satisfaction) per unit declines. This is the principle of **diminishing marginal utility** with added consumption (see Chapter 1).

By summing the quantities of a good consumed by each individual at each price, we can get the *market demand curve,* as shown in the third panel of Figure 2-1, for the two-person market. In general, a demand curve for any group is the horizontal sum of all the demand curves for individuals in the group. That's not normally the way demand curves are empirically derived, but it helps to explain the concept of demand. As you add millions of individual curves for a large market demand curve, you must compress the horizontal axis scale to keep the graph from running off the paper and into the next county. Thus, a region's demand curve doesn't necessarily appear less sloped as you add more consumers.

In forest economics, it's important to start with and to understand demand because *consumers' demands drive the practice of forestry:* demands for recreation, endangered plant and animal species, wood products, scenic beauty, biological diversity, soil and water protection, oxygen . . . The list is long. We can't always

FIGURE 2-1
Market demand is the horizontal sum of individual demand curves.

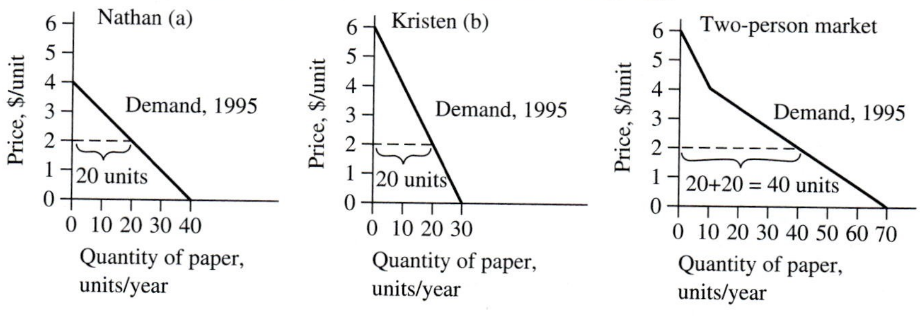

derive precise demand curves for every forest output, but the same concepts apply to all demands. Chapter 14 explores ways of deriving demand curves for unpriced goods.

In this section, *the key to looking at demand curves from the* ***consumer's*** *view is to ask, "If the price of a good or service is P, what quantity will the consumer(s) buy?"* Figure 2-2a shows a hypothetical market demand curve for softwood lumber in a state, for one year. As you'd expect for most goods, the higher the price per unit (here, dollars per board foot), the less is consumed per unit of time (here, billions of board feet per year), other things being equal. In the case of lumber, for example, if prices are high, home owners might postpone adding that new deck or room, or perhaps make the addition smaller. Those in the market for a new house might delay the purchase or opt for a smaller home. High lumber prices could also lead builders to substitute other materials like masonry. Reactions are the reverse if lumber prices drop. Note that for all demand curves, although not always specified, the horizontal axis is in units *per unit of time,* for example, per week, month, or year.

In Figure 2-2a, consumers would buy 1.75 billion board feet of lumber annually if the price were \$0.30 per board foot (BF) and 1 billion board feet if the price were \$0.60. At any given price P, and the corresponding quantity Q, consumers' total expenditure on lumber would be $P(Q)$. For example, at \$0.30/BF, total lumber sales would be $0.30(1{,}750{,}000{,}000) = \$525{,}000{,}000$ per year. The same sum is the total sales revenue to lumber retailers. For different price times quantity combinations, Figure 2-2b plots the total sales dollars based on Figure 2-2a.

How do these demand concepts relate to forestry? The higher the prices consumers are willing to pay for wood products, the more the mills are willing to pay for standing trees, given the same production costs. That's why it is so important to start with consumers and demand functions in our study of forest economics. To be efficient, we shouldn't spend more producing marketed forest outputs than they're worth to consumers. A market measure of this "worth" or value for any group of consumers can be found from its demand curve: at any given price, the market value is the price times the quantity read off the demand curve for that consumer group.

While, for now, we'll consider demand for wood products, you can think of demand curves for nontimber outputs like visits to a forest recreation area. With such a demand curve, the horizontal axis would read visits per unit of time, and the vertical axis would be the entry fee in dollars per visit. Demand concepts for nontimber outputs are the same as those for wood products and will be covered later.

Consumer Surplus

In a competitive economy, price times quantity gives the total expenditure on a good because all buyers pay the same price. But some consumers are willing to pay more than the going price for part of the year's supply. For example, in Figure 2-2a, suppose the market price is \$0.50 per board foot. Consumers buy 1.25

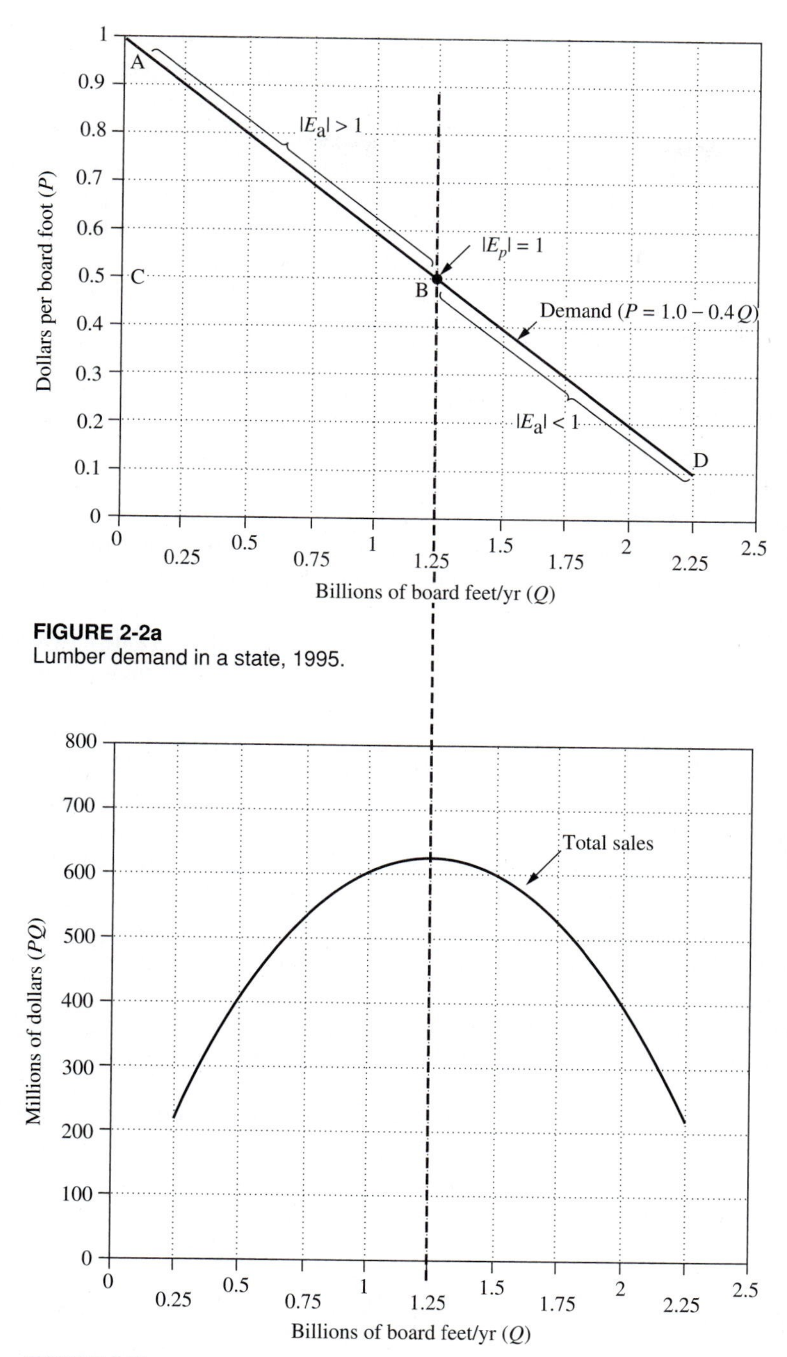

FIGURE 2-2a
Lumber demand in a state, 1995.

FIGURE 2-2b
Total sales, dollars (PQ).

billion board feet, and total expenditure is 0.50(1.25) = \$625 million. But you can see from the demand curve that at a \$0.90 price, 0.25 billion feet would be bought; at \$0.80, 0.5 billion would be bought, and so forth. So among the consumers paying \$0.50/BF, many are willing to pay more. In Figure 2-2a, consumers' extra willingness to pay above the \$0.50 market price is defined by the triangle *ABC* and is called **consumer surplus.** In general, consumer surplus is the area under the demand curve and above the price line, and should be added to price times quantity for a true estimate of social value. We will touch on this concept again later, but for now, we'll deal with market values of products consumed: the price per unit times the number of units bought.

Price Elasticity of Demand

The link between the demand curve and total sales value in Figure 2-2a and b is extremely important. As you increase quantity consumed, total sales value in Figure 2-2b may increase, decrease, or remain the same, depending on the relationship between price and quantity on the demand curve. You measure this relationship with the **price elasticity of demand:** *the percent change in quantity of a good demanded divided by the percent change in its price. A close estimate of elasticity is the percent change in quantity of a good demanded with a 1 percent change in price.* Since the demand curve slopes downward, an increase in price means a decrease in quantity, so the price elasticity of demand is negative, although the sign is sometimes omitted. In Figure 2-2a, between points *A* and *B*, the absolute value of the curve's price elasticity is greater than 1. In that range, if price decreases by 1 percent, the quantity demanded will increase by more than 1 percent, and total sales value will therefore increase. But between points *B* and *D*, the absolute value of elasticity is less than 1, and total sales value will decrease as the quantity of lumber sold increases. Thus, the total dollar sales from changes in consumption or prices are drastically affected by the elasticity of demand. That's why economists and producers are so concerned about price elasticity of demand.

Rather than memorize the formula for elasticity, it's best to know the concept, and then you can always derive the equation. Start with the following:

$$\text{Price elasticity of demand} = \frac{\%\text{ change in quantity demanded}}{\%\text{ change in price}} \tag{2-1}$$

The percent change in any quantity Q is ΔQ divided by Q. Thus, equation (2-1), for a given point on the demand curve, can be written as:

$$\text{Point elasticity} = E_p = \frac{\Delta Q/Q}{\Delta P/P} = \frac{\Delta Q}{Q} \cdot \frac{P}{\Delta P} = \frac{\Delta Q}{\Delta P} \cdot \frac{P}{Q} \tag{2-2}$$

Note that $\Delta P/\Delta Q$ is the slope of the demand curve ("rise over run" or, in calculus, the first derivative). So equation (2-2) says that **point elasticity** at a given P and Q point on the demand curve is the reciprocal of the curve's slope times P/Q.

Going back to Figure 2-2a, the demand equation is $P = f(Q_d)$. Numerically, for the linear function in Figure 2-2a,

$$P = 1.00 - 0.4Q_d$$

where 1 is the intercept, and -0.4 is the slope.[1] On this demand curve, a price-increase of \$0.10 reduces quantity demanded by 0.25 billion BF: the "rise over run" is $0.10/-0.25 = -0.4$, which is the slope. (This is a simplified lumber demand curve; normally we wouldn't expect it to be linear or to have consumption drop to zero so quickly with rising price.) You can find the point elasticity at B on the demand curve in Figure 2-2a by substituting the values of $P = 0.5$, $Q = 1.25$, and the slope of -0.4 into equation (2-2), remembering that $\Delta Q/\Delta P$ is the reciprocal of the slope:

$$E_p = (1/-0.4)(0.5/1.25) = -1.00$$

Thus, at point B, a 1 percent increase (decrease) in price would lead to a 1 percent decrease (increase) in quantity demanded.

On most demand curves, Figure 2-2a being no exception, elasticity changes from point to point, so one often figures a form of "average" elasticity or **arc elasticity** between two points. For example, we can adapt equation (2-2) to calculate the arc elasticity between points A and B in Figure 2-2a. Since there is no fixed P and Q, one approach is to use the midpoint or average P between A and B, or $(0.5 + 1)/2 = \$0.75$, and the average Q, which is $(0 + 1.25)/2 = 0.625$. Moving from point A to B, the difference in Q is $\Delta Q = (1.25 - 0) = 1.25$, and the $\Delta P = (0.5 - 1) = -0.5$. Arc elasticity is:

$$\text{Arc elasticity} = E_a = \frac{\Delta Q}{\Delta P} \cdot \frac{\text{average } P}{\text{average } Q} = \frac{1.25}{-0.5} \cdot \frac{0.75}{0.625} = -3.0 \qquad (2\text{-}3)$$

The above arc elasticity "averages" -3.0 between points A and B, but the point elasticity is -3.0 at only one spot on the Figure 2-2a demand curve. In the range where the absolute value of elasticity is greater than 1, we say the demand curve is **elastic,** as is the segment AB just examined. Demand is defined as **inelastic** where the elasticity's absolute value is < 1. Elasticity is *unitary* when its absolute value is 1. Using equation (2-3), the arc elasticity between B and D in Figure 2-2a is:

$$E_a = (1/-0.4)(0.3/1.75) = -0.43$$

which indicates an inelastic portion of the demand curve.[2]

[1] From the consumer's view, we usually think of the quantity demanded as a function of price—the independent variable—which mathematicians would usually put on the horizontal axis. By tradition in economics, however, the price variable is always on the vertical axis. Thus, for elasticity calculations that have a slope input (here, -0.4), we use the above functional form, often called an "inverse demand function."

[2] While elasticity of demand generally refers to price elasticity, other forms exist. For example, *income elasticity of demand,* a positive quantity, measures the percent change in demand with a 1 percent change in consumer income.

As you can see from Figure 2-2b (which ties into Figure 2-2a), *when demand is elastic, increasing consumption means increasing total sales value; when demand is inelastic, increasing consumption leads to decreased sales value.*

Summarizing possible price elasticities of demand:

If $E = 0$, demand is "perfectly inelastic" (vertical demand curve).
If $|E| < 1$, demand is "inelastic."
If $|E| = 1$, elasticity is "unitary."
If $|E| > 1$, demand is "elastic."
If E is infinity, demand is "perfectly elastic" (horizontal demand curve).

Remember, slope is not elasticity. For example, the slope of the straight-line demand curve in Figure 2-2a is constant, but the elasticity varies along the curve. Also, a perfectly elastic curve (horizontal with infinite elasticity) has a *zero* slope, and a perfectly inelastic curve (vertical or zero elasticity) has an *infinite* slope.

Let's look more closely at why price elasticity of demand is important to producers.

Demand Curves from the Producer's View

Initially we looked at demand curves from the consumer's view. But we can also consider the demand curve "faced by the producer." *The key to looking at demand curves from the* ***producer's*** *view is to ask, "If a firm produces quantity Q of a good, what will the price be?"* Going back to the example of Figure 2-2a, given our global lumber markets, not all lumber in a state would be produced by one firm. But if there were only one lumber company in Figure 2-2a, the curve shown would be the demand curve faced by the producer. Such a firm—a **monopolist**—could (1) affect lumber price by increasing or decreasing output or (2) change the lumber price and affect the quantity demanded. Such a firm is said to have **market power.** To see the demand curve from a producer's view, think of yourself as the only lumber producer in the market of Figure 2-2a. Reading off the demand curve, if you produced 0.5 billion board feet per year, you'd receive \$0.80/BF. If you increased output to 2 billion feet, you'd get only \$0.20/BF. What you saw before as consumers' expenditures on lumber—price times quantity—now becomes total sales *revenue* to producers. Note that this is gross revenue before subtracting any production costs.

Still imagining yourself as the only lumber producer facing the demand curve in Figure 2-2a, suppose you were producing at point B, and your analysts convinced you to expand lumber output into the inelastic portion of the demand curve, between B and D. Such analysts might well fear for their jobs, since your sales revenues would decline (see Figure 2-2b). But if you were producing between A and B, expanding output would *increase* total revenue. Thus, it's important for firms to know the price elasticity of the demand curves they face.

While no United States forest products firms are monopolists, some have large enough market shares for certain products that they face downward-sloping demand curves, as mentioned in the next chapter. These are cases of **oligopoly**—a few large firms, each facing a demand curve sloping downward but not as steeply as that which would be faced by a monopolist for the same market. Thus, the elasticity of downward-sloping demand curves is relevant for certain forest products firms.

When a large number of producers compete in the same market, they face horizontal or nearly horizontal demand curves. This is **perfect competition,** and each firm, relative to the market, is so small that no matter how much it sells, its output price is the same (or nearly so). Such producers are called **price takers;** they have to take the price offered in the market. If one firm increases its price, buyers will shop elsewhere. Let's assume that's the case for lumber producers in our hypothetical state in Figure 2-2a. With a downward-sloping market demand curve, how can separate firms face horizontal curves? The answer lies in the quantity scale differences. Figure 2-3 shows the previous demand curve for the entire market, with the horizontal axis in billions of board feet. Suppose price is \$0.50 and quantity is $Q = 1.25$ billion board feet. Now consider adding one firm producing 20 million board feet annually, or 1.6 percent of the 1.25 billion in the market, as shown beneath the market demand curve in Figure 2-3. Changing such a small percentage of the total output would have so little effect on the market price that the firm faces a roughly horizontal demand curve. Starting at point B in Figure 2-3, adding 1.6 percent to the total output yields a new output of

$$Q = 1.25 + 0.016(1.25) = 1.27 \text{ billion board feet}$$

Substituting this quantity into the given market demand equation, the market price drops from \$0.50 per board foot to

$$P = 1 - 0.4(Q) = 1 - 0.4(1.27) = \$0.492$$

The single firm above faces a nearly horizontal demand curve: if it produces a minimal output, the lumber price received is \$0.50 per board foot, and by expanding output to its 20 million board foot annual capacity, price drops by only $\frac{8}{10}$ of a cent. In reality, lumber markets are national and even international, so that single producers can be much larger than in the above example and still face nearly horizontal demand curves.

Figure 2-4 shows three other types of demand curves that firms could face for goods produced. Below each are the total revenue curves for output levels. In panel (a), the highly elastic demand causes total revenue to nearly double when output doubles. If the demand curve were horizontal (*perfectly elastic*), doubling the firm's output would truly double its total revenue. In (b) where demand is inelastic, increasing output will decrease total revenue in the range graphed. For (c), price elasticity is unitary, and total revenue remains constant for output levels shown (in such cases the demand function is a hyperbola).

This section shows how important it is to know the relevant price elasticities of demand when you want to measure the impacts on revenues from actions that change outputs of goods and services. A forestry example comes to mind: A Pacific Northwest industry forester once told me about a large purchase of old-growth timber his firm made with the intent of quickly harvesting the inventory

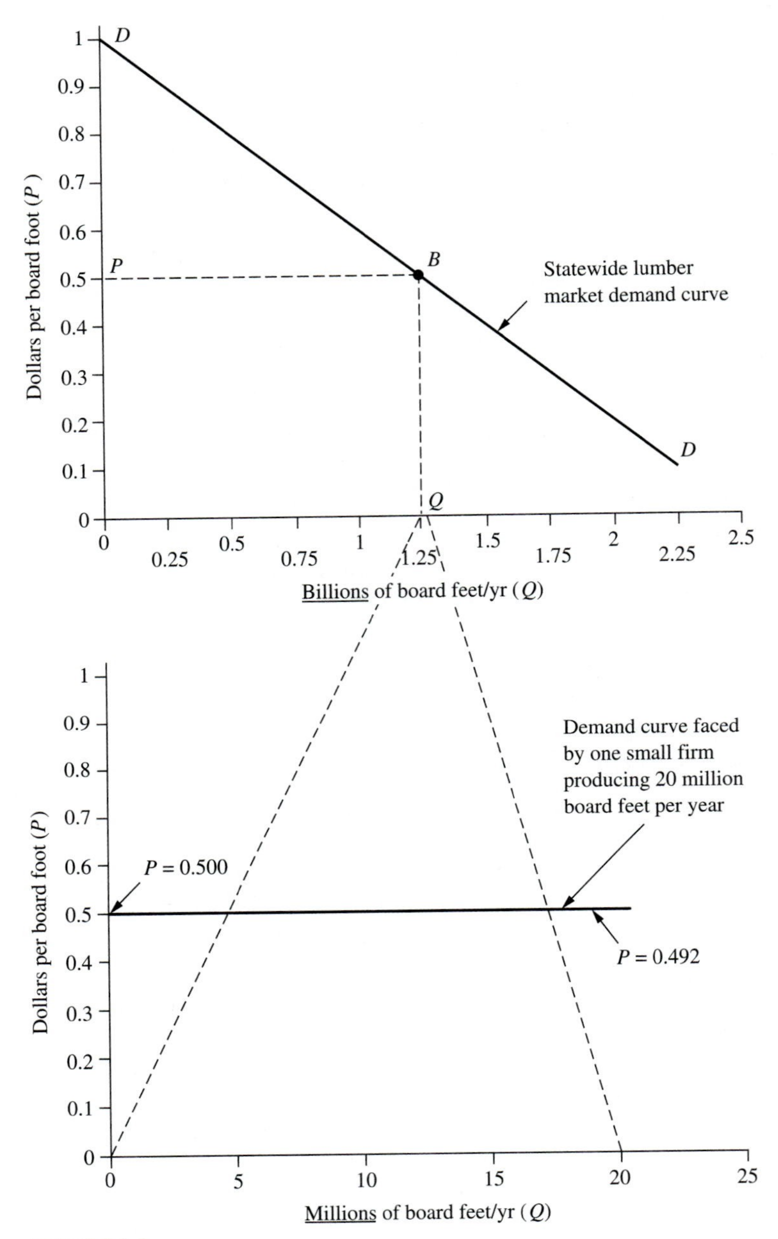

FIGURE 2-3
Statewide market demand and demand curve faced by small firm.

and selling many of the logs on the open market. At that time, export markets were not well developed, and high transportation costs limited log transport distances. Thus, after liquidation began, the firm's annual cut became such a large percentage of the local log market that prices for logs and stumpage (standing timber) dropped far more than was expected. Thus, revenue from the tract was much

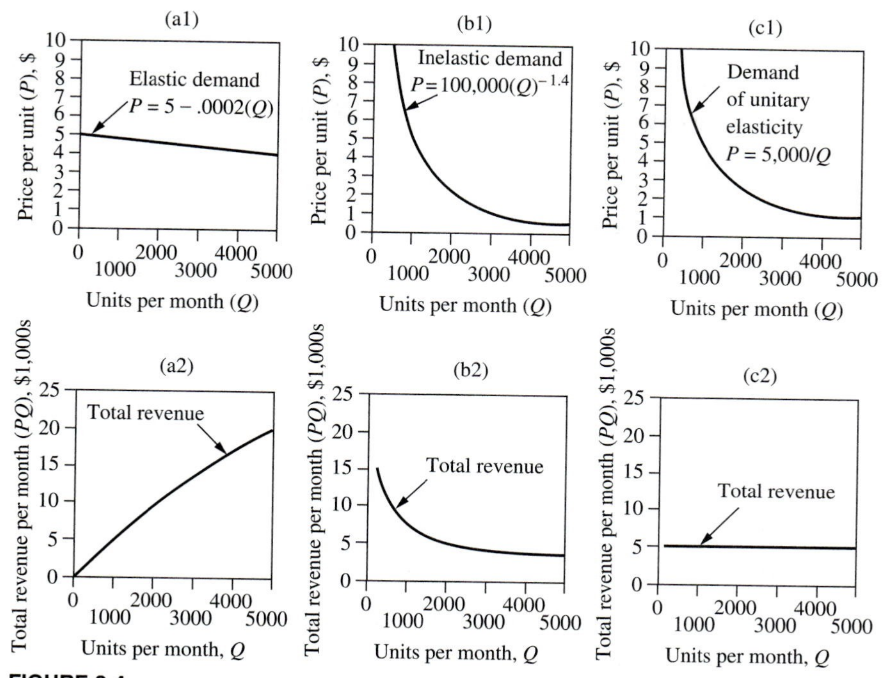

FIGURE 2-4
Demand and total revenue curves when demand is elastic (a), inelastic (b), and of unitary elasticity (c).

less than projected. Had the firm known the elasticity of the log demand function it faced, it could have more accurately estimated harvest revenues. Later sections will cover empirical estimates of elasticities and difficulties in measuring them.

Cross-Price Elasticity of Demand

When goods are substitutes, such as lumber and steel for building, if the price of one increases, the quantity of the other consumed will increase. For example, when lumber prices increased sharply in 1992–1993, some builders began using more steel studs to cut costs. We measure this relationship with the **cross-price elasticity of demand** for steel with respect to the price of lumber. One study estimated this cross-elasticity as 0.85, which means that as the price of lumber increases by 1 percent, the quantity of steel used in construction will increase by 0.85 percent. Corresponding numbers for concrete and plastic are 0.60 and 0.05 (Alexander and Greber, 1991). Thus, if lumber prices increased, you'd expect to find much more substitution of steel than plastic for lumber.

In general, cross-price elasticity of demand for good X with respect to the price of good Y is:

$$\text{Cross-price elasticity}_{XY} = \frac{\Delta Q_X / Q_X}{\Delta P_Y / P_Y}$$

When goods are substitutes, as in the above examples, cross-elasticities will be positive. For complementary goods—goods consumed together, like lumber and roofing materials—cross-elasticity will be negative. If the price of lumber increases, the quantity of lumber and roofing consumed will decrease. Knowing cross-elasticities helps in predicting economic effects of policies and events.

Demand Shifters

You must always be careful to specify that a demand curve applies only to a given group at a particular time. Figure 2-1 showed that adding more consumers to the group will shift the demand curve to the right. Thus, we say that population is a **demand shifter.** Until now we've held shifters constant and expressed quantity demanded as a function of price alone. Other demand shifters include consumers' incomes, tastes, advertising, and prices of related goods, (i.e., substitutes and complementary goods mentioned above). In statistically estimating demand curves for groups, economists use variables like years of education, age, or sex to capture the taste variable. So the complete demand function is:

$$Q_d = f(P, S_1, S_2, S_3, S_4, \ldots) \tag{2-4}$$

where the S's refer to shifters. Demand curves as traditionally drawn have the following complete function, where the shifters after the vertical line are held constant:

$$Q_d = f(P \mid S_1, S_2, S_3, S_4, \ldots) \tag{2-5}$$

To see the effects of different demand shifters, consider the commodity demand curve *DD* for one region in Figure 2-5. The curve is for one point in time, holding all shifters constant. Ask yourself, "If *Q* units of a good were originally demanded at price *P*, would more ($Q+$) or less ($Q-$) be demanded *at the same price,* after some shifter changed?" For example, if the quantity axis were houses, and average consumer income decreased, we'd expect fewer houses to be bought at any given price. That implies a leftward shift in the demand curve. Often special variables affect certain demand curves. For example, suppose Figure 2-5 showed

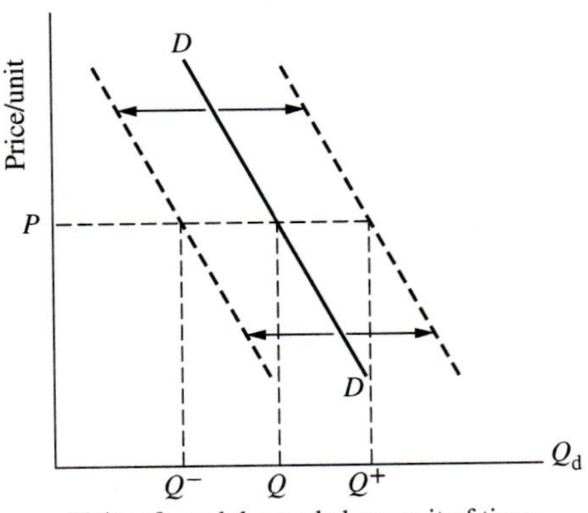

FIGURE 2-5
Shifts in the demand curve for a commodity.

visits to a forest park at different entry fees. Increased leisure time (or shorter workweek) could stimulate more visits at any price and shift the curve to the right. Try to imagine how changes in various shifters could move the demand curves for different forest outputs.

Before considering supply curves and the interaction between supply and demand, let's first derive a firm's supply curve through a simple production example. If you fully understand where the supply curve comes from, supply-demand interaction will make much more sense.

DERIVING THE SUPPLY CURVE THROUGH PRODUCTION ECONOMICS

To illustrate the economics of supply, we'll show optimum levels of fish output from Reba Pisces' Catfish Farm, given different fish prices. To supplement her tree farm income, Reba annually stocks her pond with catfish fingerlings, adds feed regularly, and sells the fish once a year. This is a simplified hypothetical example that may seem far afield from most forestland management. I chose it to review basic principles without the complexities of forestry's long production periods. Later we'll examine forestry cases that include the time element in more detail.

For simplicity, assume that Reba is one of many small producers in a large market: no matter how much fish she sells, she receives the same price (she faces a horizontal demand curve for fish). And no matter how much feed she buys, she pays the same price per pound of feed. These are two key assumptions of perfect competition. We'll also assume the **short run**—an economics term for the time period when some inputs are fixed. Here we'll hold constant the pond size and Reba's management input. In the **long run,** all inputs are variable. Assume that Reba's objective is to maximize the excess of revenues over costs—her net revenue or profit per year. Due to "market failures," covered in the next chapter, profit maximization may not necessarily lead to the best results from society's view.

Table 2-2 gives revenues and costs for the catfish farm. The annual fish harvest brings $70 per 100 lb, net of harvesting costs. The **variable costs** that increase with output are $30 per 100 lb of feed and include handling and storage costs. The remaining costs are annual **fixed costs** of $1,300 per acre of pond. Such costs are constant, regardless of output (in this case, for one year); they include payment for Reba's management time, other fixed costs such as maintenance and insurance, and a normal *return to capital* invested in land, pond, buildings, and equipment, on a per acre basis. The $400 normal annual return to capital is 10 percent of capital value, based on Reba's estimate that she could earn at least that much profit elsewhere, at the same risk, on the $4,000 market value per acre of her fish farm. Economists often call this normal return the **opportunity cost**—the cost of an alternative investment opportunity forgone. For now, let return to capital be like interest on a savings account or a bond: if a $100 bond or certificate of deposit yields you $7 per year, the $7 is your return to capital, which is a 7 percent **rate of return** on your $100 of capital. Later chapters will treat this concept in detail.

TABLE 2-2
REVENUES AND COSTS FOR CATFISH FARM

Item	Value
Revenue from fish harvested	$70/100 lb
Cost of feed (includes handling and storage) This is the "variable cost."	$30/100 lb
Fixed costs:	
Stocking fingerlings	$ 300/acre/year
Management, maintenance, interest on debt, depreciation, insurance, etc.	$ 600/acre/year
Normal return to capital at 10 percent of $4,000 fish farm value per acre*	$ 400/acre/year
Total fixed cost	$1,300/acre/year

*This is a real rate of return, excluding inflation. Inflated rates of return would be higher (see Chapter 5).

It may seem strange to list under costs the normal returns to capital, since they are actually income or profit. In business income statements, the reported profit is the total return to capital without separating the normal profit and excess profit as will be done here. However, for efficiency reasons it's important to keep track of the two to make sure that invested capital earns as much as it could earn elsewhere. This is extremely important in forestry, since capital is one of the major inputs. That's why we say forestry is "capital intensive."

Reba's economic problem is to find the optimum amount of feed or the optimum fish output. Remember that optimization implies maximizing something—in this case the net annual revenue, or the income she has left after paying all costs. Figure 2-6 shows the commercially salable fish output at different levels of feed, all in hundreds of pounds per acre per year. Such a curve is a **production function** showing the relationship between inputs and outputs. Its sigmoid shape is typical of many biological processes. The eventual flattening of the curve shows diminishing returns to added feed because the pond size is held constant (short run). The optimum amount of feed depends on prices of inputs and outputs.

Optimum Input Level

To find the optimum amount of feed, we first need to know costs and revenues for each input level, as shown in Table 2-3. Input is feed in 100-lb units per year. Where relevant, values are per acre of pond. Columns (a) and (b) are the production values plotted in Figure 2-6, and column (c) is the total revenue at $70 per 100 lb of fish. Column (e) shows total cost, which is the $1,300 fixed cost plus feed costs at $30 per 100 lb of feed. Subtracting (e) from (c) gives column (g), the **excess profit** above the normal return included in fixed costs. Most economics texts call column (g) **economic profit,** a term I avoid because it's too easily confused with profits as reported by most businesses. Business profit or net revenue would be reported as the $400 normal return included in column (e) plus the excess profit in column (g).

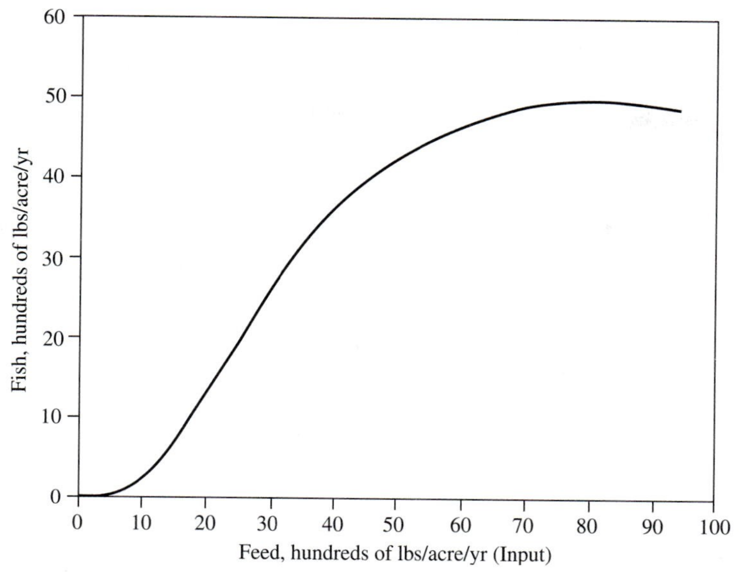

FIGURE 2-6
Catfish farm production function.

The *optimum input* is 55 hundred lb of feed/year because it yields the maximum excess profit of $225.20 in column (g). At that point, the total profit is the sum of normal and excess profits, or 400 + 225.20 = $625.20 per year. Column (d) of Table 2-3 is the **marginal revenue product:** *the addition to total revenue from adding another 100-lb unit of feed per year.* (You can think of this as a marginal revenue per unit of input. But normally, marginal revenue refers to added revenue per unit of *output.*) *Since not all rows in the table are in 100-lb feed increments, you need to look between the dashed lines to see that the marginal revenue product for any row is that row's total revenue minus the previous row's total revenue.* Similarly, the **marginal resource cost** in column (f) *is the addition to total cost when adding another 100-lb unit of feed*—the extra cost or price per 100 lb of feed. This is the difference between total costs in one row and the previous row, between the dashed lines. (You can think of marginal resource cost as a marginal cost per unit of input. But normally, marginal cost refers to added cost per unit of *output.*)

Reba should keep adding 100-lb units of feed per year as long as the extra revenue from doing so exceeds the cost per unit. If she goes 100 lb beyond the optimum of 5,500 lb, the extra 100 lb of feed costs $30 [column (f)] but only brings an added revenue of $28.70 [column (d)]. So she stays at 55 hundred lb of feed per year. This type of *marginal analysis* or incremental thinking is tremendously important and will be encountered again and again in forest economics. In columns (d) and (f), at the optimum 55 hundred lb of feed we can see *the marginal conditions necessary for profit maximization: added revenue and added cost per unit of input must be equal.* (We don't have exact equality here because the input units were not split finely enough.)

TABLE 2-3
CATFISH FARM REVENUES AND COSTS FOR ONE ACRE OF POND AT EACH *INPUT* LEVEL

(a) Feed (input) 100s lb/yr*	(b) Fish 100s lb/yr*	(c) Total revenue 70(b) $/ac/yr	(d) Marginal revenue product† $/100 lb feed	(e) Total cost 1,300 + 30(a) $/ac/yr	(f) Marginal resource cost‡ $/100 lb feed	(g) Excess profit§ (c) − (e) $/ac/yr
0	0	0	0	1,300	0	−1,300.00
10	1.12	78.40	26.05	1,600	30	−1,521.60
20	10.21	714.70	89.06	1,900	30	−1,185.30
30	24.01	1,680.70	94.33	2,200	30	−519.30
40	35.49	2,484.30	67.32	2,500	30	−15.70
50	42.91	3,003.70	40.54	2,800	30	203.70
53	44.47	3,112.90	34.20	2,890	30	222.90
54	44.93	3,145.10	32.20	2,920	30	225.10
55¶	45.36	3,175.20	30.10¶	2,950	30¶	225.20¶
56	45.77	3,203.90	28.70	2,980	30	223.90
57	46.16	3,231.20	27.30	3,010	30	221.20
60	47.18	3,302.60	22.04	3,100	30	202.60
70	49.47	3,462.90	11.58	3,400	30	62.90
80	50.50	3,535.00	3.63	3,700	30	−165.00
90	50.00	3,500.00	−12.20	4,000	30	−500.00

Note: Letters in parentheses refer to columns. *Fine-tuning by 100-lb units of feed shown within dashed lines.*
*From Figure 2-6 (production function).
†Added revenue from the last 100-lb unit of feed. Between the dashed lines, this is the difference between total revenues in one row and the previous row.
‡Added cost (or price) per 100 lb of feed. Between the dashed lines, this is the difference between total cost in one row and the previous row.
§Since normal profit is included in column (e), column (g) shows the excess above normal profit. Most economics texts call this column "economic profit."
¶Optimum where net revenue is maximized.

Figure 2-7a and b plots the values from Table 2-3; letters in parentheses refer to columns in the table. Figure 2-7a shows that the excess profit (the shaded area) is maximized at the optimum feed input of 55 hundred lb. As in the table, total profit is the normal return to capital plus the excess profit. Figure 2-7b uses the same horizontal scale as Figure 2-7a and shows that *the optimum input is where marginal revenue product and marginal resource cost are equal.* In plain English, that's where added revenue and added cost per unit of input are equal. Try, without using any jargon, to explain in simple terms why the optimum is 55 hundred lb of feed. (Unit by unit, increase the feed input up to 55 hundred lb. Look at extra revenues and costs for each unit increase. Now go one unit farther, and look at extra revenues and extra costs. What happens?) This gives us a general optimization principle in using any variable input: *go to the point where the extra cost of using the last unit of input just equals its contribution to total revenue.*

In Figure 2-7b, marginal resource cost is the slope (rise over run) of the total cost curve in Figure 2-7a: let the run be squeezed down to one unit of input (100 lb of feed), and the rise will be the $30 price for one unit of input—the

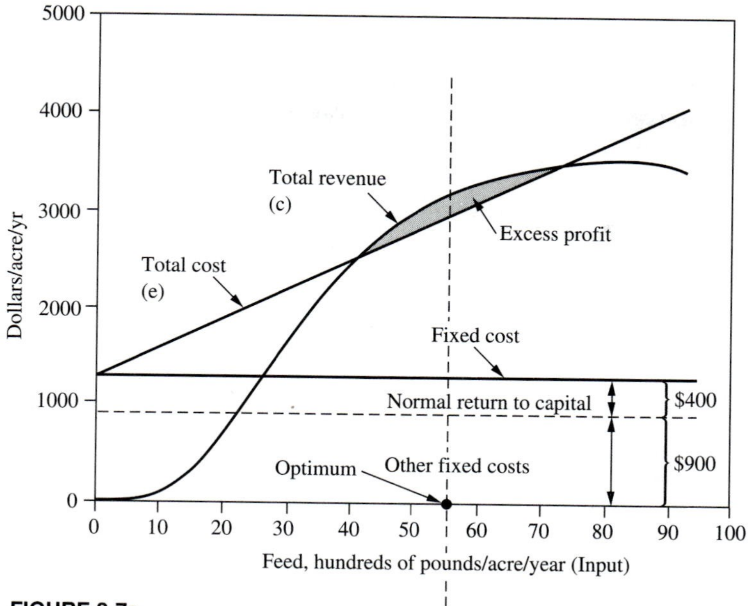

FIGURE 2-7a
Catfish farm – total costs and revenues over input.

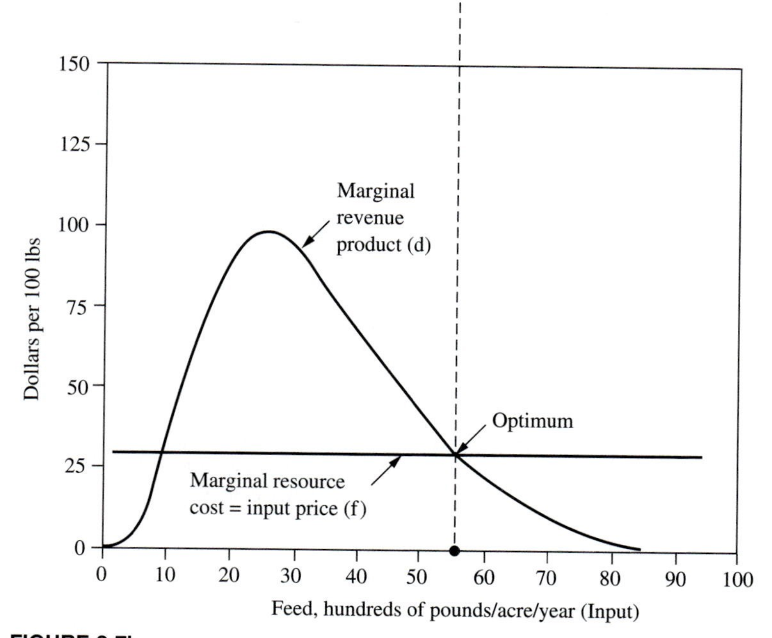

FIGURE 2-7b
Added costs and revenues per 100 pounds feed. (Letters in parentheses refer to columns in Table 2-3.)

marginal resource cost under perfect competition. By the same reasoning, the marginal revenue product is the slope of the total revenue curve. You can see in Figure 2-7a and b that the optimum is where the slopes of the total cost and revenue curves are equal (the calculus fans will recognize this as the place where the first derivatives of the two curves are equal). If feed price decreases, the total cost curve's slope declines, the marginal resource cost curve is lower, and optimal input increases. The reverse is true for a higher feed price. Changing the feed price affects *variable* costs.

Higher or lower *fixed* costs in Figure 2-7a will raise or drop the total cost curve and decrease or increase the profit. But this won't change the slope of the total cost curve. Therefore, *a change in fixed costs doesn't affect the marginal resource cost curve in Figure 2-7b and leaves the optimum input level unchanged.* But if fixed costs rise enough to make excess profits negative in the long run, the decision could be to cease production because returns to capital would be subnormal.

In Figure 2-7b, under perfect competition, the downward-sloping part of Reba's marginal revenue product curve is her demand curve for feed because it shows the amounts of feed she'd buy at different prices. This is demand from a consumer's view: Reba as a consumer of feed.

Allocation of Total Revenue

Before deriving the supply curve, note in Table 2-4 how the total revenue pie is divided. Together, the last two items in the table [(ii) and (iii)] are the after-tax return to capital, or profit that Reba is trying to maximize. Remember that the analysis here separates profit into normal and excess to make sure Reba is earning as

TABLE 2-4
CATEGORIES WHERE A FIRM'S TOTAL REVENUE (UNITS OF OUTPUT TIMES PRICE PER UNIT) IS SPENT ANNUALLY

Total Annual Revenue Goes to:

(1) Paying employees.
(2) Materials and supplies.
(3) *Depreciation* (a charge for the wearing out of assets such as buildings and machinery; *this is spent on replacing worn assets*).
(4) Other items such as research, administration, insurance, leases, royalties, etc.
(5) Remainder is *return to capital.* (Capital includes the market value of all durable assets like machines, real estate, financial assets, and inventories including raw materials such as timber.)

Return to Capital Goes to:

(a) Interest on debt (a return to capital loaned to the firm).
(b) Remainder is *return to equity* (ownership), or profit before taxes. In a firm's annual report, this is shown as income before taxes.

Return to Equity, before Taxes, Goes to:

(i) Income taxes and other taxes.
(ii) Dividends to stockholders.
(iii) Retained earnings (later, much of this may be spent on new assets for expansion and improvements).

much as she could elsewhere with the same amount of capital. The simple catfish farm example omits or combines some of the above categories; for example, it incorporates wages in the feed price which includes costs of storage and handling. Payment for Reba's management is in the fixed costs, since she is the only manager at all output levels for a fish farm of this size.

An important point is the division of revenue between labor and capital. For simplicity, let labor be all employees, and wages be all compensation. The return to capital can be expressed as an interest earning rate on invested capital in the same way you earn interest on savings, but at a higher rate in Reba's case because of risk (she aims for at least 10 percent). In the short run, with fixed technology and holding labor and capital constant, for a given amount of revenue, if labor gets more, capital gets less. This tugging and pulling between labor and capital occurs in the forest products industry as well as elsewhere. Note that the competition isn't necessarily between two different groups: most people who own capital also earn wages or salaries. It's important to express capital income as a *percentage of capital value* or a *rate of return.* Workers sometimes point to total dollar profits in an industry as a justification for higher pay. But the crucial measure is profit as a percentage of invested capital. For example, a $50 million annual profit for a firm sounds high but is low if it's only 1 percent of the invested capital value. Stockholders and lenders wouldn't be satisfied with a 1 percent rate of return, and such a firm would have problems raising capital for operation and expansion. You wouldn't be satisfied either with a 1 percent interest rate on a savings account. Chapter 5 discusses normal earning rates on different types of investments.

The concept of a rate of return on capital is vital in forestry and is a key element throughout this book. Details are in Chapter 4. You'll see that timber capital values can be high, and it's important to measure forestry income as a percentage of this capital value. For example, raving about a terrific timber growth of, say, 800 board feet per acre per year doesn't mean much from an efficiency view. You also need to know the value of this growth as a percentage of the capital value in land and timber per acre. A seemingly high annual volume growth may or may not represent an acceptable rate of return on invested capital. Later you'll learn how to incorporate nondollar values into this framework.

Measuring returns to capital as a percentage is like measuring wages as a *rate* in dollars per hour. A firm's total annual wage bill can mask a high or a low wage rate, just as the total profit per year says nothing about the rate of return on capital. Other things equal, given fixed productivity and technology, there's a short-run trade-off between wage rates and rates of return: higher wage rates bring lower rates of return to capital and vice versa.[3] Moving too far toward high wages and low rates of return causes problems in raising capital, thus reducing output, employment, and wages. Moving too far in the other direction of high rates of return to capital and low wages leads to an insufficient supply of labor, or dissatisfied workers, or strikes, all of which will reduce output and eventually lower returns to capital. These trade-offs hold for separate firms as well as the

[3]For a summary of literature on this point, see Hagemann (1990).

whole economy. Over time we reach a dynamic equilibrium. Now let's return to deriving supply curves.

Optimum Output Level

To derive a supply curve for the catfish farm, let's conduct a marginal analysis per unit of *output* (fish) rather than input. For each output level, the required amount of feed is found from the production function and recorded in columns (a) and (b) of Table 2-5.[4] As in Table 2-3, the values for total revenue and total cost are entered in each row of columns (c) and (e) in Table 2-5, but now for different output levels shown in column (a). **Marginal revenues** (added revenues) and **marginal costs** (added costs) in columns (d) and (f) are the differences in total revenues and costs for each 100-lb increase in fish, the output.[5] Since not all rows in Table 2-5 are in

---

[4]For most production functions [$y = f(x)$] it is difficult to mathematically find a function g such that $x = g(y)$. In the fish example, for each output y, the inputs x were found numerically from the production function.

[5]By convention, in economics, marginal revenue and marginal cost are always per unit of *output.* That's why columns (d) and (f) of Table 2-3 (the input table) are labeled "marginal revenue product" and "marginal resource cost." However, columns (d) and (f) in Tables 2-3 and 2-5 are conceptually the same: one is per unit of *input,* and the other is per unit of *output.*

TABLE 2-5
REVENUES AND COSTS FOR ONE ACRE OF POND AT EACH *OUTPUT* LEVEL

(a) Fish (output) 100s lb/yr*	(b) Feed 100s lb/yr*	(c) Total revenue 70(a) $/ac/yr	(d) Marginal revenue† $/100 lb fish	(e) Total cost 1,300 + 30(b) $/ac/yr	(f) Marginal cost‡ $/100 lb fish	(g) Excess profit§ (c) − (e) $/ac/yr
0	0	0	0	1,300.00	0	−1,300.00
10	19.83	700	70	1,894.90	23.40	−1,194.90
20	27.07	1,400	70	2,112.10	21.30	−712.10
30	34.78	2,100	70	2,343.40	25.50	−243.40
40	45.45	2,800	70	2,663.50	40.20	136.50
43	50.16	3,010	70	2,804.80	51.30	205.20
44	52.05	3,080	70	2,861.50	56.70	218.50
45¶	54.18	3,150	70¶	2,925.40	63.90¶	224.60¶
46	56.61	3,220	70	2,998.30	72.90	221.70
47	59.43	3,290	70	3,082.90	84.60	207.10
50	73.89	3,500	70	3,516.70	196.20	−16.70

Note: Letters in parentheses refer to columns. *Fine-tuning by 100-lb units of feed shown within dashed lines.*

*From Figure 2-6 (production function).

†Added revenue from the last 100 lb of fish (the price of fish). Between the dashed lines, this is the difference between total revenues in one row and those in the previous row.

‡Added cost of the last 100 lb of fish. Between the dashed lines, this is the difference between total cost in one row and that in the previous row.

§Since normal profit is included in column (e), column (g) shows the excess above normal profit. Most economics texts call this column "economic profit."

¶Optimum where net revenue is maximized.

100-lb increments of fish, *you need to look between the dashed lines to see that the marginal cost for any row is that row's total cost minus the previous row's total cost.* For revenues, the same relationship holds between totals and marginals, between the dashed lines.

In Table 2-5, *the optimum output* is 45 hundred lb of fish/ac./yr, where excess profits are maximized in column (g) ($224.60). At that point, total profit is the sum of normal and excess profits, or 400 + 224.60 = $624.60 per acre per year. This maximum is in theory exactly the same as in Table 2-3 but differs slightly because the discrete units in each table are not small enough. In seeking the optimum, Reba keeps producing more fish until the marginal cost (MC) per added 100 lb of fish exceeds the marginal revenue (MR): that occurs at 46 hundred lb of fish, where the extra cost of producing the last 100 lb is $72.90 [column (f)] compared with the extra revenue of $70 [column (d)]. Thus, she stays at 45 hundred lb where MR = $70, the price of fish, and MC = $63.90. If the analysis were fine-tuned to the nearest pound, the optimum would be where the MR per pound of fish exactly equaled the MC per pound—a point between 45 and 46 hundred lb of fish, as shown in the continuous graphic solution in Figure 2-8. Looking in column (b) of Table 2-3, that's the same output found when optimizing the feed input.

Figure 2-8a and b graphs columns from Table 2-5 in a way similar to Figure 2-7, except now values are plotted over units of *output.* When approaching the maximum fish production (50.5 hundred lb) in Figure 2-8a, note how total costs rise sharply because no matter how much feed is added, production can't increase beyond the maximum—diminishing returns again. The vertical distance between the total and fixed cost curves is the *total variable cost* at any output level. Following the same reasoning as in Figure 2-7a and b, the marginal curves are the slopes of the corresponding total curves.[6] Looking at Figure 2-8b, you can see that it wouldn't pay to produce beyond the optimum because the resulting marginal cost would exceed the marginal revenue (the $70 price per unit of output). Figure 2-8a shows that the excess profit (the shaded area), is maximized at the optimum output. Remember, in both Figures 2-7a and 2-8a, the shaded "excess profit" (often called "economic profit") is an amount *above normal profits* which are included in the fixed cost curve. So total profit is the normal return to capital plus excess profit.

Figure 2-8a and b again illustrates *the necessary conditions for profit maximization, but now on the basis of output units: increase production until the added cost and added revenue per unit of output are equal (marginal cost = marginal revenue).* Since marginal revenue equals output price under competitive conditions, the firm should produce where marginal cost = output price. This relationship defines the firm's supply function, as explained below. In Figure 2-8a and b, note that changing the fixed cost simply raises or lowers the total cost curve and doesn't affect marginal cost or the optimum production point, as was noted for Figure 2-7. But altering fixed costs will change profits.

---

[6]In graphs such as Figure 2-8b, we often see curves of average fixed cost, average variable cost, and average total cost. I have omitted these in order to concentrate on the importance of marginal cost and revenue.

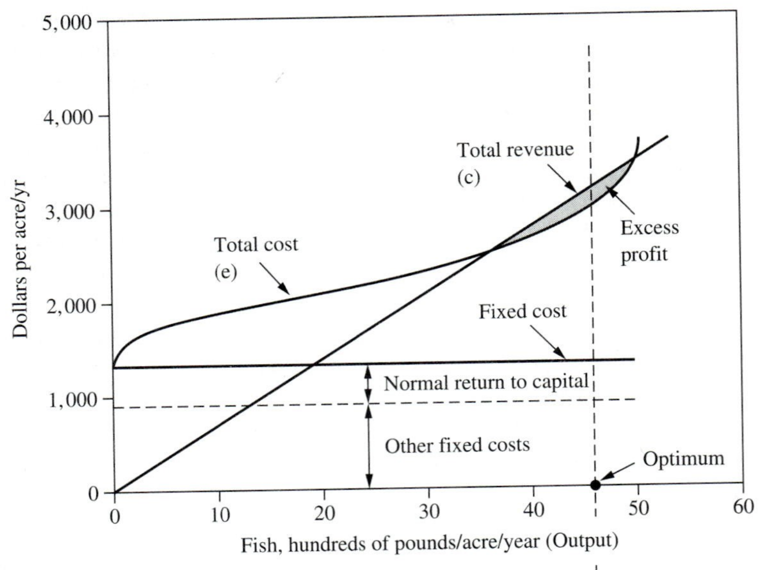

FIGURE 2-8a

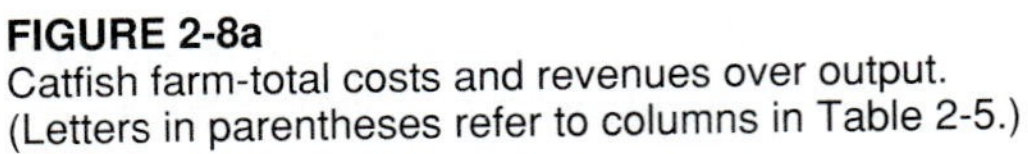

Catfish farm-total costs and revenues over output.
(Letters in parentheses refer to columns in Table 2-5.)

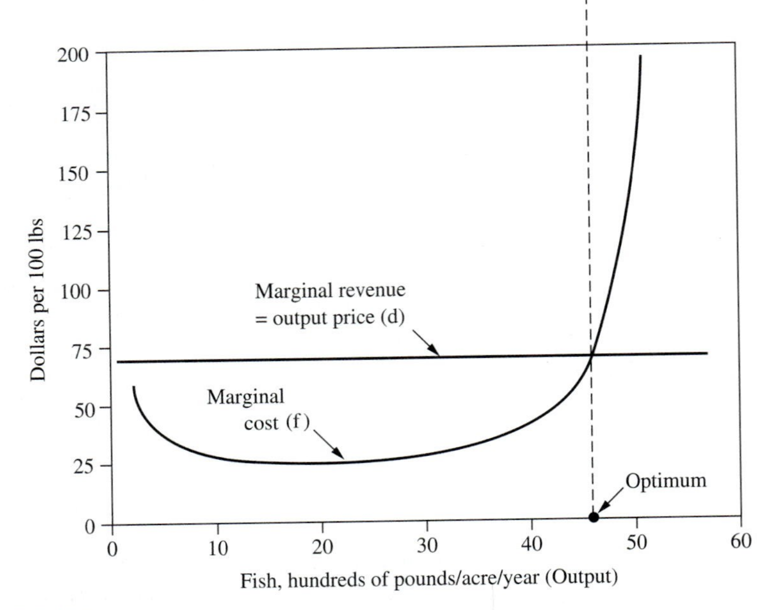

FIGURE 2-8b
Marginal costs and revenues per 100 pounds of fish. (Letters in parentheses refer to columns in Table 2-5.)

Supply Curves from the Producer's View. In Figure 2-8b, *the marginal cost curve defines the firm's supply curve because it shows optimal output levels at different output prices.* In other words, the marginal cost curve shows *how much of a good would tend to be supplied at different prices, which is the definition of a supply curve.* Figure 2-9b illustrates this point for Reba's catfish farm, using three different fish prices, or marginal revenues (MR): $64, $100, and $150 per 100 lb. At each of these prices, optimum supply levels are shown as (a), (b), and (c). Think of the three price lines in Figure 2-9b as possible fish demand curves that Reba could face, given our initial assumptions about perfect competition. In general, the higher a firm's output price, the more will be produced.

Figure 2-9a shows the total revenue (TR) curves for each fish price in parentheses. The higher the fish price, the steeper the total revenue curve because higher price makes the curve rise more per 100 lb of fish output. The $64 price was chosen to show the point (*E*) where excess profit is zero: only the normal return to capital is earned. This is the competitive equilibrium. Over the long term, firms are unlikely to stay in business if total revenues are lower, because the returns to capital would then be below normal. Owners of capital and lenders wouldn't support such ventures. Thus, the firm's long-term supply curve is the heavy portion of the marginal cost curve in Figure 2-9b.[7]

An industry supply curve is the horizontal sum of the individual firms' supply (or marginal cost) curves, as shown in Figure 2-10 for an "industry" of two firms producing units of some output. This is the same concept as summing individuals' demand curves to get the market demand curve in Figure 2-1. Intuitively the supply curve's upward slope makes sense: if a commodity's price increases, producers are likely to supply more.

In economics we need to constantly bridge the gap between theory and practice. In the catfish farm case, for example, you wouldn't expect Reba Pisces to be drawing production functions, sketching marginal revenue and marginal cost curves, and carefully plotting out her optimum feed level. In many cases, production functions aren't known, so that you can't optimize precisely. But the theory provides useful ways of thinking. For example, without drawing any graphs, Reba could keep records of different feed levels and the resulting fish production over time. She could use marginal analysis and determine the extra production gained when adding an extra unit of feed. Knowing the costs of adding feed and the prices of fish, she could keep adding feed until the extra benefit just equaled the extra cost. In some types of production, the principles of profit maximization can be

[7]Because the total revenue curve is a straight line out of the origin, the point where it's tangent to the total cost (TC) curve is the point of minimum average cost (*E* in Figure 2-9a). Why? Because the slope of a line from the origin to the TC curve is the average cost: total cost divided by units of output (rise over run). The tangency point *E* is minimum average cost because it marks the lowest possible slope of a line from the origin to the TC curve. For short periods of time, the firm can operate if the output price falls below $64, and the total revenue curve is below the TR(64) curve in Figure 2-9a. In that case, total cost exceeds total revenue. The firm may operate temporarily if this excess of total cost over total revenue is less than its fixed cost, since the latter is incurred whether or not the firm shuts down. But such a position is not viable over the long term. If total cost exceeds total revenue by an amount greater than fixed costs, the firm loses less by shutting down.

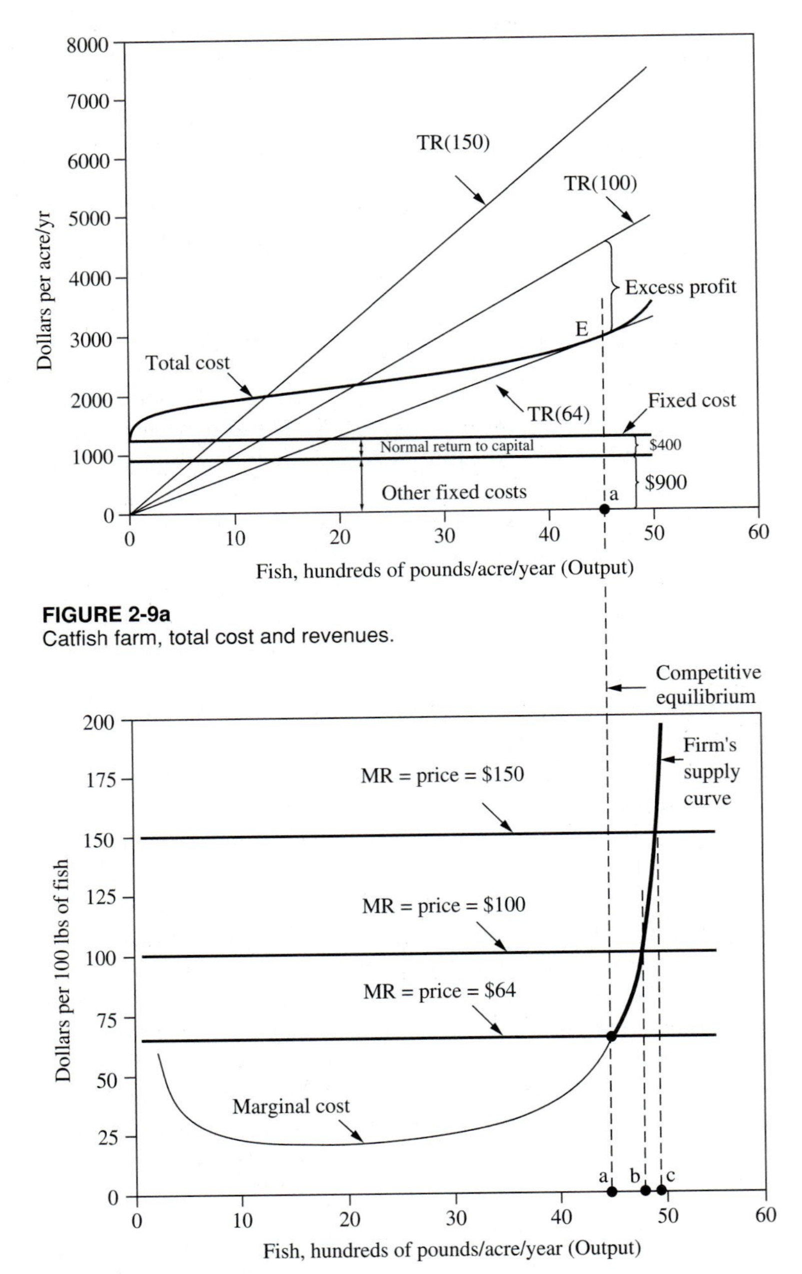

FIGURE 2-9a
Catfish farm, total cost and revenues.

FIGURE 2-9b
Marginal costs and revenues per 100 pounds of fish.

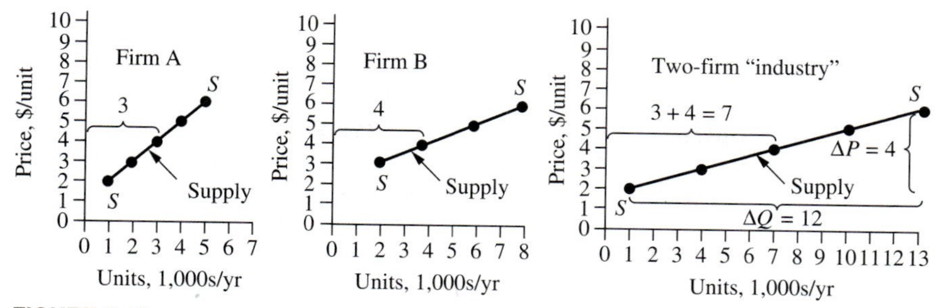

FIGURE 2-10
Industry supply curve is the horizontal sum of firms' supply curves.

applied only in the broadest sense. In other cases, the marginal conditions are applied quite precisely, for example, in computer-based models for finding optimal timber harvest schedules over time.

Price Elasticity of Supply

Price elasticity of a supply curve is the percent change in quantity of a good supplied divided by the percent change in its price for a given range of output. Or you can think of supply elasticity as the *percent change in quantity supplied with a 1 percent change in price.* The concept and the equations are the same as those for point and arc elasticity of demand [see equations (2-1) to (2-3)]. But with supply, the elasticity is positive, since the quantity response moves in the same direction as the price (the curve is positively sloped).

For the "industry" supply curve of Figure 2-10, you can find the *arc elasticity of supply* between points S and S, using equation (2-3). The midpoint or average price is $(2 + 6)/2 = 4$, and the average quantity is $(13 + 1)/2 = 7$. As marked on the curve, the changes in price and quantity are 4 and 12. Arc elasticity of supply is:

$$\text{Arc elasticity} = E_a = \frac{\%\text{ change in quantity supplied}}{\%\text{ change in price}}$$

$$E_a = \frac{\Delta Q/\text{average } Q}{\Delta P/\text{average} P} \tag{2-6}$$

$$E_a = \frac{\Delta Q}{\Delta P} \cdot \frac{\text{average } P}{\text{average } Q} = \frac{12}{4} \cdot \frac{4}{7} = 1.71$$

On average, over the range SS in Figure 2-10, if price increases (decreases) by 1 percent, quantity supplied will increase (decrease) by 1.7 percent. Because elasticity averages over 1, the supply curve is elastic over that range. Since the range SS is quite large, point elasticity can vary greatly at different places on the curve and can be figured with equation (2-2). Price elasticity of supply tells us how responsive supply will be to changes in price in a given market. For example, timber supply is often said to be fairly inelastic in the short run, so that one cannot expect dramatic changes in quantity supplied when stumpage prices change, other things being equal. In the section on supply-demand interaction, we'll look at the difference between short- and long-run supply curves.

You can calculate an arc elasticity for the inelastic portion of Reba's fish supply curve to the right of the point *a* in Figure 2-9b. Using Table 2-5, let's calculate arc elasticity between fish outputs of 46 and 50 hundred lb. Then $\Delta Q = 4$, and average Q between the two points is 48 hundred lb. Reba would tend to produce where price = marginal cost. Since the marginal cost curve is the supply curve, in column (f) you can read the marginal cost values of \$72.90 per 100 lb of fish at a fish output of 46 hundred lb and \$196.20 for an output of 50 hundred lb. You can think of these marginal costs as the fish prices per 100 lb that would bring forth outputs of 46 and 50 hundred lb of fish. Thus, for the arc elasticity, $\Delta P = (196.20 - 72.90) = 123.30$, and average P is $(72.90 + 196.20)/2 = 134.55$. Substituting these values into equation (2-6), arc elasticity of Reba's fish supply curve in this range is:

$$(4/123.3)(134.55/48) = 0.091$$

On average, for each 1 percent increase in fish price Reba would increase output by less than 1/10 percent. This indicates a fairly inelastic supply curve, since elasticity is substantially below 1. Although the inelastic fish supply curve "looks" inelastic, because it's steep, and the elastic industry supply curve in Figure 2-10 "looks" elastic because it's fairly flat, appearances aren't always helpful. By scaling the axes differently, you could make either of the curves look steeper or flatter without changing the elasticities.

Supply Curves from the Consumer's View

So far we've looked at supply curves from the producer's view. You can also think of a supply curve faced by a consumer. Under perfect competition, individual consumers are such a small percentage of the market that they face horizontal (perfectly elastic) supply curves. For example, you face a horizontal supply curve for lumber: no matter how much you buy for personal use, you pay the same price. Or Reba "faces" a horizontal supply curve for fish feed, which is her *marginal resource cost* per unit of input in Figure 2-7b (although she's a producer, she's also a consumer of inputs).

If we derive upward-sloping regional supply curves, as in Figure 2-10, how can individual buyers face horizontal supply curves? The answer is like the explanation on the demand side. Under perfect competition, consumers each buy such a tiny sliver from the upward-sloping market supply curve that they don't change the commodity price. Another example: a small sawmill in a large timber market will face a roughly horizontal supply curve for logs, as Reba faced for fish feed (see her marginal resource cost curve in Figure 2-7b). Such mills can get as much of the input as they need at the market price and would be foolish to offer more.

Without any competition in a resource market, only one firm buys the resource. Such a firm is called a **monopsonist** and faces an upward-sloping supply curve for the resource or input. A small group of resource buyers may also face an upward-sloping resource supply curve—a situation called **oligopsony.** Large wood-processing mills often face upward-sloping supply curves for logs. The next chapter treats these issues further.

This section is a mirror image of the demand side. We derive demand curves from the consumer's view, but producers face demand curves. We derive supply curves from the producer's view, but consumers face supply curves.

Supply Shifters

As with a demand curve, a supply curve applies only to a given group or "industry"—of suppliers rather than consumers—at a given time. You can represent the industry supply curve with the following equation where quantity supplied is a function of price, holding constant the shifters S_n, shown after the vertical line:

$$Q_s = f(P \mid S_1, S_2, S_3, S_4, \ldots) \qquad (2\text{-}7)$$

Figure 2-10 shows that adding more firms will shift the industry supply curve to the right. Other **supply shifters** include prices of inputs, changes in production technology, natural disasters, the cost of borrowing funds, prices of substitute inputs, and government regulations. To think about effects of supply shifters, look at the supply curve *SS* in Figure 2-11. Ask yourself, "If Q units of the good were originally supplied at price P, would more (Q^+) or less (Q^-) be supplied after some shifter changed?" For example, if the x axis is cords of pulpwood in a region and a new low-cost thinning technology is developed, it's likely that more pulpwood would be supplied at any given price: the supply curve would shift to the right. Or suppose Figure 2-11 shows regional sawtimber stumpage, and government regulations impose logging restrictions for preserving endangered species or scenic values. At any given stumpage price, less timber would be supplied: the

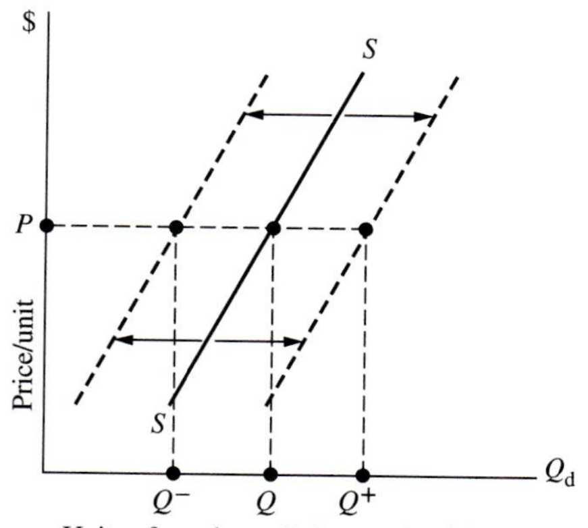

FIGURE 2-11
Shifts in the industry supply curve for a commodity.

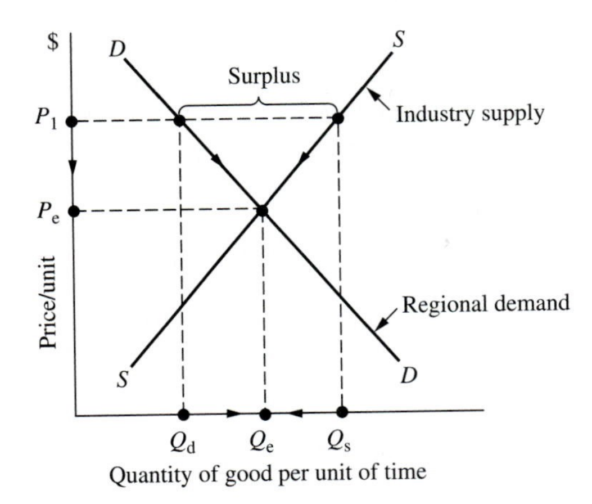

FIGURE 2-12
Surplus from an economic view—one region.

supply curve would shift to the left. Try to think of other commodities and effects of shifters on their supply functions.

SUPPLY-DEMAND EQUILIBRIUM

Figure 2-12 shows hypothetical demand (DD) and supply (SS) curves for a commodity in a large market with many buyers and sellers. Suppose price is initially at P_1. Then consumers would tend to demand Q_d, and producers would supply Q_s, with the **surplus** indicated on the graph. With a surplus, sellers cut prices to reduce inventories. As prices fall, the quantity demanded increases, and the quantity supplied decreases until equilibrium is reached at price P_e, and supply equals demand at Q_e.

In Figure 2-13, which also represents the above market, suppose price is at P_2. In that case, producers would supply Q_s; consumers would demand Q_d, and we'd have the indicated **shortage.** Buyers would then bid up prices, causing Q_d

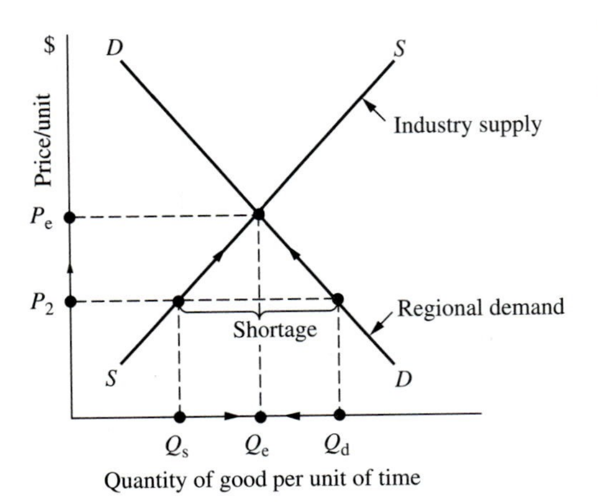

FIGURE 2-13
Shortage from an economic view—one region.

to decrease and Q_s to increase. Equilibrium would again be at P_e with supply equaling demand at Q_e. Given a well-functioning market, in the long run, these mechanisms prevent us from running out of salable commodities that can be produced. A key variable is price. Under the foregoing conditions, there always exists some price at which supply eventually equals demand. But we may not like the price, and we may not like lags that occur.

For a competitively produced consumer good, Figure 2-14 shows how individual firms are tied into the above equilibrium process. The quantity axes in Figures 2-12 through 2-14 are left general so that you can think of any forest or non-forest output (e.g., plywood, waferwood, wheat, beans, catfish, skier days, etc.).

FIGURE 2-14
Process of market equilibrium for one good.

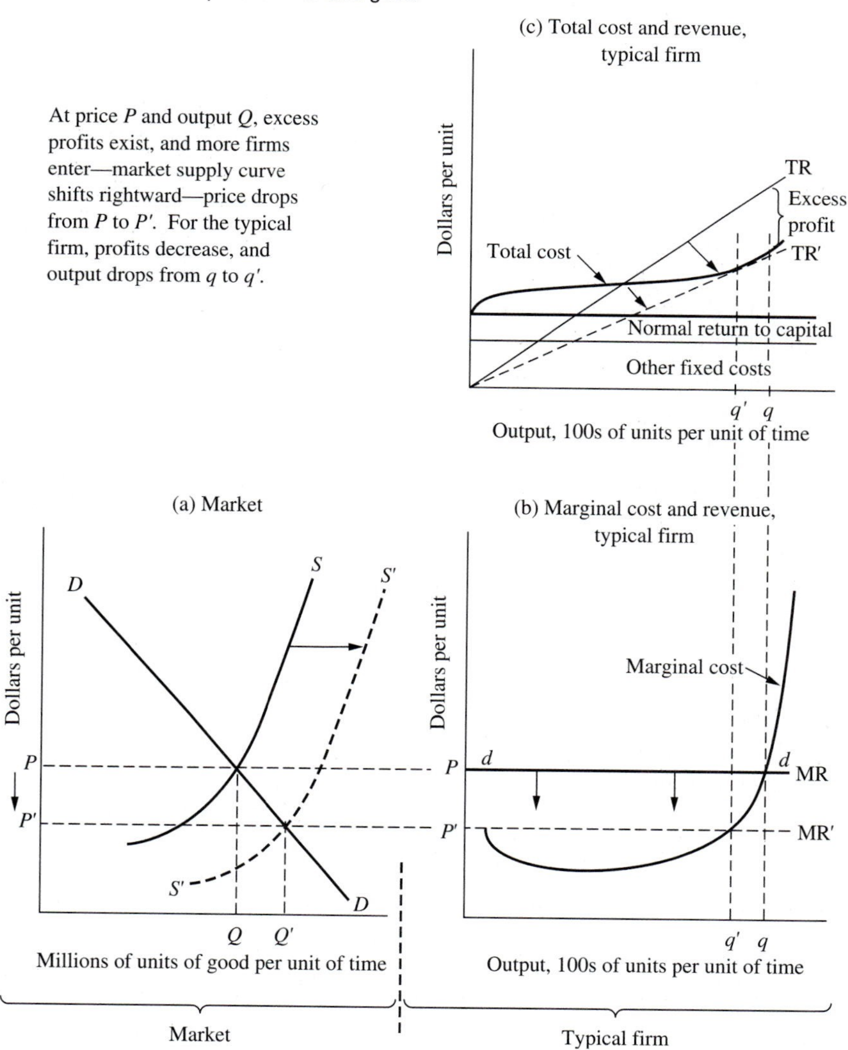

In Figure 2-14, start in panel (a), which represents a large market with the demand and supply curves *DD* and *SS* and the equilibrium price (*P*) and quantity (*Q*). Panel (b) shows the horizontal demand curve (*dd*) faced by one typical, competitive firm, given price *P*, which is also the firm's marginal revenue (MR) (see Figure 2-8b). This price generates the total revenue, or price times quantity, which is the TR curve in panel (c). Panels (b) and (c) are analogous to Figure 2-8, and when price = *P*, they show optimum output at quantity *q*, where excess profit is maximized. As before, normal profits (returns to capital) are included in fixed costs.

Now, what would you expect to happen if, as pictured, higher-than-normal annual profits—measured as a percentage of invested capital—were being earned by typical firms in this industry? Excess profits would attract investors, who would build more production facilities, and more of the good would be produced. Then, in panel (a) of Figure 2-14, at any given product price, more would be supplied, leading to a series of points to the right of *SS* that would form a new supply curve, $S'S'$. So investing more in production will shift the supply curve to the right, creating a surplus at the original price *P*. Product prices therefore fall to P', yielding lower MR′ and TR′ curves in panels (b) and (c) for the typical firm. This is a "long-run" response that involves actions like plant expansion, new production facilities and processes, new firms, or more land brought under cultivation for the case of fiber and food crops. (Remember that in the short run, capital and plant size were kept constant.) In panels (b) and (c), equilibrium would be reached at output q', when the lower TR′ curve eliminated excess profits for the typical firm, permitting no more than the normal return to capital.[8] Incentive would no longer exist for rapid industry expansion.

Figure 2-15, starting at the upper left, outlines verbally what's going on in Figure 2-14. The end result is that the consumer receives the lowest prices *consistent with a fair return to capital.* Obviously things don't work this smoothly in the real world, and in the next chapter, we'll examine problems with free markets. Firms don't always face horizontal demand curves, there are ambiguities in measuring rates of return (especially considering differences in rates of return on a firm's total capital and on added capital), and problems exist in predicting prices and other variables. Nevertheless, there's evidence that in many sectors, mechanisms like those in Figures 2-14 and 2-15 are occurring in a rough sense, given large numbers of firms. In the United States we've seen this happen in the southern pine plywood industry, which expanded rapidly between 1965 and 1980 when rates of return looked promising in such ventures. New plant construction dropped as rates of return on southern pine plywood production declined. At the same time, investment in producing products such as oriented strand board and waferwood appeared more promising and began to increase. Similar dynamics have occurred in areas like electronic calculators, video game rooms, TV video rentals, compact

[8]Economics texts often call this competitive equilibrium a point of "zero economic profit" for the firm. This can cause confusion, since normal profits (which are included in the fixed cost curve) are still being earned.

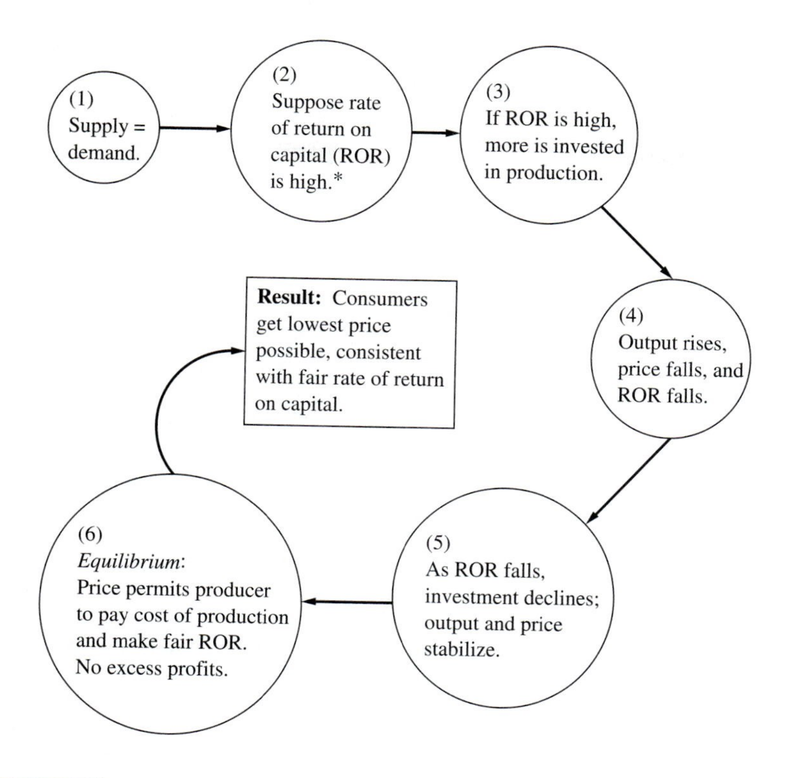

* You could start with a low rate of return, say due to increased competition from low-priced substitutes. In that case, the stages would read: (2) Suppose rate of return on capital (ROR) is low. (3) If ROR low, less invested in production. (4) Output falls and price rises. (5) As price rises, investment-decline tapers off; output and price stabilize. The equilibrium results are the same.

FIGURE 2-15
Pricing results of a competitive market for one good.

discs, and personal computers. Expected rates of return in these areas seemed high at first; entrepreneurs rushed in; output increased, and prices fell.

Perhaps one of the biggest dangers in studying market equilibrium theory is that we become so enthralled with the logic that we think it works precisely in practice. It doesn't always. Sticky prices, fluctuations, and lags sometimes occur. Pockets of market power exist, and even with large numbers of buyers and sellers, the competitive result doesn't necessarily always occur. For example, credit card interest rates, which spiked in the United States during the 1970s inflation, didn't settle down when inflation subsided. Thus, real interest rates on such loans remained abnormally high for many years, in spite of competition among thousands of banks serving millions of customers (Ausubel, 1991). But such examples, when there are many buyers and sellers, are the exception rather than the rule.

There's little doubt that a well-functioning market is extremely efficient: it can avoid shortages, get goods and services where they're needed, assure acceptable

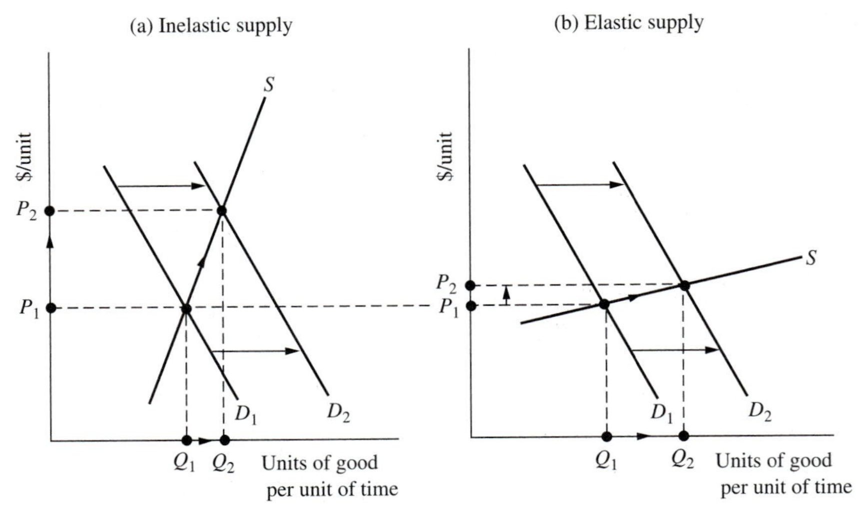

FIGURE 2-16
Responses to increased demand—short run.

returns to capital, and stimulate employment. But markets aren't perfect, and there are real dangers in blindly worshiping unregulated markets as the solution to all of society's ills. The next chapter notes some of these imperfections or "market failures" in a forestry context and what we can do about them.

Elasticity Interactions

Supply and demand elasticities can influence the outcome of forest policies. Figure 2-16 shows the importance of supply elasticity when demand shifts in the short run—a time period not long enough for major changes in capital investment. Suppose the *x* axis is cords of firewood per year, and demand is shifted rightward from D_1 to D_2 by government consumer education programs and income tax credits for woodstove purchases in order to lessen oil consumption. With inelastic supply (*S*) in panel (a), this increase in demand causes a relatively large increase in price per cord, compared with the increase in quantity of firewood consumed. But in panel (b), with more elastic supply, the same policy causes a relatively large increase in quantity of firewood consumed, compared with the price-increase.

Figure 2-17 shows how demand elasticity can affect policy outcomes. Let the *x* axis be visitor days per year at private campgrounds. Suppose the government policy is educational programs and tax incentives for private forest owners in order to stimulate campground construction and lessen pressure on public facilities. With inelastic camping demand (*D*) in panel (a), the resulting supply shift from S_1 to S_2 causes a large drop in the fee per visit but a fairly small increase in consumption. However, in panel (b), with elastic demand, the same supply shift causes a large increase in campground visits and a relatively small drop in price. Policy makers should know about the relevant elasticities before implementing programs.

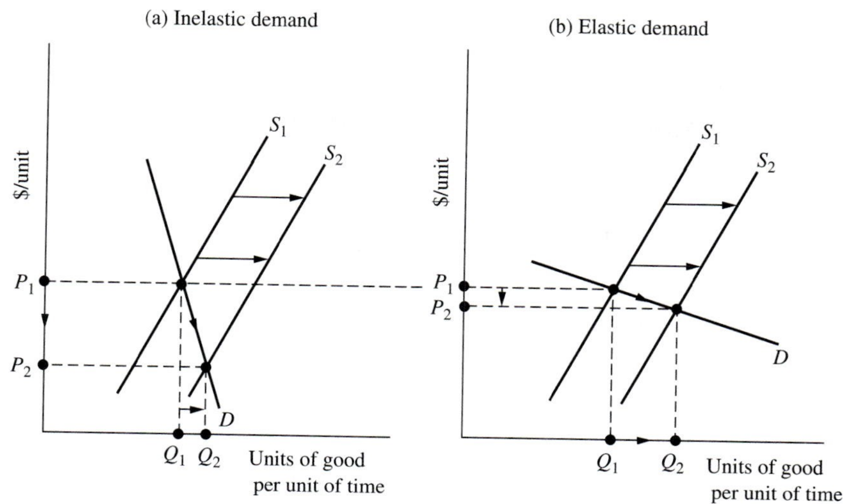

FIGURE 2-17
Responses to increased supply—short run.

Both Figures 2-16 and 2-17 show the important distinction between the movement along a curve and a shift in the curve. For example, Figure 2-16 shows a shift in the demand curve but a movement along the supply curve in response to a higher price. In Figure 2-17 we have a shift in the supply curve but a movement along the demand curve in response to lower prices. In Figures 2-16 and 2-17, the *x* axes are left general, so that you can construct scenarios for a wide variety of forest outputs.

Long-Run Supply

Figure 2-18 shows short- and long-run adjustments to an increase in demand for a good. Suppose the *x* axis is private campground visits, and years are assigned to demand and supply curves (*D* and *S*). Starting with 1990, the fee per visit is P_{1990}, with consumption at Q_{1990} visits per year. Suppose lower gasoline prices shift the demand curve rightward to $D_{1991-95}$ for four years. For 1991, the resulting higher *P* and *Q* are shown in Figure 2-18 without a supply shift, since, in the short run, the number of campgrounds is fixed. However, by 1995, in response to higher prices per visit, suppose more capital is invested in campground construction, shifting supply to S_{1995}. The resulting price is now lower and quantity higher. If we envision further rightward shifts in demand due to factors like increased population and higher income, we'd have a series of points in line with points *a* and *b*, thus defining a **long-run supply curve.** You can think of the dashed curve as a long-run output supply response to changes in output price. Similar mechanics would occur in the opposite direction if demand shifted leftward.

A long-run supply curve may be upward-sloping, horizontal, or downward-sloping, depending on trends in resource costs and production technology. In

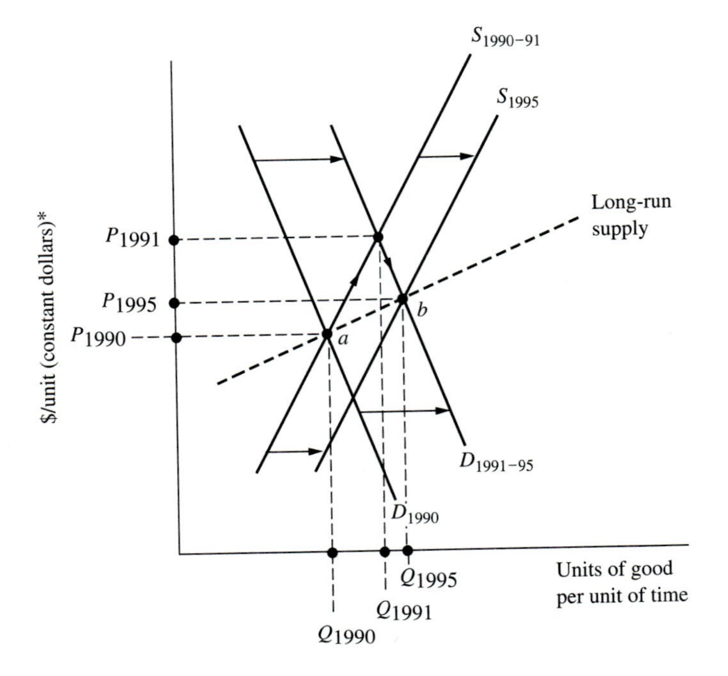

*Excludes inflation.

FIGURE 2-18
Long-run supply response when demand shifts rightward.

general, as a result of rightward shifts in the short-run supply curve, long-run supply is more elastic than short-run supply. Forestry needs long production periods to shift the short-run timber supply curve rightward—a fact posing interesting policy questions and technical problems in timber supply analysis. These are considered in the chapter on timber supply.

Long-Run Demand

Long-run demand mechanics are a mirror image of the supply side. Figure 2-19 shows short-run demand and supply curves (*S* and *D*) for a good in 1990, with price and quantity at P_{1990} and Q_{1990}. Suppose the good is lumber, and the supply curve for 1991–1995 shifts leftward due to environmental restrictions on timber cutting. For 1991, lumber price rises to P_{1991}, with quantity dropping to Q_{1991}. The short-run demand curve for lumber ($D_{1990-1991}$) tends to be inelastic because building commitments have already been made and lumber-using technology is already in place. But in the long run, say, by 1995, in response to higher lumber prices, we'll see new lumber-saving techniques for construction and substituting other materials, and the short-run lumber demand curve can shift leftward to D_{1995} (at any given price, less lumber is consumed). We then have a reduction in price and quantity to P_{1995} and Q_{1995}. Repetitions of this pattern would give a series of points in line with *a* and *b* in Figure 2-19, tracing the **long-run demand**

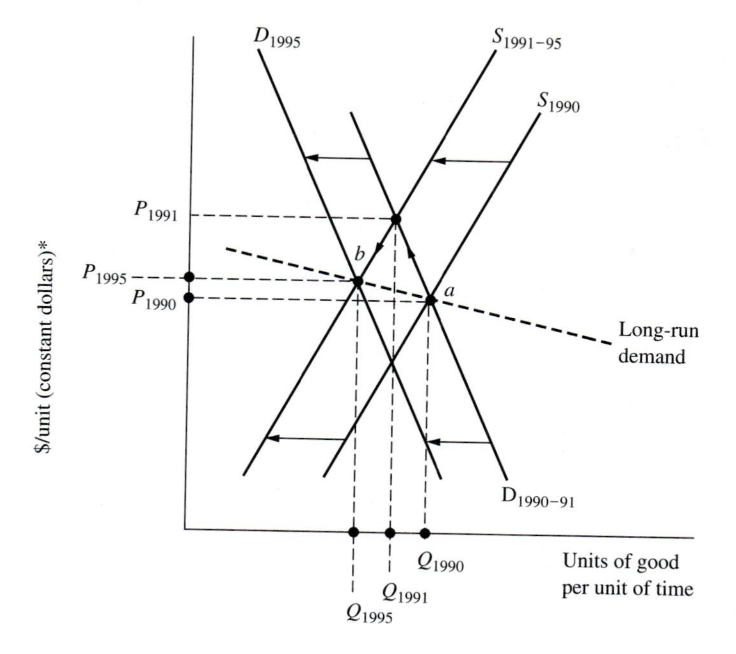

*Excludes inflation.

FIGURE 2-19
Long-run demand response when supply shifts leftward.

curve. You can think of the dashed curve as a long-run output demand response to changes in output price. Analogous mechanics would occur in the opposite direction if supply shifted rightward.

As with supply, long-run demand tends to be more elastic than short-run demand. Although the lumber example is not the final consumption stage, consumers also make long-term adjustments in response to higher prices of many goods, for example, by substituting other goods or by starting conservation measures.

Figures 2-18 and 2-19 explain why prices often spike up and later settle back somewhat (after correcting for inflation). In the short run, changes in demand or supply yield greater price-changes than in the long run. But the resulting long-run quantity-changes exceed the short-run changes. These graphs, as well as others, change only selected variables to illustrate a point. In reality many factors change at once, creating challenging problems in statistically estimating supply and demand functions.

THEORY VERSUS PRACTICE

This chapter has covered mainly the theory of supply and demand. Chapter 12 introduces ways of deriving timber supply and demand functions for groups of producers and consumers. You can statistically estimate these functions from historical data on prices, quantities, and the shifters mentioned earlier. Chapter 14 explains how demand functions can be roughly estimated for some forest outputs that have no market price. Throughout the book you'll find many examples of how supply-demand concepts and specific functions can help guide public and private policy decisions. Chapter 11 mentions ways to predict forest output prices that result from supply-demand interactions.

While it's important to understand the theory of profit maximization underlying supply curves, you should also realize that firms can rarely apply the theory as precisely as shown in this chapter's graphs. They must deal with multiple inputs and outputs, poor data, constantly changing scenarios, incredible complexity, and uncertainty. Thus, as Herbert Simon (1979) suggests, firms are usually "satisficing," or aiming for satisfactory earnings rather than maximum profits in any precise sense—they are "sub-optimizing," surviving, increasing sales, and doing as well as they can in an uncertain world. You also need to remember that maximizing profits won't necessarily maximize society's well-being—the thrust of the next chapter.

After the next chapter's review of forestry and the market, several chapters deal with forest investment analysis, which is crucial to understanding Chapter 12 on timber supply and demand and the rest of the book. But the investment analysis doesn't make much sense without the basics of price determination covered in this chapter. Thus, after what seems like a detour, you'll tie back into this chapter (so don't forget it!).

KEY POINTS

- The demand curve for a good shows the quantities of the good that will be bought at different prices.
- A demand curve applies to a specific group of one or more people, holding shifters constant. Examples of demand shifters are population, income, tastes, advertising, and prices of related goods. Changes in such factors will shift a demand curve to the right or left.
- Price elasticity of demand is the percent change in quantity demanded with a 1 percent change in price. Elasticity has a profound impact on the way total sales revenue from a good changes as output is increased.
- A market demand curve, from the consumer's view, will slope downward. For the same commodity, the demand curve faced by a producer may be horizontal or downward-sloping, depending on the size of the firm's output relative to the entire market.
- A firm's profit-maximizing output (input) level is where the added cost per unit of output (input) equals the added revenue per unit of output (input).
- For any enterprise, it is important to express returns to capital as a percentage of capital value—the rate of return.
- A firm's output supply curve plots the added cost per unit of output (the marginal cost) and shows profit-maximizing levels of output at different output prices.

- The industry supply curve is the horizontal sum of the individual firms' supply curves. Holding shifters constant, it shows the amount of a good the industry will produce at different prices. Changes in shifters like the following can move the supply curve to the right or left: number of and size of production facilities, technology, prices of inputs, cost of borrowing, and natural disasters.
- Price elasticity of supply is the change in quantity of a good supplied with a 1 percent change in its price.
- Equilibrium price and quantity in a market for a good is at the intersection of the supply and demand curves. The elasticities of demand and supply can sharply affect the price and quantity-changes when either demand or supply curves shift.
- Under competitive conditions, when excess profits exist for firms producing a good, more is invested in production; the industry supply curve shifts to the right, and price falls until normal returns to capital are earned.
- Long-run supply is a dynamic construct and shows intersection points of short-run supply and demand curves over time as demand shifts rightward (or leftward).
- Long-run demand is a dynamic construct and shows intersection points of short-run supply and demand curves over time as supply shifts rightward (or leftward).
- Both long-run demand and long-run supply tend to be more elastic than short-run demand and supply. Thus, shifts in supply or demand curves in the short run cause relatively large price-changes and small quantity-changes but in the long run cause smaller price-changes and larger quantity-changes.
- Due to imperfect information, rather than maximizing profits in the precise sense of this chapter's graphs, firms are more often seeking satisfactory profits. The next chapter explains why profit maximization doesn't necessarily maximize our collective well-being.

QUESTIONS AND PROBLEMS

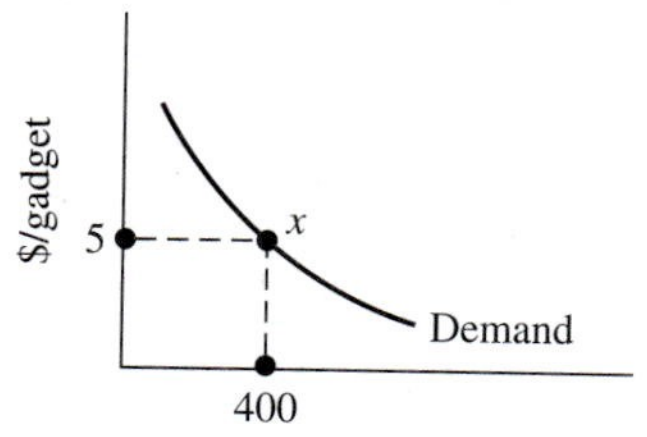

2-1 Given that the slope of the above demand curve is -0.025 at x, what is the price elasticity of demand at that point?

2-2 **a** Why is it important for a firm to know the elasticity of the demand curves it faces for its products?

b Why is it important for policy analysts to know the regional demand and supply elasticities of outputs affected by proposed policies?

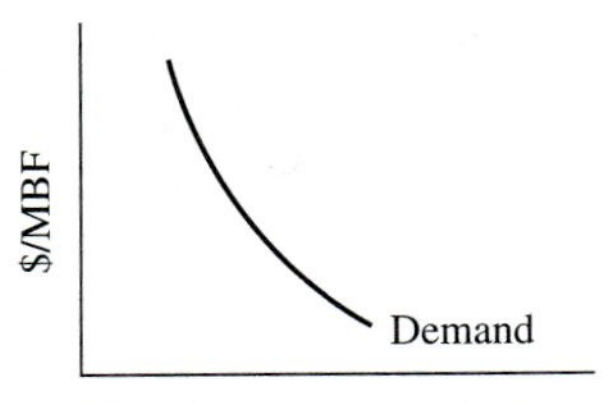

2-3 What will happen to the above regional demand curve for lumber if:

a An increased supply of lumber of the same kind is imported?

b Real per capita income increases (other things being equal)?

c Lumber manufacturing costs increase?

2-4 A firm increases output of its product and finds that total (gross) revenue has also increased.

a What can you say about the arc elasticity of the demand curve faced by this firm over the above range of output?

b Is the firm selling in a perfectly competitive market?

2-5 A firm doubles output of its product and finds that total (gross) revenue also doubles. In this range of output, what can you say about (a) elasticity of the demand curve faced by the firm and (b) the structure of the market in which these products are sold?

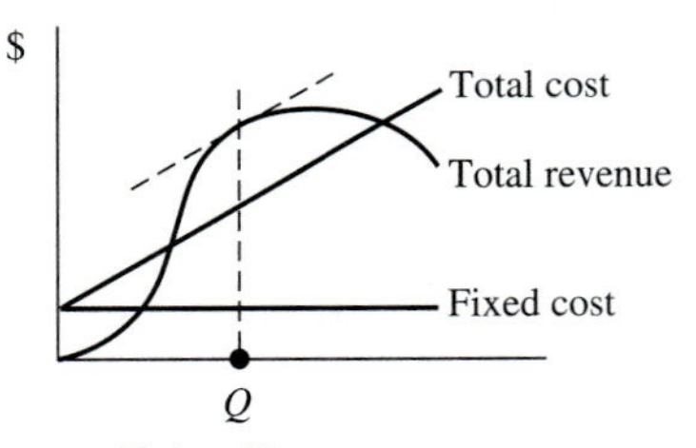

2-6 In the above graph, Q shows the input level where net revenue is maximized. If the price per unit of input is reduced by 30 percent at all input levels, will the point of maximum net revenue change? If so, in what direction?

2-7 What happens to the point of maximum net revenue if fixed costs increase in the above graph?

2-8 Assume the following production function for total gross revenue per day as a function of variable input I:

$$\text{TR} = 500.3I - 0.157I^2$$

Assume that fixed costs are $5,000 per day and that variable inputs cost $300 per unit used each day.

You have been asked to write a computer program that calculates net revenue per day for the above production process at various levels of input. Write out the net revenue equation that you would use in the program.

(For you calculus fans: What's the net revenue-maximizing level of input, i.e., the best combination?)

2-9 **a** On the lower set of axes, sketch in the marginal revenue product and marginal resource cost curves implied by the upper graph. (Concentrate on the shape of curves and intersection point, but don't worry about actual values.)

b In the upper graph, sketch in the *net* revenue curve.

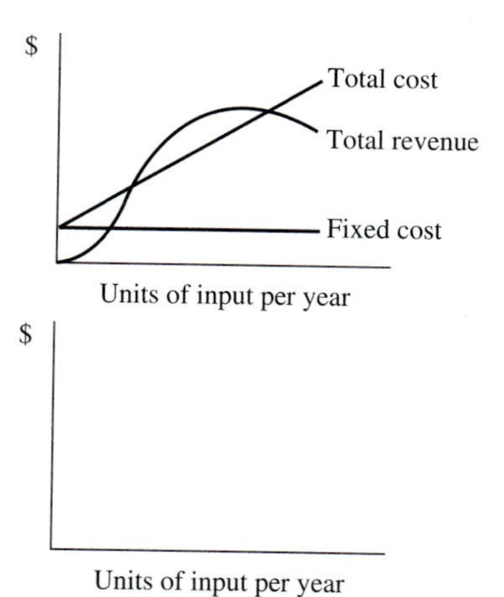

2-10 Calculate arc elasticity between *A* and *B* for the given demand relationships:

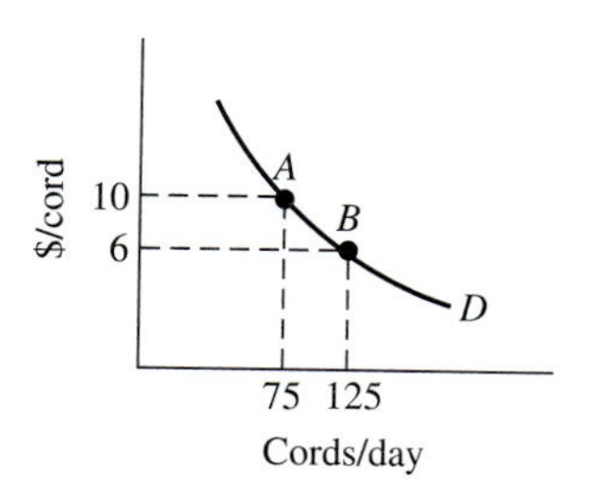

2-11 Given the regional supply and demand curves below for softwood plywood, if the price per square foot is at *P*, what is likely to happen to the price? Why?

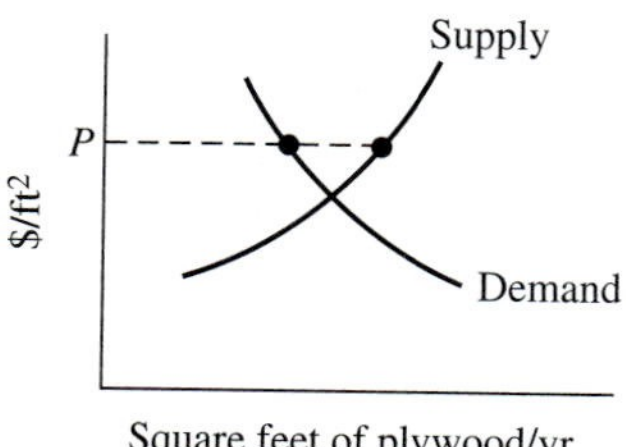

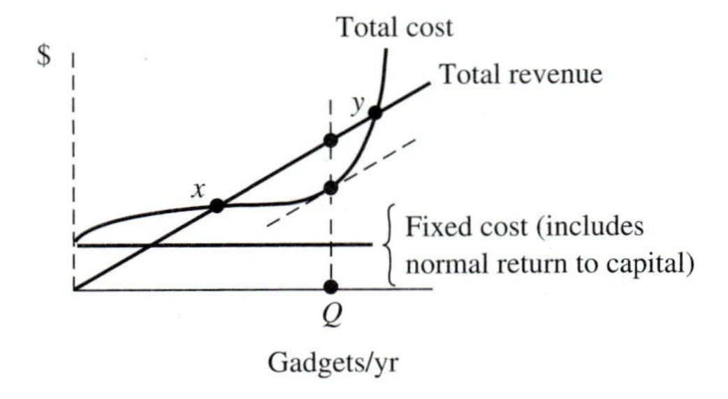

2-12 Suppose the above graph represents the typical factory in the gadget industry. If output is at Q, what would you expect to happen in the future, given a competitive economy?

2-13 At point x in the above graph, which is greater, marginal cost or marginal revenue? And at point y? Where does marginal cost equal marginal revenue?

REFERENCES

Alexander, S., and B. Greber. 1991. *Environmental Ramifications of Various Materials Used in Construction and Manufacture in the United States.* USDA Forest Service Report PNW-GTR-277, Corvallis, OR. 21 pp.

Ausubel, L. M. 1991. The failure of competition in the credit card market. *American Economic Review.* 81(1):50–81.

*Fischer, F., R. Dornbusch, and R. Schmalensee. 1988. *Economics,* 2nd edition. McGraw-Hill, New York. 813 pp.

Hagemann, H. 1990. Internal rate of return. In *The New Palgrave: Capital Theory.* W. W. Norton and Co., New York, pp. 195–199

Mansfield, E. M. 1985. *Microeconomics—Theory and Applications.* W. W. Norton and Co., New York. 590 pp.

*McConnell, C. R. and S. L. Brue. 1990. *Economics—Principles, Problems, and Policies,* 11th edition, McGraw-Hill, New York. 866 pp.

Pappas, J. L., and M. Hirschey. 1990. *Managerial Economics.* Dryden Press, Chicago. 826 pp.

Pindyck, R. S., and D. L. Rubinfeld. 1989. *Microeconomics.* Macmillan Publishing Co., New York. 668 pp.

*Samuelson, P. A., and W. D. Nordhaus. 1989. *Economics,* 13th edition. McGraw-Hill, New York. 1013 pp.

*Schiller B. R. 1989. *The Economy Today,* 4th edition. Random House, New York. 935 pp.

Simon, H. A. 1979. Rational decision making in business organizations. *American Economic Review.* 69(4):493–513.

*Stiglitz, J. E. 1993. *Economics.* W. W. Norton and Company, New York. 1142 pp.

*Introductory economics texts.

3

FORESTRY AND THE FREE MARKET

How much government intervention in private forestry is needed in a market economy? At one pole are those believing that the best interests of society are met if forests are managed in an unfettered free market. Extremists in this camp argue that all public forests should be sold to private owners, and government intervention in private forestry should be eliminated. At the other pole are those who distrust the way markets allocate resources to forestry and who suggest that most forests should be publicly held. Others support the current situation in the United States: a mix of public and private forests, the latter often under some regulation. A 1983 University of Washington conference, "Selling the Federal Forests," spawned serious discussions on the issue (Dowdle, 1984; Gamache, 1984). Privatization of federal forests is often debated in the United States Congress.

At some time most foresters are involved in the debate about the best role for government in forestry. To better understand these issues and to cut through emotionalism on both sides, this chapter will review how a free market could in theory maximize "social welfare," or the well-being of people. We'll see how the theoretical **welfare maximum** may not necessarily occur in a free market and what forms of government action might make us better off. The emphasis will be on forestry; more extensive analyses are in the references. I'll assume that the United States will remain committed to a market economy with some government action to improve welfare. The thorny questions are what kind of intervention, how much, when, and where.

A welfare maximum is an elusive allocation of resources—land and natural resources, people, and capital—such that no reallocation could yield a net gain. At this optimum, the losses from any change would exceed the accompanying gains. You must be at a welfare maximum if no change can bring net gains. "Reallocation" could be more or less money, human effort, land, or other resources devoted to some aspect of forestry or any other enterprise. A narrower view of welfare maximization is called the **Pareto optimum,** after the Italian economist Vilfredo Pareto (1848–1923). At the Pareto optimum, no resource reallocation could make anyone better off without making someone else worse off. While that sounds fair, seldom are there policies where no one loses. Thus, we usually take the less restrictive approach and allow a gain to occur at the expense of someone's loss, as long as gains exceed losses, or as long as gainers could overcompensate losers. Kaldor (1939) has called this the "compensation principle." Welfare maximization isn't as simple as it sounds because gains and losses aren't always easy to measure, and each person's evaluation of gains and losses will differ.

RESOURCE ALLOCATION IN A COMPETITIVE FREE MARKET

To see what's behind arguments for unregulated free markets in forestry and elsewhere, let's review the conditions under which a market economy could, in theory, maximize social well-being. When these conditions are not met, which is often the case, we have **market failures** preventing the free market from reaching a welfare maximum. These failures will be reviewed shortly, with reference to forestry. Why set up a straw man—welfare maximization in a market economy—and then knock him down? The reason is to more clearly understand the frequent arguments about the degree to which markets should allocate resources in forestry. Listed briefly below are (1) the conditions required for a market to maximize welfare and (2) the results under welfare maximization.[1] Both (1) and (2) will be discussed in more detail under market failures.

The starting point presumes private ownership of most resources and the existence of a functioning government, a monetary system, and a central bank.

Conditions Required for a Free Market to Maximize Welfare

Property Rights to Resources Are Enforced Ownership of all land and resources such as timber is clearly defined so that free access to any scarce resource isn't allowed.

Firms and Consumers Are Maximizers Firms attempt to maximize net revenues, and consumers wish to maximize utility.

Perfect Competition Under perfect competition, firms are **price takers;** each is so small that it faces a horizontal demand curve for its product and cannot

[1]This discussion of welfare maximization is fairly intuitive and condensed; more rigorous treatments are found in the references and include concepts such as marginal rates of substitution and transformation, and price ratios. One objective here is that you learn to explain these concepts to noneconomists, using a minimum of jargon.

affect the market price (see lower panel of Figure 2-3). Price-increases by one firm will cause buyers to shop elsewhere. Consumers and firms face horizontal supply curves for inputs and can't affect their prices.

Free Entry of Firms If higher than normal rates of return can be earned in one industry, new firms can enter to increase output, leading to lower prices (see Figures 2-14 and 2-15).

Perfect Information Consumers know about available outputs and their prices, and firms have information about production technologies, inputs, and prices of all inputs and outputs over time.

Mobility of Labor and Capital Labor is available at the competitive wage, and capital comes forth at competitive interest rates.

No Unpriced Negative Side Effects In production processes, there are no costs imposed on others for which the producer needn't pay. Examples are various forms of pollution.

Priced Inputs and Outputs All products, services, and resources can be sold at a price, so there's an incentive for the market to provide them.

Satisfactory Income Distribution In some sense, the distribution of income among citizens, current and future, must be satisfactory.

Results When Welfare Is Maximized

Now let's review the theoretical outcomes of a competitive market under the above conditions.

Consumers' Needs Are Met Consumers vote with dollars to receive desired outputs. Prices will be as low as possible, consistent with acceptable rates of return to capital (see Figures 2-14 and 2-15).

Consumers Maximize Their Satisfaction by Applying the Equi-Marginal Principle A rational consumer allocates funds so that the last dollar spent per unit of time on every good or service brings the same satisfaction. If this point isn't clear, look at Figure 1-2, and let the horizontal axis to each panel read, "Dollars spent per week on good A, B, etc." Any reallocation from the illustrated optimum results in a net utility loss, following the same reasoning given in Chapter 1's discussion of the equi-marginal principle.

Capital Will Be Allocated Efficiently Investors will place funds where they will earn the best returns. At any given risk level, interest rates or rates of return on added investment will move toward an equilibrium (see Figures 2-14 and 2-15). We will consider this in more detail in Chapter 4.

Labor and Other Resources Are Allocated Efficiently Employees move to jobs where the pay is best. Other resources, for example, timber, are used in enterprises that can afford to pay the most for them. Employers hire more workers until the added revenue per worker per day equals the added cost (wage) per day. Other resources (inputs) are similarly employed (see the optimum feed input in Figure 2-7).

Producers Seek Efficient Levels of Output This occurs where the added revenue (price) per unit of output equals the added cost (where net revenue is maximized). (See Figure 2-8.)

Land Is Allocated Efficiently Land will go to the highest bidder who uses it for the most profitable enterprise (under the above conditions, this would be socially the most desirable land use). Chapter 7 discusses the economics of land use decisions.

No Redistribution of Income Could Yield Any Net Benefit The losses from redistributing income would exceed the benefits.

If all the above results occurred, welfare would be maximized because no reallocation of resources could yield a net gain. Remember, the above model is outlined, not because it's believed to exist in any real economy but because it forms the theoretical basis for many arguments to limit the government's role in market economies. Also, because our United States society has many of its roots in beliefs about individual freedom, private property rights, consumer sovereignty, and limited government interference, the competitive market model appeals to many of us, whether or not we completely understand its nuances.

The above conditions and results don't always hold in market economies, and therefore welfare is not maximized in a completely free market. When welfare maximization conditions are not met, we have *market failures.* These are reasons for government action to increase total satisfaction above the level attained by the free market. Let's now examine market failures, cases where they might occur in forestry, and examples of government action. These are extremely important points in a society committed to welfare maximization in a market economy. A market failure justifies some collective action, *but only if the resulting benefits exceed the costs.* However, *in a society committed to a market economy, if you cannot demonstrate market failure, there is little justification for government action in the marketplace.*

MARKET FAILURES

While we often think of market failures occurring only in private enterprises, the same problems can also occur in government activities. For example, public agencies often cause pollution and other environmental damage—cases of **government failure.** Bear this in mind when reviewing the following market failures.

Property Rights Not Enforced

If we allow unlimited access to resources, welfare will not be maximized. Examples are open access fisheries or grazing areas. A user is unlikely to invest in improving the resource, since others will reap most of the gains. Users, if numerous enough, can eventually destroy an *open access resource* unless they develop regulations. Hardin (1968) called this "the tragedy of the commons."[2] An example is serious depletion of certain ocean fisheries (Howe, 1979). Champions of the market say that we avoid such problems when resources are privately owned and sold at a price under enforced property rights. But, as we'll see, questions remain about other market failures.

For timber production in the United States, the open access problem is virtually nonexistent. Most property boundaries are respected, stumpage is sold at a price, and timber thievery is rare. But in parts of countries with extreme population pressures like Haiti and Costa Rica, open access to some areas has led to widespread forest destruction (Foy and Daly, 1992). A similar situation existed in parts of medieval Europe. On many forests, access wasn't limited, and people simply cut wood as they needed it. The resulting forest destruction spawned the forestry profession in Europe together with detailed government regulation of forestry practices. Many of the German foresters who brought forestry to the United States therefore came with a feeling that government regulation of private forestry and even government ownership of major forest areas were needed to assure successful timber production. But since the United States doesn't have open access problems in timber production, some analysts say that the European brand of regulations isn't necessary to assure wood output on private forests. The chapter on timber supply and demand addresses these issues further.

Contrary to the case of timber production in the United States, open access problems sometimes *do* occur with forest resources like fish, game, scenic beauty, water, and hiking in natural areas. For example, without high enough fees or some other form of rationing, free access can damage the character of hiking areas and can deplete fish and game populations. In the case of fishing and hunting, the problem is eased by licenses, by fish and game laws, and by the increasingly common practice of charging hunting fees on private land, which gives an incentive for owners to manage for game. Higher fees could also be charged on overused public lands.

---

[2]Without communal constraints, individuals keep using an open access resource until their costs of consuming the last unit exactly equal their added benefits. However, since users don't bear the costs of production or the cost to other users as the resource is depleted, the result can be overuse to the point where added cost per unit exceeds added benefit. Bromley (1992) notes that the term "common property resource" shouldn't be used to denote an open access resource. History abounds with examples of groups owning resources collectively as common property and enforcing rules for restricted use. Thus, I interpret Hardin's use of the term "commons" as an open access resource.

Imperfect Competition

Monopoly A **monopolist** is a single firm producing all output of one product or service in a market. As shown in Figure 3-1, such a firm faces a downward-sloping demand curve. As a monopolist changes output, product price will change. We say that such a firm has **market power,** as opposed to the competitive firm, which is a price taker and faces a horizontal demand curve for its output. Although, in the United States forest products industry, pure monopoly doesn't exist, it's useful to understand the theory because some forest products firms having a large share of a market do face downward-sloping demand curves for their output. Figure 3-1 shows that marginal revenue for a monopolist will lie below the demand curve, because whenever output increases, the resulting decreased price is received on *all units sold,* not just the last one. Such firms tend to produce at Q_1 and P_1 in Figure 3-1, where marginal revenue equals marginal cost. But the socially optimal output in the figure is at output Q_2 and price P_2, at point A, where the marginal cost of the last unit equals its price (as you saw in the last chapter's Figure 2-8b, under perfect competition). That's where the extra cost to the firm equals the added benefit to those last consumers near point A (i.e., the last consumers' willingness to pay, which is the price). Thus, *a monopolist's output tends to be lower and price higher than at the social optimum, and profits are likely to be higher than under competition.*

An unregulated monopolist earning excess profits can continue to do so, since new firms aren't entering to increase output and drive down prices. The more

FIGURE 3-1
Monopolist's optimum output vs. social optimum.

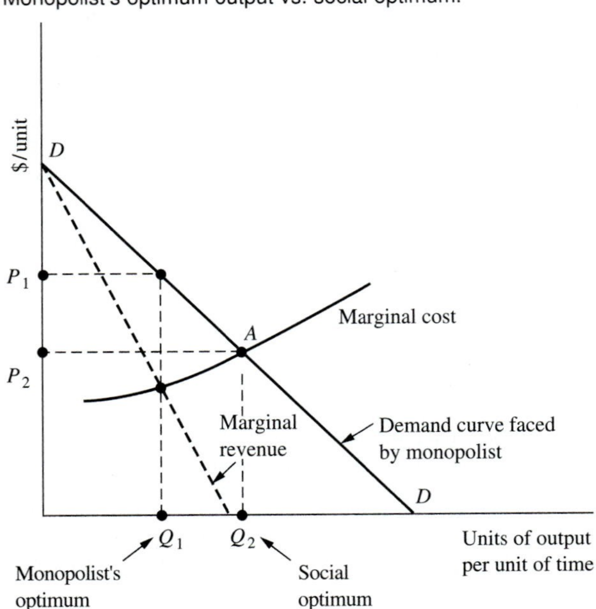

common case of market power is that of a few firms **(oligopoly),** each facing a somewhat downward-sloping demand curve. So, other things being equal, oligopoly prices and profits would be lower than under monopoly, but not as low as under perfect competition. Imperfect competition keeps us from getting around the circular diagram in Figure 2-15, where the outcome should be the lowest prices possible, consistent with a fair rate of return to capital.

Monopsony Figure 3-2 shows a market failure when an input is bought by only one firm, a **monopsonist.** The monopsonist's marginal revenue product curve shows the firm's added revenue for the last unit of input, for example, inputs such as hours of labor or cubic feet of logs for a sawmill. In Figure 3-2, the monopsonist does not face a horizontal supply curve for the input, as the competitive firm does in Figure 2-7b. Since the monopsonist is the only buyer, it faces an upward-sloping input supply curve showing the quantities supplied in the whole market at various prices. In that case, the monopsonist's **marginal resource cost** will lie above the supply curve (or above the price) because to increase its input supply, the firm must offer a higher price not only on the last unit but on *all units* previously bought at lower prices. Thus, the monopsonist will maximize profits by using Q_1 units of input in Figure 3-2, where its marginal revenue product

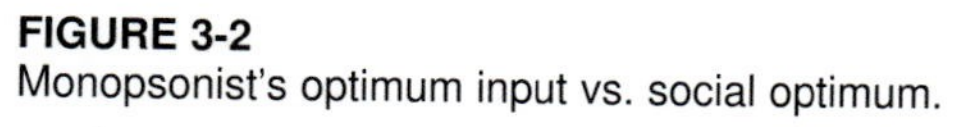

FIGURE 3-2
Monopsonist's optimum input vs. social optimum.

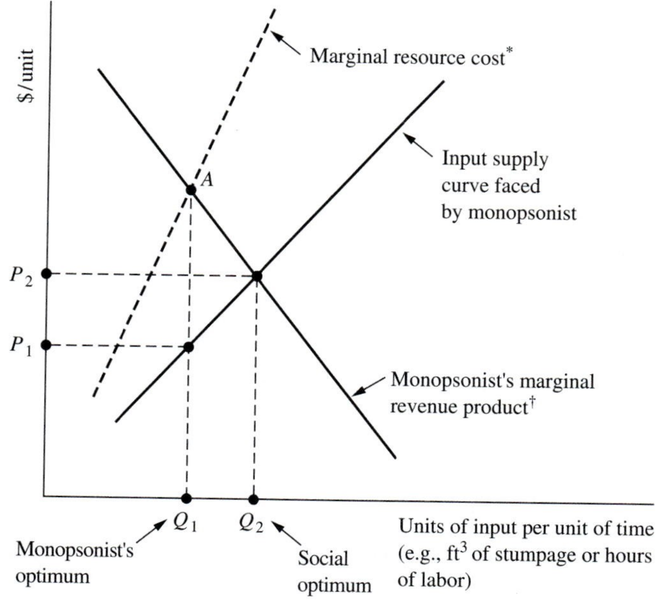

* Added cost per added unit of input.

† Added revenue per added unit of input.

equals the marginal resource cost—at point A—yielding an input price of P_1.[3] Using one more unit of the input will cost the firm more than that unit contributes to total revenue. *But Q_1 is a socially suboptimal input because the last unit of input receives a price (P_1) less than its contribution to the firm's total revenue* (read off the marginal revenue product curve). The social optimum under competition (as shown in the last chapter's Figure 2-7b) is to employ inputs to the point where the marginal revenue product equals input price, or at output Q_2 and price P_2 in Figure 3-2. Thus, *under monopsony, resource use will be at lower levels and lower prices than under the competitive optimum.*

Monopsony is the mirror image of monopoly. A monopolist is a single seller of a good, while a monopsonist is a single buyer of an input. The extreme case of market power in buying resources is monopsony; more common is **oligopsony**—a limited number of resource buyers, each facing a somewhat upward-sloping supply curve for the input. In United States forestry we sometimes find one or a small number of firms buying private or public timber; the result can be stumpage prices below the competitive optimum and thus not enough incentive for wood production.

Government Action to Improve Competition Although pure monopoly isn't a problem in the United States forest industry, some oligopoly problems have occurred, and Mead (1966) has documented oligopsony among timber buyers. Thus, competition issues are important. Federal antitrust laws discourage activities that could reduce competition and lead to higher output prices, lower input prices, reduced output, and excess profits. Such activities include:

- Collusion between firms to raise product prices.
- Collusion to reduce input prices, for example, stumpage prices bid for public or private timber—in which case timber growers receive less than the true value to society and are not stimulated to provide optimal amounts of timber.
- Price discrimination, where firms sell the same product at different prices to separate groups for reasons unrelated to supply costs.
- Predatory pricing, or cutting product prices below production costs in order to prevent competing firms from entering the market.

To lessen imperfect competition, governments have sometimes required firms to break up into smaller separate companies if they have enough market power to control prices. What does all this imply for forestry? Compared to some industries, concentration in most of the United States forest products industry is not high (see Chapter 13). But the U.S. Department of Justice in the late 1960s alleged that the Georgia Pacific Corporation was large enough to have too much control over

[3]This explains why, in Figure 3-2, you can't call the dashed curve an "input supply curve" faced by the producer. The dashed curve doesn't show how much input is supplied at different prices: that's what the upward-sloping solid curve shows. Likewise, with monopsony, you can't call the downward-sloping curve in Figure 3-2 a "demand curve" because it doesn't show how much input the monopsonist demands at different prices. For example, at point A, the input price is way down at P_1.

the price of certain solid wood products. Backed by clauses in the 1914 Clayton Antitrust Act, the U.S. Department of Justice in 1972 required Georgia Pacific to spin off a number of its wood-processing plants into an independent corporation (Louisiana Pacific). The Federal Trade Commission has levied fines on several paper and plywood producers for alleged price-fixing (see Chapter 13).

There are other signs that forest products firms don't always face horizontal demand curves for their products. For instance, at a seminar I attended, a representative of a major plywood producer stated that his firm had a large enough share of the market to depress plywood prices by increasing output. As another example, major forest landholders selling stumpage are well aware that harvesting their timber too rapidly can dampen stumpage prices. Note that such market power in stumpage markets also occurs with public timber owners such as the U.S. Forest Service, especially in the West.

To lessen oligopsony problems, the U.S. Department of Justice has prosecuted firms alleged to be colluding to keep bid prices low on federal timber sales. But without informers, such cases are hard to prove. To lessen monopsony or oligopsony problems, stumpage sellers can set minimum acceptable bid prices and can encourage competitive bidding. Governments publish stumpage price reports and have helped landowners organize marketing cooperatives to bargain with timber buyers.

Although we know that perfect competition doesn't always exist, the road to social bliss will not necessarily lie in breaking the economy into many small producers. In some sectors, including forest products and timber management, a certain degree of bigness helps to foster the price stability, capital reserves, and long-range planning required for today's massive investments in production, research, and development. Also, for industries such as utilities, a monopoly may provide more efficient service. In the latter case, governments usually set prices lower than the monopoly price but high enough to guarantee the firm a competitive rate of return on capital. Recognizing these economies of scale in many industries, regulatory policy in the United States hasn't sought to break up all industries into tiny fiercely competitive units but has rather sought to stimulate reasonably competitive behavior.

Antitrust laws make employees of forest products firms reluctant to discuss prices even if the firm doesn't intend to limit competition. You can see this concern at informal and formal meetings attended by forest industry representatives.

Imperfect Information

The market cannot function efficiently if consumers don't have enough information about products and prices or if producers have insufficient knowledge about production processes or about resources and their prices. Most consumers seem to have enough information about wood products and other forest outputs to make choices that maximize their satisfaction. Truth-in-advertising laws are an example of government action to assure that consumers have good information. Major forest products firms likewise appear to have adequate information or know where to find it. But many foresters and lawmakers feel that nonindustrial private forest (NIPF) owners can't find enough information in a free market to carry out utility-maximizing forest management. Federal and state extension and technical assistance programs thus help to educate NIPF owners about forestry and forest products marketing. Others argue that if NIPF management were really so lucrative, private forestry consultants and industry assistance and lease programs would enhance NIPF management in a free market, which has in fact occurred in many areas. Chapter 12 further addresses issues about nonindustrial private forestry.

In deciding how much to invest in timber production, landowners have trouble projecting long-term wood prices. Although government has the same problem, public research on long-term timber supply, demand, and prices is often better than that available to many smaller investors.

Unbridled competition can also lead to firms too small to develop sophisticated research programs, and where private research is developed, results are often not shared. This justifies state and federally funded research programs in many areas, including agriculture and forestry. Examples are research at universities and at the forest experiment stations operated by the U.S. Forest Service. Private research can also be encouraged by tax benefits and direct subsidies.

Immobility of Labor and Capital

If labor or capital can't readily move to where earnings are best, welfare is not maximized. Capital immobility is generally not a problem in United States forestry. Capital markets are truly international in forestry as well as in other enterprises. If expected rates of return are attractive in forestry or forest products manufacturing in the United States, investment funds appear from domestic or foreign sources. On the labor side, forest products firms offering competitive wages appear to have few problems attracting workers. But we do worry about labor immobility in areas where mills have closed. For example, in Pacific Northwest towns with mill closures, unemployment has often remained high. Economic theory suggests that unemployed or underemployed workers would move to better jobs. But often people can't afford to leave their homes and move elsewhere, can't afford retraining if needed, or stay because of emotional or family ties. Governments in such cases can provide relocation assistance, job retraining, unemployment insurance, or economic development activities.

Unpriced Negative Side Effects

Another market failure is unpriced negative side effects of production. Examples are all forms of pollution: air, water, noise, and visual. A key point is that the cost is external to the perpetrator, who doesn't have to pay it—hence the term "unpriced." (In economics jargon, such damages are called "negative externalities" or **external diseconomies.**) Negative side effects are considered market failures because, being unpriced, there's no market incentive for the firm to eliminate or

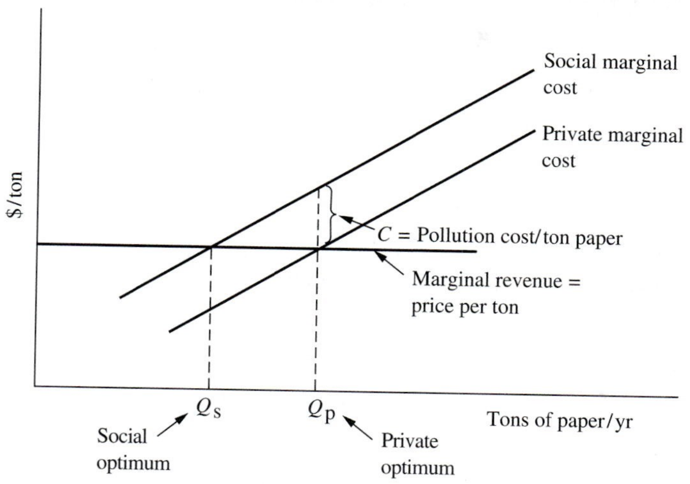

FIGURE 3-3
Divergence between the private and social optimum with water pollution.

at least abate them, even though the benefits of abatement could often exceed the costs.

Figure 3-3 shows the nature of the problem for a hypothetical paper mill causing downstream water pollution. From the firm's view, the private optimum is at Q_p tons of paper per year, where the private marginal cost equals the marginal revenue or price per ton. If the firm produced less than Q_p, it would have an incentive to increase output because the extra revenue from producing one more unit would exceed the extra cost. But from society's view, if the pollution cost is $\$C$ per ton of paper, the social optimum is at Q_s tons per year. At that point, the social marginal cost (the marginal cost to the firm plus the $\$C$ cost to the rest of society) equals the marginal revenue. So in this case, the social optimum is at a lower output of paper and pollution than the private optimum. Producing more than Q_s would bring extra costs (including the damage costs) that exceeded the extra benefits. At the socially optimal output Q_s, environmental damage still exists at $\$C$ per ton of paper, but total damage is less than at the private optimum because paper output is lower. Figure 3-3 is a highly simplified case where the only way to reduce pollution is to reduce output. A more general framework follows.

Unpriced negative side effects in forestry are common. Logging can cause damage to scenery, soil, water quality and regimen, and fish (through decreased shade over streams, damage to spawning beds, or reduced water quality). Slash burning and forest products manufacturing can cause air pollution. Mills often pollute water.

Most government actions to correct these market failures in the United States have been requirements to maintain certain emissions standards from factories and, in some states and counties, to require selected logging and forest management practices. Other actions are to give tax incentives or direct payments for damage control measures or to levy fines on damage caused.[4] Publicly funded education and assistance programs encourage environmentally responsible actions among farmers, forest owners, and loggers. One study showed that education was effective in getting farmers to adopt soil conservation practices that were economically appealing (Feather and Amacher, 1994). In forestry, a problem is that many practices for enhancing the environment reduce the landowner's income (examples are given later). There's little disagreement about the need to reduce environmental damage. Most arguments center on how and how much to intervene. But as you'll see later in the chapter, the optimal damage level is a slippery concept.

A related issue is voracious consumption in industrial nations. Many economists have questioned the notion that more is better (Daly and Cobb, 1989; Galbraith, 1958; Mishan, 1967; Schumacher, 1973). Not only do we have the resulting environmental damage—ozone depletion, solid waste problems, pollution—but poorer nations are pinched by the resulting rise in resource prices.

Outputs Not Easily Priced

If a good isn't priced, there is no incentive for its production, and the market fails us. Unpriced goods fall into several categories.

Public Goods Goods or services that can't be parceled out to each consumer and sold at a price are called **public goods.** Examples are national defense and street lighting. You can't exclude those who don't pay for the good from receiving its benefits; more for one person doesn't mean less for another. A free market will not produce enough public goods. "Not enough" means that the benefit of producing more exceeds the cost, but firms won't invest in production because they can't reap the benefits. Perhaps the prime example of a public good in forestry is scenic beauty along public roads. You can't sell units of scenic beauty, so firms don't go out of their way to enhance scenery on their lands or to avoid scenic damage that may occur from practices like clear-cutting in certain locations. If firms could capture the willingness to pay for scenic beauty, we'd no doubt find more effort among landowners to produce it. Note that public ownership doesn't necessarily solve the problem. Some of the most flagrant examples of damage to scenic beauty occur on public forestlands (Anderson and Leal, 1991). Examples are clear-cutting on some U.S. Forest Service lands near recreation areas and

[4]For certain polluters, mainly utilities, the federal and some state governments in the United States have issued marketable pollution permits such that fines are levied on levels of pollution above those allowed by the permits. Firms with high pollution control costs can buy added pollution permits from those with low costs. Thus, in theory, a market develops where total pollution control costs are minimized with the least government regulation (see Pearce and Turner, 1990, and see Goodstein, 1995, pp. 278–280, who questions how effective these markets have been in practice). Although the solution is cost-effective, the result can be fairly heavy pollution around older, less efficient factories. The approach hasn't been widely applied to forest products manufacturing and is less applicable to damage like water siltation from forest land management because it's often hard to tell how much pollution each landowner causes.

well-traveled scenic highways. Bureaucrats are often as likely to follow market-oriented behavior as businesspeople.

Other public goods are research results that are often diffuse and difficult to predict and sell to all users. Thus, amounts of private research in some areas may be less than optimal—another justification for publicly funded research. Forestry examples are tissue culture, genetics, climatology, fire control, and satellite-based mapping.

Some outputs are not pure public goods that are impossible to price in units, but they may be hard to sell. In forestry, high-quality water flowing evenly throughout the year is an example. In theory, landowners could sell water to consumers, but in most cases that's not practical. If firms could collect the value of improved water quality, they'd tend to produce it to the point where the added benefit just equaled the added cost, and society's welfare could be increased.

Timber is certainly *not* a public good. It is easily sold to individual buyers for a price. Thus, it's not valid to use the public goods argument to justify government actions strictly to increase timber supply. In that case, imperfect information may be a more relevant market failure. What about government subsidies for tree planting to gain outputs not easily sold, like scenic beauty, water quality improvement, and oxygen production? The rub here is that foresters' planting proposals are often costly efforts to grow more valuable commercial timber, while much less costly reforestation would usually yield the above nontimber values. Think of what happens on most logged areas in the United States with minimal site preparation but no planting. As long as soils are stable and sites can support vegetation, the result is usually brush, hardwoods, or mixed hardwoods and softwoods through natural regeneration—not usually what warms the heart of a forester. But such vegetation provides abundant beauty, soil and water benefits, and oxygen; and it isn't very costly. But on unstable soils, failure to plant can often cause major damage to nonmonetary or **nonmarket** values.

Outputs Traditionally Free When outputs that could be priced have been traditionally free, consumers resist efforts to charge for them. Examples are hunting and fishing in many areas. Efforts to post lands and charge for hunting have sometimes been met with arson and vandalism by irate hunters. But such acts are becoming rarer as hunting and other recreation gradually move into the marketplace on private lands. As fee hunting becomes accepted—and it has in many regions—landowners make more effort to produce and protect game values. But they won't spend more producing such outputs than users are willing to pay for them—an efficiency advantage of the market.

Governments often provide underpriced outputs like hunting, hiking, fishing, and camping. Then private landowners can't collect high enough fees for recreation to encourage its production. This isn't a market failure but simply the result of an underpriced public supply. Where this problem is absent, private markets often supply outdoor recreation (for examples, see Anderson and Leal, 1991).

Unpriced Positive Side Effects Unpriced positive side effects (the economics jargon is "positive externalities" or **external economies**) are a subset of public goods, discussed above. The key point here is that these desired outputs occur as an *unmarketed by-product* of producing marketed goods. A forestry example is the open space benefit to urban populations when nearby lands are managed for timber production. Figure 3-4 shows that the private and social optima in such a case can differ. The horizontal axis is acres forested around an urban area. The market solution or private optimum is at Q_P, where the private marginal cost for the last forested acre equals the private marginal revenue. The rising private marginal cost reflects the increasing cost of land as more is bought, and the private marginal revenue is based on the market price of products sold from the land. Both of these can be thought of in **present value** or capital terms, a concept discussed in the next chapter. If there's a positive unpriced side effect—the open space benefit, $\$B$ per forested acre—the social marginal revenue lies above the private marginal revenue, and the socially optimal acreage is Q_S, yielding more open space benefits than the private optimum. Private landowners won't move to Q_S on their own because they can't cash in on the open space benefit. The key word here is "unpriced"; if such positive side effects could be sold at a price reflecting the marginal benefit to recipients, the private and social marginal curves in Figure 3-4 would coincide, and the market would provide these benefits in optimal quantities, assuming no other market failures.

Option Demand Another demand not easily satisfied in a free market is the willingness to pay for the option to use some resource in the future; this is called **option demand.** An example would be recreationists' willingness to pay for the option to use some unique park area like Sequoia National Park in the future, even if they don't currently visit the park. Since such options are difficult to sell, private park operators might find it unprofitable to maintain some areas that the public would be willing to pay for if option demand were included. You must be

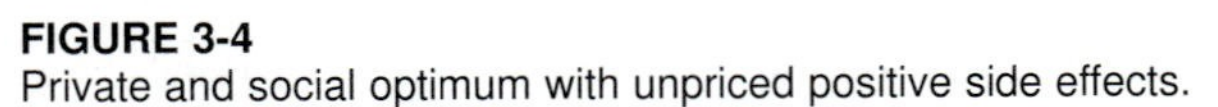
FIGURE 3-4
Private and social optimum with unpriced positive side effects.

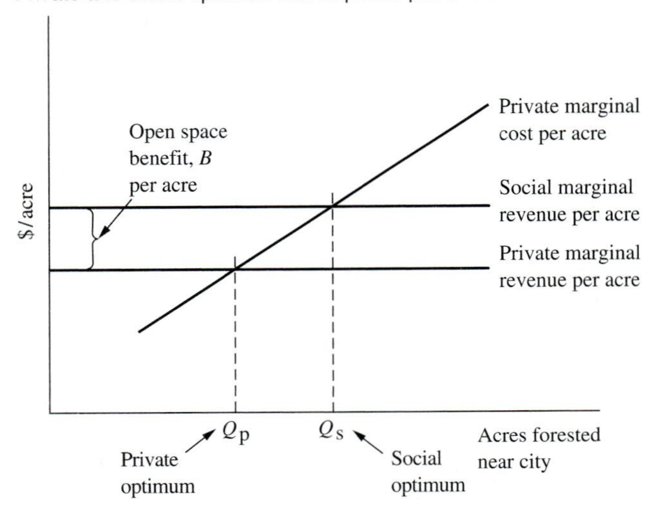

careful in using this argument only for nontimber goods because it also applies to timber. There is a demand for the option to use reasonably priced wood in the future. But the case for option demand is strongest with unique resources having few substitutes, for example, Yellowstone National Park. Note that option demand can be partially exercised in the market: some private groups raise funds to buy unique lands for preservation. In this way The Nature Conservancy has protected over 7.5 million acres in North America.[5]

An analogous concept is **existence demand,** which is the willingness to pay for the assurance that something remains in existence, even though the person will probably never use or consume it. For example, many people are willing to pay to preserve African parks, although they never plan to visit one. Organizations do in fact collect some of this willingness to pay, but the market is far from perfect. Related to these benefits is **bequest value:** the satisfaction we receive from leaving resources to future generations. As with option demand, the above values are most valid for truly unique resources without substitutes. Such values are not fully registered in market decisions.

Government Action to Provide Unpriced Goods We can consider the problem of unpriced goods only in general terms because their values are so difficult to measure in dollars. The main point is that free markets produce less than the optimal amount of unpriced benefits. In forestry, most government actions to provide or protect unpriced goods fall in five areas: (1) tax incentives or direct subsidies for providing unpriced goods, (2) educational programs about conservation practices, (3) taxes or fines to discourage certain types of environmental damage, (4) regulation of private forests, (5) government ownership. For example, governments can limit landowners' liability and give financial incentives to protect or encourage outputs that are difficult to sell, such as scenic beauty and hiking trails. Public programs subsidize publications, classes, and technical assistance stressing conservation practices in land use. Fines can be levied on landowners causing excessive siltation in streams. Some states and counties regulate logging practices on private land to protect unpriced environmental values. Zoning laws can require certain lands to remain undeveloped. Public forests are set aside to provide unpriced outputs like scenic beauty and open space at higher levels than the unregulated market would supply.

Economic Instability

In a free market's equilibrium processes, like those shown in Figure 2-14, lags often lead to boom and bust cycles accompanied by swings in prices and employment. A prime example in the forest products industry is housing cycles. When housing markets are strong, overbuilding sometimes occurs, after which housing sales slump, together with employment in construction and wood products

[5]The Nature Conservancy fund-raising letter CR-10-EM (December 1993).

industries. And the cycle starts anew. In Chapter 5 you'll see how rising inflation can increase interest rates on new home loans and dampen housing demand. These cycles cause unstable wood products prices and fluctuations in employment and economic activity. Avoiding the unemployment of downswings could increase social well-being and lessen income inequality.

Forestry students have felt the cycles of free market decisions: in the early 1970s, forestry school enrollments skyrocketed with growing interest in the environment. But it took several years for the word to spread that the bulge in graduates led to meager job opportunities and fairly low pay. By the mid-1980s forestry enrollments had plummeted.

Because prices are a market's major production signals, legislators have usually been reluctant to impose price controls in the United States. The exception is agriculture, which has a long history of federally controlled price stabilization and support programs. But these efforts have often come under attack for sometimes causing overproduction and inefficient resource allocation. Other government stabilization programs are price reporting systems, loan guarantees, job retraining, unemployment insurance, and monetary and fiscal policies designed to stimulate sluggish demand, dampen excess demand, stimulate investment, increase employment, and stabilize interest rates.

Compared to market systems, centrally planned economies tend to have fairly stable prices and employment, but they've been much less responsive to consumer demands and not noted for providing high living standards. Perhaps some degree of instability is a price we pay to reap the efficiency benefits of markets. But government can do much to at least dampen the seemingly inherent cycles in market economies.

Unsatisfactory Income Distribution

Suppose all the above market failures were corrected, and it seemed like no reallocation could bring gains that exceeded the resulting losses. We still might be dissatisfied with the income distribution. Market economies tend to yield large disparities in individual income and wealth—one of the things that worried Karl Marx, the father of communism. However, as many communist nations have discovered, Marx's solution, as practiced to date, seems to breed low productivity and economies that respond poorly to consumer demands.

What do we do about excessive income inequality in market economies? Government actions have been programs like welfare and food stamps, certain free public services, assistance to low-income students, low-income housing subsidies, social security, health programs, and progressive income taxation where wealthier people pay higher tax rates than the poor. Economics can't be very precise in defining the best income distribution. It seems obvious that if you take $100 from someone very wealthy and give it to a beggar, the beggar's utility gain will exceed the wealthy person's utility loss. But how do you compare these utilities? Do the rich have a greater capacity for joy than the poor, or vice versa? For this reason, economists say you can't make "interpersonal utility comparisons." But such a statement shouldn't let us completely ignore income inequality. When parting with a given sum of money, it's hard to argue that very poor people don't suffer a greater utility loss (starvation in the extreme case) than the rich. Thus, most of us tend to support programs that shift at least some revenue from higher to lower income groups. But creating too much income equality can lessen incentives to be productive. Income distribution questions are usually addressed in the political sphere, with legislators trusting their intuition and somehow muddling through, but not without lots of arguing.

Again you might think, "What does this have to do with forestry?" Many policy issues in forestry have an income distribution side. For example, some will argue for free public forest recreation areas to help low-income groups. Others cite studies showing that most users of federal forest recreation areas aren't from low-income groups; in fact, the poorest can't even afford the cost of getting to many public forestlands.

In forestry we're concerned about the way markets distribute benefits between generations. Unborn citizens have no voice in today's market decisions that affect their future timber supplies. We also worry about the way markets distribute nontimber benefits to future generations: will they have enough wilderness, species diversity, and other nonwood outputs? Noting that most offspring have eventually been better off than their parents, some policy makers have argued that shifting income from current to future generations is distributing assets from the poor to the rich. But today we're not so sure that our descendants will be better off than we are.

Another issue: the U.S. Forest Service worries about rising stumpage prices and favors government action like reforestation assistance to increase timber supplies, which would ease the upward pressure on prices. (Can you use the supply and demand curve framework to illustrate this point? What does more reforestation do to the future supply curve for stumpage?) Some argue that lower wood prices will make housing cheaper for the poor. Others point out that it would be more efficient to help the poor directly rather than to lower wood prices for everyone, the rich included. (Chapters 12 and 16 address this point further.)

The sections on economic instability and optimal environmental damage also have important income distribution aspects.

Other Market Failures?

Given some commitment to a market economy, market failure is the major justification for government actions. Without such failure, government programs could reduce welfare by distorting the necessary conditions for welfare maximization. For example, suppose the government forced more investment in timber production in the absence of market failure. One necessary marginal condition for a welfare maximum is for the last dollar spent in forestry to yield the same return as in any other enterprise (assuming correction for risk)—another application of the equi-marginal principle. This would occur naturally in a market economy, given no market failure. If the government forced more timber investment, the forestry rate of return would drop—through eventual wood price-declines caused by higher output—and the above marginal condition wouldn't be met.

Can you see why that would reduce welfare (i.e., reduce total returns from society's capital)? If the government program made forestry rates of return lower than those available elsewhere, welfare could be increased by shifting capital from timber to higher return investments until returns were again equated at the margin.

The above shows why market failure should be documented before recommending government actions. Without such failures, intervention could *decrease* social welfare, rather than increase it. Another example of a shaky defense for intervention is a U.S. Forest Service rationale sometimes given for government programs to increase forestry investment: Short-run timber supply functions tend to be inelastic, which implies that as stumpage prices rise by a certain percentage, timber supply doesn't increase by as large a percentage (Cubbage and Haynes, 1988). This simply means that timber supply elasticity is less than 1, or that short-run supply isn't highly responsive to price. But that's not a market failure preventing welfare maximization: many products have supply elasticities less than 1, and the market can continue to provide them in optimal quantities (for example, see the last chapter's hypothetical inelastic fish supply curve and the equilibrium analyses based on it).

Some analysts feel that forestry's long production period is a form of market failure hindering private timber investment. Others point out that investors in reforestation needn't wait until rotation-end to cash in; they can (and do) sell immature forests or sell cutting rights. Too, when small woodlands are not reforested, the forest industry often stimulates timber output on these lands by leasing, buying, or providing reforestation assistance, *if* the expected rate of return is adequate. If returns are not competitive, and *if* no market failure exists, reforestation may not be a wise expenditure.

The above list of market failures and justifications for government programs is certainly not all-inclusive. But it does cover some of the more important failures related to forestry. There are reasons to be wary of arguments for government programs based on "market failures" unrelated to the above list.

Existence of market failures doesn't necessarily argue for abandoning markets. The alternative of centralized planning hasn't been impressive. Market failures simply point to areas where efficient government action can improve welfare within a market system. By "efficient," I mean ensuring that the costs of added government actions don't exceed the resulting benefits (marginal analysis again!). We also have to guard against assuming that we'll always have "good" government countering "bad" effects of markets. Government bureaucrats can often be empire builders, self-interested, and power hungry, the result being "government failure" where program costs exceed benefits. Thus, watchdog groups need to closely monitor government programs. Examples of such organizations in forestry are Cascade Holistic Economic Consultants (O'Toole, 1988), the Political Economy Research Center, the Native Forest Council, the American Forest and Paper Association, the Association of Forest Service Employees for Environmental Ethics, Public Employees for Environmental Responsibility, and a host of other groups.

OPTIMAL LEVELS OF ENVIRONMENTAL DAMAGE

We've briefly looked at the notion of optimal damage under "Unpriced Negative Side Effects." But due to the increasing public attention to environmental damage from timber harvesting, let's examine the issue in greater detail. In widely distributed publications like *Time, Sports Illustrated, The Wall Street Journal, Newsweek,* as well as in virtually all environmental magazines, recent articles have highlighted damage to owls, woodpeckers, fish, scenic beauty, soil, and water quality as a result of certain timber harvesting practices. Foresters don't always come across as the resource stewards we think we are. Our approach has often been to say, "If we could only educate the public about what we're doing, they'd be more understanding." That tactic hasn't been especially fruitful. Let's consider other views and how economic thinking can shed some light on environmental damage problems. These will be some of the major issues affecting forestland use in the twenty-first century.

Damage Liability Rules

Those causing damage will be termed "damagers," while those first harmed by damage will be "victims."[6] The damager is termed a "firm," which includes any property owner—*public or private*—doing anything from backyard gardening to tree farming or paper manufacturing. In a sense, roles of damager and victim are reversible. For example, if a firm kills fish through water pollution, the damager is the firm, and victims are anglers (and fish!). If regulations require costly pollution controls, the firm could consider itself a victim. Here the party first causing damage will be the damager; the one harmed by this damage will be the victim.

When one party imposes damage upon another, two bargaining approaches or *damage liability rules* exist:

Victim liability: Start with damage allowed. Make victims liable for damage reduction, and determine whether they'd be willing to pay the costs of reducing damages.

Damager liability: Start with no uncompensated damage allowed, and place the liability on the damagers—public or private—to compensate damaged parties. Compensation will be the minimum monetary payment that victims would require to willingly endure the damage. With compensated damage, victims would be just as satisfied as they were before the damage occurred.

These damage liability rules focus on two types of rights. Allowing damage to start and then placing liability for its reduction on the victims (victim liability) stresses *private property rights* or individuals' rights to do with their property

[6]Unfortunately, the terms "damager" and "victim" are emotionally charged. I'm not implying that damagers are willfully torturing victims. I assume damages here are unintended. I tried calling damagers "producers" or "operators" and victims "recipients" or "receivers," but sometimes the discussion became so bland that it was hard to see who was doing what to whom. Thus, I returned to the more descriptive terms, recognizing the risk of misunderstandings.

what they wish. Refusing to allow uncompensated damages (damager liability) emphasizes what Mishan (1967) calls **amenity rights:** the community's right to enjoy amenities like clean air and water, attractive landscapes, and peace and quiet. Our founding fathers placed great emphasis on private property rights. And this made sense, given the vast amount of land and a fairly small population. But crowding and development have led to more concern about amenity rights, and conflicts abound. You could view this as tension between community values and individual freedom.

Let's now consider how we might conceptually use bargaining frameworks in trying to reach "optimal" environmental damage levels, under both damage liability rules. It's important to remember with each rule that payments don't actually have to be made to show an improvement in welfare; one only needs to show that, if the payment were made, gains would exceed the cost.

Optimal Solutions with Monetary Damages

Figure 3-5a and b explains the concept of optimal environmental damage imposed by one firm. Compared with Figure 3-3, this more general analysis allows damage abatement that doesn't necessarily reduce the firm's output. The horizontal axes, units of damage, could be, for example, units of sediment or turbidity per month discharged in a river by a mill or by logging. *This section assumes that the only damage is reduced income to victims,* for instance, reduced income of commercial fishers or increased water purification costs for water companies. The next section will cover damage to nonmarket values like scenic beauty.

In Figure 3-5a, XP is the firm's maximum profit (total revenue minus total cost) with uncontrolled environmental damage at X units. The dashed curve shows the firm's profit at different damage abatement levels. The difference between XP and the profit (dashed curve) is the firm's total damage control cost at any damage level. The solid curve is the total damage cost to victims, in terms of lost income. The distance between the firm's profit curve and the damage curve (the dashed and solid curves) is the net social revenue. Thus, the optimal damage level is at D where net social revenue is maximized. That may sound strange; if environmental damage is bad, how can we have an optimal amount of it? Figure 3-5b helps clarify the concept.

Victim Liability In Figure 3-5b, start at uncontrolled damage (X) and move from right to left (r. to l.): *this follows the victim liability rule, stressing private property rights.* The dashed curve is the firm's loss in profit per unit of damage reduced (marginal cost) and is the slope of the dashed curve in Figure 3-5a. The solid curve in Figure 3-5b is victims' added revenue from each unit of damage reduced, or maximum **willingness to pay** to avoid the last unit of damage (or to gain more fishing income). This is the slope of the solid curve in Figure 3-5a. Reducing damage, unit by unit, up to D, the increased income to victims exceeds the added cost to the firm. Using the commercial fishing example, from their higher income, fishers could actually afford to pay the firm not to pollute.

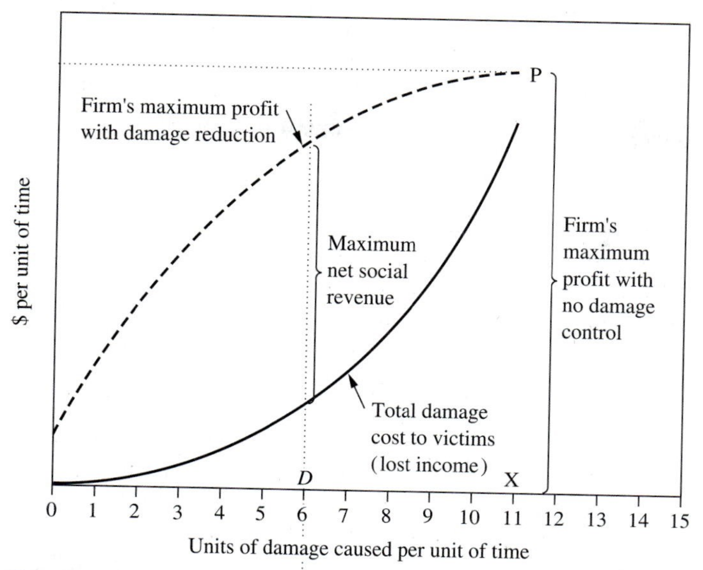

FIGURE 3-5a
Environmental damage costs and firm's profit.

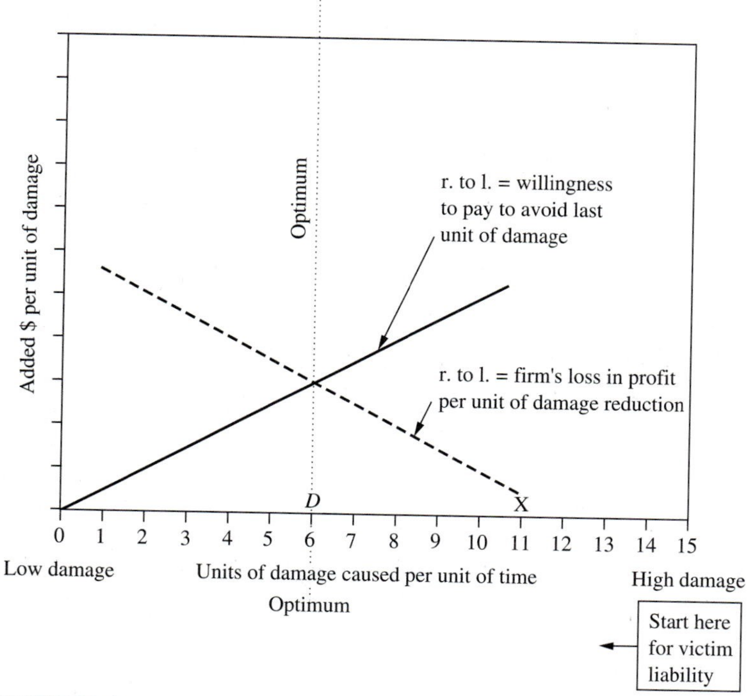

FIGURE 3-5b
Optimizing monetary damage—stressing property rights.

Reducing damage below D, the extra revenue to victims wouldn't be enough to cover the added abatement cost to the firm. So D is the optimal damage level. Note the economic thinking in incremental units. Figure 3-5b shows that, without bargaining between victims and damagers, a free market tends to produce too much environmental damage—at X, where the firm's profit is maximized. "Too much" (for example, at any level above 6 units of damage in Figure 3-5b) means that the added benefits of eliminating one more unit would exceed the added costs.

Damager Liability An alternative view is to start at zero damage in Figure 3-6: *this is following the damager liability rule and stresses public amenity rights.* Figure 3-6 has the same curves as Figure 3-5b, but they are now labeled to read from left to right (l. to r.). The dashed curve is now the firm's added profit or cost-saving from being allowed to cause one extra unit of damage. The solid curve in Figure 3-6 is the marginal damage cost, or victims' income loss, per unit of damage. Starting at no damage and moving rightward in Figure 3-6, it's efficient to allow added damage as long as the added profit to the firm is enough to pay the added cost to victims. In this bargaining framework, you can think of the solid curve as the minimum compensation victims require to willingly endure one more unit of damage (often called the **willingness to sell**—in this case, sell catchable fish). You can see that up to point D, the firm's added profit exceeds the victims' added cost. But if we allow damage beyond D, the firm's

FIGURE 3-6

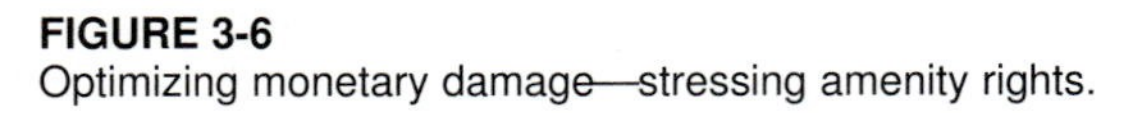
Optimizing monetary damage—stressing amenity rights.

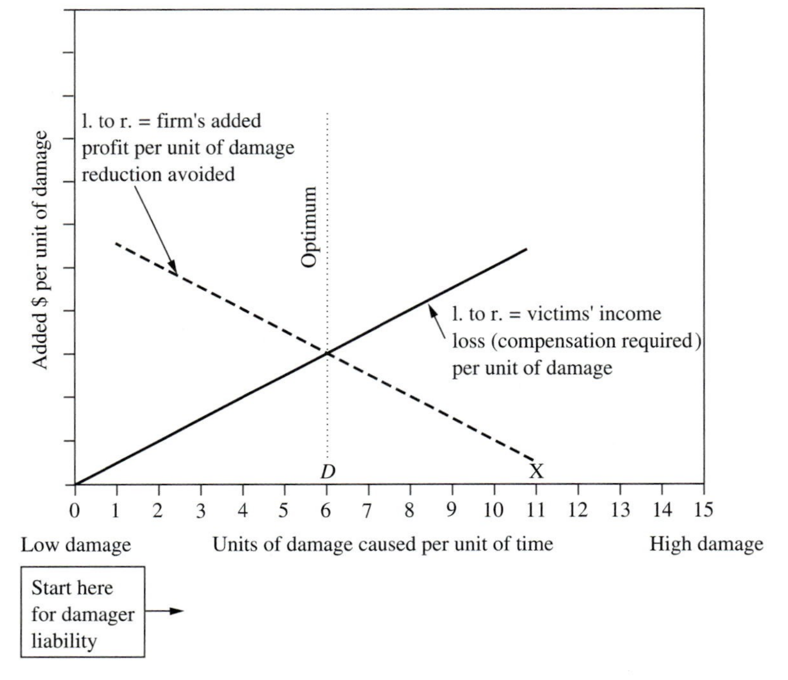

added profit wouldn't be enough to compensate victims for their costs. Thus, the optimum damage level is still at *D*, the same level as in Figure 3-5b.

When the only environmental damage is a reduction in the victims' income, the optimal damage level is exactly the same under either bargaining framework (victim or damager liability). This results because victims' income losses from one more unit of damage exactly equal their maximum willingness to pay to prevent that damage. If some impending damage would cost you $100—for example, in lost commercial fishing income—you wouldn't pay more than $100 to prevent it. Likewise, you'd ask for $100 as compensation if the damage occurred. But this symmetry, called the Coase theorem after the 1991 Nobel prize winner Ronald Coase (1960), doesn't necessarily hold for the case of nonmonetary damage, which is common in forestry.

Optimal Solutions with Nonmonetary Damages

Now let's consider environmental damage that does *not* reduce victims' incomes but causes a nonmonetary loss like reduced air quality, reduced scenic beauty, or loss of recreational fishing or swimming due to water siltation. In forestry, harvesting and wood-processing activities can cause such damage. Now the two bargaining frameworks don't necessarily give the same optimal damage solution because the amount victims are willing to pay to reduce damage is often less than the compensation they'd require to willingly endure that same damage. (That's the same as saying your willingness to pay to gain a nonmonetary amenity is less than your willingness to sell it.) The case is irrefutable for very poor victims: they have no ability to pay to reduce environmental damage, but they would demand some compensation for such losses. This asymmetry is shown in Figure 3-7, which demonstrates nonmonetary damage optimization under both damage liability rules—stressing private property rights (r. to l.) or public amenity rights (l. to r.). The figure shows the same curves as in Figures 3-5b and 3-6, but now the victims' compensation curve, the dotted curve (l. to r.), lies above the solid curve of their willingness to pay (r. to l.). *Thus, with nonmonetary damage, stressing amenity rights gives a lower optimal damage level (D'), than stressing private property rights (D).* This makes sense: if you initially allow nonmonetary damage and then seek an optimum by asking what people would be willing to pay to reduce damage (victim liability), you're likely to end up with more damage than if you don't permit any uncompensated damage (damager liability).[7]

Stressing amenity rights (damager liability), you start at zero damage in Figure 3-7 and allow added units of damage as long as the added profit to the firm (dashed curve) exceeds the victims' compensation required per unit of damage (the dotted curve, which includes monetary and nonmonetary damage). To the right of *D'*, the firm's added profit from allowing one more unit of damage isn't enough to pay the required compensation, so further damage reduces social

[7]Several authors have noted this difference in optimal nondollar damage, depending on assignment of property rights (contrary to the Coase theorem); I believe Mishan (1967: 57–66) was the first. Also see Klemperer (1979), Knetsch (1990), and Kahneman et al. (1990).

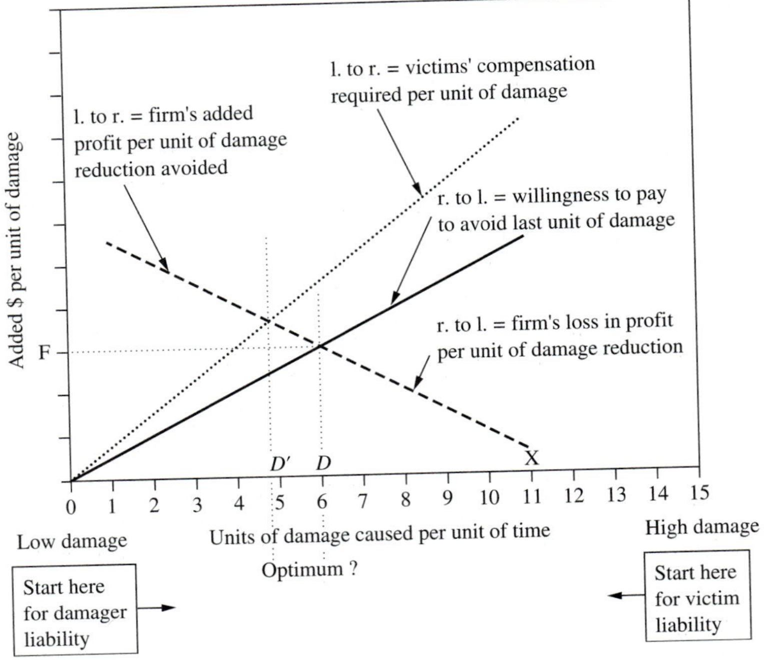

FIGURE 3-7
Optimizing nonmonetary damage: victim and damager liability.

welfare. In Figure 3-7, victim liability is interpreted as in Figure 3-5b, reading from right to left, arriving at the optimum damage of *D*. As you'll see shortly, arriving at a monetary value of willingness to pay to reduce nonmarket damage isn't as easy as the case where the only damage is lost income.

Your required compensation for accepting a given amount of nonmonetary damage will exceed your willingness to pay to avoid that damage if you have a declining marginal utility of money, as shown in Figure 3-8. With increasing money, the total utility curve's slope (the marginal utility) declines. Suppose your initial wealth is at M, and some pollution exists. Consider a possible pollution reduction of P that would increase your utility by U (the pollution brings a disutility of U). Placing liability for damage reduction on the victim (you), you'd be willing to pay up to M_1 to eliminate P. You wouldn't pay more, because that would bring a utility loss greater than U.

Now place the liability on the damager, and ask what compensation you'd require to willingly endure the last increment of pollution (P, above). To receive a utility gain of U to offset the pollution disutility of U, Figure 3-8 shows you'd need a compensation payment increasing your wealth by M_2. Due to the declining marginal utility of money, M_2 exceeds M_1. This explains how the compensation you'd require to willingly endure damage could exceed the amount you'd be willing to pay to reduce that same damage. It explains how the dotted curve can lie above the solid curve in Figure 3-7. In that case, the optimal nonmonetary damage would be lowest under damager liability (stressing amenity rights over

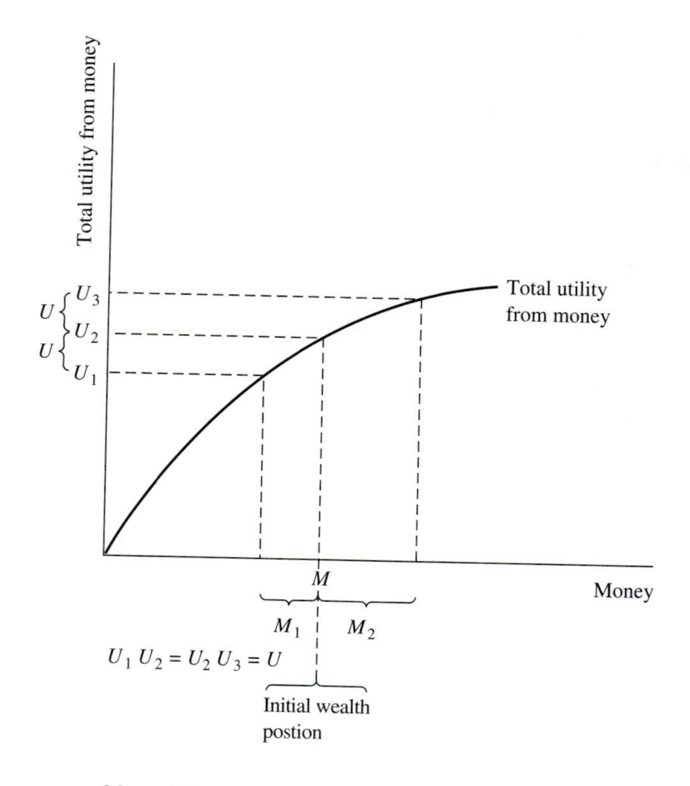

M_1 = willingness to pay to avoid nonmonetary loss of U or to gain U
M_2 = compensation required to endure loss of U (i.e., willingness to sell nonmonetary benefit worth U)

FIGURE 3-8
Declining marginal utility of money.

private property rights). If you imagine a nonmarket good that is being damaged (e.g., recreational fishing), the foregoing implies that your willingness to pay to attain one more unit of the good is less than your willingness to sell the same good (the compensation required to accept the loss of the good).

Although not everyone has a declining marginal utility of money, empirical research suggests that most people do.[8] Most folks whose wealth and income have increased over time will find that an added amount of money today brings less satisfaction than the same amount did when their income was much lower, even if we correct for inflationary loss in buying power. Also, the income distribution section noted that at very low income levels, declining marginal utility of money seems irrefutable, since one's initial amounts of income sustain life itself. Added income must be less important.

Chapter 14 reviews empirical studies consistently showing that people's required compensation when losing a nonmarket amenity exceeds their willingness

[8]See Halter and Dean (1971); and Wilson and Eidman (1983).

to pay to attain it or save it from damage, the difference averaging in the three- to fourfold range but often being much more.[9] Several of these studies point out other factors—beyond declining marginal utility of money—that make willingness to sell exceed willingness to pay by more than Figure 3-8 suggests. One of these is the observation that people hate to give up what they already have (Kahneman et al., 1990).

Damage Fines Some have suggested charging damagers a fine per unit of damage caused per unit of time, in cases where damage like water pollution can be measured. In Figure 3-7, if the optimal damage level is at D, the idea is to set the fine at $\$F$/unit of damage, which is the dollar level where dashed and solid curves cross. That presumes victim liability; with damager liability, the fine is set where the dashed and *dotted* curves cross. You could plot the fine as a horizontal line going through this intersection point, as shown for F in Figure 3-7. In that case the firm would tend, of its own accord, to seek the optimal damage level D at 6 units of damage. Why? Because, starting at 0 damage, the firm would create added units of damage and pay the fine per unit as long as its added profit from avoiding pollution control costs (dashed curve) exceeded the fine. Beyond 6 units of damage, the $\$F$ fine per unit would exceed the firm's added profit. At that point, it would no longer pay for the firm to cause more damage. So the firm would cause 6 units of damage, pay the total fine of $6(F)$, and pay the cost of controlling damage beyond 6 units.[10]

While the damage fine or tax is theoretically attractive, it really doesn't help in finding the optimum damage level, for example, D in Figure 3-7, unless you know the social cost per unit of damage. Then, as with regulations, you have to monitor damage levels to enforce the fine system. But there is the advantage that governments could use fine receipts to cover administrative costs and compensate damaged parties. Note that a fine system would allow a firm with high costs of damage reduction (a high dashed curve in Figure 3-7) to pollute more than a firm with lower costs—an efficient result, but not necessarily intuitively appealing, especially for those receiving the heavier damage. For many of the broad forestry side effects like scenic damage or nonpoint source water pollution, where you can't always pinpoint units or sources, the fine system may not be practical.

Transactions Costs

In theory, under either liability framework, bargaining markets could form, and one party could "bribe" the other either to reduce or to accept environmental damage, thus reaching an optimal damage level. The problem is the excessive "transactions costs" of organizing these markets, especially when millions of people are

[9]Hanemann (1991) cites several empirical studies showing that willingness to pay for nonmarket benefits is less than the required compensation when losing the same benefit, and he provides theoretical defense for such results. His work supports findings that this divergence can be significant for the case of nonmarket environmental amenities for which there are few substitutes.

[10]Ideally, the fine per unit of damage should follow either the solid or the dotted curve, varying with the damage level, but that would be almost impossible to administer.

involved. Such markets don't develop effectively—a form of market failure—and thus bargaining solutions are mainly a theoretical way to view environmental problems.

Other transactions costs are those of implementing any government actions, such as fines, taxes, incentives, regulations, or educational programs to lessen market failures. Governments should consider such costs when comparing the benefits and costs of any intervention.

Income Effects

Using the *same* damage liability rule, nonmarket damage optima could differ, depending on the income levels of victims (Mishan, 1967). Consider the victim liability case. The higher the victims' incomes, the greater their willingness to pay to reduce a given amount of damage. And the poorest can't afford to pay anything to reduce damage. That makes it likely for a wealthy group of victims to have a higher solid curve in Figure 3-7 than poorer groups of the same size, in which case optimal damage levels for high-income groups would be lower.

Now consider placing the liability for damage reduction on the damager. The sum demanded by wealthy groups as compensation for a given amount of nonmarket damage is likely to exceed that demanded by poor groups as compensation for the same amount of damage. Again the result is that the Figure 3-7 framework would yield lower optimal damage levels for wealthy groups than for poor ones, given the same number of people.[11]

You can see the above income effects in the fact that you'll often find higher environmental damage levels in areas where incomes are low. This doesn't imply that firms intentionally impose more pollution on the poor: it's just a natural consequence of market forces. Land where low-income groups live is cheaper for firms to buy, and areas already damaged are more affordable residential areas for the poor. And such residents, compared to wealthier groups, have less money and time to influence damage control policies in their areas.

Where to Place the Liability for Damage Reduction

A major body of research shows that the monetary value of people's willingness to pay for nonmarket environmental amenities is far less than their willingness to sell the same amenities (see Chapter 14). If decision makers acknowledge this, the question of where to place the liability for nonmonetary damage reduction is vital, since it can sharply affect the optimal damage level. Those who rally to the

[11] What do the above two paragraphs require us to assume about the marginal utility of increasing income? Only if the marginal utility of money is decreasing can we have wealthy people willing to pay more than the poor to remove a given quantity of damage (or likewise to demand a larger compensation to willingly accept that damage). Such reasoning also requires the heroic assumption that the utility from removing a unit of damage (or disutility from accepting it) is comparable for people of all income levels—the heresy of interpersonal utility comparisons. While many economists shy away from such assumptions, they can't be avoided when analyzing issues of optimal environmental damage.

cry of "private property rights!" say that costs incurred by private landowners to reduce damage to others should be reimbursed by the public. Others stress that all costs imposed on others should be compensated for or stopped.

At this point two questions should be clarified:

1 Where should we place the liability for damage reduction?
2 Should compensation payments actually be made? (That is, should we compensate damagers under victim liability or compensate victims under damager liability?)

The second question is independent of the first. Under either damage liability rule, any action would increase welfare if the resulting gainers *could* overcompensate losers. Whether or not compensation actually occurs is a question of ethics and income distribution.

Placing Liability for Damage Reduction on Victims Starting with uncontrolled damage, you could estimate costs (out-of-pocket costs and income forgone) to the damager for various levels of damage reduction. By various means—education, fines, or tax policies, payments to damagers, regulations—a government could encourage increasing levels of environmental improvement as long as the rest of society was willing to pay the cost. Willingness to pay might be gauged by voting for increased taxes to cover the costs of damage reduction or by conducting surveys to consider the likelihood of such a vote. Some surveys directly ask people how much they're willing to pay for environmental improvements. Respondents should be told by how much these improvements will increase prices, which in turn could reduce their willingness to pay. For example, harvest restrictions can increase wood products prices by shifting regional timber supply curves to the left (can you sketch this scenario?). Note that if respondents felt they wouldn't actually have to pay, they could declare a damage reduction expenditure worthwhile when they might not have done so, had they actually been required to pay the costs.

In surveying someone's hypothetical willingness to pay, even if you somehow eliminate overstatement, you can't keep adding many independent evaluations and assume that the same person would willingly pay the sum of all the separate amounts. Eventually there wouldn't be enough income. Thus, major problems exist in summing hypothetical willingness-to-pay values over a large number of amenities for one population.

One philosophy says that we should collect the public's willingness to pay for damage reduction and use the funds to compensate landowners for the resulting loss in property income. Such income reduction will actually reduce property values, which is an implicit confiscation: lawyers call this "inverse condemnation." Many of the arguments in this area hinge on how long people have owned their property and how its value has increased. For example, suppose you bought land for $50/acre 20 years ago, and now it's worth $150,000/acre, mainly because of development by others on neighboring lands. Now suppose the government imposes regulations that decrease environmental damage but reduce your property value by $50,000/acre (e.g., maximum building height, minimum lot size, or requirements for drainage and scenic buffer strips mean that potential buyers would pay only $100,000/acre for the land). Some maintain that you should be compensated for the loss in property value—that individual freedom, liberty, and property rights entitle you to all increases in your property value. Others argue that part of your increase in property value was a windfall gift from society—you couldn't have received it without the development on neighboring lands—and now society shouldn't feel sorry about removing some of that value to prevent you from damaging the environment. But suppose you paid $150,000 for the same property *last year,* and new *unanticipated* regulations reduced your property value by $50,000. One could no longer claim that you had a gift from society. Some argue that the case for compensation might be stronger if your property value increase is due only to *your* investment in improvements like drainage, roads, structures, and plantings rather than from others' development. The courts are full of cases on these issues, and there's little consensus on how they should be resolved.

A 1994 bill introduced in the Senate, "The Private Property Rights Restoration Act," illustrates a property rights view of environmental regulations. The proposed legislation would have compensated landowners when federal regulations reduced their property value by more than a specified percentage. Neither the Senate nor the House took further action on this bill in 1994, but similar proposals were being drafted again in 1995.[12]

An objection to the victim liability approach is that it implies the public should pay to clean up someone else's mess. Placing liability on the damagers, while eliminating this problem, raises others.

Placing Liability for Damage Reduction on Damagers Initially forbidding landowners to impose costs on others, you could estimate extra income accruing to landowners if added damage were allowed. Unit by unit, more damage would be allowed as long as the resulting increase in income to landowners exceeded society's required compensation to willingly endure that damage (their willingness to sell the environmental benefits to be lost).

Asking what compensation society would require to tolerate given levels of damage is fraught with problems because people aren't used to selling amenity values. Also, some groups might state unrealistically high compensation sums in an effort to prevent any environmental damage, especially if no one believed that compensation would actually be paid. Aside from such problems, we'd need detailed surveys and realistic payment vehicles like lower taxes and reduced commodity prices. In a realistic survey of this type, a group of electric customers was asked their willingness to trade varying increases in air pollution (from coal-fired power generators) for different reductions in their electric bills (Randall et al., 1974). When being asked about required compensation to accept increased damage, people should be told of the expected decrease, if any, in certain prices. For example, accepting more environmental damage from higher timber harvests will result in lower timber prices (timber supply curve shifts rightward) and could decrease the compensation required.

[12]For a review of recent property rights legislation in the states, see Lund (1995).

Class action suits, where a group seeks compensation for some form of damage, are an example of applying the damager liability framework. When large groups are involved, the values can quickly escalate. Simple arithmetic tells us that a million people seeking $100 each amounts to $100,000,000.

It may be that among some groups, only very high payments could compensate for certain environmental damage. To use an extreme analogy, consider a mother threatened with a kidnapping. The amount she could pay to prevent the kidnapping is finite. Yet the sum she would demand as compensation for losing her child could well be infinite. While environmental damage isn't like stealing children, it's extremely difficult for some people to estimate a payment that would compensate them for the loss of certain environmental qualities. Given our increasing affluence, there may be a growing trend toward guarding some types of environmental quality in the same way that children are protected.

The damager liability framework is rife with income distribution problems. For example, in the case of forestry-caused environmental damage, wealthy groups could conceivably make compensation requirements high enough to cause heavy timber-cutting restrictions. Some see this as providing playgrounds for the rich at the expense of high housing costs for the poor. (Can you use your knowledge of supply and demand curves for timber and housing to explain the reasoning behind this statement?[13] Note that the statement is emotionally loaded because the increased housing costs would occur for all income groups, not just the poor.) In spite of problems with the damager liability framework, we need more research in this area due to the significant effect that damage liability rules can have in finding optimal levels of environmental damage.

Changing Attitudes In the United States we seem to be gradually placing more emphasis on amenity rights and community values at the expense of individual freedom and private property rights. For example, as a former smoker, I now shudder at my past behavior of smoking in buses, restaurants, and other public places without considering the costs to nonsmokers. Earlier views favored a smoker's individual freedom, and smokers usually held the property rights to clean air. But today most people are equally concerned with amenity rights. Under damager liability, imagine a case where smokers would have to compensate people annoyed by smoke. I can picture enjoying an exquisite meal in a smoke-free restaurant and feeling that $5 wouldn't be nearly enough compensation for me to willingly accept smoke from someone nearby. A smoker might have to pay over $50 to bribe 10 people to accept smoke: that could make smoking outdoors a better alternative.

Most cities now regulate activities like leaf burning, building design, waste disposal, animal husbandry, and landscaping. Many landowners originally cursed these limits to property rights but now accept them for the common good. As population density increases, freedom to manage private and public forests is being

[13]Draw a regional timber supply and demand curve showing a price and quantity equilibrium. Harvest restrictions would shift the supply curve to the left, giving higher wood prices and lower output.

limited, often voluntarily, by concern about negative effects on others. These effects include siltation and turbidity of water, undesirable changes in water flows, scenic damage from certain clear-cuts, damage to endangered species, loss of soil, and fish kill. The key point is that attitudes are changing: landowners as well as smokers are becoming more sensitive to the effects of their actions on others. While some "damagers" are being dragged kicking and screaming into this new ethic, a shift in attitudes is obvious.

In a 1992 Tennessee Valley survey of nonindustrial forest owners and others, over 80 percent of both groups agreed with the statement, "Private property rights are important, but only if they don't hurt the environment." Agreement among forest owners was only slightly behind the general public. About 69 percent of the forest owners and 77 percent of the general public agreed with the statement, "Private property rights should be limited if necessary to protect the environment" (Bliss et al., 1994). This report cites other studies reaching similar conclusions about nonindustrial forest landowners' apparent willingness to change profit-maximizing management practices for environmental reasons.

Note that willingness to limit property rights doesn't necessarily mean support for costly government regulatory programs. Forest landowners and loggers often voluntarily adopt "best management practices" (BMPs) like erosion control that soften the environmental impacts of harvesting but reduce landowners' profits. (AFPA (1994) notes the forest industry's commitment to BMPs.) Private forest owners usually become more interested in voluntary BMPs when they observe states like California, which has the strictest private forestry regulations in the nation (McKillop, 1993). Another sign of change: The entire February, 1995 issue of the *Journal of Forestry* is devoted to forest aesthetics. An added example of changing forest management is the U.S. Forest Service's "new perspectives" program of the early 1990s, which emphasized environmentally more sensitive forest management. Be sure to remember that the notion of optimal damage levels applies to both private and public lands.

Avoiding Polarization In studying the theory of optimal environmental damage, you might mistakenly think that liability rules must be all one way or the other. That's unnecessarily polarizing. We can apply both rules at once. For example, a society can decide not to allow tractors to be driven through streams because the harvesters' gain from the practice wouldn't be enough to compensate losers (damager liability). The same society might use tax revenue to pay part of forest owners' reforestation costs to prevent erosion on unstable soils (victim liability).

Many liability rule applications are voluntary, without any government action: using the victim liability view, we send money to conservation groups to help reduce environmental damage, or we donate money and time to elect environmentally conscious legislators. From the damager liability view, some private firms voluntarily give the public free hunting and other recreation, picnic sites, and campgrounds and make other goodwill gestures that may soften public criticism of business practices. Again, these are examples of both liability rules operating at once.

Concerns about appropriate damage liability rules can't be solved with "hard science" or some technique of economic analysis. Foresters and economists can tell us about benefits and costs of alternatives and what different liability rules imply. But the final decisions will be made in legislatures, in courts, and in the hearts and minds of citizens. The most productive approaches will involve negotiated solutions sensitive to the rights of all parties.

Theory versus Practice

In Figure 3-7, note the slope of the firm's dashed curve (right to left) of loss in profit, or added cost per unit of damage reduction. This curve wouldn't necessarily have to slope upward from the right, but it's likely to. Consider a pulp mill operating at 11 units of damage and dumping all of its waste in a river. To remove initial units of pollution is fairly inexpensive; the firm needs merely to install a simple screen to eliminate the largest chunks of waste. As the pollution control progresses leftward in the graph, added costs to the firm become higher. At the extreme left, it becomes very expensive to remove the last traces of pollutant, making the water clean enough to drink. Similar cases exist in forest management: the first efforts at cleansing water runoff from a logging area aren't expensive; they are actions like making roads less steep, keeping tractors and debris out of streams, and adding water bars on steep roads. It's easy to support low-cost measures that yield major benefits. But the last efforts at minimizing site disturbance involve more costly practices like switching to a less profitable uneven-aged management without clear-cutting on some areas, seeding cut-and-fill slopes to grasses, and extensive culvert installations.

Why does the dotted curve of compensation required per unit of damage (victims' added cost) in Figure 3-7 slope upward from the left? Again, it wouldn't have to, but usually does. Picture unpolluted rivers and lakes not being used for drinking water. Initial increases in turbidity or sediment may hardly be noticed by boaters, swimmers, or anglers. But as we move rightward in the graph and more pollution is added, costs to users per unit of pollution grow higher as waters become unsuited for recreation.

Does the optimal damage framework of Figure 3-7 suggest that all environments should become partially damaged with no areas remaining pristine at zero damage? Definitely not! In some unique areas like the Yosemite Valley, Sequoia National Park, or certain wilderness areas, even very slight environmental damage is so unspeakable that the marginal damage cost curve measured by the dotted or solid curves is so high that it would lie above the firm's curve of added profit at all points. In such cases, the optimal damage level is zero. (Try sketching such a situation.) Conversely, in areas where, for example, temporary scenic damage from logging is seen by few people, and there's little **option demand** or **existence demand** for the scenic value, marginal damage or compensation curves could be low, and optimal damage levels might be relatively high. (Again, picture how such a case would look in Figure 3-7.) Such a view suggests that optimal damage levels depend on how many people are affected and that a single environmental damage standard won't generally be suitable for all areas. But others would disagree. In any case, the analyses should include future impacts as well as current effects of actions.

Upward pressure on optimal damage levels occurs when damage control could sharply reduce employment and income, with limited chances for reemployment elsewhere. Examples are heavy timber-cutting restrictions in areas where logging and wood processing are major sources of employment. For instance, in the early 1990s many jobs were lost in the Pacific Northwest due to old-growth harvest constraints to prevent clear-cutting damage to spotted owls and scenery. In Figure 3-7, such losses could be added to the dashed curve of the firm's loss in profit: as you move from right to left and impose more damage controls, the added costs are now greater, and the optimal damage level moves to the right at a higher level.

Also, moving from right to left in Figure 3-7 (victim liability), the resulting price-increases of the firm's output could lower people's willingness to pay to reduce damage, thus lowering the solid curve and increasing the optimal damage level. Or under damager liability, allowing more business output and damage can decrease certain prices and lower the damage compensation curve (dotted), thus increasing optimal damage levels.

Interesting compensation issues are raised if jobs are lost when reducing environmental damage. In this case, the general public is the damager, and the unemployed (a subset of the public) are the victims. If we set environmental standards that cause job losses, should we also use our tax dollars as compensation to fund job retraining and relocation programs? This is not without precedent: after setting aside the Redwoods National Park in 1968 and 1978, the federal government gave such help to forest industry workers who lost jobs as a result of the reduced regional harvest.[14] Similar programs were started in the mid-1990s to aid unemployed forest industry workers after U.S. Forest Service harvest reductions in the Pacific Northwest.

Starting at low damage in Figure 3-7, employment and income benefits from development can lead society to accept more damage than it otherwise might. Such benefits lower the marginal damage curve (solid or dotted curves), thus moving optimal damage to a higher level. On public lands, such thinking has influenced decisions like those to put an oil pipeline through the Alaska Wildlife Refuge and a section of Interstate Highway 87 through part of the "forever wild" Adirondack Forest Preserve, although not all would agree that the resulting damages were optimal.

The Figure 3-7 decision framework is a typical example of economic theory being far more precise than actual practice. In most cases it would be hard to precisely estimate the dotted or dashed curves and locate the exact optimum damage level. But that doesn't mean the framework is useless. The important thing is to ask about incremental costs and benefits as we move, for example, from more to less environmental damage, even if all units aren't in dollars. In a given river, one could determine the industry costs of reducing pollution by a set increment and list benefits like increased swimming, fishing, boating, and reduced municipal

[14] Public Law 90-545. October 2, 1968. 82 Stat. 931 et seq. Amended in Pub. L. 95-250. March 27, 1978. 92 Stat. 163 et seq.

TABLE 3-1
EXAMPLES OF FORESTRY-CAUSED ENVIRONMENTAL DAMAGE

Practice	Damage	Solutions
Streamside clear-cutting	Increased water temperature, fish kill, siltation and turbidity in streams	Leave streamside strips of trees and other vegetation
Careless logging, especially on hillsides and unstable soils	Reduced water quality, fish kill, scenic damage, damage to some endangered wildlife	Water bars and culverts in roads, limit road steepness, more bridges, scatter slash, seed vegetation on exposed soil, cable logging on steep areas, equipment restrictions (also see below)
Clear-cutting in scenic areas	Scenic damage	Limit clear-cut areas, partial cutting, convert to tolerant species, and never clear-cut
Slash burning and prescribed burning	Air pollution	Limit burning to specific days, or eliminate burning

water treatment costs. Under the victim liability approach, legislators could ask, step by step, "Are the benefits worth the costs?" Having gauged optimal damage in this way, you could assume that under damager liability, optimal damage would be even less.

Graphs like Figure 3-7 assume constant added benefits and costs per unit of time at any given damage level. We should really consider the "present value" of all added future costs and benefits over time because costs and benefits per year can change over time. Present value will be discussed in the next chapter.

Forestry Applications Admittedly, we often don't have very precise information about the relationships between forestry practices and degrees of damage like fish kill or siltation and turbidity levels in water. But continuing research will let us better estimate environmental impacts of different forestry operations.[15] Table 3-1 gives examples of forestry practices, types of damage, and possible solutions. Each solution will reduce the forest owner's income through direct costs or lower harvest revenues. For each damage liability rule, the optimal combination of practices will be where the extra costs and benefits of added or deleted practices are equated.

Figure 3-9 illustrates this line of thinking for optimizing logging damage from a given forest, using the optimal damage framework of Figure 3-7. The horizontal axis would be a series of environmental descriptions—air and water quality, soil, scenery visible to the public, fish, wildlife—starting with a fairly undisturbed environment at the origin and moving rightward toward more damage. Beginning

[15]For example, see Brown, 1969; Fredericksen, 1970; Gibbons and Silo, 1973; Hall and Lantz, 1969; Meehan, 1991; Rothacher et al., 1967; USEPA, 1976.

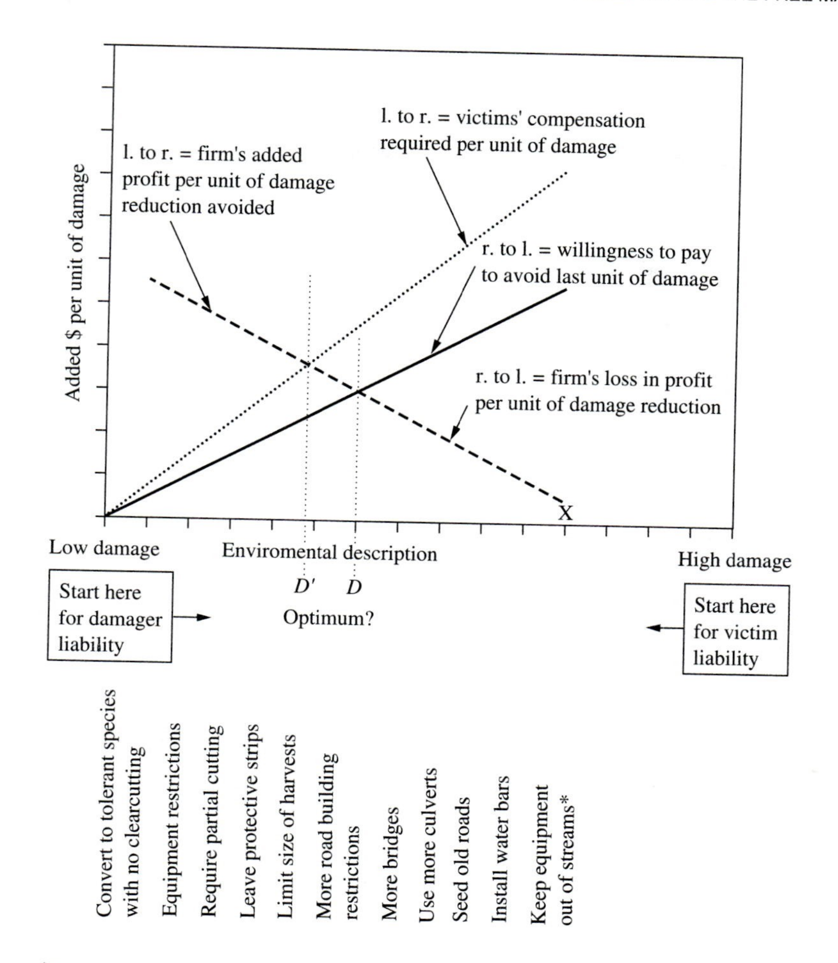

FIGURE 3-9
Optimum damage from logging.

at the right on the horizontal axis (victim liability), you can think of lessening damages by cumulatively adding practices shown below the graph. Proceed leftward until the victims would no longer be willing to pay the added cost to forest landowners (point *D*). Under damager liability, you'd start with low damage, at the origin in Figure 3-9, with all the practices in effect. Then you'd eliminate practices until the added gains to forest landowners were no longer enough to compensate victims for the resulting damage (point *D′*).

It's easiest to think of Figure 3-9 in terms of a public forest where citizens, in theory at least, have some control over management. The problem for private landowners is that they don't receive most of the benefits from the practices shown, so there's no incentive to implement them. A role for government here can be to provide incentives for practices, for example, direct payments or tax reductions—a victim liability approach. Educational programs can encourage voluntary best management practices, which have been implemented in several states. A more extreme tactic is for governments to impose regulations requiring certain practices, moving stepwise from right to left in Figure 3-9—a controversial and costly route. Such systems often require permits, large bureaucracies, and inspections to assure compliance, all costing money for both sides. Also, cumbersome permit requirements hamper quick harvesting decisions needed when weather and markets are best. Rather than regulate practices, some argue that it would be more efficient for governments to impose optimum regional environmental standards for, say, quality of water, air, or scenery and to let landowners choose the lowest-cost means of achieving the standard (again moving from right to left in Figure 3-9). Note that the practices shown are just examples, which aren't necessarily in a least-cost sequence.

In all cases, you'd want to consider the costs of administering government programs. For example, when trying to reach optimum environmental standards, extra administrative costs to the government and landowners should be added to the dashed curve in Figure 3-9 (moving from right to left). Or when relaxing standards, the administrative cost-savings should be added to the dashed curve (left to right). In either case, the effect is to relax environmental standards and accept more damage. The same reasoning applies to the cost of private landowners losing individual freedom as environmental standards are tightened (see "Individual Freedom and Multiple-Use" in Chapter 15).

Within each one of the practices listed in Figure 3-9, you should also optimize. For example, forested streamside strips have an optimum width. Each added foot of strip width decreases timber income and increases the quality of water, fishing, and scenery: these are marginal costs and marginal benefits per year. The marginal costs of added strip width (lost income) are fairly stable in this case. But added benefits of wider strips will eventually decline; for example, once a strip is wide enough to shade a stream and provide cooler water and better trout habitat, added width won't reduce water temperature by much more. And beyond a point, added width does little to decrease water siltation. So the optimum width is where the extra cost of the last foot of width just equals the extra benefit. The added benefit of making the strip still wider would be less than the added cost. (A question at the chapter's end asks you to graph this concept.)

We should avoid knee-jerk reactions like "All clear-cutting causes scenic damage." I was recently cross-country skiing through a heavily wooded area in Vermont, and our group came upon a clearing exposing open sky and a pleasant vista. We all paused and enjoyed the scene. I couldn't resist noting that this was a timber harvesting site where little slash was left. So generalizations are dangerous. Most of us don't want uninterrupted forest cover everywhere; nor do we want total deforestation. It's a question of optimum landscape diversity. A further problem with optimizing forestry practices is that one group's gain may be another's loss.

In this vein, we need to consider that damage from any given practice often varies with the location as well as the numbers of people affected. For example, logging damage to water or scenic resources tends to be greater in mountainous areas than on flat plains. This again argues for flexible environmental standards. One set of forestry practices can't be optimal for an entire country. None of these guidelines can be implemented precisely, but forest owners and governments should continue to refine data so that the key tradeoffs can be quantified. The added benefits and costs needn't all be measured in dollars; they just need to be compared. Without such information, we can never hope to maximize benefits from forestry. For a good review of forest policy and regulation issues, see the June 1995 *Journal of Forestry.*

CONCLUSIONS

Theories of welfare maximization in a free market are grounded in Adam Smith's (1776) idea that a society's well-being would be greatest in a competitive market with everyone pursuing their self-interests. But many have questioned that self-interest and dog-eat-dog competition will necessarily make us all as well off as possible. These concerns spawned the literature on so-called market failures. Today, many economists have further questioned the me-first model, and we find their writings sprinkled with words like "communal, fairness, cooperation, altruism, commitment, community, public interest, morality, ethics, and honesty" (see Daly, 1980; Daly and Cobb, 1989; Dixit and Nalebuff, 1991; Frank, 1988; Mishan, 1967; Schumacher, 1973). These are not the stuff of which Adam Smith's philosophy was woven. While most of us in the United States still believe strongly in market economies, more and more questions are being raised about how to tinker with the free market, in forestry and elsewhere, to provide more of the natural amenities the public craves.

KEY POINTS

- Arguments abound about the appropriate amount of government involvement in private forestry and about the amount of forestland that should be publicly owned.
- The chapter briefly outlines conditions required for an unregulated market to maximize social well-being and gives theoretical results when these conditions are met.
- When the above conditions are not met, we have "market failures," which are the justification for government action to improve social welfare. Such actions include financial assistance, taxes and tax reductions, fines, education, research, regulations, and government ownership of some resources.
- In forestry, market failures that could exist (but are not necessarily widespread) include poorly defined property rights, imperfect competition, imperfect information, immobility of labor, unpriced negative and positive side effects, and outputs not easily priced.
- On both public and private land, one of the major problems in timber production is to find the optimal level of negative side effects such as damage to water and air quality, soil, fish, certain wildlife, and scenery.
- In theory, the optimum damage level is where the extra costs of moving from the optimum exceed the extra benefits.

- A complicating problem is that the "damage liability rule" applied in aiming for environmental standards can affect the theoretically optimum level of damage from any activity, when the damage doesn't reduce victims' incomes.
- In thinking of bargaining solutions to damage problems, a theoretically optimum damage level can be found under two liability rules: (1) start with existing damage, and keep reducing damage as long as victims are willing to pay the costs of damage reduction; or (2) start with no damage, and allow added damage as long as damagers can compensate victims for damage imposed. The latter approach is likely to result in less nonmonetary damage than the former.
- Regardless of the damage liability rule, optimal damage levels will vary depending on people's evaluation of benefits and costs in different localities.
- Today more and more questions are being raised about whether the competitive self-interest model can lead to welfare maximization or whether it even reflects how a large number of people behave. But if governments try to lessen market failures, we must assure that the extra benefits of actions exceed the extra costs.

QUESTIONS AND PROBLEMS

3-1 In terms of resource allocation, what is a welfare maximum?

3-2 List the market failures that can prevent a free market from attaining a welfare maximum. Give forestry examples to illustrate each failure.

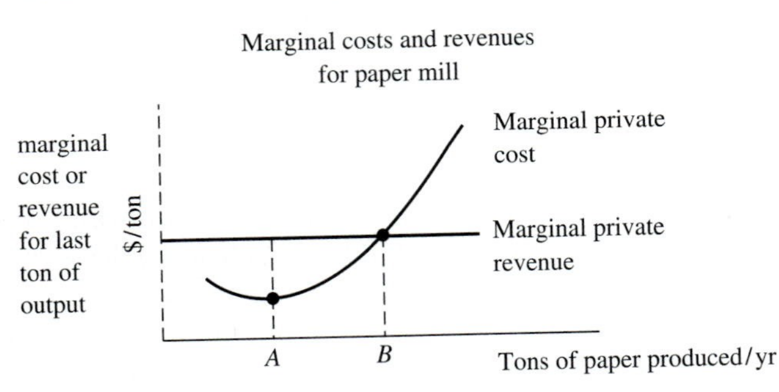

3-3 In the graph above, the firm's net income is maximized at output level *A* where marginal cost is minimized. (Explain your answer.)

a True

b False

3-4 In the previous question's graph, assume that we're showing private marginal costs and revenues for thousands of board feet (MBF) produced annually on a private tree farm. Assume that negative side effects external to the firm (e.g., water siltation) are costing society $20 per added MBF produced annually. Assume further that positive side effects external to the firm (e.g., open space and employment stability) are worth $10 per added MBF produced annually.

a How does the socially optimal tree farm output compare with the private optimum?

b Is it always necessary to change product output in order to reduce environmental damage? Explain your answer.

3-5 If policy makers always followed the analyses in Figure 3-7 of the text, would we have a partially damaged environment everywhere and no relatively undisturbed environments anywhere? Explain and justify your answer.

3-6 In the above-mentioned graph, sketch a situation where zero environmental damage would be optimal. What do you mean by optimal?

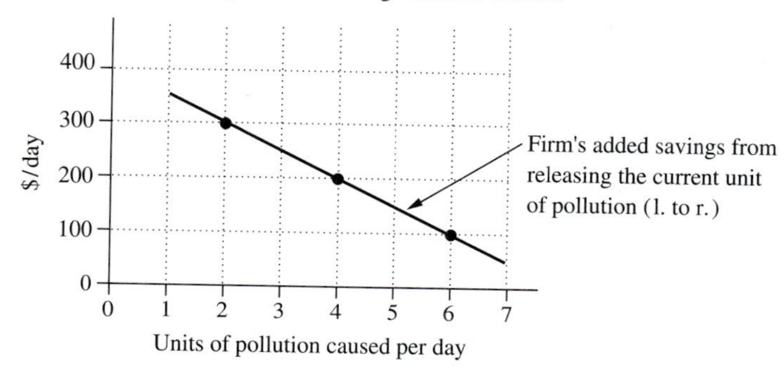

3-7 In the above graph, suppose the firm is fined $200 per day per unit of pollution caused daily. Starting at zero pollution, what is the greatest amount of pollution the firm could release without reducing its net income (this would be the firm's cost-minimizing level of pollution, including fines paid)?

3-8 At the firm's cost minimization point in the preceding question, what is the firm's total fine per day?

3-9 If the daily social damage external to the firm exceeds $200 per unit of pollution, will the socially optimal pollution level be greater than, less than, or equal to the firm's cost-minimizing level in question 3-7?

3-10 Is all this graphing of "optimal" damage levels so theoretical as to be meaningless in terms of real-world applications? If your answer is no, discuss some possible applications of this decision framework. What are some of the problems in such applications?

3-11 If firms are required to reduce negative side effects, the public will benefit.

a What might be some associated costs to society?

b How could these costs be lessened?

3-12 Suppose you expect temporary reductions in employment and regional income as a result of logging regulations. How could you incorporate such costs into Figure 3-7? How would that affect the optimal damage level under either liability framework?

3-13 Discuss the difference between damage to the environment and damage to people (let the latter be a loss of satisfaction in some sense). What does this difference, if any, imply for environmental policy?

3-14 Some will argue that government regulation to reduce environmental damage is an infringement on private property rights. Such regulation is sometimes viewed as a "partial taking" or confiscation of private property. In what sense is that true? What is the other side of the private property rights issue?

3-15 Use a "production economics" and supply-demand framework to illustrate possible price and output effects from environmental regulations in areas such as forestry, manufacturing, and agriculture. Use graphs to illustrate points (see Figure 2-14).

3-16 The section on "Forestry Applications" discusses the idea of optimal width for streamside strips to protect fish, water, or scenic values. Draw a graph that illustrates this concept, using marginal costs and benefits. Can you extend this notion to show the optimal intensity of the other practices listed in the chapter's last figure? The above-referenced discussion implied a "victim liability" view (start with stream damage and without streamside strips and decide how wide to make the strip). Can you discuss and graph the notion of optimal streamside strip width from the "damager liability" view?

REFERENCES

American Forest and Paper Association. 1994. *Sustainable Forestry Principles and Implementation Guidelines.* AFPA, Washington, DC. 11 pp.

Anderson, T. L., and D. R. Leal. 1991. *Free Market Environmentalism.* Westview Press, Boulder, CO. 192 pp.

Bator, F. M. 1957. The simple analytics of welfare maximization. *American Economic Review.* 47(1): 22–59.

Bliss, J. C., S. K. Nepal, R. T. Brooks, Jr., and M. D. Larsen. 1994. Forestry community or granfalloon? Do forest owners share the public's views? *Journal of Forestry.* 92(9): 6–10.

Bromley, D. W., Editor. 1992. *Making the Commons Work.* Institute for Contemporary Studies Press, San Francisco. 339 pp.

Brown, G. W. 1969. Prediction temperature of small streams. *Water Resources Research.* 5:68–75.

Coase, R. H. 1960. The problem of social cost. *Journal of Law and Economics.* 3:1–44.

Cowen, T. 1988. *The Theory of Market Failure—A Critical Examination.* George Mason University Press, Fairfax, VA. 384 pp.

Cropper, M. L., and W. E. Oates. 1992. Environmental economics: A survey. *Journal of Economic Literature.* 30(2):675–740.

Cubbage, F. W. 1991. Public regulation of private forestry. *Journal of Forestry.* 89(12): 31–35.

Cubbage, F. W., and R. W. Haynes. 1988. *Evaluation of the Effectiveness of Market Responses to Timber Scarcity Problems.* U.S. Forest Service Marketing Research Report No. 1149, Washington, DC. 87 pp.

Daly, H. E. (Ed.). 1980. *Economics, Ecology, Ethics—Essays toward a Steady State Economy.* W. H. Freeman and Co., San Francisco. 372 pp.

Daly, H. E., and J. B. Cobb. 1989. *For the Common Good.* Beacon Press, Boston. 482 pp.

Dixit, A. K., and B. J. Nalebuff. 1991. *Thinking Strategically.* W. W. Norton & Co., New York. 393 pp.

Dowdle, B. 1984. The case for selling federal timberlands. In Gamache (1984), pp. 21–46.

Feather, P. M., and G. S. Amacher. 1994. The adoption of best management practices for water quality and the role of information. *Agricultural Economics.* 11:159–170.

Foy, G., and H. Daly. 1992. Allocation, distribution and scale as determinants of environmental degradation: Case studies of Haiti, El Salvador and Costa Rica. In *Environmental Economics: A Reader.* A. Markandya and J. Richardson (Eds.), St Martin's Press, New York. pp 294–315.

Frank, R. H. 1988. *Passions within Reason.* W. W. Norton & Co., New York. 304 pp.

Fredericksen, R. L. 1970. *Erosion and Sedimentation Following Road Construction and Timber Harvest on Three Small Western Oregon Watersheds.* USDA Forest Service PNW For. and Range Exp. Sta. Res. Paper PNW-104, Portland, OR. 15 pp.

Galbraith, J. K. 1967. *The New Industrial State.* Houghton Mifflin, Boston. 427 pp.

Galbraith, J. K. 1958. *The Affluent Society.* Mentor Books, New York. 286 pp.

Gamache, A. E. 1984. *Selling the Federal Forests.* Proceedings of a symposium at the Univ. of Washington. Institute of Forest Resources no. 50, Seattle. 271 pp.

Gibbons, D. R., and E. O. Silo. 1973. *An Annotated Bibliography of the Effects of Logging on Fish of the Western United States and Canada.* USDA Forest Service Gen. Tech. Rept. PNW-10, Portland, OR. 145 pp.

Goodstein, E. S. 1995. *Economics and the Environment.* Prentice Hall, Inc., Englewood Cliffs, NJ. 575 pp.

Hall, H. D., and R. L. Lantz. 1969. Effects of logging on the habitat of coho salmon and cutthroat trout in coastal streams. In *Symposium on Salmon and Trout in Streams.* (T. G. Northcote, Ed.). University of British Columbia, Vancouver. pp. 355–375.

Halter, A. N., and G. W. Dean. 1971. *Decisions under Uncertainty, with Research Applications.* South-Western Publishing Co., Cincinnati. 226 pp.

Hanemann, W. H. 1991. Willingness to pay and willingness to accept: How much can they differ? *American Economic Review.* 81(3):635–647.

Hardin, G. 1968. The tragedy of the commons. *Science.* 162(3859):1243–1248.

Howe, C. W. 1979. *Natural Resource Economics—Issues, Analysis, and Policy.* John Wiley & Sons, New York. 350 pp.

Jones, G. T. 1993. *A Guide to Logging Aesthetics.* Cooperative Extension. NRAES-60, Ithaca, NY. 27 pp.

Kahneman, D., J. L. Knetsch, and R. H. Thaler. 1990. Experimental tests of the endowment effect and the Coase theorem. *Journal of Political Economy.* 98(6):1325–1348.

Kaldor, N. 1939. Welfare propositions of economics and interpersonal comparisons of utility. *Economic Journal.* 49(195):549–552.

Klemperer, W. D. 1979. On the theory of optimal timber harvesting regulations. *Journal of Environmental Management.* 9(1):253–262.

Knetsch, J. L. 1990. Environmental policy implications of disparities between willingness to pay and compensation demanded measures of values. *Journal of Environmental Economics and Management.* 18:227–237.

Lerner, Abba P. 1947. *The Economics of Control.* Macmillan, New York. 428 pp.

Lund, H. L. 1995. *Property Rights Legislation in the States: A Review.* PERC Policy Series. No. PS-1. Political Economy Research Center, Bozeman, MT. 28 pp.

McKillop, W. 1993. Economics of forest practice regulation in California. Proceedings of the Southern Forest Economics Workshop. Duke University, Durham, NC. pp. 13–17.

Mead, W. J. 1966. *Competition and Oligopsony in the Douglas Fir Lumber Industry.* University of California Press, Berkeley. 276 pp.

Meehan, W. R. (Ed.). 1991. *Influences of Forest and Rangeland Management on Salmonid Fishes and Their Habitats.* American Fisheries Society, Bethesda, MD. 751 pp.

Mishan, E. J. 1967. *The Costs of Economic Growth.* Praeger, New York. 190 pp.

O'Toole, R. 1988. *Reforming the Forest Service.* Island Press, Washington, DC. 248 pp.

Pearce, D. W., and R. K. Turner. 1990. *Economics of Natural Resources and the Environment.* The Johns Hopkins University Press, Baltimore. 378 pp.

Randall, A., B. Ives, and C. Eastman. 1974. Bidding games for valuation of aesthetic environmental improvements. *Journal of Environmental Economics and Management.* 1:132–149.

Reder, M. W. 1947. *Studies in the Theory of Welfare Economics.* Columbia University Press, New York. 208 pp.

Rothacher, J., C. T. Dyrness, and R. L. Fredericksen. 1967. *Hydrologic and Related Characteristics of Three Small Watersheds in the Oregon Cascades.* USDA Forest Service PNW For. and Range Exp. Sta., Portland, OR. 54 pp.

Schumacher, E. F. 1973. *Small Is Beautiful.* Harper & Row, New York. 304 pp.

Scott, A. 1973. *Natural Resources: The Economics of Conservation.* McClelland and Stewart Limited, Toronto. 313 pp.

Smith, A. 1776. *The Wealth of Nations.* Republished by Modern Library, Inc., New York.

Thaler, R. 1980. Toward a positive theory of consumer choice. *Journal of Economic Behavior and Organization.* 1:39–60.

U.S. Department of Agriculture, Forest Service. 1988. *The South's Fourth Forest: Alternatives for the Future.* Forest Resource Report No. 24, Washington, DC. 512 pp.

U.S. Environmental Protection Agency. 1976. *Forest Harvest, Residue Treatment, Reforestation and Protection of Water Quality.* EPA 910/9-76-020, Seattle, WA. 273 pp.

Wilson, P. N., and V. R. Eidman. 1983. An empirical test of the interval approach for estimating risk preferences. *Western Journal of Agricultural Economics.* 8:170–182.

4

THE FOREST AS CAPITAL

In this chapter we'll initially consider the forest as a store of wealth, or capital. In that sense, a forest is like a certificate of deposit or a stock you buy in the hope that, over time, it will return more money than you paid for it. Obviously, forests are much more than that and yield other benefits. But the financial view is a useful framework into which we can later weave nonmonetary aspects. Furthermore, many investors own forestland mainly for its financial returns.

It may seem like a detour to deal with forest finance for the next few chapters and interrupt the discussion of supply, demand, and market forces, to which we'll return in Chapter 12. But forestry is so capital-intensive that thoroughly understanding capital theory applied to forestry is vital before delving deeper into other aspects of forest economics. It's a bit like the chicken-and-egg dilemma: you can't grasp market equilibria in depth without understanding capital theory, but you can't comprehend capital allocation without knowing how prices are determined in the market. I've chosen to first cover the basics of resource allocation and pricing in markets, to inject capital theory and applications in forestry, and then return to more detail on short- and long-run supply and demand in a forestry context.

In a financial sense, if you consider trees and land as capital, two of the most important inputs into forestry are capital and time. How can we allocate these in a way that maximizes satisfaction to society? Let's now see how investors can use standard tools of financial analysis to evaluate forestry decisions—to decide how much to pay for forest properties and management practices and to measure how profitable forest investments are. In making such financial calculations, you'll need a calculator that will raise a number to a power and take the *n*th root of a number x (the latter can also be calculated as $x^{(1/n)}$). A good business calculator has many of the standard financial equations preprogrammed, but to avoid making mistakes, you need to be aware of certain nuances.

To simplify, we'll make some assumptions that will later be relaxed:

1 Taxes aren't specifically considered in this chapter. You could assume either that all income is before taxes or that taxes have already been subtracted. Taxes will be covered in Chapter 9.

2 Inflation, or the general rise in prices, will be zero. Thus, the average cost of a fixed typical market basket of goods and services will be constant over time. But prices of individual goods could change relative to one another. Unless otherwise stated, all interest rates and price-changes will be considered "real," excluding inflation. Inflation will be covered in Chapter 5.

3 Investment risk won't be considered explicitly except through increasing the rate of return investors demand. Risk in a future revenue occurs when you expect variable returns compared to very predictable yields from, say, government bonds. For example, due to risks, suppose the future harvest revenue from a newly planted Christmas tree farm ranges from $0 to $20,000 per acre in eight years. The greater the spread, the greater the risk. The average value or **expected value** of repeated plantings and harvests might be $12,000. In this chapter, future costs and revenues will be given as such average values, realizing that averages can mask large expected variation (high risk) or small variation (low risk). The chapter on risk covers these issues in more detail.

4 To demonstrate principles, this chapter will assume that all satisfactions from capital can be measured in dollars. Chapters 6, 14, and 15 show ways to incorporate nonmonetary benefits into investment analyses.

CAPITAL AND INTEREST

Classical economic theory defines **capital** as durable goods produced by people and used in production. In the broadest sense, this chapter will consider capital as any store of wealth yielding satisfaction to its owner. Under this broad definition, three types of capital assets are:

1 *Durable goods* such as machinery, equipment, tools, works of art, and buildings. Also included are inventories of goods.

2 *Financial assets* such as savings accounts, bonds, stocks, and certificates of deposit.

3 *Land and natural resources* such as coal, oil, and timber.

Because we'll evaluate these assets in the same way, let's look at their similarities. They're all assets that can be bought and sold. Also, *over time, the buyer expects to receive more than was paid for the asset.* Something for nothing? Not really. If you buy an asset, you give up the chance to spend that money now on goods and services. So your reward for postponing expenditure is the extra value you receive from an asset above its purchase cost. The simplest example is a savings account: If you put $100 in the bank for a year and earn 7 percent interest, you can withdraw $107 at the end of the year. People don't normally invest money (capital) without expecting some extra return—**interest** or a **rate of return.**

There's another reason why investors expect more revenue over time from an asset than they paid for it: In a well-functioning economy, such gains are widely available. For example, at the end of a growing season, farmers can usually receive more money for a crop than they spent on planting and tending it. Otherwise, why make the investment? The same prospect exists for many forms of capital, including forests. The income over time, whether in one lump sum or as a flow, can far exceed the original outlay. And even for nondollar returns from assets like art or scenic areas, you can imagine many cases where the satisfaction gained over time exceeds the satisfaction lost in creating or setting aside the asset.

A key point is that, *in this discussion, timber and forestland are considered capital.* Buyers of a forest or those who invest in forest management expect to eventually get more than they gave up for it, in dollar and nondollar terms. How much more? At least as much as they could reap over the same period on an equal amount invested in their best alternative at a similar risk. When investing, you give up an opportunity for earnings elsewhere on that same capital: this is what economists call **opportunity cost,** or the cost of an opportunity forgone. To willingly invest more, some people need a **minimum acceptable rate of return** (MAR) even *higher* than available alternatives. This doesn't necessarily mean that you must always sell assets for more than their cost. For example, you can sell, say, land or buildings for amounts equal to or less than costs and still make acceptable returns if your income (or "rent") over time is high enough.

Below are methods of measuring investment values and earnings by means of future value, present value, and rate of return. Table 4-1 explains all notation in this chapter.

TABLE 4-1
NOTATION FOR CHAPTER 4

ARR = alternative rate of return
C_y = cost in year y
MAR = minimum acceptable rate of return
n = number of years of compounding or discounting
NPV = net present value or present value of revenues minus present value of costs
p = amount of fixed payment occurring regularly in a series
PV = present value
r = real annual interest rate, percent/100. (Excludes inflation—Chapter 5 discusses the difference between real and inflated interest rates.) In equations, interest will be in decimals, so that 8 percent will be 0.08.
R_y = revenue in year y
t = number of years between periodic occurrences of a payment, p
V_0 = initial value at the start of an investment period (may sometimes refer to "present value")
V_n = future value in year n (excludes inflation)
y = an index for years

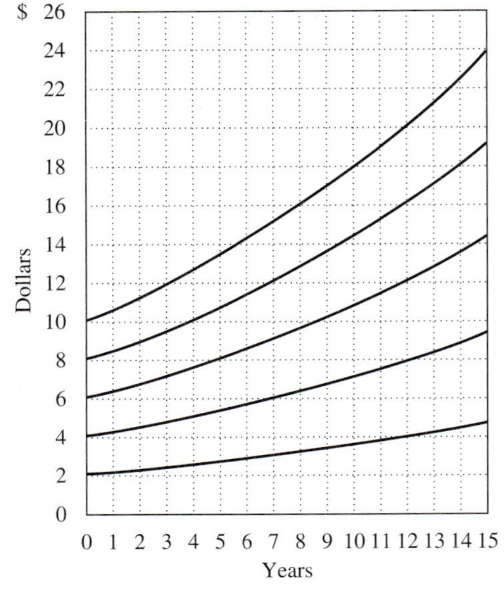

Each curve is:
$V_0(1+r)^n$ = value in year n
= V_n = future value
V_0 = year *0* value
r = interest rate/100
n = years

FIGURE 4-1
Family of 6 percent interest lines.

Future Value of a Single Sum

Figure 4-1 shows the concept of capital growing at a compound interest rate without any income withdrawn. The interest rate is 6 percent per year, and the vertical axis shows value in dollars, while the x axis is the number of years an initial investment of V_0 grows (here, starting at 2, 4, 6, 8, or 10 dollars). To compute any **future value** V_n from an initial investment of V_0, use the following formula:

$$V_n = V_0(1 + r)^n \tag{4-1}$$

For example, if you put $10 in the bank at year 0 at 6 percent interest ($r = 0.06$), it grows to $V_n = 10(1.06)^{12} = \$20.12$ in $n = 12$ years, as shown in Figure 4-1. So at 6 percent interest, you'll double your money in 12 years, since $(1.06)^{12} = 2.0$, a multiplier of 2. As defined in Table 4-1, the interest rate, as well as all other rates, will be divided by 100 when used in equations. For this example, the 6 percent interest is your "rate of return."

Equation (4-1) defines the curves in Figure 4-1, V_0 being the point where each curve starts on the y axis. Compound interest curves are exponential, and a family of such curves for one interest rate will diverge over time as shown. The higher the interest rate, the steeper the curves.

Below are values over time when V_0 is $1.00 invested at 6 percent interest:

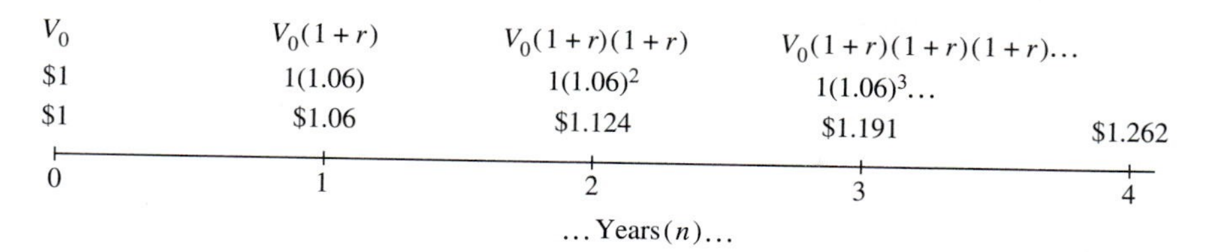

Each year's value is the previous year's value times $(1 + r)$. Therefore, any future value is the initial value (V_0) times $(1 + r)^n$, as shown in equation (4-1). This growth is **compound interest** because you earn interest not only on the principal of $1.00 but also on the accumulated interest. If this were 6 percent **simple interest,** the interest would always be $0.06 each year, and investment growth curves would be straight lines rather than the exponential functions shown in Figure 4-1. However, investments don't normally grow at simple interest. In the above example, annual interest after the first year rises above $0.06 by an increasing amount each year. What is the interest earned in the third year? [The value at the end of the third year minus the value at the end of the second is $1.191 - 1.124 = \$0.067$. Or multiply the second year value by the interest rate: $1.124(0.06) = \$0.067$.]

Table 4-2 is a compound interest table showing values of $(1 + r)^n$ for different interest rates r and different time periods n. The tabulated values are also the future value of $1.00 compounded for n years at an interest rate of r, since $V_n = \$1(1 + r)^n$. For instance, if you put $1.00 in the bank at 8 percent interest for 30 years, it would grow to $10.06. Or if it was $100 invested, the future value would be $1006.27: you just multiply the initial value by the tabulated value to get the future value for any r and n. Foresters used to consult books of interest tables to find future values and values for the formulas discussed below. But now you can do this so easily with handheld calculators and spreadsheets on personal computers that such tables are no longer common. But Table 4-2 is useful for seeing the power of compound interest over time. Look at the tabulated value for 10 percent interest and $n = 100$. One dollar grows to $13,780.61, or if it were $600 per acre spent by a public agency to buy land and plant oaks on a 100-year rotation, in order to earn a 10 percent rate of return, the acre of land and timber would have to be worth $600(13{,}780.61) = \$8{,}268{,}366$ in 100 years. This is a simple example that ignores other costs and revenues, but it shows why foresters spend lots of time discussing appropriate interest rates and rotation lengths.

Both Figure 4-1 and Table 4-2 are vital concepts because they show how capital investments grow and how forest investments must perform if they are to earn given rates of return. Table 4-2 is a printout from a personal computer spreadsheet program with equation (4-1) in each cell; in this case, $V_0 = 1$. The spreadsheet automatically changes values for n and r in each cell and computes the future value. You can easily construct similar spreadsheets for most of the equations in this chapter. Chapters 5 and 7 show other useful ways to apply spreadsheets.

TABLE 4-2
VALUES OF $1 INVESTED FOR *n* YEARS AT AN INTEREST RATE OF *r*(100). VALUES ARE $(1 + r)^n$

Number of years, *n*	Interest rate						
	2	**4**	**6**	**8**	**10**	**12**	**14**
0	1.00	1.00	1.00	1.00	1.00	1.00	1.00
1	1.02	1.04	1.06	1.08	1.10	1.12	1.14
2	1.04	1.08	1.12	1.17	1.21	1.25	1.30
3	1.06	1.12	1.19	1.26	1.33	1.40	1.48
4	1.08	1.17	1.26	1.36	1.46	1.57	1.69
5	1.10	1.22	1.34	1.47	1.61	1.76	1.93
6	1.13	1.27	1.42	1.59	1.77	1.97	2.19
7	1.15	1.32	1.50	1.71	1.95	2.21	2.50
8	1.17	1.37	1.59	1.85	2.14	2.48	2.85
9	1.20	1.42	1.69	2.00	2.36	2.77	3.25
10	1.22	1.48	1.79	2.16	2.59	3.11	3.71
15	1.35	1.80	2.40	3.17	4.18	5.47	7.14
20	1.49	2.19	3.21	4.66	6.73	9.65	13.74
30	1.81	3.24	5.74	10.06	17.45	29.96	50.95
40	2.21	4.80	10.29	21.72	45.26	93.05	188.88
50	2.69	7.11	18.42	46.90	117.39	289.00	700.23
60	3.28	10.52	32.99	101.26	304.48	897.60	2595.92
70	4.00	15.57	59.08	218.61	789.75	2787.80	9623.64
80	4.88	23.05	105.80	471.95	2048.40	8658.48	35676.98
90	5.94	34.12	189.46	1018.92	5313.02	26891.93	1.32×10^5
100	7.24	50.50	339.30	2199.76	13780.61	83522.27	4.90×10^5
150	19.50	358.92	6250.00	1.03×10^5	1.62×10^6	2.41×10^7	3.43×10^8
200	52.48	2550.75	115125.90	4.84×10^6	1.90×10^8	6.98×10^9	2.40×10^{11}

Present Value of a Single Sum

You can solve equation (4-1) to get V_0, the **present value** of any single future value:

$$V_0 = \frac{V_n}{(1 + r)^n} \tag{4-2}$$

For example, at 6 percent interest, the above-calculated value of $20.12 expected in 12 years has a present value of $20.12/(1.06)^{12} = \$10$, which is the same value you started with to get the future value. This present value computation says that *if you need to earn 6 percent interest,* the promise of $20 in 12 years is worth a maximum of $10 today. If you spend more than $10 to get $20 in 12 years, your investment will grow at an unacceptable rate (less than 6 percent). That's why the interest rate used to calculate present values and future values will be called the "minimum acceptable rate of return (MAR)." *Equation (4-2) is the foundation for finding your maximum willingness to pay for any asset yielding future income.* Since so much of the income from forest properties and practices occurs in the distant future, equation (4-2) is an indispensable tool for evaluating forestry investments.

Looking at equation (4-2) you can tell that at any interest rate greater than 0, present value will be less than future value. And that makes sense: Wouldn't you rather have \$100 now than receive it in 10 years, even if there were no inflation? That's why we **discount** future values to arrive at present values. Thus, the interest rate used for discounting is sometimes called the **discount rate.** Often we don't use equation (4-2) for this discounting, but do it intuitively. For example, if a monthly magazine is worth \$1 an issue, we're not usually willing to pay \$60 now for a 60-month subscription. The value would be discounted to, say, \$50.

Again looking at equation (4-2), *note that the farther in the future a value is (larger n), the lower its present value will be.* Figure 4-2a shows this graphically: using 7 percent interest, a future income of $V_n = \$5{,}000$ occurring in year 5 has a present value of $5{,}000/(1.07)^5 = \$3{,}565$. But if the same \$5,000 doesn't occur until year 40, its present value would be $5{,}000/(1.07)^{40} = \$334$. Given forestry's long production periods, this impact of time depresses investors' willingness to pay for many timber properties and forest management practices.

Equation (4-2) also shows that *the higher the interest rate, the lower the present value will be.* For example, in Figure 4-2b, the future value of \$5,000 to be received in year 40 has a present value of \$2,264 at 2 percent interest. But present value drops to \$110 at 10 percent interest. Thus, high minimum acceptable rates of return will decrease the amount investors are willing to pay to generate future income. This effect is especially dramatic in certain forestry ventures with long payoff periods such as planting Douglas-fir, which is grown for 40 years or more.[1]

Rate of Return

You can solve equation (4-1) or (4-2) for r, which would be the interest rate or the average annual rate of return on an investment for which you paid V_0 and received V_n, n years later. Rearranging (4-1):

$$(1 + r)^n = \frac{V_n}{V_0} \qquad (4\text{-}3)$$

Taking the nth root of both sides and solving for r:

$$r = \sqrt[n]{\frac{V_n}{V_0}} - 1 \qquad (4\text{-}4)$$

[1]Figure 4-2b illustrates a confusion that can occur between an "initial value" and "present value," both of which are defined as V_0. The figure shows that an initial value of \$110 would grow to \$5,000 in 40 years at 10 percent interest (read from left to right), but at 2 percent interest, that future value has a present value of \$2,264. For a single-input, single-output investment, initial value—say, the purchase cost of an investment—will equal present value only if the investment growth rate equals the discount rate.

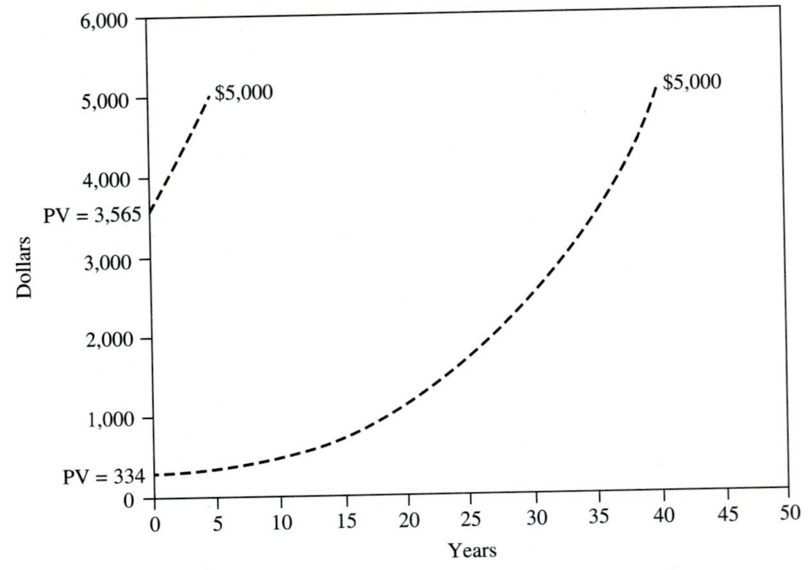

FIGURE 4-2a
Effect of time on present value of income at 7% interest.

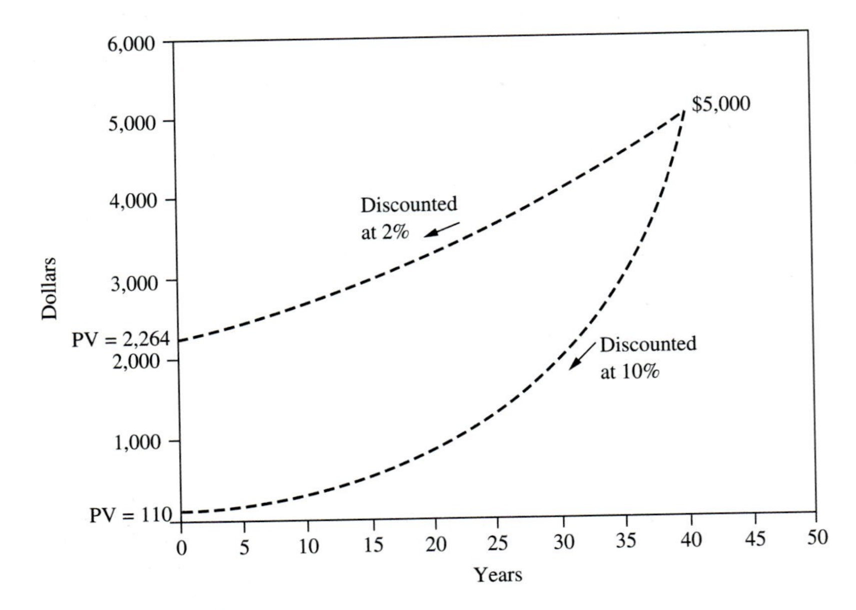

FIGURE 4-2b
Interest rate effect on present value of income in 40 years.

This r is often called the **internal rate of return.** In the example from Figure 4-1, \$10 grew to \$20.12 in 12 years. Using equation (4-4), taking the twelfth root of 20.12/10 and subtracting 1 gives 0.06, the 6 percent interest rate in Figure 4-1. This formula is very useful for measuring performance of simple investments with one input and one output, for example, stocks, real estate, or timber that you might buy at one date and sell some years later. It's not enough to say, for instance, that you doubled your money in the stock market; your rate of return will vary depending on the number of years it took you to do it. Noting that doubling your money means that $V_n/V_0 = 2$, you can quickly figure your rate of return on such an increase over, say, 4 years versus 20 years by using equation (4-4): Doubling your money over 4 years, $r =$ (the fourth root of 2) $- 1 = 0.189$, or 18.9 percent. Doubling over 20 years, $r =$ (the twentieth root of 2) $- 1 = 0.035$, or 3.5 percent. Equation (4-4) is useful for figuring average annual percentage rates of change in any quantities such as population, volume, or temperature over an n-year period.

Before continuing, let's note some conventions and terms.

Conventions

Unless otherwise specified, the following conventions will hold throughout the book:

- Interest rates will be yearly percentage rates of change. You often hear of monthly or daily interest. A daily interest rate, for example, is the annual rate divided by 365, in which case the time units are days. Here, our units will be years, unless otherwise stated. Percentage inflation rates, value growth rates, and other growth rates will also be annual.
- For any investment, costs and revenues are assumed to occur at the same time of year.
- In the term present value, "present" means now, if not otherwise clarified. You can compute present values for other dates, but these will be specified.
- Year 0 means now.
- In this book, any rate such as interest, inflation, growth rates, or tax rates will be expressed in a formula as a decimal (the rate divided by 100). For example, 5 percent interest would be $r = 0.05$, or a 30 percent tax rate would be $T = 0.30$.
- Interest will be compound rather than simple, unless stated otherwise.

Terms for the Interest Rate

In different contexts, several terms are used for the earning rate of capital. It's important to be familiar with these terms and to realize that mathematically they are the same thing: an annual percentage rate of change. Below are several of these terms and notes on how they are used.

- **Interest rate**—Generally refers to earning rates on financial instruments like bank accounts, bonds, certificates of deposit; or the rate charged on loans;

or the rate used in discounting future values to arrive at present values. It is also referred to as *compound interest.*

- **Discount rate**—The interest rate used in discounting future values to present values.
- **Rate of return**—The earning rate on investments to produce goods and services. A synonym is *internal rate of return* (IRR). [A term often used in business is "return on investment" (ROI), but it's not necessarily the same as the internal rate of return, as shown in Chapter 5.]
- **Alternative rate of return** (ARR)—The earning rate available on an investor's best alternative, at a risk similar to new ventures being considered. Ambiguity surrounds the ARR. It's often suggested that new projects should earn at least the ARR. But if a high ARR is available on just a few small projects, we can't expect all new ventures to earn this much.
- **Minimum acceptable rate of return** (MAR)—The minimum earning rate a decision maker requires on invested capital. *This term will be used for the investor's discount rate throughout the book.* Engineering economists often refer to the MAR, sometimes calling it the "minimum attractive rate of return." As suggested above, the MAR may sometimes be less than an ARR. On the other hand, some people with a great need for current consumption may discount future income at an MAR higher than available alternative rates of return.
- **Guiding rate of interest**—Another term for the MAR: a minimum rate of return that projects must earn to be acceptable to an investor.
- **Hurdle rate**—Same as above; refers to a minimum acceptable rate of return or "hurdle" that new investments must clear before they are acceptable to a business.
- **Reinvestment rate**—The rate of return at which you assume future income from a project could be reinvested. This rate should be at least as high as the minimum acceptable rate, and sometimes, for limited funds, it may be some available rate higher than the MAR.
- **Opportunity cost**—The cost of an opportunity forgone. It is the highest rate of return given up on an alternative venture when you make a new investment—the same concept as the alternative rate of return. Opportunity cost isn't always expressed as an interest rate; it could also be the income or satisfaction given up in one area as a result of an expenditure in another.
- **Cost of capital**—When raising capital for a new investment, a corporation can borrow money or sell shares of stock. Borrowing money requires paying interest, which is called the **cost of debt.** On stock issued, a firm eventually will pay dividends, which are the **cost of equity.** Equity refers to the share of ownership that a stockholder has in the firm. The cost of debt and equity combined is the **cost of capital.** This can be expressed as a weighted average interest rate on capital obtained, or the **"weighted average cost of capital"** (WACC). Weighting is by the percentages of a firm's debt and equity. Firms sometimes use the WACC as a minimum acceptable rate of return that new projects should earn. But when the cost of capital is below widely available

alternative rates of return, it would be inefficient to invest in projects yielding less than available alternatives.[2]

- **Capitalization rate**—Same as above. The term derives from the fact that discounting future values is sometimes called "capitalizing." Property appraisers often refer to this rate as the "cap rate."
- **Rate of time preference**—The rate at which someone discounts future values. Such discounting is often with seat-of-the-pants judgment rather than a given rate as in equation (4-2). An example is the magazine subscription discounting mentioned earlier. Knowing how much someone has discounted a future value, you can calculate their rate of time preference with equation (4-4). If Karen is willing to pay a maximum of $60 today for the promise of $100 in 3 years, what is her rate of time preference? [Using equation (4-4), subtracting 1 from the third root of 100/60 gives 0.186, or 18.6 percent.]

The next chapter will discuss how interest rates are chosen for financial calculations and the effects of inflation on these rates.

Terms for Future Value and Present Value

Several terms exist for *future value:* accumulated value with interest, compounded value, and principal with accumulated interest. These all refer to the future value at one date for one or more cash flows received at earlier dates. The procedure for finding future values is called **compounding** or *accumulating,* as shown in equation (4-1).

Likewise, there are several terms for *present value:* present worth, year 0 value, discounted value, capitalized value, **discounted cash flow (DCF),** and **net present value.** These are all the result of finding the present value of future cash flows, a procedure called **discounting** or **capitalizing,** as shown in equation (4-2). Net present value is explained in more detail below.

Net Present Value

Net present value (NPV) is the present value of revenues minus the present value of costs. You can compute present values of negative cash flows (costs) as well as positive ones. Symbolically, NPV is:

$$\begin{aligned} \text{NPV} = R_0 &+ \frac{R_1}{(1+r)^1} + \frac{R_2}{(1+r)^2} + \frac{R_3}{(1+r)^3} + \ldots + \frac{R_n}{(1+r)^n} \\ &- C_0 - \frac{C_1}{(1+r)^1} - \frac{C_2}{(1+r)^2} - \frac{C_3}{(1+r)^3} - \ldots - \frac{C_n}{(1+r)^n} \end{aligned} \tag{4-5}$$

[2]Several studies have found business hurdle rates three or four times the interest cost of borrowing funds. Thus the WACC is unlikely to represent typical business MARs, especially with risky investments (Dixit, 1992). As an example of a WACC, suppose 60 percent of funds raised were from stock on which a 6 percent dividend was paid, and 40 percent were borrowed at 9 percent interest. The WACC would be $0.60(6) + 0.40(9) = 7.2$ percent.

where R and C are revenues and costs in the subscripted years. Note that R_0 and C_0 are not discounted, since they are already at year 0 (or they are divided by $(1 + r)^0$, which is the same as dividing by 1). In more general terms:

$$\text{NPV} = \sum_{y=0}^{n} \left[\frac{R_y}{(1 + r)^y} - \frac{C_y}{(1 + r)^y} \right] \tag{4-6}$$

which says that NPV is the sum of revenues in each year, y, discounted to year 0 minus the sum of costs in each year discounted to year 0. This is simply an expansion of equation (4-2) to include negative as well as positive cash flows at more than one point in time. Your willingness to pay for an asset is reduced by the present value of costs (such as taxes) and is increased by the present value of revenues. *Net present value defines an investor's willingness to pay for an asset based on estimated benefits, costs, and the desired rate of return. Thus, NPV is a powerful tool in valuing forest properties.* Other chapters apply this concept. Chapters 10 and 11 note when you should discount costs and revenues with different interest rates, something not done in the introductory examples. Later you'll also see how nonmonetary values can be included in NPV calculations.

You can make an NPV calculation from the following **time line** showing expected costs and revenues on 1 acre of bare forestland. Revenues are above the line; costs are below.

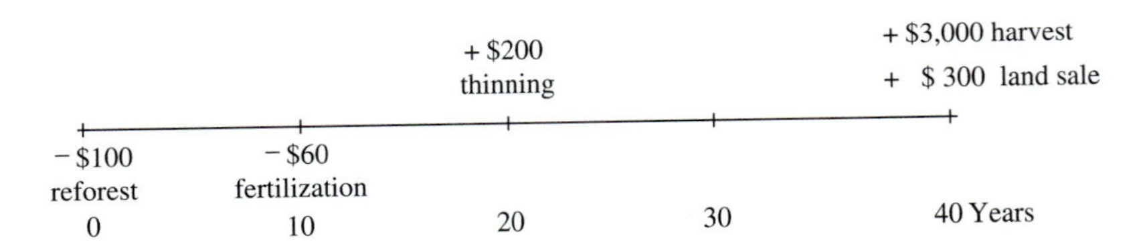

Using x percent interest, in year 0, the NPV of the above projected cash flows would be the maximum an investor could pay for the acre of bare land if an x percent rate of return were desired, assuming no other costs or revenues. If the desired rate of return were 6 percent, the net present value, using equation (4-5) or (4-6), would be:

$$\begin{aligned} \text{NPV} &= \frac{200}{(1.06)^{20}} + \frac{3000 + 300}{(1.06)^{40}} - \frac{60}{(1.06)^{10}} - 100 \\ &= \frac{200}{3.207} + \frac{3300}{10.286} - \frac{60}{1.791} - 100 \\ &= \$249.69 \text{ per acre} \end{aligned} \tag{4-7}$$

If all costs and revenues were as projected, a buyer who paid \$249.69 for the bare acre would earn a 6 percent rate of return on the investment. If the NPV were computed just after reforestation, the negative \$100 reforestation cost wouldn't appear in the equation, and the NPV would be the willingness to pay for land *plus* reforestation, or \$349.69. These NPVs provide a way to compare investment alternatives and to compute maximum bid prices for assets.

Don't be fooled by the seeming precision of calculating present values to the nearest penny. While the formulas are very precise, the inputs such as incomes, costs, and the interest rate are approximations. Usually analysts will compute a range of present values based on different scenarios. Decision makers can then choose which value is most reasonable, according to their view of the future.

In the broadest sense, an asset's value is the net present value of *satisfactions* it yields. If, for example, the aesthetic pleasure from owning the above acre of forestland was worth at least \$10 per year to an owner, the property's value to that owner would be increased by at least the present value of \$10 per year. The next section shows how to quickly compute present values of such repeating series of payments.

DISCOUNTING AND COMPOUNDING PAYMENT SERIES

So far we've discounted and compounded single or irregular cash flows. Now let's consider easier ways to discount series of equal payments occurring at regular intervals, since these are quite common in forestry. The present and future value equations are in Table 4-3 and are explained below. Derivations are given in Appendix 4A. It's important to remember five conditions that must be met for the equations to be valid:

1 Payments must be equal.
2 Payments must occur at regular intervals called "periods."
3 No payment occurs at year 0.
4 The first payment occurs at the end of the first period.
5 Payments must be all of the same sign, positive or negative.

If a payment series does not meet these conditions, formulas in Table 4-3 must be altered. If you can recognize and name each of the series described below, it's easy to look up the right formula in the decision tree of Table 4-3: just start at the left side of the tree and work your way through the branches to the correct formula, based on the formula names described below.

Perpetual Annual Series

The time line below diagrams a *perpetual annual series* of payments at \$$p$ each.

	p	p	p	p	p	p	p	p ...
0	1	2	3	4	5	6	7	8 ... ∞ Years

To find the *present value of a perpetual annual series,* use the following formula from Table 4-3:

$$V_0 = \frac{p}{r} \tag{4-8}$$

TABLE 4-3
DECISION TREE FOR PRESENT VALUE AND FUTURE VALUE FORMULAS
(First Payment at End of First Period. No Payment in Year 0.)

Number of payments	Time between payments	Evaluation period	Time of value	Formula	Formula name	Formula no. in text
One		Terminating	Future	$V_n = V_0(1 + r)^n$	Future value* of an amount	(4-1)
			Present	$V_0 = V_n/(1 + r)^n$	Present value of an amount	(4-2)
Series	Annual	Terminating	Future	$V_n = p\left[\frac{(1 + r)^n - 1}{r}\right]$	Future value* of a terminating annual series	(4-11)
			Present	$V_0 = p\left[\frac{1 - (1 + r)^{-n}}{r}\right]$	Present value of a terminating annual series	(4-10)
		Perpetual	Future	$V_n =$ infinity		
			Present	$V_0 = \frac{p}{r}$	Present value of a perpetual annual series	(4-8)
	Periodic	Terminating	Future	$V_n = p\left[\frac{(1 + r)^n - 1}{(1 + r)^t - 1}\right]$	Future value* of a terminating periodic series	(4-14)
			Present	$V_0 = p\left[\frac{1 - (1 + r)^{-n}}{(1 + r)^t - 1}\right]$	Present value of a terminating periodic series	(4-13)
		Perpetual	Future	$V_n =$ infinity		
			Present	$V_0 = \frac{p}{(1 + r)^t - 1}$	Present value of a perpetual periodic series	(4-9)

(Decision tree begins at START.)

r = annual interest rate/100. (If payments are fixed in real terms, r is real; if payments are fixed in nominal terms, r is nominal. See Chapter 5.)
V_0 = present value (or initial value)
V_n = future value after n years (including interest)
n = number of years of compounding or discounting
p = amount of fixed payment each time in a series (occurring annually or every t years)
t = number of years between periodic occurrences of p

*The future value of any terminating series is its present value formula times $(1 + r)^n$.
Adapted from Gunter and Haney (1978), by permission.

For example, suppose a firm is committed to perpetual timber management and wishes to compute the present value of annual hunting lease income at $3 per acre in perpetuity, the first payment due at the end of the first year. Using equation (4-8), the present value at 5 percent interest is 3/0.05 = $60/acre. If the payments were costs, the computed value would be negative.

This formula makes sense if you think of leaving $100 in the bank forever. The $100 is your present value V_0. If the interest rate is 5 percent, you can withdraw an income of $p = 0.05(100) = \$5$ per year in perpetuity. So $p = rV_0$. Solving for present value, $V_0 = p/r$, which is formula (4-8).

Suppose in the hunting lease example the first income occurred in year 0, as shown below:

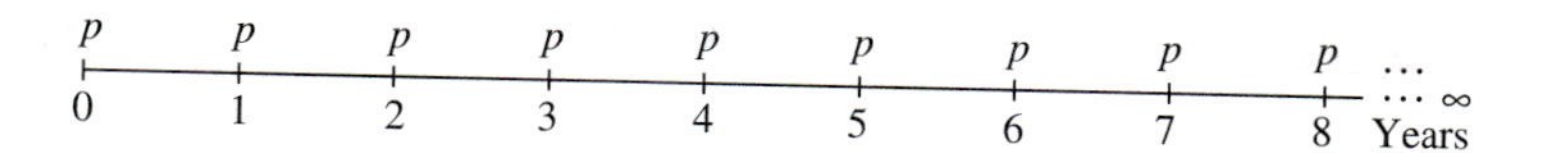

Since the formula assumes no payment in year 0, you must add it, thus making the present value $3 + 3/0.05 = \$63$. You needn't discount the first payment, since it's already at year 0. *This same principle applies to any of the series present value formulas in Table 4-3: if payment p occurs in year 0 as well as every t years, simply add p to the equation.*

In the previous example, what is the present value if the first lease income occurs at the end of year 4, as shown below?

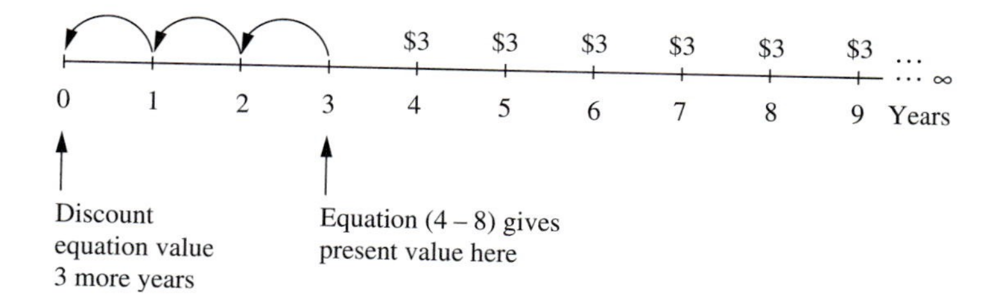

Equation (4-8) will compute present value at the end of year 3, since it assumes the first payment is 1 year after the present. Therefore you need to discount the equation (4-8) result 3 more years, and the correct present value in year 0 is $(3/0.05)/(1.05)^3 = \$51.83$, still using 5 percent interest.

The future value of a perpetual series needn't concern us because it is infinite.

Perpetual Periodic Series

Perpetual regular payments more than 1 year apart are called a *perpetual periodic series,* as shown below for the case where the period is $t = 10$ years.

p p p p ...
0 10 20 30 40 ... ∞ Years

Table 4-3 shows that the formula below will give the *present value of a perpetual periodic series:*

$$V_0 = \frac{p}{(1 + r)^t - 1} \tag{4-9}$$

This is an important series in forestry, since in long-term management, harvest incomes and many costs occur at regular intervals. As an example, at 6 percent interest, what is the present value of a $3,000 Christmas tree harvest income occurring in 10 years and every 10 years thereafter, in perpetuity? Using equation (4-9), $t = 10$, $r = 0.06$, and present value is:

$$3{,}000/[(1.06)^{10} - 1] = \$3{,}793$$

That means, given no other costs or revenues, if you paid $3,793 today for the above expected series of incomes, you would earn a 6 percent rate of return. In other words, $3,793 should be your maximum willingness to pay if you want to earn 6 percent. Another view is to think of putting $3,793 in the bank at 6 percent interest. You could then withdraw $3,000 every 10 years, forever, without adding any more funds.

If, in the previous example, the trees are now 10 years old, and the first harvest income occurs right after purchase and every 10 years thereafter, you need to add the first $3,000 to the above result, since the formula doesn't include it. The present value would then be $6,793.

Suppose, for the same harvest income series, the first harvest occurs in 4 years, so that the trees are now 6 years old. What is the present value (in year 0) at 6 percent interest? As shown on the time line below, this series doesn't quite meet the conditions required by equation (4-9), so some adjustments are needed.

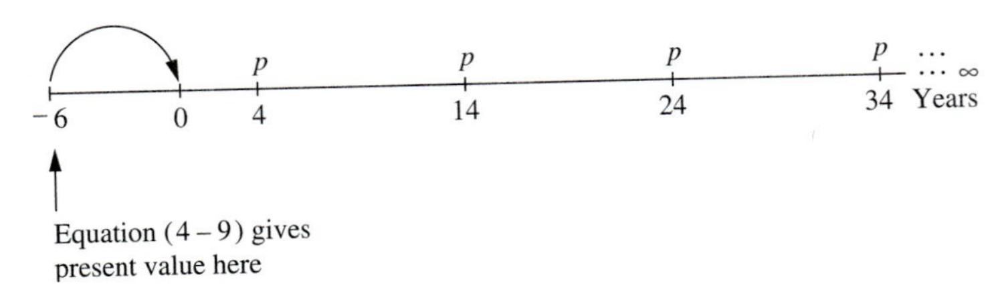

Equation (4-9) computes present value in year minus 6, since that's the only way you'll get exactly 10 years between all payments. Remember, the equation assumes that the first payment is at year t and every t years thereafter: here $t = 10$. Compounding the result from equation (4-9) for 6 years will give the present value today (year 0). So the answer is:

$$\{3{,}000/[(1.06)^{10} - 1]\}(1.06)^{6} = \$5{,}381$$

Check for logic: The answer should lie between the value obtained when the first harvest was 10 years away and when it was now—between $3,793 and $6,793. [You may often find more than one way to compute present values. Here you could have used equation (4-9) to obtain a value in year 4, added the $3,000 for that year, since the formula assumes no income at the start, and discounted the result 4 years to get to year 0.]

As in the previous case, equation (4-9) is for a perpetual series, so future value isn't relevant.

Terminating Annual Series

Equal annual payments that stop at some date are a *terminating annual series,* as shown below for $n = 8$ years.

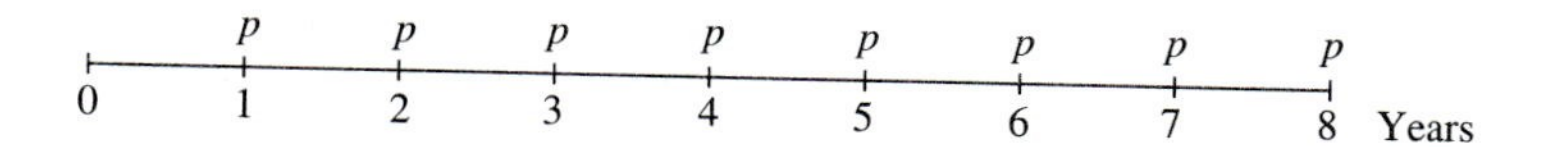

Table 4-3 gives the following formula for the *present value of a terminating annual series:*[3]

$$V_0 = p\left[\frac{1-(1+r)^{-n}}{r}\right] \tag{4-10}$$

Suppose you own land that a hunting club wishes to lease for \$5 per acre per year for 15 years. If your minimum acceptable rate of return (MAR) is 7 percent, what is the present value of this lease to you? Using the above formula with $r = 0.07$, $n = 15$, and $p = \$5$, present value is:

$$(5/0.07)[1-(1.07)^{-15}] = \$45.54/\text{acre}$$

Assuming you don't care when the income arrives, you'd be indifferent between the lease payment schedule and the full payment of \$45.54 now, on a per acre basis. They both have the same present value. You can also think of the lease as increasing your property value by \$45.54.

Be careful when keying into a calculator the negative exponent in equation (4-10) and elsewhere. *Immediately after pushing the exponentiation key, you must enter the exponent and then change its sign with the +/– key. If you enter a negative sign before the exponent, your answer will be wrong.* You can easily adapt equation (4-10) to compute a future value.

Future Value of a Terminating Annual Series *The future value of any terminating series is its present value formula times* $(1 + r)^n$. Thus, multiplying equation (4-10) by $(1 + r)^n$ gives *the future value of a terminating annual series* shown in Table 4-3:

$$V_n = p\left[\frac{(1+r)^n - 1}{r}\right] \tag{4-11}$$

[3] An equivalent form of equation (4-10) is:

$$V_0 = \frac{p}{r}\left[1 - \frac{1}{(1+r)^n}\right]$$

A conventional but more complex form of equation (4-10) is:

$$V_0 = p\left[\frac{(1+r)^n - 1}{r(1+r)^n}\right]$$

In the lease example above, if you put each $5 lease payment in the bank at 7 percent interest, what would the accumulated value be in 15 years? Using the above equation, future value is:

$$5[(1.07)^{15} - 1]/0.07 = \$125.65$$

As a check, you can see that compounding the above present value of $45.54 for 15 years gives the same result:

$$45.54(1.07)^{15} = \$125.65$$

Equation (4-11) is handy for calculating personal savings. For example, if you save $1,000 yearly at 8 percent interest, in 30 years you'll have:

$$1{,}000[(1.08)^{30} - 1]/0.08 = \$113,283$$

Considering that you put in only $30,000, this again shows the power of compound interest over long periods. But this isn't quite as good as it looks; the only way you'll consistently make 8 percent on fairly safe investments is if annual inflation stays in the 4 to 6 percent range. That eats into the purchasing power of your payoff, making it worth less than it appears. If inflation were lower, interest rates would tend to be lower. The next chapter covers inflation in more detail.

Capital Recovery Formula Solving equation (4-10) for p gives the *terminating annual payment with a present value equal to* V_0*:*

$$p = V_0\left[\frac{r}{1 - (1 + r)^{-n}}\right] \tag{4-12}$$

You can use this **capital recovery formula** to figure the terminating equal annual revenue required to justify paying V_0 for an investment, given that you want to earn a rate of return of r. For example, suppose the 10-year cutting rights on a mature timber tract are offered for $V_0 = \$1{,}000{,}000$. In order for an investor to earn a 9 percent rate of return, what equal annual harvest income p must be received?

$$p = 1{,}000{,}000(0.09)/[1 - (1.09)^{-10}] = \$155{,}820/\text{year}$$

In other words, the present value of $155,820 per year for 10 years is $1,000,000, given $r = 0.09$. The same equation is used to figure fixed payments on loans, where V_0 is the loan amount, but since inflation adds a wrinkle here, that topic is covered in the next chapter.

Terminating Periodic Series

Terminating regular payments more than 1 year apart are called a *terminating periodic series.* The example below is for $t = 10$ and $n = 40$.

0	10	20	30	40 Years
	p	p	p	p

Table 4-3 gives the following equation for the *present value of a terminating periodic series:*

$$V_0 = p\left[\frac{1-(1+r)^{-n}}{(1+r)^t - 1}\right] \tag{4-13}$$

Suppose a Christmas tree farmer plans to grow four crops of small trees on a 5-year rotation. If each harvest brings $1,000, and the first is due in 5 years, what is the present value of these yields to a farmer whose MAR is 8 percent? Using equation (4-13), for $t = 5$ and $n = 20$, present value is:

$$1{,}000\left[\frac{1-(1.08)^{-20}}{(1.08)^5 - 1}\right] = \$1{,}673.57$$

Since equation (4-13) is terminating, it can be changed to compute future value.

Future Value of a Terminating Periodic Series Multiplying equation (4-13) by $(1 + r)^n$ yields the *future value of a terminating periodic series* shown in Table 4-3:

$$V_n = p\left[\frac{(1+r)^n - 1}{(1+r)^t - 1}\right] \tag{4-14}$$

In the above Christmas tree example, what would be the future value of the harvest incomes in year 20 if they were invested at 8 percent interest? Using equation (4-14), future value is:

$$1{,}000[(1.08)^{20} - 1]/[(1.08)^5 - 1] = \$7{,}800.42$$

As a check, this equals the above present value of $1,673.57 times $(1.08)^{20}$.

Equation (4-14) is also useful in personal finances. For example, if you invested, say, $1,000 every 3 years, the equation tells you how much you'd have after a given number of years n at different interest rates.

General Procedures

When you encounter a regular series of equal payments to be discounted or compounded, follow this procedure:

- Name the series.
- Find the formula in Table 4-3 by following the appropriate branches, depending on whether the series is annual or periodic, terminating or perpetual, and whether you want present or future value.
- Make sure the series meets the five conditions required by the series formulas, as outlined at the start of this section. Ask, *"Does the series fit the formula?"* If not, make adjustments to the formula.
- Find the decision maker's real minimum acceptable rate of return, which becomes r in the equation. The next chapter tells how to choose r.
- Substitute the appropriate values for the remaining variables in the equation, and compute the answer.

Appendix 4B shows ways to adapt many of the Table 4-3 equations if payments are increasing or decreasing at a fixed geometric rate.

FUTURE INCOME NEEDED TO OFFSET COSTS

Sometimes, instead of estimating very uncertain forestry revenues in the distant future, we can ask, "How much income will be needed at rotation-end to offset production costs plus interest?" Then we can judge the probability of receiving that income. Suppose it would cost you $400 to buy an acre of land, $200 to plant it to trees, and (10 years later) $75 for brush control and thinning. If expected harvest is in 30 years, what is the minimum income "needed"? For simplicity, other costs and incomes are ignored. The "needed" income, or *future value*, is what you could have earned if the same capital had been invested at your minimum acceptable rate of return. For a 7 percent MAR, the future values of costs are shown below.

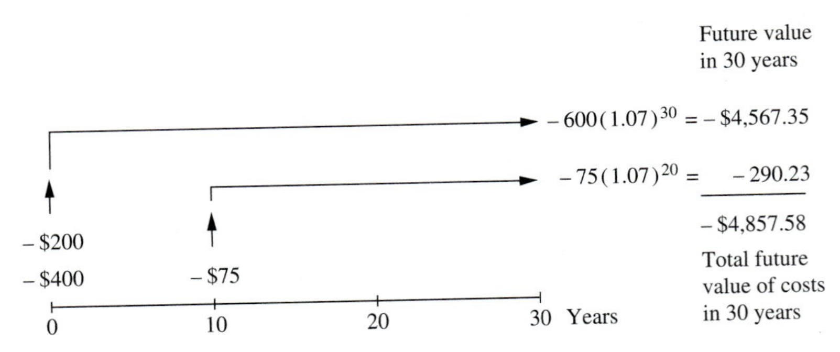

The initial cost of $600 is accumulated for 30 years, and the $75 cost in year 10 is accumulated for the remaining 20 years, summing to $4,857.58. This means that if these costs were invested at your MAR, instead of in the forest, they could accumulate to $4,857.58 by the time the forest was mature (30 years). So the expected value of your land and timber in 30 years needs to be at least that amount if you are to earn at least a 7 percent rate of return.

Foresters often speak of the "interest cost" of waiting long periods before receiving harvest income. This is the opportunity cost mentioned earlier: the cost of forgone earning opportunities. But it's important to know that this "cost" is

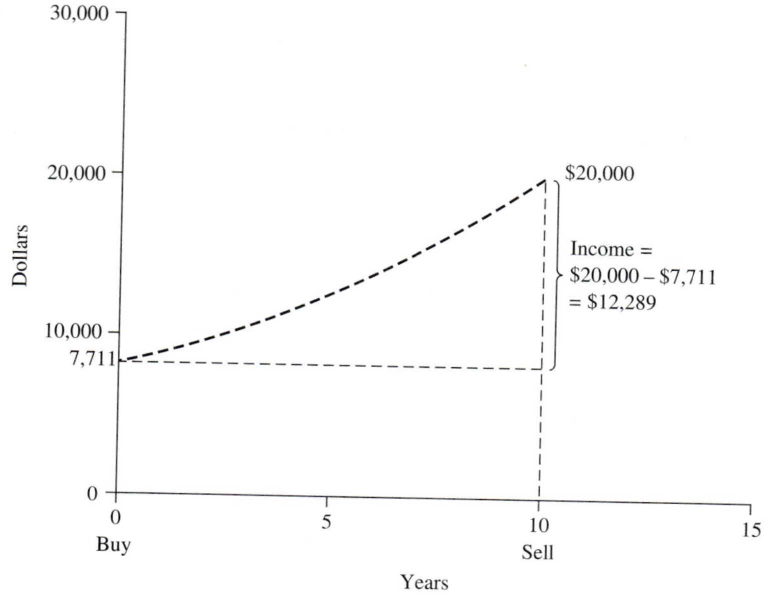

FIGURE 4-3
Income to capital.

also the income to capital. Figure 4-3 clarifies this concept. Let the projected sale price of an asset like a forest be $20,000 in 10 years. Suppose Mark's MAR is 10 percent, and he pays $20{,}000/(1.10)^{10} = \$7{,}711$ for the asset—the discounted value of the future sale price. Now assume he sells the asset in 10 years for $20,000, in which case his rate of return is 10 percent. The difference between his purchase price and sale price is the interest he could have earned elsewhere: $12,289. While it's important to keep track of this opportunity cost to make sure earnings are as good as they could be elsewhere, you should realize that *the $12,289 is an income,* or **capital gain.** The Internal Revenue Service also considers this as income for tax purposes. The opportunity cost is not a tax-deductible expense and does not reduce income. However, interest paid on business debt *is* a legitimate tax-deductible expense that reduces income.

OPTIMAL ALLOCATION OF CAPITAL

Under an ideal allocation of capital to different investments, total satisfaction from all the nation's capital would be maximized over time. In that case, no reallocation could bring any net gains: we would have an *optimal allocation of capital.* For the moment, assume the same risk level in all investments. Applying the **equi-marginal principle,** the optimal allocation of capital occurs when the last dollar invested in each activity earns the same rate of return. Figure 4-4 illustrates this concept in a manner parallel to Figure 1-2. The x axes in the panels are billions of dollars invested per unit of time in separate industries like wheat, forestry, compact discs, automobiles, etc. The vertical axes are the **marginal rate**

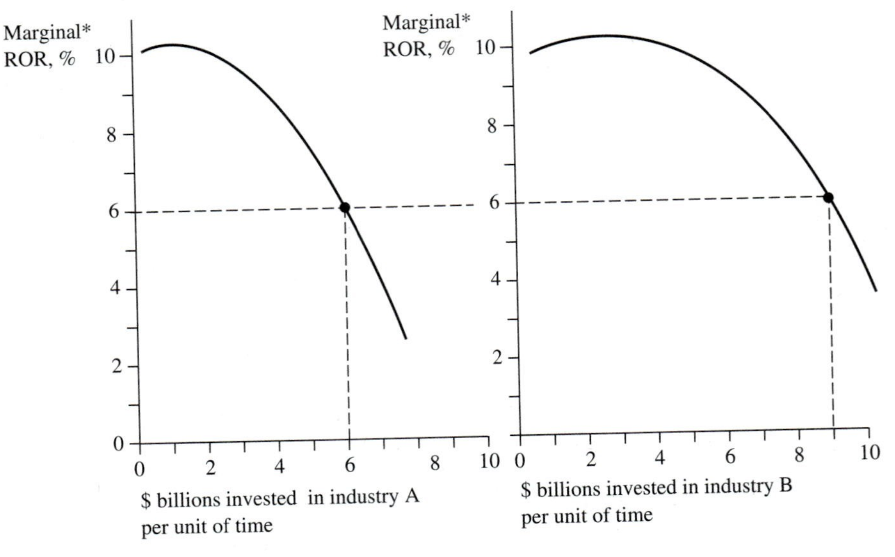

Assumptions: Total budget is $15 billion per unit of time.
Risk is the same in each industry.
What is meant by "optimal"?
Why is the above allocation optimal?
Sketch in a suboptimal allocation of $15 billion.

*Rate of return on the last billion dollars invested.

FIGURE 4-4
Optimal allocation of capital among industries.

of return—the rate of return on the last dollar invested in each industry (*not* the average rate of return on all capital). The marginal rate of return is shown to eventually fall as more is invested in each industry. You have to think big here: if you, as an individual, invest more in one large industry, there's no impact on the rate of return because your added capital is relatively small. But if billions of dollars are added to one sector, the rate of return eventually declines. Why? Because increased output drives down output prices and with it the rate of return. Think back to Chapter 2, and sketch a downward-sloping regional demand curve for the sector's output. Draw in the original regional supply curve and the new supply curve that is shifted to the right because the increased investment boosts the sector's output. At the new intersection of supply and demand, output is higher and product price is lower, so the rate of return on new investment declines.

Assume that a full $15 billion is to be allocated between industries A and B in Figure 4-4. The optimum allocation is $6 billion in A and $9 billion in B. At that point, and at no other, the rate of return on the last dollars invested (the marginal rate of return) in each industry is 6 percent. The equi-marginal principle has been met. You can see that any reallocation—more invested in A and less in B, or vice versa—will reduce total returns to the $15 billion. For example, take a billion out of A, where it was earning 6 percent, and put it in B, where the marginal rate of return will drop below 5 percent. Not a good move![4] If any reallocation brings a net loss, we must be at the optimum where total returns to capital are maximized.

In Figure 4-4 you could imagine hundreds of graphs for different industries and billions of dollars allocated optimally so that the marginal rates of return were equal. In a free market there is a tendency for this to happen over time, but not in a precise sense at any given moment. Although initially risk (variance in returns) was ignored, risk differences cause marginal rates of return to vary between industries. But this causes no problem because the disutility of added risk is offset by a higher average rate of return. What we're really trying to equate at the margin is **utility** from the last dollar invested rather than the financial rate of return. With differing risk among industries, the marginal rates of return will vary in order that the marginal utilities can be closer (see Chapter 10).

Without monopolies, it is highly unlikely for widely differing rates of return among *similar-risk* industries to persist in a competitive market. Investors are generally savvy, and capital tends to flow to sectors where rates of return are highest. The increased investment causes output to rise and rates of return to drop in those sectors (because their output prices fall). Examples are mentioned under "Supply-Demand Equilibrium" in Chapter 2. This is nothing terribly precise: examples of stickiness in prices and rates of return do exist (again see Chapter 2). Nevertheless, allowing for differing risk, in a very rough and average sense over time, markets appear to be meeting the equi-marginal principle for optimizing investment, as economic theory suggests they should (look at Figure 2-14 to review the mechanics).

Some allege that inefficient capital allocation is more likely in government because agencies may not always ferret out the best investments as effectively as businesspeople. To address this problem, the U.S. Office of Management and Budget suggests a 10 percent minimum acceptable rate of return for federal agencies. However, because of differing risk, not all projects should earn the same expected rate of return. And, as noted, not all agencies use the same MAR, although not necessarily for reasons of differing risk. Chapter 10 will consider these issues further.

DOES THE INTEREST RATE SHORTCHANGE FUTURE GENERATIONS?[5]

At reasonable interest rates, the present value of future timber harvests from long rotation species can sometimes be less than costs of intensive site preparation and planting, especially on poorer sites. In such cases, intensive reforestation won't yield an acceptable rate of return for timber production alone, and the economical decision may be to spend less on forestry or change land use. Forest owners often

[4]Chapter 6 points out that a firm wouldn't always find it optimal to invest first in projects with the highest rates of return. But it would not invest in projects with returns below the equilibrium market rate of return. Thus, Figure 4-4 is still a useful explanation of the equilibrating process for rates of return.

[5]This section borrows heavily from Klemperer (1976), by permission.

make such decisions, and you can argue that if their capital earns more elsewhere, society is better off (see "Optimal Allocation of Capital"). But some foresters curse the interest rate because it can discourage certain investments that could provide future generations with more timber.

Typical of such concerns is Strong's (1973) statement that "on a purely economic basis, in fact, it would just not pay to save the planet earth!" This is an overstatement. Preserving the earth is saving humanity, so the stakes are infinitely high. Since the costs of saving the earth are less than infinite, the investment is obviously worthwhile.

The anxiety about discounting future values is perhaps best illustrated by an experience I once had with one of the early electronic calculators. I was toying with the machine's capacity to discount future values at the push of a button. One startling result was that the promise of a million dollars in 260 years discounted at 6 percent interest is worth only about $0.25 today.

Twenty-five cents? At first this seemed to imply, forget about future generations! Their demands discounted to the present are insignificant. But the flip side is that you need to invest only twenty-five cents today at 6 percent interest in order to give future generations a million dollars in 260 years. Admittedly, such long-term compounding doesn't apply to society's total capital stock, which in the long run never accumulates as fast as mathematical compounding might suggest (otherwise we'd be inundated with wealth). In the first place, we siphon off part of our investment returns for consumption, simply to continue existing. Furthermore, investment cycles are much shorter than 260 years. We can seldom, if ever, grace our heirs with investments that expand 4 million-fold over 260 years. But for smaller cycles, whether a year or a few decades, continuous compounding really can occur. We see it in tree value growth and in series of shorter investment cycles where part or all of the proceeds are reinvested. Thus, future citizens are taken care of with a continuing chain of such ventures, the yields from which are partially consumed along the way.

So the interest rate cuts both ways. True, future benefits are reduced in present value terms, but the compounding of current investments works in the opposite direction. What seems like a curse to our successors from one view is a boon from another.

Note that this potential boon doesn't depend on a growth-mania philosophy. Even in a no-growth and zero-population-growth economy, possibilities for continued investment growth would exist. Farm crops would still grow and so would trees. Capital invested in buildings and machinery could still bring returns exceeding their cost; people would still pay interest on loans. The compounding of investment growth is here to stay, as long as people remain enterprising and innovative. Thus, rather than shortchanging future generations, the interest rate can provide abundantly for them, *if* we invest enough today.

And the foregoing principle isn't confined to easily measured monetary returns. For example, we often see continuing nondollar returns to education and research far beyond the value of initial inputs. The same can hold for strategically located parks, wildlife reserves, and activities in the arts.

What Does All This Imply for Forestry?

Applying an interest rate yardstick to forest management means that land, labor, and capital would be allocated to wood production as long as expected rates of return were equal to or above those available elsewhere. It means that not much capital would be invested on poor timber sites, while better sites would be cultured more intensively. As you'll see in Chapter 7, it also means that in young-growth timber production, forests would be cut when the value growth rate of trees and land combined fell below alternative earning rates, even though that's usually before the forest reaches its maximum volume growth per year.

Beyond a certain point, if we keep sinking more money into producing wood (or any other output), the returns to each added dollar will fall below benefits available in other pursuits. Then it's time to cease additional wood production and put our resources to work elsewhere.

Under this efficiency approach, timber output wouldn't be maximized on commercial timberlands. Nor should it be, if our objective is to maximize total social well-being from all our activities combined. The efficient farmer doesn't maximize production of corn or wheat per acre, and the factory manager doesn't try to maximize physical output per day.

If efficient resource allocation results in less than the maximum wood output for future generations, that doesn't necessarily mean less satisfaction. Down to a certain point, the resources freed by producing less-than-maximum timber output would be available to yield greater returns in other pursuits like developing air-entrained concrete slabs, new stone building and insulation technology, energy harnessing, education, electronics, services, and recreation.

What we ultimately want is an optimal mix of outputs without maximizing any particular one. In allocating resources over time, an interest rate performance indicator helps us to do just that, by making sure added resources in one activity would work at least as effectively as in any other.

Therefore, economists needn't be viewed as ogres who don't care about future generations. On the contrary, they care so much about them that they want the sum of all invested resources to bring the highest possible aggregate benefits over time.

THEORY VERSUS PRACTICE

We know that not all people use discounted cash flow to see what they're willing to pay for assets. But it's clear that nearly all of us at least do the seat-of-the-pants discounting mentioned earlier. For any financial asset, we tend to pay less today than the sum of its future net revenues. The net present value equation simulates this behavior. And such discounting also applies to nondollar returns like benefits from memberships in clubs: If one year's membership costs $25 and is worth that much to you, you wouldn't normally pay $100 today for a four-year membership; you'd discount the benefits to something less than $100 today. Sophisticated investors such as major corporations, banks, and real estate firms conduct very detailed NPV calculations before making large investments. They'll rarely rely on one answer but will calculate several NPVs under different assumptions about interest rates, future revenues, and costs. Furthermore, NPV is a maximum willingness to pay, given certain assumptions. It's not necessarily what buyers of an asset pay. If they can buy it for less, they usually will.

For the present, we've only scratched the surface of discounting and compounding. The chapters on inflation, appraisal and valuation, capital budgeting, and risk will go into more depth.

KEY POINTS

- Forests are capital assets. From an efficiency view, they should yield at least as much satisfaction (a rate of return) as that same capital value could yield in other uses (the alternative rate of return). Some people require a rate of return greater than available alternative rates.
- The above line of reasoning applies for nondollar values as well as financial returns.
- Over time, productive capital will yield satisfactions that exceed the original cost. Thus, the future value of capital benefits exceeds the present value. Conversely, the present value must be less than the future value.
- The further in the future a value occurs, the lower its present value. Discounting formulas can give present values of single future payments or regular series of future payments (revenues or costs). Compounding formulas give the future values if payments are not perpetual.
- A buyer's maximum willingness to pay for an asset is the net present value of the asset's yields, using the buyer's discount rate.
- Capital is allocated efficiently if the last unit invested yields the same rate of return in all pursuits, ignoring corrections for risk—the equi-marginal principle again.
- The discount rate may seem to make the needs of future generations appear minuscule today. On the other hand, the power of compound interest lets us make major contributions to future citizens by investing fairly small amounts now.

QUESTIONS AND PROBLEMS[6]

4-1 If you invest $2,200 in a stock that grows at 11 percent annually, how much would it be worth in 22 years?

4-2 If you expect a timber harvest to yield $10,000 in 25 years, and your minimum acceptable rate of return is 7 percent, what is this harvest worth to you today (what's its present value to you)?

4-3 Suppose you invest $100 for two years at 5 percent interest. What is the interest earned during the second year?

4-4 Using a 6 percent discount rate, what is the present value of 15 annual hunting lease revenues of $200 each, the first due in one year?

[6]For more exercises, see Bullard and Straka (1993) and Gunter and Haney (1984).

4-5 The formula for the present value of a terminating annual series of payments (of $\$p$ each) is:

$$V_0 = p\left[\frac{1-(1+r)^{-n}}{r}\right]$$

Derive the formula for the future value of this series in year n (without looking it up in Appendix 4A!). *Hint:* How do you get a future value from a present value?

4-6 In question 4-4, if you invested each of the $200 revenues at 6 percent interest, what would be the total future value in 15 years?

4-7 At 8 percent interest, what is the present value of annual tax payments of $10, beginning in one year and continuing in perpetuity?

4-8 In the previous question, what would the present value be if the annual tax payments didn't start until four years from now?

4-9 Answer question 4-7, making the first tax payment now.

4-10 Find the present value of timber harvest income that will be $3,300 in 40 years and $3,300 every 40 years thereafter, in perpetuity. Use 6 percent interest.

4-11 How much of the present value in the previous question results from the harvests *after* 40 years?

4-12 Assume the following expected incomes and costs from an acre of bare forestland:

$ 125	Initial reforestation cost today and every 40 years thereafter
$ 50	Brush control cost in 5 years and every 40 years thereafter, in perpetuity
$ 75	Thinning cost in 10 years and every 40 years thereafter, in perpetuity
$ 2	Annual taxes in perpetuity, starting in one year
$ 1.25	Annual hunting revenues in perpetuity, starting in one year
$ 200	Pulpwood harvest revenue in 20 years and every 40 years thereafter, in perpetuity
$3,000	Final harvest every 40 years, in perpetuity

If your minimum acceptable rate of return is 6 percent, what is the maximum you'll pay for the acre of bare land, assuming no other costs or revenues?

4-13 You are planning a forest management program for even-aged loblolly pine to be planted this year and harvested in 35 years. You plan to spend $100 per acre to fertilize the trees at age 20. What **increase** in the harvest income per acre (age 35) must the fertilization produce in order for you to earn 6 percent interest on your fertilization expenditure? (*Hint:* After spending the $100, how long will you have to wait until you get your return?)

4-14 Assume that you are planning to buy a 15-year-old pine plantation, which you intend to grow to age 30. At that time you will cut the timber and sell the land for $300 per acre. You expect to harvest 60 cords of pulpwood per acre at age 30, and you will have annual costs of $3 per acre for the 15 years. The plantation will cost you $600 per acre. How much will you have to get for the pulpwood per cord to make a 7 percent return on your investment?

4-15 If Karen bought a house for $90,000 in January 1986 and sold it for $192,900 in January 1994, what was her rate of return on the investment?

4-16 If an investment value doubles over a nine-year period, what is the rate of return?

4-17 If an investment value increases at 12 percent annually, how long will it take to triple in value?

4-18 You have a 10-year-old, 1-acre, even-aged forest. You estimate that if you spend $35 per acre fertilizing this forest every 5 years, you can receive $1,000/acre net stumpage returns 20 years from now. In 20 years, you predict a potential land sale price of $300/acre. Four fertilizer applications will be made, the first one now. Based only on the above value and cost projections, what is your net present value per acre now for this forest, if your minimum acceptable rate of return is 8 percent?

4-19 What kinds of information and analyses would you need in order to determine whether it would be worthwhile, from a financial standpoint, to practice forestry on the area described in the last question?

APPENDIX 4A: Derivations of Compound Interest Formulas

TERMINATING PERIODIC SERIES—PRESENT VALUE

Start with equation (4-13) in Table 4-3:

$$V_0 = p\left[\frac{1-(1+r)^{-n}}{(1+r)^t-1}\right] \tag{4A-1}$$

where r = interest rate, percent/100
V_o = initial value = present value
V_n = value in year n = future value
p = value (revenue or cost) occurring annually or every t years (no payment occurs in year 0)
t = number of years between periodic occurrences of p
k = number of periods
n = number of years of compounding or discounting (kt)

The basic formula for this series is:

$$V_0 = \frac{p}{(1+r)^t} + \frac{p}{(1+r)^{2t}} + \frac{p}{(1+r)^{3t}} + \ldots + \frac{p}{(1+r)^{kt-t}} + \frac{p}{(1+r)^{kt}} \tag{4A-2}$$

where $kt = n$ or the total number of years.

Multiply both sides of (4A-2) by $(1 + r)^t$:

$$V_0(1+i)^t = p + \frac{p}{(1+i)^t} + \frac{p}{(1+i)^{2t}} + \ldots + \frac{p}{(1+i)^{kt-t}} \tag{4A-3}$$

Subtract (4A-2) from (4A-3), remembering $kt = n$:

$$V_0(1+r)^t - V_0 = p - \frac{p}{(1+r)^n} \tag{4A-4}$$

[Note that on the right side of (4A-2) and (4A-3), most of the terms are the same, so they fall out when subtracting one formula from the other, simplifying the equation.]

Factor V_0 out on the left side and p on the right side:

$$V_0[(1 + r)^t - 1] = p\left[1 - \frac{1}{(1 + r)^n}\right] \tag{4A-5}$$

Isolate V_0 on the left side of the equation, and reformulate the right side:

$$V_0 = p\left[\frac{1 - (1 + r)^{-n}}{(1 + r)^t - 1}\right] \tag{4A-6}$$

which is the formula for the present value of a terminating periodic series [equation (4-13) in Table 4-3].

TERMINATING PERIODIC SERIES—FUTURE VALUE

Multiply equation (4A-6) by $(1 + r)^n$ to get its future value, which is equation (4-14) in Table 4-3:

$$V_n = V_0(1 + r)^n = p\left[\frac{1(1 + r)^n - (1 + r)^{-n}(1 + r)^n}{(1 + r)^t - 1}\right] \tag{4A-7}$$

$$V_n = p\left[\frac{(1 + r)^n - 1}{(1 + r)^t - 1}\right] \tag{4A-8}$$

PERPETUAL PERIODIC SERIES—PRESENT VALUE

The difference between this formula and the present value of a terminating periodic series (4A-6) is that n goes to infinity. Therefore (4A-6) can be converted to a perpetual case by letting n go to infinity:

$$V_0 = \lim_{n \to \infty} p\left[\frac{1 - (1 + r)^{-n}}{(1 + r)^t - 1}\right] \tag{4A-9}$$

Separate the terms containing n:

$$V_0 = \frac{p}{(1 + r)^t - 1} \cdot \lim_{n \to \infty} 1 - (1 + r)^{-n} \tag{4A-10}$$

In this case, as n goes to infinity, the term $1 - (1 + r)^{-n}$ approaches 1. Therefore,

$$V_0 = \frac{p}{(1 + r)^t - 1} \tag{4A-11}$$

is the present value formula for a perpetual periodic series [equation (4-9) in Table 4-3].

PERPETUAL ANNUAL SERIES—PRESENT VALUE

Convert equation (4A-11) to a perpetual *annual* series by letting $t = 1$:

$$V_0 = \frac{p}{(1 + r)^1 - 1} \tag{4A-12}$$

Simplify the denominator to get equation (4-8) in Table 4-3:

$$V_0 = \frac{p}{r} \tag{4A-13}$$

TERMINATING ANNUAL SERIES—PRESENT VALUE

To convert equation (4A-6) to an annual series, let $t = 1$:

$$V_0 = p\left[\frac{1 - (1 + r)^{-n}}{(1 + r)^1 - 1}\right] \tag{4A-14}$$

Simplify the denominator to get equation (4-10) in Table 4-3:

$$V_0 = p\left[\frac{1 - (1 + r)^{-n}}{r}\right] \tag{4A-15}$$

TERMINATING ANNUAL SERIES—FUTURE VALUE

Multiply equation (4A-15) by $(1 + r)^n$ to find future value, which is equation (4-11) in Table 4-3:

$$V_n = V_0(1 + r)^n = p\left[\frac{1(1 + r)^n - (1 + r)^{-n}(1 + r)^n}{r}\right] \tag{4A-16}$$

$$V_n = p\left[\frac{(1 + r)^n - 1}{r}\right] \tag{4A-17}$$

APPENDIX 4B: Discounting and Compounding Geometrically Changing Payments

Often in forestry we project series of equally spaced payments that, instead of being constant, are increasing or decreasing at some fixed real rate. Examples are values of equal annual harvest volumes when stumpage prices increase at a fixed annual percentage rate, or real loan payments decreasing when inflation is projected at some average rate. Also, we often see discounted cash flow analyses where different revenues, prices, and costs are increasing or decreasing at separate rates. As explained here, such payments can be discounted with the standard formulas in Table 4-3 if an adjustment is made to the real interest rate. All values and rates are in real terms, excluding inflation.

In addition to the notation in Table 4-1, the following notation is used:

g_r = real annual payment growth rate, percent/100

r_d = downward adjusted real interest rate/100 for capitalizing rising payments with a fixed payment formula

r_u = upward adjusted real interest rate/100 for capitalizing declining payments with a fixed payment formula

n/t = number of payments, where n = number of years in series, and t = number of years between payments

p = the amount the first payment would be if it occurred in year 0. Since it occurs at the end of the first period, the first payment is $p(1 + g_r)^t$ for rising payments and $p(1 - g_r)^t$ or $p/(1 + g)^t$ for declining payments.

Except that payments will not be fixed, the same constraints for the Table 4-3 formulas will apply: no payment in year 0, first payment at the end of the first period, and payments equally spaced over time.

Capitalizing Real Payments Increasing at Rate g_r

When payments rise at an annual rate g_r, the first payment is $p(1 + g_r)^t$, the second is $p(1 + g_r)^{2t}$, etc. We know that lower discount rates increase present values; therefore, to capitalize a series of rising payments with any of the traditional fixed payment formulas, use a *downward adjusted interest rate* r_d as follows:

$$r_d = \frac{1 + r}{1 + g_r} - 1 \qquad r \neq g_r \tag{4B-1}$$

Qualifications If $g_r = r$ for terminating series, the rate of payment growth and the discount rate exactly cancel each other out, and the present value is simply the number of payments times p:

$$V_0 = (n/t)p \qquad \text{when } r = g_r \tag{4B-2}$$

For perpetually increasing series of payments, the rate of payment growth must be less than the interest rate ($g_r < r$). If payments rise at a rate equal to or greater than the discount rate, the present value of an infinite series of payments will also be infinite.

A Rising Payments Example Suppose we expect real timber harvest incomes from a forest to increase at 2 percent annually. At 6 percent interest, what is the present value of 15 such annual harvests, the first of which would be worth \$1,000 today? The initial harvest would occur next year at 1,000(1.02) = \$1,020. The downward adjusted discount rate using equation (4-B1) is $r_d = (1.06)/(1.02) - 1 = 0.03922$. Using this rate in the fixed payment equation (4-10), present value is:

$$V_0 = 1{,}000\left[\frac{1 - (1.0392)^{-15}}{0.0392}\right] = \$11,181$$

With rising payments, the same type of downward adjusted interest rate can be substituted into any of the standard discounting formulas for fixed payments, annual or periodic, terminating or perpetual.

Declining Real Payments

If payments decline, present value will be less than that for the fixed payment case. Thus, when discounting declining payments with a fixed payment formula, the interest rate must be adjusted upward. Two cases are common.

Payments Declining at a Percentage Rate When payments decline annually at a rate g_r, the first payment is $p(1 - g_r)^t$, the second is $p(1 - g_r)^{2t}$, etc. The *upward adjusted interest rate* r_u for this case is:

$$r_u = \frac{1 + r}{1 - g_r} - 1 = \frac{r + g_r}{1 - g_r} \tag{4B-3}$$

Consider, for example, a pulp mill projecting a real 1.5 percent annual decline in its recycled paper costs, which are $100,000 in the current year. Given a 7 percent real discount rate, what is the present value of these annual costs for the next 10 years? The upward adjusted interest rate would be $r_u = (0.07 + 0.015)/(1 - 0.015) = 0.0863$. Using this rate in equation (4-10), present value is:

$$V_0 = 100{,}000\left[\frac{1 - (1.0863)^{-10}}{0.0863}\right] = \$652{,}351$$

Deflating Payments When payments decrease annually by a factor of $1/(1 + g_r)$, the rate of decline is slightly less than in the previous example. In that case, the upward adjusted interest rate is:

$$r_u = (1 + r)(1 + g_r) - 1 \tag{4B-4}$$

In such series, the first payment is $p/(1 + g_r)^t$, the second is $p/(1 + g_r)^{2t}$, etc. As you'll see in the next chapter, loan payments that remain fixed in inflationary times decline in this pattern in real terms. The g_r in such cases is the inflation rate.

Future Values

At the end of any terminating series of geometrically changing payments, you can calculate the future value as follows: Compute the present value as shown above, and simply multiply the result by $(1 + r)^n$. For example, the future value of the above declining recycled paper costs would be $652{,}351(1.07)^{10} = \$1{,}283{,}273$. In other words, if the firm did not spend the money but invested it at 7 percent interest, it would accumulate to that value in 10 years.

Summary

With the interest rate adjustments shown above, you can adapt any of the six series formulas in Table 4-3 to compute present or future values of equally spaced payments increasing or decreasing at a fixed geometric rate. These are useful techniques, since cash flows in the same forestry project often change at different rates. Remember, the price-changes considered here do *not* include inflation; if all prices rise collectively, we have inflation, which involves no real change in value but just a reduction in purchasing power of the dollar (see Chapter 5).

The above discounting approach for increasing payments is not advised for perpetual series. Rarely can we expect real values or quantities in forestry to increase exponentially forever. For example, if timber prices increased relative to other prices over long periods, wood substitutes would be more widely used, mitigating the rise in wood prices. Thus, if you are discounting a perpetual series of payments and real prices are projected to increase, it's reasonable to allow the increase to occur for a projected period and then let it remain constant thereafter.

For example, consider a perpetual series of equal annual harvests from a sustained yield forest, the first cut being in one year. Suppose today's value of one harvest is $1,000, and real stumpage values are projected to increase at 2 percent per year for 30 years, remaining constant thereafter. What is the present value (PV) of these harvests, at a 5 percent real discount rate? Compute the PV of the first 30 harvests as an increasing series, using $r_d = (1.05/1.02) - 1 = 0.0294$ in equation (4-10). After 30 years we have a perpetual annual series of constant harvest values starting at $1{,}000(1.02)^{30} = \$1{,}811$, which includes the 2 percent annual price-increase for the first 30 years. In year 30, the PV of this perpetual series is calculated with equation (4-8) and then discounted 30 years. These two PV segments are as follows:

$$V_0 = 1{,}000\left[\frac{1-(1.0294)^{-30}}{0.0294}\right] + \frac{\dfrac{1{,}000(1.02)^{30}}{0.05}}{(1.05)^{30}} = \$28{,}135$$

The PV of *perpetually* increasing harvest values calculated with equation (4-8), using the above r_d, is likely to be unrealistically high:

$$V_0 = 1{,}000/0.0294 = \$34{,}014$$

REFERENCES

Brigham, E. F., and L. C. Gapenski. 1991. *Financial Management—Theory and Practice*, 6th ed. The Dryden Press. 995 pp.

Bullard, S. H., and T. J. Straka. 1993. *Forest Valuation and Investment Analysis.* GTR Printing, Starkville, MS. 70 pp.

Dixit, A. 1992. Investment and hysteresis. *Journal of Economic Perspectives.* 6(1): 107–132.

Goforth, M. H., and T. J. Mills. 1975. Discounting Perpetually Recurring Payments under Conditions of Compounded Relative Value Increases. USDA Forest Service Research Note WO-10, Washington, DC.

Gunter, J. E., and H. L. Haney. 1978. A decision tree for compound interest formulas. *Southern Journal of Applied Forestry.* 2(3):107.

Gunter, J. E., and H. L. Haney. 1984. *Essentials of Forestry Investment Analysis.* Oregon State University Bookstores, Inc., Corvallis, OR. 337 pp. (Second edition in press.)

Klemperer, W. D. 1976. Economic analysis applied to forestry: Does it short-change future generations? *Journal of Forestry.* 74(9):609–611.

Klemperer, W. D. 1979. A simplified technique for capitalizing geometrically changing cash flows. *The Real Estate Appraiser and Analyst.* 24(4):11–13.

Merrett, A. J., and A. Sykes. 1973. *The Finance and Analysis of Capital Projects.* Longman Group Ltd, London. 573 pp.

Strong, M. 1973. One year after Stockholm: an ecological approach to management. *Foreign Affairs.* 51:690–707.

5

INFLATION AND FOREST INVESTMENT ANALYSIS

We've all experienced rising prices—**inflation**—at one time or another. When inflation occurs, an average market basket of goods and services gets more expensive. At the same time, rising prices tend to make interest rates increase. In most market economies, inflation has been a fact of life in recent decades, and it has profound effects on investments. In explaining how inflation affects forestry decisions, this chapter ties in very closely with the last one. If you don't correctly handle inflation in financial analyses, you cannot make valid comparisons between investment alternatives. You could seriously over- or underestimate asset values. Such errors can cause rejection of projects that are worth more than ones accepted. Because of its long payoff periods, forestry is very sensitive to these potential errors. Also, inflation, by boosting interest rates, can sharply reduce housing demand, causing serious boom and bust cycles in the forest products industry.

In this chapter you'll learn how inflation is measured, how and why it affects interest rates, how inflation is correctly handled in evaluating forest investments, the types of errors that can be made in inflationary times, and the impacts of inflation on forest products demand. You'll see that correct handling of inflation is important when evaluating investments not only in timber but also in recreation, wildlife, water, fish, and other forest outputs. This may seem like a detour, but knowing about inflation is crucial to understanding the investment analysis in Chapter 6.

Usually inflation is caused by several things at once: one example is too many people chasing too few goods, in which case, buyers will bid up prices. This is called **demand-pull inflation.** Examples are excessive government spending, too rapid increase in the money supply, or buying-frenzies, whatever the cause. The opposite is **cost-push inflation** where certain production costs increase, thus causing general price increases. Examples are wage-increases exceeding productivity gains and increased oil prices, which occurred in the 1970s. These higher costs work their way through the economy, causing most prices to rise. Our concern here is less with the causes of inflation than with learning how it affects forest investments.

Table 5-1 defines all notation used in this chapter. As in Chapter 4, unless otherwise stated, analyses will be before taxes, future values will be considered as average or **expected values** and any risk is assumed to be compensated for by an increased discount rate, and all satisfactions will initially be measured in dollars.

MEASURING INFLATION

The trouble with trying to measure inflation is that all prices don't increase at the same rate. At times, some prices are rising (like gasoline during the mid-1970s) while others are falling (for example, calculators during the same period). So inflation can be measured only in an average sense. To do this, the U.S. Department of Labor follows the cost of an average "market basket" of about 360 goods and services in 91 urban areas in the United States, weighting the costs by population in each area. This market basket includes goods and services in the proportions consumed by the average urban consumer. Now, suppose the weighted average cost of the market basket was $7,000 for 1988 and $7,336 for 1989. The percentage rate of change in this cost is the annual inflation rate for 1989: subtract 1 from

TABLE 5-1
NOTATION FOR CHAPTER 5

C	= purchase cost of an asset
CPI_n	= consumer price index in year n
f	= annual inflation rate, percent/100 (for example, for 7 percent inflation, $f = 0.07$)
g_n	= nominal rate of annual price growth, percent/100
g_r	= real rate of annual price growth, percent/100
i	= inflated or nominal interest rate, percent/100; ($i = r + f + rf$). This also refers to nominal rates of value growth in assets.
I_n	= inflated or current dollar value in year n
MAR	= minimum acceptable rate of return
MBF	= thousand board feet
n	= number of years in investment period
NPV	= net present value (PV revenues minus PV costs)
PV	= present value
PPI_n	= producer price index in year n
r	= real interest rate, percent/100; excludes inflation. This also refers to real rates of value growth in assets.
ROI	= return on investment [see equation (5-11)]
T_i	= income tax rate, percent/100 (for example, for a 30 percent tax rate, $T_i = 0.30$)
V_0	= initial value at the start of an investment period (also referred to as present value)
V_n	= future value in year n, in constant dollars of year 0

the 1989 cost divided by the 1988 cost, or (7,336/7,000) − 1 = 0.048, which is 4.8 percent inflation between 1988 and 1989. (You can also get the percent change by dividing the actual change by the beginning value: 336/7,000 = 0.048.)

In practice, the market basket cost is translated into an index—the consumer price index (CPI)—which is arbitrarily set to 100 in a base year. One base year had been 1967, but recent revisions set the average CPI for 1982–1984 at 100. Thus, you'll sometimes see different CPIs for the same year. This is no problem, since we're mainly interested in the percent change in the CPI from year to year, which is the inflation rate. For example, the CPI was 118.3 in 1988 and 124 in 1989. That indicates an inflation rate of 4.8 percent for 1989 [or (124/118.3) − 1 = 0.048]. The CPI is actually computed monthly, and you'll often see it published that way together with monthly rates of change. Here, we'll consider only annual rates. Unless otherwise stated, inflation rates are almost always annual.

Columns (2) and (3) of Table 5-2 show consumer price indexes for the years 1960 to 1993 and the percentage changes or inflation rates from year to year. You can figure the average annual rate of change in the CPI the same way you computed annual rates of change in asset values. For example, to find the *average annual inflation rate f over an n-year period,* you can adapt equation (4-4) by substituting f for r and CPI for V, since r and f are both percentage rates of change:

$$\text{Average annual inflation rate} = f = \sqrt[n]{\frac{\text{CPI}_n}{\text{CPI}_0}} - 1 \qquad (5\text{-}1)$$

What is the average annual inflation rate between 1979 and 1989? Using equation (5-1) and CPIs from Table 5-2, taking the tenth root of 124.0/72.6 and subtracting 1 gives 0.055, or 5.5 percent. Remember that this is an annual rate. Over the 10-year period, the CPI increased about 1.7 times, but the increase averaged 5.5 percent per year. This inflation rate means that in 1989 it took 1.7 times as much income to buy the same market basket of goods and services as it took in 1979. So people whose annual spendable incomes increased from $10,000 in 1979 to $17,000 in 1989 would have found their purchasing power unchanged. But this would apply only to an average consumer in the average urban area in the United States who consumed exactly the average market basket of goods and services. Inflation figures rarely pertain well to an individual, since they're based on averages. Nevertheless it's important to correct for inflation so that we can see what incomes really yield in terms of average purchasing power.

It's more difficult to measure impacts of inflation on businesses as compared to consumers, because of the widely differing mixes of inputs that various industries use. The U.S. Department of Commerce maintains a **producer price index (PPI),** which shows the increases in prices for an average mix of industrial outputs, excluding services, from year to year. As with the CPI, annual percent changes in the PPI are also a measure of inflation—a measure that is sometimes more relevant for figuring inflationary impacts on businesses. As shown in Table 5-2, over a number of years the CPI and PPI tend to move together but aren't always the same.

TABLE 5-2
CONSUMER PRICE INDEX (CPI) AND PRODUCER PRICE INDEX (PPI)*

Year	CPI 1982–1984 = 100	% change in CPI†	PPI 1982 = 100	% change PPI†
1960	29.6	1.7	33.4	0.9
61	29.9	1.0	33.4	0.0
62	30.2	1.0	33.5	0.3
63	30.6	1.3	33.4	−0.3
64	31.0	1.3	33.5	0.3
65	31.5	1.6	34.1	1.8
66	32.4	2.9	35.2	3.2
67	33.4	3.1	35.6	1.1
68	34.8	4.2	36.6	2.8
69	36.7	5.5	38.0	3.8
1970	38.8	5.7	39.3	3.4
71	40.5	4.4	40.5	3.1
72	41.8	3.2	41.8	3.2
73	44.4	6.2	45.6	9.1
74	49.3	11.0	52.6	15.4
75	53.8	9.1	58.2	10.6
76	56.9	5.8	60.8	4.5
77	60.6	6.5	64.7	6.4
78	65.2	7.6	69.8	7.9
79	72.6	11.3	77.6	11.2
1980	82.4	13.5	88.0	13.4
81	90.9	10.3	96.1	9.2
82	96.5	6.2	100.0	4.1
83	99.6	3.2	101.6	1.6
84	103.9	4.3	103.7	2.1
85	107.6	3.6	104.7	1.0
86	109.6	1.9	103.2	−1.4
87	113.6	3.6	105.4	2.1
88	118.3	4.1	108.0	2.5
89	124.0	4.8	113.6	5.2
1990	130.7	5.4	119.2	4.9
91	136.2	4.2	121.7	2.1
92	140.3	3.0	123.2	1.2
93	144.5	3.0	124.7	1.2

*Data from CEA (1994).
† Percent change from previous year.

Recognizing that individuals and businesses use different products, I'll assume that the CPI gauges the impact of inflation on consumers and the PPI on producers. So, when analyzing investments, you need to estimate how investors will spend their income in order to choose the best measure of inflation. For example, if you're evaluating investments for a nonindustrial private forest owner who is a consumer like you and me, use the CPI to measure inflation; if the investor is a forest products firm, use the PPI. Some firms construct their own price index that reflects their mix of inputs. For government agencies, the choice of price index isn't always so clear. In any case, since our data in forest investment analyses are usually so uncertain and since the CPI and PPI tend to move together over long periods, the choice of inflation measure usually isn't crucial. Unless otherwise mentioned, here the inflation rate will be based on the consumer price index.

Most of us aren't hurt as much by inflation as the statistics suggest because we change buying habits and substitute cheaper items for more expensive ones. For example, when beef prices are up, you might switch to chicken. That tactic isn't always possible in the short run; for example, firms using oil-based products suffer more than others when oil prices are high. But in the long run, we conserve the costliest items. Thus, the standard ways given here to correct for inflation may often overcompensate buyers and investors.

CURRENT DOLLARS AND CONSTANT DOLLARS

Monetary values such as costs, prices, or incomes are usually given in **current dollars**—the values as they actually occur in a given year, including inflation. But during inflationary times, current dollar figures can be deceiving. For example, we noted that if someone's income rises at exactly the inflation rate, their purchasing power doesn't increase—in other words, their **real** income, excluding inflation, has not increased. Likewise, if you own a forest yielding harvest income that increases over time, you'll want to know how much of that increase, if any, is a real increase in purchasing power. Imagine an initial value (either an income or an asset) V_0 that grows to an inflated value I_n, n years later (in current dollars). If you know the average rate of inflation f, you can use the following equation to express the later current dollar value I_n in **constant dollars** V_n of a base year n years earlier:

$$\frac{\text{Current dollar value, year } n}{(1+f)^n} = \frac{I_n}{(1+f)^n} = V_n \tag{5-2}$$

$$= \text{value in constant dollars of base year } n \text{ years ago}$$

This procedure is sometimes called **deflating** a value. Comparing V_n with V_0 allows you to see the real change in value over the n-year period. *Note that constant dollars are always referenced to some base year, and to make valid comparisons between constant dollar values, the base year must be the same.* Figure 5-1 graphs these relationships for an appreciating asset like a growing forest. The rates of growth will be discussed shortly. These principles are relevant in evaluating personal finances as well as forest income. Suppose someone bought a home for $V_0 = \$70{,}000$ in 1979 and sold it for $I_n = \$115{,}000$ in 1989. While that looks like a 64 percent increase in value, due to inflation it's not the real increase in purchasing power. Remembering that annual inflation averaged 5.5 percent over that period, equation (5-2) will give the 1989 home sale price in 1979 dollars:

$$V_n = 115{,}000/(1.055)^{10} = \$67{,}325, \text{ in 1979 dollars}$$

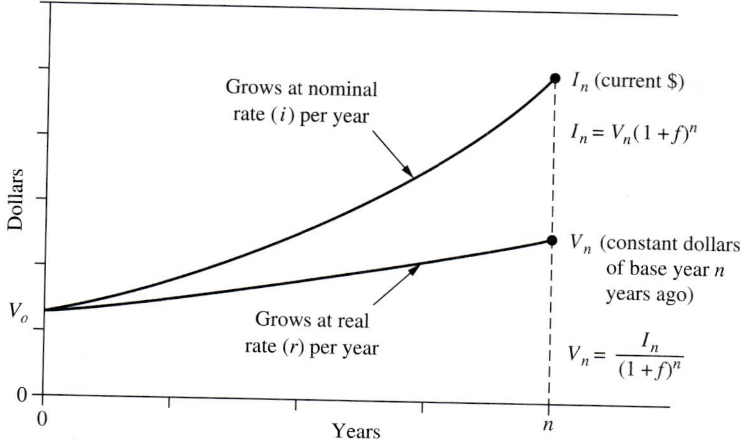

FIGURE 5-1
Asset values—current dollars and constant dollars.
(For example, a forest, stock profolio, or real estate investment. The figure shows an appreciating asset, but cases exist where V_n or even I_n could be less than V_0, for example, a stock which decreased in value. If $f > 0$, $I_n > V_n$. However, during deflation when average $f < 0$, $I_n < V_n$.)

In real terms, the sale price was *below* the purchase price. In inflationary periods, looking only at current dollar figures masks the real values and can cause people to be misled by the so-called **money illusion.** But remember, as noted under "Measuring Inflation," inflation often doesn't bite as hard as the raw numbers suggest.

Rearranging equation (5-2) tells us that a value in current dollars is its constant dollar value (base year n years earlier) multiplied by $(1 + f)^n$, where f is the average inflation rate over the n-year period:

$$V_n(1 + f)^n = I_n \tag{5-3}$$

For example, suppose in today's dollars you expect real Douglas-fir lumber prices to be $300 per 1,000 board feet in 5 years, regardless of the inflation rate. If you project 4 percent inflation, the current dollar value 5 years hence[1] is $300(1.04)^5 = \$365$.

While you can use equation (5-2) to express current dollar values in constant dollars of any chosen base year, the more common practice, especially for annual series of values, is to compute constant dollar values by using the consumer (or producer) price index as follows:

$$\text{Constant dollar value} = \frac{\text{current dollar value}}{\text{CPI/100}} \tag{5-4}$$

[1]This doesn't mean that stumpage by definition *must* rise at the inflation rate: the $365 could be above, equal to, or below today's price, depending on the projected *real* rate of price change.

TABLE 5-3
LOUISIANA SOUTHERN PINE PULPWOOD STUMPAGE PRICES, DOLLARS PER CORD, 1960–1987

Year	$/cord* (current dollars)	CPI†	1967 dollars per cord (constant dollars)‡
1960	4.40	88.7	4.96
61	4.25	89.6	4.74
62	4.25	90.6	4.69
63	4.30	91.7	4.69
64	4.30	92.9	4.63
65	4.40	94.5	4.66
66	4.55	97.2	4.68
67	4.60	100.0	4.60
68	4.65	104.2	4.46
69	4.65	109.8	4.23
1970	4.70	116.3	4.04
71	4.75	121.3	3.92
72	4.75	125.3	3.79
73	5.20	133.1	3.91
74	6.05	147.7	4.10
75	6.40	161.2	3.97
76	6.70	170.5	3.93
77	7.10	181.5	3.91
78	7.80	195.4	3.99
79	9.30	217.4	4.28
1980	10.30	246.8	4.17
81	12.65	272.4	4.64
82	14.30	289.1	4.95
83	14.85	298.4	4.98
84	17.65	311.1	5.67
85	15.20	322.2	4.72
86	12.05	328.4	3.67
87	13.85	340.4	4.07

*Ulrich (1990).
†Consumer price index, 1967 = 100 (CEA, 1988).
‡(Current dollars)/(CPI/100).

The result will be in constant dollars of the base year in which the price index equals 100. For example, column (2) of Table 5-3 shows southern pine pulpwood stumpage prices in Louisiana in current dollars. These are the average prices buyers paid for private standing timber to be cut in the near future. Column (3) shows the consumer price index with 1967 = 100. Using equation (5-4), column (4) gives the column (2) prices in 1967 dollars (constant dollars of the base year 1967). (Sometimes current dollars are referred to as "actual," and constant dollars as "relative.") The constant dollar prices indicate real values in terms of relative purchasing power, while the current dollar prices include inflation. For example, in column (2) of Table 5-3, note how the stumpage price more than doubled from $4.65 per cord in 1969 to $10.30 in 1980. But in constant dollars, the price actually *decreased* slightly from $4.23 to $4.17 over the same period, as shown in column (4). Thus, in terms of purchasing power, these landowners received about the same value per cord for their pulpwood in 1969 and 1980. On the other hand, real sawtimber stumpage prices increased in many parts of the United States.

Note that the base year for the CPI in Table 5-3 is earlier than that in Table 5-2. Had a later base year been used in Table 5-3, the constant dollar figures would be higher, but the *relative* values between years would be the same as that for the constant dollar series in Table 5-3. With constant dollar values, the important thing is the percent change from year to year, which shows the change in terms of purchasing power.

Whenever you see series of values over time, make sure you know whether they're in current or constant dollars. *Unless otherwise stated, values and prices you read or hear about are almost always in current dollars (nominal).* If you don't deflate such values to a common year, you won't be able to compare them over time, in real terms. These relationships are very important in forestry where we deal with long time periods over which inflation can cause current dollar values to increase greatly. The *real* value-increase will usually be far smaller.

Several terms exist for describing values with and without inflation. Table 5-4 lists the more common terminology you should know.

NOMINAL AND REAL RATES OF RETURN

The rate of return earned on investments measured in current dollars (including inflation) is typically called the **nominal** rate of return or nominal interest rate. The return expressed in constant dollars is the *real* rate of return. Just as values over time are nearly always expressed in current dollars (unless otherwise stated), **rates of return** and **interest rates** are usually expressed in *nominal* terms (again, unless otherwise stated). Thus, these nominal rates include inflation and are not real earning rates in terms of purchasing power. Let's now see why nominal rates of return tend to rise and fall with inflation.

Suppose you loan someone $100 now, to be returned in one year in one payment, and you wish to earn a 5 percent real rate of return r beyond the inflation rate. In other words, with the money you receive at year's end, you'd like to be able to purchase 5 percent more than you could buy with $100 now. If inflation were 0, you would charge 5 percent interest, and the borrower would pay back $105. If you expect 4 percent inflation over the next year, how much interest

TABLE 5-4
INFLATION TERMINOLOGY RELATING TO VALUES, INTEREST RATES, AND RATES OF RETURN

Inflation is included	Inflation is not included
Nominal	Real
Current dollars	Constant dollars
Inflated	Uninflated or deflated
Actual (prices)	Relative (prices)

should you charge? Just to keep up with inflation, you need to receive \$104 after one year. That's $100(1 + f)^n = 100(1.04)^1 = \$104.00$, according to equation (5-3), or 4 percent on the \$100 loan. But, due to 4 percent inflation, the \$104 allows you to buy no more than \$100 would at the beginning of the year. If you also want 5 percent more at year's end, you need to multiply the \$104 by $(1 + r)^n$ or $(1.05)^1$ to get a loan repayment of \$109.20. So you must charge 9.2 percent interest (nominal). That was obtained, in this one-year example, by combining the inflation and real growth factors: $(1.04)(1.05) - 1 = 0.092$. As a check, with equation (5-2), express the current dollar repayment in constant dollars of the year before: $109.2/(1 + f)^n = 109.2/(1.04)^1 = \$105.00$, which is exactly the 5 percent real rate of return you wanted.

Generalizing the above example, the equation below gives the required single income I_n in current dollars, n years after an investment of V_0, in order to earn a real rate of return r during a period with an average annual inflation rate f:

$$I_n = V_0(1 + r)^n(1 + f)^n = V_0(1 + r + f + rf)^n = V_0(1 + i)^n \qquad (5\text{-}5)$$

The last two segments of equation (5-5) give the solution for the *nominal rate of return or interest rate* i:

$$r + f + rf = i \qquad (5\text{-}6)$$

A simpler way to punch this out on a calculator is:

$$i = (1 + r)(1 + f) - 1 \qquad (5\text{-}7)$$

mentally subtracting the 1. For example, in the loan above, the nominal interest rate was $(1.05)(1.04) = 1.092$, implying 9.2 percent. Often nominal interest rates are estimated by adding the inflation rate to the real interest rate and ignoring the cross product rf. But the more precise formulation will be used here. Failure to do so can result in discounting errors, especially with high rates of inflation and real interest. For example, if $r = 0.08$ and $f = 0.13$, the nominal interest rate is 22.04 percent, more than a point above the simple addition of $r + f$.

The foregoing illustrates an extremely important point: *Nominal interest rates tend to rise and fall with inflation* in the fashion shown by equation (5-6). The higher the expected inflation rate, the higher the interest rate that lenders will charge in order for them to still earn an acceptable real rate of return. That also applies to buyers of government bonds who are lending money to the Treasury. As inflation increases, to attract investors, the government must advertise higher nominal interest rates on newly offered bonds. And likewise for all investors, if they wish to earn a given real rate of return, the nominal rate must be higher, if inflation is greater than 0.[2]

[2]One might think that interest rates are driven by the Federal Reserve Board setting the rate it charges on loans to banks (the "discount rate"), and that's sometimes the case. But typically, changes in the discount rate follow changes in Treasury bill interest rates (Fischer et al., 1988; Ritter and Silber, 1974).

If you know the nominal rate of return, just solve equation (5-7) for r to get the *real rate of return:*[3]

$$(1 + r)(1 + f) = 1 + i$$

$$1 + r = \frac{(1 + i)}{(1 + f)} \tag{5-8}$$

$$r = \frac{(1 + i)}{(1 + f)} - 1$$

To the extent that the published inflation rate f is overstated (see "Measuring Inflation"), real rates of return from equation (5-8) are understated in terms of purchasing power or utility. But if all analysts use the same inflation data, these understated real rates of return are still consistent measures for comparing investments.

Figure (5-1) graphs the relationships between V_0, V_n, I_n, r, and i. As an example of finding a real rate of return, suppose you bought timberland 10 years ago for $40,000 and sold it today for $101,047 (that's in current dollars, since nothing was stated otherwise). If annual inflation averaged 6 percent over that period [calculated with equation (5-1)], what was your real rate of return? First figure your nominal rate of return:

$$i = \sqrt[n]{\frac{I_n}{V_0}} - 1 \tag{5-9}$$

This is a variation on equation (4-4), where I_n is substituted for V_n, in which case r becomes the nominal rate i. Substituting the above values, the nominal rate of return is:

$$i = \sqrt[10]{\frac{101{,}047}{40{,}000}} - 1 = 0.0971$$

or 9.7 percent. This return includes 6 percent inflation; so to compute the real rate of return, use equation (5-8):

$$r = (1.097/1.06) - 1 = 0.035$$

or 3.5 percent. (Note that this is slightly less than the real rate minus the inflation rate.)[4] A vital lesson here is that, *unless you know the average inflation rate over*

[3]Values given here for i and r can also be derived from Figure 5-1, noting that $V_n = V_0(1 + r)^n$ and $I_n = V_0(1 + i)^n$. Substituting these values for V_n and I_n into equation (5-2) yields an equation that can be solved for i and r.

[4]You could also compute the real rate of return by first deflating the sale price with equation (5-2): $101{,}047/(1.06)^{10} = \$56{,}424.12$ in constant dollars of the purchase year. Since values are now in real terms, use equation (4-4) to find r:

$$r = \sqrt[10]{\frac{56{,}424.12}{40{,}000}} - 1 = 0.035$$

an investment period, you have no idea what most reported rates of return are in real terms. Nearly all rates of return reported by financial publications, newspapers, magazines, firms' annual reports, banks, and the U.S. Treasury are nominal, and you must translate them to real terms with equation (5-8) in order to make valid comparisons. The same goes for interest rates on loans. If you borrow money at 12 percent interest and repay it when inflation averages 7 percent, you will pay (and the lender will earn) only 4.7 percent real interest [using equation (5-8)]. This occurs because the loan is repaid in current dollars that are worth less each year.

The foregoing examples calculated real rates of return based on nominal rates *already earned.* What about projecting real rates of return on *new* investments? For example, if you just bought a government bond yielding a fixed 8 percent interest, what will your real rate of return be? You need to consider various projections of future inflation and calculate r with equation (5-8), based on your best guess at f. If expected inflation is 5 percent, your projected real return on the bond would be $(1.08)/(1.05) - 1 = 0.0286$, or 2.86 percent. But this is just an estimate. You can only figure your true real rate of return after you know what inflation rate actually develops.

There is much misunderstanding about the difference between real and nominal interest rates, and it's important, whenever you hear of a historic or desired rate of return, to ask whether it is real or nominal. Don't be surprised if those involved aren't certain what you're talking about. But with a few questions about the rate's origin, you should be able to tell if it's real or nominal.

Rates of Price Growth

You can use equations (5-7) through (5-9) to figure annual rates of increase in prices, which involve the same arithmetic as rates of return. Substitute the nominal rate of price growth g_n for i and the real price growth rate g_r for r. Let nominal and real prices in year n be I_n and V_n, and the year 0 price be V_0. For example, column (2) of Table 5-3 shows Louisiana pulpwood stumpage prices increasing from \$7.10/cord in 1977 to \$13.85 in 1987 in current dollars. To find the *nominal* rate of price growth g_n, use equation (5-9). That gives 6.9 percent per year (subtracting 1 from the tenth root of 13.85/7.10 yields $g_n = 0.069$). This is a nominal rate of increase and includes inflation, since prices are in current dollars. To find the *real* rate of price growth g_r, first find the inflation rate or rate of increase in the CPI between 1977 and 1987 in column (3) of Table 5-3, using equation (5-1). That yields $f = 0.065$, or 6.5 percent inflation. Using equation (5-8), $g_r = (1.069)/(1.065) - 1 = 0.004$, or a 0.4 percent annual real rate of price increase. You could get the same real rate of increase by applying equation (4-4) to the constant dollar prices for 1977 and 1987 in column (4) of Table 5-3. These are the same relationships graphed in Figure 5-1, except that the above values are prices rather than the asset values shown in the figure. Prices are sometimes called "actual prices" if they are nominal, and "relative prices" if they are real.

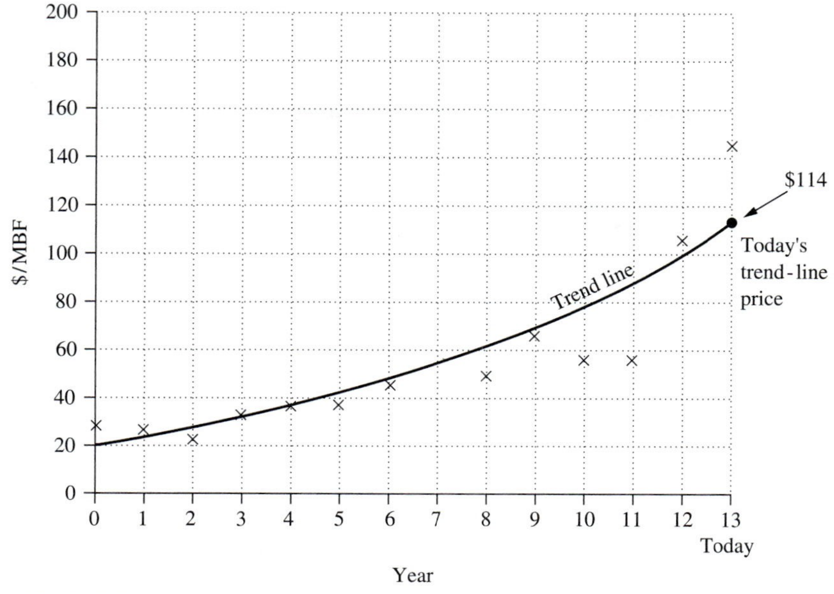

FIGURE 5-2
Trend line through fluctuating nominal stumpage prices. (Exponential regression line fitted to data points. x = actual prices; $r^2 = 0.872$).

As with rates of return, whenever you encounter rates of increase in prices, be sure you know whether they are real or nominal. Until you translate nominal rates into real terms, they are meaningless in terms of purchasing power. If you want to project real prices incorporating a real rate of price-increase at a rate g_r for n years, multiply today's trend-line price by $(1 + g_r)^n$ to get the estimated price in n years expressed in constant dollars of today:

$$\text{Constant dollar price, year } n = (\text{today's trend-line price})(1 + g_r)^n \qquad (5\text{-}10)$$

Figure 5-2 shows the concept of **trend-line** prices. The best fit regression curve through past prices is the trend line from which projections should be made if prices fluctuate. In the case of Figure 5-2, the best fit is an exponential curve rising at about 14.5 percent per year (that's nominal, since the prices are in current dollars). Using the Figure 5-2 example, suppose a landowner wanted to project today's trend-line stumpage price of \$114/MBF 12 years into the future in constant dollars, using a forecasted 2.4 percent real yearly increase. The estimated future price would be $114(1.024)^{12} = \$152$/MBF in constant dollars. Such projections are highly uncertain, but the above approach at least avoids projecting from points that are likely to be peaks or troughs. When figuring average past percentage rates of price-increase, you should also use trend lines when actual prices have fluctuated. For example, in Figure 5-2, the annual nominal rate of increase between the \$50 price 5 years ago and today's price of \$143 is 23.4 percent per year. [Can you reproduce this number? Use equation (5-9).] A more realistic average trend would be 14.5 percent per year, based on the trend line.[5]

If you want to project prices in current dollars, just substitute the nominal rate of price increase into equation (5-10). But be forewarned: That's a fairly useless number unless you know the projected inflation rate. It's best to use real value projections.

HISTORICAL RATES OF RETURN

Given some equilibrium between wage demands of workers and interest demands of capital owners, over time you might expect fairly stable real interest rates, at a given risk level. Then nominal interest rates would fluctuate with inflation according to equation (5-6). As shown in Figure 5-3, things aren't that neat in reality, since inflation is volatile and hard to predict. The figure shows that the **prime interest rate** on new loans varies with inflation as we'd expect. This "prime rate" is the interest rate paid by businesses with the best credit rating. The vertical distance between the prime and the inflation rate—a rough estimate of the real prime rate—is not constant over time. While real interest rates tended to be fairly stable during the 1960s, they were sometimes negative in the mid-1970s—at times when the inflation rate exceeded the interest rate. Later gyrations in the inflation rate caused lenders and other investors to demand higher nominal rates of return, giving rise to unusually high real interest rates for the 1980s, as shown

[5]For information on availability of historical stumpage prices, see Chapter 11, "Appraising Stumpage Price by Comparable Sales."

FIGURE 5-3

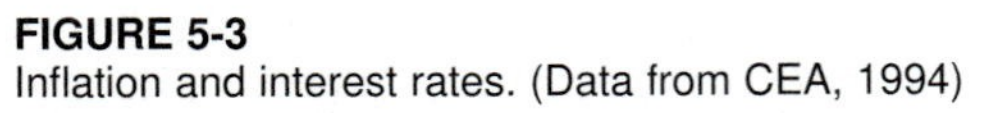
Inflation and interest rates. (Data from CEA, 1994)

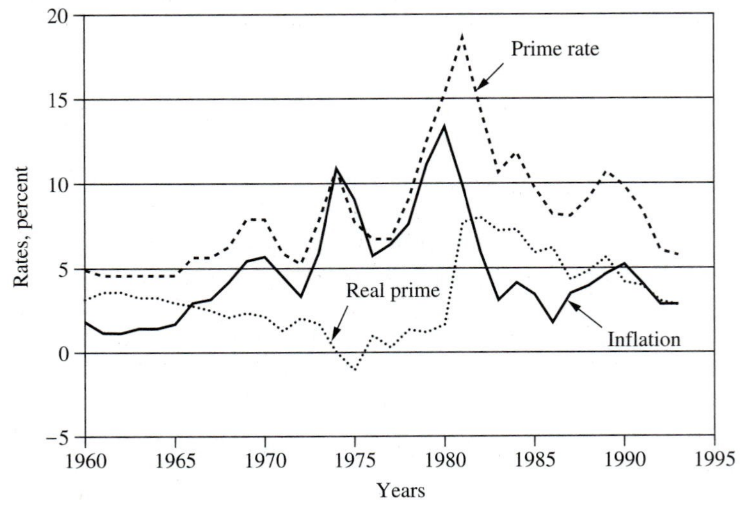

in Figure 5-3 for the prime. That same pattern is seen in interest rates on all loans and bonds—probably because, in the 1980s, investors perceived higher risk, having been burned by unpredictable inflation during the 1970s and early 1980s.

On fixed-interest loans and bonds *already issued,* interest rates will not vary with inflation. Thus, graphs like Figure 5-3 refer only to interest rates on financial instruments *newly issued* in each year.

From Figure 5-3 and similar graphs of other interest rates, we can see that, over time, real earning rates are generally much more stable than nominal rates. But during times when inflation is volatile, real earning rates will also gyrate. Thus, real nonforestry rates of return for a given year won't always be a valid hurdle rate for long-term forestry investments to meet. And averages for several years may also not be valid; for example, the high real prime for the early 1980s is probably an anomaly from a long-term forestry view. When estimating the long-term real earning rate for an investment type and risk level (for example, U.S. bonds, corporate bonds of a given rating, or loans), *you should generally use long-term averages extending several decades back.* For example, in seeking hurdle rates for 40-year investments, look at 40-year historical average real earning rates. And if you're evaluating a prospective 1-year investment, don't look only at last year's rates of return on alternatives. Let's now consider average earning rates for different periods.

Alternative Rates of Return

What are typical real rates of return that you could earn? Table 5-5 lists returns for different investment categories, in order of increasing risk. The rates range from about 1 to 3 percent for low-risk investments to 10 percent or more for higher-risk ventures, before taxes. Note that the higher the risk (variability in returns), the higher the expected rate of return will be. Think of these returns as average rates. Thus, for a risk-free investment, you'll always get the listed return; but on risky ventures, with repeated investment, sometimes you'll get less than the listed rate and sometimes more. *All the Table 5-5 rates are real, excluding inflation; the nominal rates you usually see are higher.* Also, these rates are *before taxes* (after-tax returns would be lower).

The sampling period for rates of return is crucial. In Table 5-5, the long-term real rate of return on a diversified stock portfolio averaged about 5 percent between 1871 and 1973. But between 1983 and 1993 it averaged around 10 percent. From 1960 to 1993, the table shows average real returns to corporate Aaa bonds of around 3 to 3.5 percent, but real returns on Aaa bonds averaged about 5.8 percent for 1983–1993. In the latter period, investors demanded high nominal returns because they were afraid the previous decade's high inflation would return to eat away their earnings. Inflation turned out to be modest from 1983 to 1993, so real rates of return were fairly high.

Bear in mind that the Table 5-5 rates of return are averages and can fluctuate over time. In nominal terms, variations would be even greater, because of

TABLE 5-5
AVERAGE REAL BEFORE-TAX RATES OF RETURN BY CATEGORIES
(Listed in Order of Increasing Risk)

Investment category	Average real before-tax rate of return
U.S. government bonds—These are considered risk-free. For many years, earning rates had been in the 1% range, but recently they have been higher (CEA, 1994; Ibbotsen and Sinquefield, 1982). Simon (1990) suggests 3% as a reasonable long-term expected real risk-free interest rate.*	1 to 3%
Corporate Aaa bonds—These are rated as fairly low-risk bonds but still carry more risk than government bonds.	
1960–1993 (CEA, 1994; Yohe and Karnosky, 1969)	3 to 3.5%
1983–1993 (CEA, 1994)	5.8%
Diversified portfolio of common stocks—For the period 1871–1973 (Bernstein, 1974).	~ 5%
1983–1993 (S&P 500 index, Schwab, 1994)	~ 10%
Returns to U.S. corporate investments—See Stockfisch (1969), Seagraves (1970), Hanke and Anwyll (1980). Nordhaus (1970) estimates this as about 6% *after* taxes.	~ 10%

*The nominal interest rate on long-term U.S. government bonds is guaranteed, and on newly offered issues, the rate rises and falls with the inflation rate. However, since the interest rate is locked in at purchase date, unpredictable changes in inflation will make the *real* rate vary somewhat. Thus, U.S. bonds with maturities of several years have some risk. Three- and six-month Treasury bills are the closest to risk-free, since short-term inflation is more predictable. Government bonds indexed to inflation would be totally risk-free, but these are not offered by the U.S. government.

changing inflation. And there are problems in measuring these rates and gathering the data, especially for corporate investments. So what's the appropriate real rate for evaluating forestry investments? Unfortunately there's no easy answer. The proper discount rate depends in part on the investor's objectives and alternatives and also on the project's risk, or variability in returns. The U.S. Office of Management and Budget recommends that most federal investments should earn the estimated 10 percent real before-tax rate of return in private industries, shown in the last row of Table 5-5 (USOMB, 1972). But as we'll see in Chapter 10, for any given amount of risk (or variability) in expected returns, the proper addition for risk in the discount rate becomes less as the investment period lengthens. So the 10 percent real discount rate may be too high for some long-term forestry investments. The U.S. Forest Service deems investments acceptable down to a 4 percent real rate of return. You can find examples of investors using a wide range of minimum acceptable rates between and beyond these two.

Some economists note that the real 10 percent industrial rate of return is not what the public gives up (the **opportunity cost**) when it spends tax dollars. That's because not all taxes come from industries that would have invested at the 10 percent rate. Thus, some suggest that the federal government's discount rate should be a weighted average of returns on individual savings and business investments. For individuals, the reasoning is that you're indifferent between spending your last dollar or saving it where you'll earn, say, the six-month U.S. Treasury bill rate. The weights would be the proportions of federal taxes raised from individual and corporate income taxes—81 percent and 19 percent for 1993 (excluding levies like social insurance, excise, and death taxes). So, based on an average real six-month T-bill rate of 2.8 percent for 1983–1993 as a personal savings rate and the above 10 percent corporate earning rate, the long-term weighted average opportunity cost of tax revenues would be $0.81(2.8) + 0.19(10) = 4.2$ percent. This is probably a lower limit since many people aren't willing to save much at T-bill rates and prefer higher but more volatile returns on mutual funds. Another suggestion is that the U.S. government's discount rate should be the rate of interest the government pays on borrowed funds—the long-term government bond rate (Lind, 1990), which averaged a real 3.8 percent between 1977 and 1993, for 30-year bonds (CEA, 1994). Over this period, the government's long-term borrowing rate was abnormally high; in earlier years it was lower.

INTERPRETING RATES OF RETURN

In doing discounted cash flow analyses or evaluating rates of return for forest owners, one of the first questions to ask the decision maker is, "What *real* rate of return do you wish to earn?" This becomes the real **minimum acceptable rate of return** (MAR) in present value calculations. Many investors will cite published returns available on opportunities they feel have risks similar to the venture in question. For example, they might aim for the current return on Baa rated bonds, an average risk rating, not the best (the highest rated bonds, Aaa, would offer a lower interest rate). First you must realize that's a nominal rate based on people's expectations about inflation. You can only speculate what the real rate of return will be. If the return on Baa bonds or any other type of investment is to be the MAR, it would be best to translate *historic* nominal returns to real rates using past inflation rates [equation (5-8)]. Although most investors' discount rates are based on available rates like those in Table 5-5, some investors will have MARs *exceeding* widely available rates.

The "Return on Investment"

A major problem in translating some desired earning rates to real terms is that often you don't know how much inflation is in the cited rate. One example is the *return on investment* or ROI often calculated by businessmen. The ROI, sometimes called the "return on total assets" or "accounting rate of return," is a firm's net income for a given year divided by the value of its assets:

$$\text{ROI} = \frac{\text{firm's net income for year}}{\text{asset value, including debt}} \tag{5-11}$$

The ROI is designed to be a rate of return in the same way that the rate of return on money left in a savings account is a year's interest divided by that year's account value. Debt is included in the denominator to reflect all assets, in the same way that a home value includes the value of the mortgage. The ROI only gives a meaningful real rate of return if income is fairly stable and the denominator is current **market value** (the likely price received if the assets were offered for sale). The problem with the above ROI formula is that asset value is sometimes understated, thus making the ROI too high. Firms usually report asset values as *original purchase cost* (or so-called **book value**) of capital such as buildings, land, timber, and other resources. But often such assets were purchased many years ago and are increasing in value. To account for the wearing out of certain assets, **depreciation** is deducted in the denominator of equation (5-11). But depreciation is sometimes deducted faster than it actually occurs, again understating asset value.

Equations for ROI don't always understate asset values, but they often do, especially with forest products firms owning large timber volumes purchased many decades ago at prices far below current market value. An extreme example is Weyerhaeuser Company's timber that was listed at a book value of $612 million in 1989, while one market value estimate was $5.5 billion (Zinkhan et al., 1992). Another study estimated that, for 20 large United States forest products firms, estimated market value of timber was 9.5 times the stated book value (Clephane, 1978). In such cases, since annual income is in current dollars, the numerator in equation (5-11) contains more inflation over time than the denominator, and the ROI can greatly overstate the real return to capital. For timber companies, the overstatement is larger than that caused by inflation alone because usually timber value has also risen in real terms.

You can't easily translate an ROI [equation (5-11)] into real terms. A firm's assets are acquired over a number of years, so you cannot know by how much the denominator understates current market value. Thus, you can't tell how much inflation, if any, is in the ROI. If assets in equation (5-11) are far enough below market value, the ROI could include much more inflation than the projected annual inflation rate. Nevertheless, managers in the forest industry sometimes suggest ROIs from annual reports as hurdle rates that forestry investments should meet. These rates are often 10 percent or higher. Foresters dealing with real rates of return often have trouble showing that forestry investments can meet such hurdle rates. You can't solve the problem by simply adding some projected inflation rate to a real forestry rate of return, or subtracting inflation from the ROI, because the amount of inflation in the ROI is unknown. An added problem is that the ROI chosen as a hurdle rate may be from a year with high income and may not be a typical average.

Foresters tend to deal with **internal rates of return** [as defined in equation (4-4) and treated in more detail in the next chapter]: these are fundamentally different from ROIs quoted by accountants. The route to reconciling the two measures lies not in massaging data and assumptions until forestry returns look higher. Both groups need to realize that sometimes they're dealing with profit measures that aren't comparable. In making comparisons, analysts should give forestry and nonforestry returns as true internal rates of return. Furthermore, to compare forestry and nonforestry rates of return, either both must be real or both must include the same amount of inflation, and risks must be similar. This is easier said than done, since misunderstanding over the issue abounds.

Related Ratios Measures such as the ROI above are often called "profitability ratios." Some analysts will compute an ROI where the net income includes interest on debt. The reasoning is that interest is an income to the debt portion of the assets. Although not typically shown in firms' annual reports, this form of ROI would therefore exceed that shown in equation (5-11), for the usual case where funds have been borrowed.

Another return ratio is the *return on equity* (ROE), where the denominator in equation (5-11) includes only ownership and not debt. Thus, ROEs, which are often 15 percent or more, generally exceed ROIs. You can see why profitability ratios must be carefully interpreted and usually aren't suitable guidelines for investment hurdle rates.

COMPUTING PRESENT VALUES IN INFLATIONARY TIMES

If you discount a future nominal value with your nominal MAR, you should get the same **present value** as you would by discounting the real value with your real MAR. In nominal terms, your future value is I_n, and your interest rate is i. Calculating in current dollar terms, the present value of a future payment is:

$$\text{PV} = \frac{I_n}{(1+i)^n} = \frac{V_n(1+f)^n}{(1+r+f+rf)^n} = \frac{V_n(1+f)^n}{(1+r)^n(1+f)^n} = \frac{V_n}{(1+r)^n} \quad (5\text{-}12)$$

In the second segment of equation (5-12), both numerator and denominator are in nominal terms. In the third segment, the numerator is the same as I_n from equation (5-3), and $r+f+rf = i$ from equation (5-6), given inflation at f. In the fourth segment, the terms $(1+f)^n$ cancel out, making equation (5-12) equivalent to the present value of a real payment as given in Chapter 4:

$$\text{PV} = \frac{V_n}{(1+r)^n} \quad (5\text{-}13)$$

To illustrate, suppose you project a real timber revenue of $15,000 in 15 years. If your real MAR is 5 percent, your present value, calculated in real terms, is:

$$\text{PV} = 15{,}000/(1.05)^{15} = \$7{,}215.26$$

Now assume you project 6 percent annual inflation. To calculate present value in nominal terms, first figure your nominal MAR with equation (5-7):

$$i = (1.05)(1.06) - 1 = 0.113, \text{ or } 11.3\%$$

Then inflate the real income to current dollars in 15 years, given $f = 0.06$, using equation (5-3):

$$I_n = 15{,}000(1.06)^{15} = \$35{,}948.37$$

Calculating present value in nominal terms, using the left side of equation (5-12):

$$\text{PV} = 35{,}948.37/(1.113)^{15} = \$7{,}215.26$$

This is exactly the same as the present value computed in real terms. And this makes sense because inflation represents no real gain in purchasing power.

The foregoing leads to two important guidelines for computing **net present value (NPV):**

1 *If future cash flows are in constant dollars (in real terms), compute NPV with a* ***real*** *interest rate.*
2 *If future cash flows are in current dollars (nominal, or including inflation), compute NPV with a* ***nominal*** *interest rate. But make sure you use the same inflation rate in the cash flows and the interest rate.*

You must never mix constant dollars and current dollars in the same equation. Whether you use guideline (1) or (2), present value should be exactly the same, given the same real discount rate. Usually it's simpler to work in real terms (method 1). But one warning applies: For present values after income taxes, the same guidelines hold, but you need to project a reasonable inflation rate because inflation will reduce after-tax present values.

Inflation and Income Taxes

Chapter 9 covers taxes, but this section explains why inflation must be included in discounted cash flow analyses after income taxes. Figure 5-4 shows the growth, over time, of an asset like timber purchased for $C = \$2{,}000$ and sold $n = 8$ years later for $I_n = \$8{,}000$ in current dollars. Under United States income tax laws, at sale date the seller must pay taxes on the **capital gain**, which is the difference between the sale price and the purchase cost, or **basis** in tax lingo. The tax, called a capital gains tax at a rate T_i, is $T_i(I_n - C)$. We say the basis is *deducted* from asset sale price. Regardless of the inflation rate, the deductible basis C remains the same, under current law. The effect is to make the income tax more burdensome as inflation increases. For example, in Figure 5-4, annual inflation at about 9.05 percent doubled the asset value from a real $V_n = \$4{,}000$ to $I_n = 4{,}000(1.0905)^8 = \$8{,}000$, but it tripled the taxable capital gain from \$2,000 to \$6,000. This causes the present value to decrease as inflation increases.

To show the above effect numerically, the following equation gives a buyer's present value of after-tax income from the asset in Figure 5-4, assuming only the

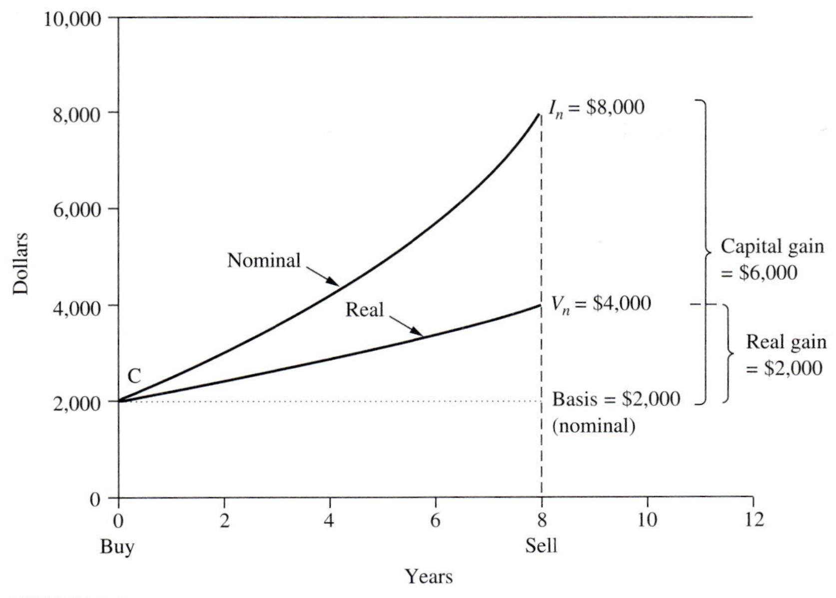

FIGURE 5-4
Inflation and taxable capital gain.

given revenue and cost. Following guideline (1) above, calculations are in real terms. Since the tax is a cost to the buyer, it is entered as a negative value in the numerator.

$$\text{After-tax net present value} = \text{NPV}$$

$$\text{NPV} = \frac{V_n - T_i\left(V_n - \dfrac{C}{(1+f)^n}\right)}{(1+r)^n} \tag{5-14}$$

$$\text{NPV} = \frac{V_n - T_iV_n + T_i\dfrac{C}{(1+f)^n}}{(1+r)^n}$$

Although C is given as the purchase cost, NPV is the maximum amount a buyer could pay for the asset and still earn a rate of return of r after taxes. The NPV may equal or exceed C; if NPV $< C$, the asset will not earn the rate of return r. Note that the basis C in this equation is divided by $(1 + f)^n$ to deflate it to constant dollars. You might think that since C occurred in year 0, it already was in constant dollars. But when C is deducted in year n, its value is in current dollars of that year. Thus, to keep all values in constant dollars, C must be deflated. Inflation cannot cancel out of equation (5-14) as it did before taxes in equation (5-13).[6] Thus, present value will decline with increasing inflation because the tax savings from deducting C become less in real terms [the real savings are $T_iC/(1 + f)^n$].

Let's assume the numerical example in Figure 5-4 is timberland that you plan to purchase. Suppose you want to earn a 5 percent real rate of return after taxes, you project inflation at the indicated 9.05 percent, and you'll have to pay combined state and federal income taxes at 30 percent of the capital gain when the timber is sold in 8 years. What would be your maximum willingness to pay for the property? Substituting these values into equation (5-14), calculating in real terms:

$$\text{NPV} = \frac{4{,}000 - 0.30[4{,}000 - 2{,}000/(1.0905)^8]}{(1.05)^8} = \$2{,}098$$

This means you'd be willing to pay up to $2,098 for the property, if all your projections were correct, given no other costs or revenues.[7] If you projected no inflation, present value would be $203 higher:

$$\text{NPV} = \frac{4{,}000 - 0.30(4{,}000 - 2{,}000)}{(1.05)^8} = \$2{,}301$$

Thus, for present values after income taxes, projecting too little inflation will overvalue the asset and could lead to bidding too much for properties. The reverse is true if you overestimate inflation. On large timberland purchases that often run into millions of dollars, such errors can be major. The percentage of error depends on the real discount rate, the number of years until income occurs, the size of the basis, and the amount by which inflation is misstated. No one can predict inflation perfectly, but the key issue is that, after income taxes, present value calculations should include a reasonable inflation projection. Without taxes, say, for government investments, it doesn't matter how much inflation is projected; as long as guideline (1) or (2) above is followed, present value will be the same.

The above negative impact of inflation on present values occurs whenever a time lapse exists between the occurrence of a cost and the date it is deducted for

[6] You can reach the same present value by computing in nominal terms, but it is more complex:

$$\text{NPV} = \frac{V_n(1 + f)^n - T_i[V_n(1 + f)^n - C]}{(1 + r)^n(1 + f)^n}$$

Dividing both numerator and denominator by $(1 + f)^n$ will result in equation (5-14). To avoid arbitrary tax-increases caused by inflation, some have proposed that the deductible basis be increased to account for the inflation occurring over the holding period, i.e., multiplying C by $(1 + f)^n$ in the above equation. In that case inflation would cancel out of the equation and could not affect the present value. This is called "basis indexing," which has been proposed several times in Congress but never passed.

[7] This is a slight underestimate, since paying that much would give you more than the $2,000 basis to deduct and would reduce your taxes. The correct present value can be reached by replacing the basis with the net present value and iteratively solving until net present value equals the basis.

income tax purposes. Another example is deducting depreciation—the allowance for wearing out of certain assets like logging equipment, buildings, bridges, culverts, and temporary roads. In such cases the depreciation deduction is a portion of a depreciable asset's purchase cost C deducted for a given number of years. In a constant dollar present value calculation, the tax savings from depreciation deducted in any year must be deflated and entered as $+T_i(\text{depreciation})/(1+f)^n$, where n is the number of years since the asset was purchased. That's the same procedure explained above for treating the tax savings from basis deductions.[8]

For most types of taxes, other than the income tax, inflation will have no impact on present value calculations. This holds for property taxes, yield taxes on harvest revenue, and death taxes, all of which are discussed in the chapter on taxation.

Potential Present Value Errors

Present values after income taxes will be overstated if you fail to include inflation when a tax deduction is taken later than the expense occurred. An example is given above.

One of the most common discounting errors is to project cash flows in constant dollars and to use a nominal interest rate. In that case the interest rate is too high and present value too low. Suppose Ms. Bois projects a sawtimber harvest of 100,000 board feet in 10 years from her woodland. Starting with today's trend-line stumpage price in her area and adding a U.S. Forest Service projection of a 1 percent per year real price increase, she projects a stumpage price of $100/MBF. [She multiplied today's trend-line price by $(1.01)^{10}$.] This means a real harvest revenue of $100(100 MBF) = $10,000 in 10 years. Suppose she chooses a current 9.5 percent interest rate on Baa bonds as her minimum acceptable rate of return on an investment of similar risk. She incorrectly computes the present value of the harvest as:

$$\text{PV} = \frac{10{,}000}{(1.095)^{10}} = \$4{,}035$$

Since the interest rate is from the market, it's nominal and includes a projected inflation rate that we'll assume to be 4 percent. But her price projection is in real terms. Thus, the denominator in the above equation has more inflation than the numerator, and her present value is too low. To correct the error, translate Ms. Bois' nominal 9.5 percent MAR to real terms with equation (5-8): $1.095/1.04 - 1 = 0.053$, or 5.3 percent. Now, assuming this real interest rate is acceptable, the

[8]If inflation is high enough, an income tax can leave you with no real income. For example, suppose a corporation paying a 38 percent combined state and federal income tax rate invests $1 for one year in an asset that shares in the inflation rate. With a real 5 percent growth rate and 10 percent inflation, the asset is sold for $1(1.05)(1.10) = \$1.155$ at year's end. The tax is $0.38(1.155 - 1) = \$.0589$, leaving $1.155 - 0.0589 = \$1.0961$ after taxes. That's a 9.6 percent nominal after-tax rate of return—less than the inflation rate, so there's no real gain! Without inflation, the after-tax rate of return is always positive, as long as the before-tax rate of return is positive.

correct present value is 48 percent higher:

$$PV = \frac{10{,}000}{(1.053)^{10}} = \$5{,}966$$

This is a simple example that excludes taxes and several other factors, but it shows an error that can be significant when investors use present values to evaluate projects or compute bid prices for properties, especially when millions of dollars are involved.

Consider another present value error: Suppose Mr. Holtz is growing pulpwood in Louisiana. He multiplies a projected harvest volume by today's trend-line stumpage price to estimate a $10,000 harvest income in 10 years. He is aware that both income and the interest rate must be in nominal terms. So, from the current dollar stumpage prices in Table 5-3, which had been increasing at 6.56 percent per year between 1970 and 1987, he projects the same nominal rate of increase into the future. Using a current Baa bond rate of 9.5 percent (nominal) based on economic projections of 4 percent inflation, he computes the following present value:

$$PV = \frac{10{,}000(1.0656)^{10}}{(1.095)^{10}} = \$7{,}617$$

His $10,000 harvest, which is based on today's stumpage prices, is multiplied by $(1.0656)^{10}$ to project the value in nominal dollars. Although everything's in nominal terms, we don't know if there's the same amount of inflation in the cash flow as there is in the interest rate. The projected inflation in the 9.5 percent interest rate was 4 percent (making the real MAR 5.3 percent). But from the CPI in Table 5-2, you can compute the inflation rate at 6.5 percent between 1970 and 1987 (the sampling period for the price projection). Thus, assuming the past real price trend is reasonable to project, the inflation in the cash flow is 6.5 percent and exceeds the inflation in the discount rate. In nominal terms, that makes the discount rate too low and the present value too high. To compute the correct present value in nominal terms, incorporate the 6.5 percent inflation into the assumed real MAR of 5.3 percent. That yields a nominal rate of $1.053(1.065) - 1 = 0.12145$, or 12.145 percent. Using that interest rate in the above calculation will yield the correct present value of $6,000 for this last scenario (try it!). Values are carried to several places to show equivalence with another approach below.

It might have been less confusing for Holtz to calculate present value in real terms. He could have projected real stumpage prices by deflating the 6.56 percent historical nominal rate of increase with the 6.5 percent inflation rate over that period: $(1.0656/1.065) - 1 = 0.00056$, or a 0.056 percent real increase per year. Thus the real future harvest is $(10,000)(1.00056)^{10} = \$10,056$, which is nearly constant in real terms. Discounting that with the real 5.3 percent MAR also gives the correct $6,000 present value (try it!).

Just because both cash flows and the interest rate are in nominal terms doesn't mean that the same inflation rate is in both. For example, from 1978 to 1982, inflation averaged nearly 10 percent, so that nominal price projections based on that period would have included about 10 percent inflation. But by 1983, interest rates on short-term government bonds were less than 9 percent, including an expected inflation component of about 6 percent. This is another reason why it's better to compute present values in real terms: it forces you to address the question of inflation and get consistency between cash flows and the interest rate.

Nonmonetary Values

The above concerns about correct present values during inflationary times are just as important when dealing with outputs like hiking or bicycling on public lands where no price is charged. Public agencies often compute present values of management plans including estimates of willingness to pay for unpriced goods (later we'll note ways to make these estimates). Suppose Marilyn, a U.S. Forest Service analyst, has found that willingness to pay for hiking is $10 per visitor day. One approach would be to project this as a real value for net present value (NPV) calculations, using a real discount rate, assuming that as incomes rise with inflation, willingness to pay would also rise at the inflation rate in nominal terms. That may not be too far from the mark, but it would be best to know whether the real willingness to pay might change. Marilyn might look at historic prices for private outdoor recreation services such as fee fishing and hunting, campgrounds, or parks. Expressing these in constant dollars [with equation (5-4)], she could measure the average real rate of price-change and apply it to the hiking value projection as a best guess, say for 5 or 10 years, but not in perpetuity.

INFLATION AND PAYMENT SERIES FORMULAS

In Table 4-3, recall that the compound interest formulas for payment series require the recurring payment p to be fixed. Therefore, if nominal payments rise at exactly the inflation rate, they are fixed in real terms and you can use the present value or future value formulas with a real interest rate as shown in Table 4-3. However, *if payments are fixed in nominal terms, you must use a nominal interest rate (i) in the series payment formulas,* making sure the same amount of inflation is in the cash flows and the interest rate. This is indicated in the notes to Table 4-3 rather than rewriting all the equations, replacing r with i. Examples are given below.

If payments are not fixed in either real or nominal terms, but they increase or decrease geometrically in real terms, Appendix 4B shows how the fixed payment formulas can be adjusted to compute correct present or future values.

Payments Rising at the Inflation Rate

Consider a series of annual sawtimber harvests of equal volumes in an area where stumpage prices are increasing at the inflation rate. In that case, the increasing nominal harvest income in any year n is the beginning value times $(1 + f)^n$.

But the real value in any year n, in constant dollars of year 0, is the current dollar value divided by $(1+f)^n$, as in equation (5-2). Since we've multiplied and divided by the same factor, the series of annual incomes is fixed in constant dollars. Thus, to compute present value, you can use equation (4-10) with a real interest rate, as shown in Chapter 4. The same reasoning would hold for any of the regular series present or future value formulas in Table 4-3: *When nominal payments rise at the inflation rate, they are fixed in real terms, and you must use a real discount rate.*

Payments Fixed in Current Dollars

An example of payments remaining fixed in nominal terms would be the value of equal pulpwood harvests in an area where stumpage prices remain fairly stable regardless of the inflation rate. Suppose such harvests are projected at $8,000 per year in current dollars over a 10-year period with average inflation of 4 percent per year. What is the present value in year 0 for this terminating annual series, using a 5 percent real discount rate? Since the payments are nominal, the discount rate must be changed to nominal terms, including the 4 percent inflation, using equation (5-7): (1.05)(1.04) − 1 = 0.092 or 9.2 percent. Using equation (4-10), with the nominal discount rate, present value is:

$$8{,}000[1 - (1.092)^{-10}]/0.092 = \$50{,}892$$

Had a real discount rate been used, present value would have been overestimated, since the *real* payments are decreasing.

Another example of equal payments in nominal terms would be installments on a fixed payment loan. To calculate annual fixed payments on a loan, use the capital recovery multiplier [equation (4-12)], letting the loan amount be V_0, and making sure to use the *nominal* loan interest rate because the payments are fixed in nominal terms regardless of the inflation rate. For example, on a 10-year, $50,000 loan, at 12 percent nominal interest, annual payments would be:

$$\text{Annual loan payment} = (\text{loan payment})\left[\frac{i}{1-(1+i)^{-n}}\right] \qquad (5\text{-}15)$$

$$\text{Annual payment on a \$50,000 loan} = 50{,}000\frac{0.12}{1-(1.12)^{-10}} = \$8{,}849$$

For any payment series fixed in nominal terms, you can use the present or future value formulas of Table 4-3, making sure to use a nominal interest rate incorporating the average inflation rate for the relevant period.[9]

[9]Chapter 11 warns that the *borrower's* present value of loan payments should be calculated with a risk-free, nominal discount rate lower than the risky rate, since payments are fixed with certainty according to the loan contract.

COMPARING FORESTS WITH FIXED-INCOME INVESTMENTS

Suppose you wish to compare a forest's rate of return with that on a **fixed income investment** like a bond, where annual income is fixed in current dollars. Then the forest's earning power is often underestimated. Suppose a 10-year corporate bond yielding 9.2 percent interest is bought for $100,000. That would yield a fixed annual interest income of $9,200 for 10 years, after which the $100,000 original value would be returned to the buyer, *all in current dollars.* Assume further that this is a moderate risk bond and inflation is 5 percent, in which case the real return is $1.092/1.05 - 1 = 0.04$, or 4 percent.

Now, compare the above bond with an uneven-aged forest of similar risk, earning a real 4 percent rate of return. Suppose the forest's market value also starts at $100,000; and it can, through selective cutting, yield equal annual sawtimber harvests in perpetuity, while the standing inventory stays constant. Each year's harvest is worth $0.04(100{,}000) = \$4{,}000$ in constant dollars of year 0. Although both the forest and the bond earn the same 4 percent real rate of return, at first glance the bond looks much better with its $9,200 annual income. But this isn't necessarily true, because *the forest is not a fixed income asset.* On average, stumpage values in much of the United States have tended to at least keep up with the inflation rate (Washburn and Binkley, 1993). Suppose, in this example, stumpage prices rise at exactly the assumed inflation rate of 5 percent. That means *both the annual forest income and the entire forest value rise over time at the inflation rate (here 5 percent) in current dollars while the bond income and ending value remain fixed.* Thus, I'll call the forest a **variable income investment**: its income and value tend to rise with inflation. For this reason, many investors view timberlands as a good hedge against inflation. Other assets that tend to be "variable income investments" include agricultural properties and natural resources in general.

Figure 5-5 shows the annual incomes and ending asset values for the above bond and forest in current dollars. Now you can see how the forest makes up for initially lower income by increased nominal earnings and asset value over time. By rising at 5 percent annually, nominal forest income increases from $4,200 at the end of year 1 [or 4,000(1.05)] to $6,516 in the tenth year [or $4{,}000(1.05)^{10}$]. Over the 10 years, the forest value has increased from $100,000 to $100{,}000(1.05)^{10} = \$162{,}889$ in current dollars. In constant dollars, the bond has declining income and asset value, while the forest has constant asset value and earnings. In order to make bond-based endowment funds behave more like the forest in Figure 5-5, managers often withdraw funds at a real "take-out" rate, say, 4 percent, and reinvest the rest, thereby allowing the endowment value and income to increase at the inflation rate.

Since variable income assets such as forests compete with fixed income assets, it's extremely important to correctly interpret earnings patterns from the two types of investments. In sustained yield forests where cut equals growth, annual income divided by asset value tends to yield the *real* rate of return, *regardless of the inflation rate.* For example, for the forest in Figure 5-5, in year 1, the real rate of return is $4{,}200/105{,}000 = 0.04$; in year 10, it's $6{,}516/162{,}889 = 0.04$;

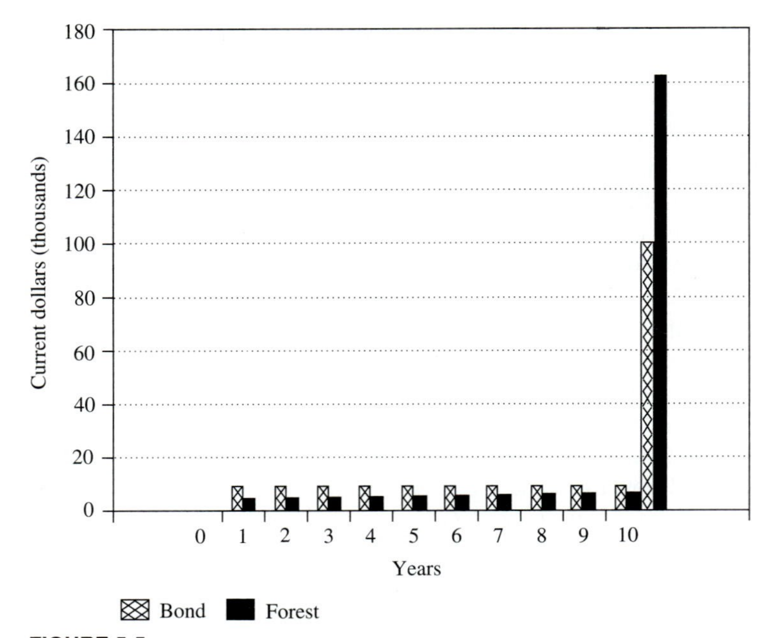

FIGURE 5-5
Bond vs. forest, nominal cash flows.
(Inflation rate is 5 percent. Both bond and forest earn a 9.2 percent nominal and a 4 percent real rate of return. In current dollars, bond revenues are fixed, and forest revenues rise at the inflation rate.)

and the same holds for other years. As already noted, nominal rates of return on newly offered, fixed income investments such as bonds will rise and fall with the expected inflation rate. Nominal and real rates are not comparable. If you consider the ending value of the forest in Figure 5-5 as an income, you'd find that incomes from the forest and the bond have the same present value, using the 4 percent real interest rate (that is one of the exercises at the end of the chapter). They are equally good investments, although more of the bond's income occurs earlier.[10]

The apparent discrepancy between bond and forest earning rates occurs only during inflationary times. If expected inflation is zero, nominal bond earning rates would tend to be the real rate, and the incomes for the bond and forest, plotted as in Figure 5-5, would be the same, given equal risk. But when inflation rates are high and interest rates on new bonds soar—in the United States, they rose well over 15 percent in the early 1980s—there may be short periods when forest outputs and values don't rise as fast as the inflation rate. Then bonds may have the edge over forests. But in the long haul, sawtimber stumpage prices have tended to rise at least as fast as the inflation rate. Make sure you understand this section, because during inflationary times, it can help to convince skeptical investors

[10]With fixed income investments (FIIs), an increase (decrease) in originally anticipated inflation can reduce (increase) the present value of an existing FII. This occurs because higher (lower) inflation will make the FIIs real income less (greater) than originally expected, thus making FII real incomes more variable than they may initially appear.

that well-managed forests often earn more than seemingly high-yield bonds of equal risk.

The above analyses also apply to forests yielding nontimber outputs. These too are "variable income assets": the willingness to pay for, say, recreation services tends to increase with the inflation rate. Thus, the recreation asset value of a forest also rises with inflation in current dollars. Using the same numbers as before, suppose the above forest yielded $4,000 worth of annual recreation benefits without harvesting. The real rate of return, using the above procedure (income divided by asset value), would also be 4 percent regardless of the inflation rate. The recreation forest earning a 4 percent real return could compete with a bond earning the same real rate but initially yielding income at a higher percentage of asset value during inflationary times.

INFLATION AND HOUSING DEMAND

By boosting mortgage interest rates, inflation can increase the cost of buying a home and can dampen housing demand. One of the main uses for solid wood products is housing. Thus, housing cycles cause wide swings in lumber and plywood manufacturing, in logging, and even in the hiring of foresters. It's important to understand how inflation influences these cycles, since they so strongly affect timber management and wood processing. Also, you, as a future or current home buyer, will probably be affected by changes in mortgage interest rates.

We can use the prime interest rate on new loans as a combined proxy for interest rates on builders' construction loans and home buyers' mortgages. Figure 5-6

FIGURE 5-6
Housing starts and interest rates. (Data from CEA, 1994)

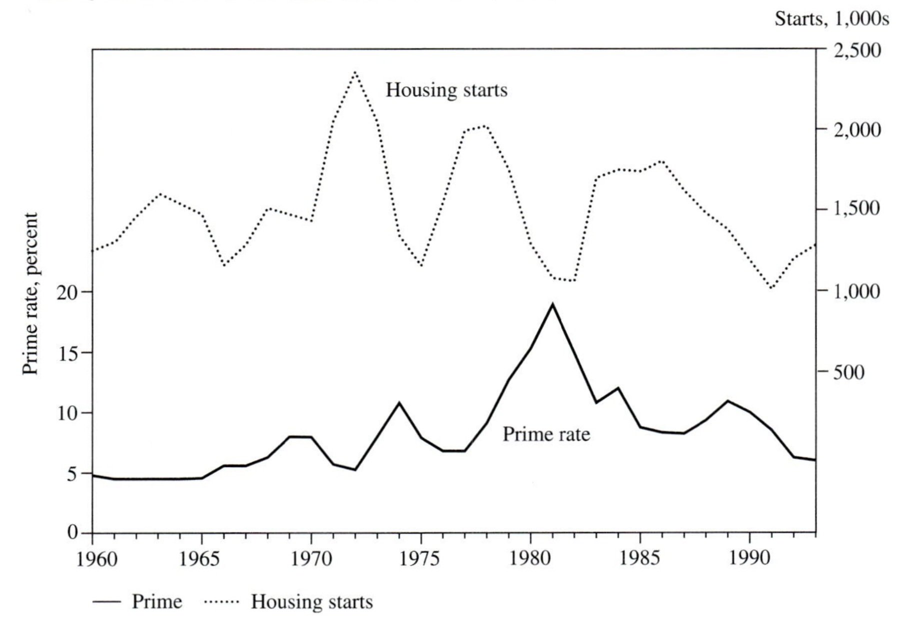

TABLE 5-6
MORTGAGE PAYMENTS, DOLLARS PER MONTH (FIXED PAYMENT LOAN)
(Mortgage = $80,000)

Loan period, years	Interest rate, %							
	4	7	10	13	16	19	22	25
5	1473.32	1584.10	1699.76	1820.25	1945.44	2075.24	2209.51	2348.11
10	809.96	928.87	1057.21	1194.49	1340.10	1493.38	1653.58	1819.94
15	591.75	719.06	859.68	1012.19	1174.96	1346.30	1524.60	1708.42
20	484.78	620.24	772.02	937.26	1113.00	1296.55	1485.65	1678.57
25	422.27	565.42	726.96	902.27	1087.11	1278.14	1472.99	1670.10
30	381.93	532.24	702.06	884.96	1075.81	1271.11	1468.79	1667.66

shows how housing starts fall when interest rates rise and vice versa. When interest rates are high, payments are greater on new home loans, and home sales are depressed. In Figure 5-3 you saw that interest rates rise and fall with inflation. Therefore, construction activity moves in the opposite direction from the inflation rate.

A large percentage of new home buyers borrow much of the cost with a **fixed payment loan** (mortgage). Based on the current interest rate, monthly payments are fixed for the duration of the loan. Table 5-6 shows monthly payments for an $80,000 mortgage at selected interest rates and loan periods (the next section gives the payment formula). Most buyers choose loan periods from 20 to 30 years in order to keep payments lower. Compared with shorter loan periods, the disadvantage is that they pay more interest over time, and it takes longer to pay off the loan. Note how monthly payments rise as interest rates increase: At 7 percent interest, the payment on a 30-year mortgage is $532.24, while at 13 percent, it is $884.96, or 66 percent higher. At shorter loan durations, the increase is less dramatic; for example, with a 5-year automobile loan, the same interest rate-increase will boost monthly payments by only 15 percent. That explains why demand for products like autos and appliances isn't as sharply affected by high interest rates as housing demand. But, in general, sharply increased inflation can severely dampen consumption and investment, economywide, because borrowing gets so much more expensive.

Table 5-6 shows that as inflation drives up interest rates, many home buyers are forced out of the market by the prospect of high monthly payments. When interest rates rise, the regional demand curve for housing shifts to the left, as shown in Figure 5-7. With higher interest rates, at any given average price per home, fewer homes are bought, because monthly payments on new mortgages are higher. Higher interest rates are also likely to shift the housing **supply** curve to the left, because construction loans become more costly. But the shift is likely to be less dramatic than the demand shift because interest rates have a smaller impact on short-term construction loan payments than on long-term mortgage payments (see Table 5-6). At the new equilibrium $P'Q'$, fewer homes are built, and *real* home prices may or may not be higher (depending on the relative demand and supply shifts and elasticities). *Nominal* home prices often rise with increasing

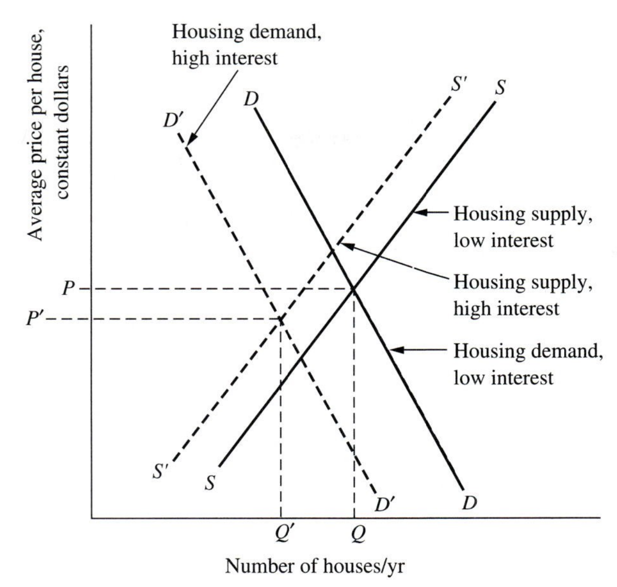

FIGURE 5-7
Hypothetical regional housing demand with high and low interest rates.

interest rates, because of the accompanying inflation, but note that the price axis in Figure 5-7 is in constant dollars.

The positive relationship between inflation and loan payments occurs mainly for *newly offered* loans. Most loans have fixed payments and fixed interest rates, so changes in the inflation rate usually won't affect payments on *existing* loans.

Two polar solutions to the inflation problem are (1) refine monetary and fiscal policy to eliminate inflation and (2) learn to live with inflation by indexing key variables to the inflation rate. The most realistic course probably lies between the two. We've been successful at avoiding persistent high inflation in the United States. But the continuing inflationary surges are enough to aggravate housing cycles, which cause economic pain in the forest products sector and elsewhere. Too, inflation tends to persist at moderate levels because it's generally written into wage contracts and budgets.

While not solving inflation, one way to dampen housing cycles is to offer **inflation-indexed mortgages** whose initial payments are based on real interest rates, with subsequent payments being adjusted upward by the inflation rate.[11] During inflationary times, initial payments are thus much lower than those on conventional loans, and housing becomes more affordable. Although not without problems, variations of such loans have been successful in dampening housing

[11] See Klemperer (1985) and Modigliani (1989). This is not the same as a "variable rate mortgage" (VRM) where the interest rate and payments on an existing loan can vary with the inflation rate. When inflation exists, initial interest rates on VRMs are lower than on fixed payment loans but not as low as on an inflation-indexed loan.

cycles in Brazil, Sweden, and elsewhere; but they've not been widespread in the United States.

Calculating Monthly Loan Payments

You can figure the monthly loan payments shown in Table 5-6 by modifying the annual loan payment equation (5-15). As before, the interest rate would be nominal (i) because loan payments are fixed in current dollars (when $f > 0$, payments decline in real terms). Since the period length is now a month rather than a year, you need a monthly interest rate, compounded monthly, or $i/12$. The number of periods and payments is $(12)n$, since n = years. Adapting equation (5-15):

$$\text{Monthly payment} = (\text{loan amt.})\frac{i/12}{1-(1+i/12)^{-12n}} \tag{5-16}$$

With equation (5-16), the bank sets loan payments so that their present value equals the amount loaned. That way the bank will earn the nominal interest rate i. For the above $80,000 mortgage, suppose the interest rate is 10 percent (monthly interest of $0.10/12 = 0.008333$), and the loan life is $n = 25$ years or 300 months. Using equation (5-16), the monthly payment is:

$$80{,}000\frac{0.008333}{1-(1.008333)^{-300}} = \$726.94$$

as in Table 5-6, accounting for rounding error. You can use this approach to figure payments on other types of borrowing like car loans. Note that, because the interest is compounded monthly rather than annually, the effective annual interest rate is $(1.008333)^{12} - 1 = 0.1047$, or 10.47 percent, not the stated 10 percent.[12] Table 5-6 is another useful spreadsheet application where each cell contains equation (5-16), which references cells containing interest rates, loan durations, and the loan amount—here set at $80,000. That way you can figure loan payments for any number of interest rates, loan durations, and loan amounts.

KEY POINTS

- Improper accounting for inflation has often caused major errors in forestry investment analysis.
- Inflation is measured by annual percent changes in the consumer price index or other price indexes.

[12]This difference between actual and effective interest is due to the incorrect but common convention of dividing the annual interest rate by the number of periods in a year to get the periodic interest rate. Given an annual interest rate of i_a, a periodic interest rate of i_p, and p periods per year, the correct periodic interest rate is found by solving the following equality for i_p (but don't expect bankers to do this):

$$(1+i_p)^p = (1+i_a)$$
$$i_p = \sqrt[p]{1+i_a} - 1$$

- Current dollar values can be deflated to arrive at constant dollar values that exclude inflation.
- Nominal rates of return and rates of value-change are based on current dollar (nominal) values; real rates of return and rates of value-change are based on constant dollar (real) values.
- Nominal interest rates tend to rise and fall with the inflation rate, while real rates of return are more stable.
- Long-term real before-tax rates of return in the United States have ranged from 1 to 3 percent on risk-free bonds to about 10 percent on corporate investments. The higher the rate, the greater the risk. One estimate of a reasonable expected real risk-free rate in industrialized countries is 3 percent.
- It's difficult to know how much inflation is in an "accounting rate of return," or ROI. The ROI is usually not comparable to the internal rate of return commonly applied in forestry.
- To correctly compute present values, express cash flows in constant dollars and use a real interest rate. Alternatively, let cash flows include projected inflation and include the same rate of inflation in the discount rate.
- If the above two guidelines aren't followed, large errors in present values can occur, especially with long-term forestry projects.
- In present value calculations after income taxes, include your best estimate of future inflation.
- Forests are "variable income investments" where income and asset value tend to increase with the inflation rate.
- When inflation increases interest rates on new mortgages, housing demand declines. This is a major cause of housing cycles that plague the forest products industry.

QUESTIONS AND PROBLEMS

5-1 Suppose you invested $100 five years ago, and it accumulated at 12 percent annual compound interest to $176.23 in current dollars today. If inflation averaged 6 percent per year, what was the *real* rate of return on this investment? Suppose you did not know the inflation rate over the last five years?

5-2 In question 5-1, what would the $176.23 payoff be in constant dollars of the date five years ago?

5-3 You are considering purchasing an asset for which you have estimated the expected positive and negative cash flows in constant dollars over the projected life of the asset. Assume your minimum acceptable rate of return is the return on corporate Bbb bonds, which has averaged 11 percent over the last five years. Assume further that inflation over the same period averaged 4 percent annually. What interest rate should you use for finding the net present value of the above cash flows to determine what the asset would be worth to you?

5-4 Over a 10-year period, while inflation averaged 5 percent/year, you've paid off a loan at 9 percent interest. Did the lender earn 9 percent real interest or 9 percent inflated interest? Specify the lender's real and nominal rate of return on the money loaned.

5-5 Corporation XYZ wants a real rate of return of 7 percent. Projecting 6 percent annual inflation, what nominal rate of return (in current dollars) must it earn to meet its goal?

5-6 A firm is considering two management plans. Management plan (1) is projected to have a 5 percent real rate of return. Management plan (2) is projected to have an 8.15 percent nominal rate of return. The firm's alternative opportunity for investment is putting its money into bonds at 9 percent nominal interest. The projected annual rate of inflation over the period is 3 percent. Which plan should the firm adopt?

5-7 **a** Suppose you borrow $100,000 today to buy a forest. The lender charges 10 percent interest and says you must pay the loan off in equal annual installments over a 10-year period, the first payment being due in 1 year. How much is each payment? (*Hint:* Since the banker earns 10 percent interest on the loan, the present value of your payments should be worth $100,000 to her at 10 percent interest.)

b Suppose inflation averages 6 percent/year over the period you pay off the loan. In constant dollars of the initial year (in real terms), what are the values of the first three loan payments? What's the value of the last payment on the same constant dollar basis?

c What is the real rate of return the banker will earn, if inflation averages 6 percent/year over the loan period?

5-8 Kurt bought some Acme Wood Products stock in 1980. At the time he sold it seven years later, its value was 95 percent higher than in 1980 (current dollars). If inflation averaged 4.7 percent/year over the period, what was his real rate of return on the investment?

5-9 Suppose over the last five years stumpage prices for one species rose at 2 percent per year *in real terms,* and during this period inflation averaged 7 percent annually. If the stumpage price was $100/MBF five years ago, what is the price today in current dollars?

5-10 **a** Suppose you bought timber for $10,000 today and project you'll sell it in 10 years for $25,000 (constant dollars of this year). Assume a 28 percent capital gains tax. Given a 7 percent real after-tax discount rate, what is the after-tax net present value (NPV) of this investment, if you project *no* inflation? Let NPV be the maximum you could have paid for the timber and still have earned a 7 percent real rate of return.

b In part (a), suppose you projected 5 percent inflation per year, and timber prices increased so that the above *real* timber sale value was still $25,000. What would your net present value be (other things being equal)?

Why is the present value different? If there were no tax, would the present value differ with and without inflation? Explain your answer.

5-11 In December 1983, Karen bought $1,000 worth of the Janus mutual fund. In December 1993, Janus reported a 15.24 percent average annual rate of return on the fund for the last 10 years. Karen found that between 1983 and 1993, the consumer price index went from 99.6 to 144.5. What was Karen's real rate of return on her Janus investment from 12/83 to 12/93? (These are real-world numbers.) In December 1993, what was her investment worth in 1993 dollars and in 1983 dollars?

5-12 Marilyn projected real net revenues from a prospective timber investment and calculated the present value using the rate of interest on corporate Bbb bonds as reported in the Wall Street Journal (this was her minimum acceptable rate of return). From her standpoint, was this present value too small, too large, or just right?

5-13 Suppose a forest owner computes the present value of a reforestation project, assuming that the MAR should be the average real return on Aaa bonds over the last seven years. Assume that during this seven-year period, inflation averaged 7 percent annually, and the nominal rate of return on Aaa bonds was 13 percent. Now suppose the forest owner computes present value in current dollar terms, projecting 5 percent annual inflation, and uses the above 13 percent interest rate. Will the present value be too high, too low, or just right?

5-14 In calculating the present value of a forestry investment, suppose the U.S. Forest Service projected all cash flows in nominal terms, using a consultant's inflation projection. Values were discounted with the real interest rate on corporate Aaa bonds, the Service's MAR. Was the resulting present value too small, too large, or correct?

5-15 Suppose you own an uneven-aged forest from which you annually harvest only the timber growth, which is constant. Thus, the standing timber volume remains the same over time. Assume that stumpage values are increasing at exactly the inflation rate, which is 7 percent (historic and projected). Suppose you bought the forest 10 years ago for $200,000, when the annual harvest income was $10,000. Let $200,000 be the forest market value 10 years ago.

a What would be this year's harvest income and total forest value?

b What would be this year's rate of return, as measured by harvest income divided by asset value? Is this a nominal or real rate of return?

5-16 Consider these newly purchased assets: (1) a corporate bond purchased for $100,000 and yielding $7,000 per year and (2) a sustained yield forest purchased for $100,000 and yielding annual harvest income starting at $7,000. Which is the better investment, from a financial view? Why?

5-17 Figure 5-5 graphs nominal incomes from a bond and a forest, both bought for $100,000 in year 0. Nominal bond income is $9,200 per year for 10 years, with the $100,000 principal returned in the tenth year in current dollars. Real forest income is $4,000 per year for 10 years (in nominal terms, rising at the inflation rate), with the real potential forest sale value in year 10 being $100,000 in year 0 dollars; count this as an income. Given projected inflation of 5 percent per year and a 4 percent real discount rate, show that both income streams have the same present value.

REFERENCES

Bernstein, P. L. 1974. Are common stocks really good investments? *Challenge.* 17(3): 56–61.

Brigham, E. F., and L. C. Gapenski. 1991. *Financial Management—Theory and Practice* (6th edition). The Dryden Press, Chicago. 995 pp. plus app.

Bullard, S. H., and W. D. Klemperer. 1986. Effects of inflation on after-tax present values where business costs are capitalized. *Mid-South Business Journal.* 6(2):11–13.

Clephane, T. P. 1978. Ownership of timber: A critical component in industrial success. *Forest Industries.* August. pp. 30–32.

Council of Economic Advisors. 1988. *Economic Report of the President.* U.S. Government Printing Office, Washington, DC. 374 pp.

Council of Economic Advisors. 1994. *Economic Report of the President.* U.S. Government Printing Office, Washington, DC. 398 pp.

Fischer, F., R. Dornbusch, and R. Schmalensee. 1988. *Economics.* (2nd edition). McGraw-Hill, New York. 813 pp.

Gregersen, H. M. 1975. Effect of inflation on evaluation of forestry investments. *Journal of Forestry.* 73:570–572.

Hanke, S. H., and J. B. Anwyll. 1980. On the discount rate controversy. *Public Policy.* 28(2): 171–183.

Hanke, S. H., P. H. Carver, and P. Bugg. 1975. Project evaluation during inflation. *Water Resources Research.* 11:511–514.

Ibbotsen, R. G., and R. A. Sinquefield. 1982. *Stocks, Bonds, Bills, and Inflation: The Past and the Future.* The Financial Analysts Research Foundation, Charlottesville. 137 pp.

Joint Economic Committee. 1959. *Employment, Growth, and Price Levels.* Hearings before the Joint Economic Committee, 86th Congress, 1st session. Part 2—Historical and Comparative Rates of Production, Productivity, and Prices. U.S. Government. Printing Office, Washington, D.C.

Klemperer, W. D. 1983. Some implications of inflation-caused changes in the timing of asset yields. *Forest Science.* 29(1):149–159.

Klemperer, W. D. 1985. Inflation-indexed loans as a means to dampen housing cycles. *Forest Products Journal.* 35(3):49–54.

Lind, R. 1990. Reassessing the government's discount rate policy in light of new theory and data in a world economy with a high degree of capital mobility. *Journal of Environmental Economics and Management.* 18(2):S8–S28.

Modigliani, F. 1989. The inflation proof mortgage: The mortgage for the young. In *The Collected Papers of Franco Modigliani.* S. Johnson (ed.). MIT Press, Cambridge, MA.

Nordhaus, W. D. 1974. The falling share of profits. In *The Brookings Papers on Economic Activity.* A. M. Okun and G. L. Perry (eds.). Washington, DC. pp. 169–217.

Ritter, L. S., and W. L. Silber. 1974. *Principles of Money, Banking, and Financial Markets.* Basic Books, New York. 546 pp.

Schwab, C. 1994. *Mutual Funds Performance Guide.* 4th Quarter, 1993. San Francisco. 47 pp.

Seagraves, J. A. 1970. More on the social rate of discount. *Quarterly Journal of Economics.* 84(3):430–450.

Simon, J. L. 1990. Great and almost-great magnitudes in economics. *Journal of Economic Perspectives.* 4(1): 149–156.

Stockfisch, J. A. 1969. The interest rate applicable to government investment projects. In Hinrich, H. H., and G. M. Taylor, (eds.). *Programming, Budgeting, and Cost Benefit Analysis.* Goodyear Rubber Co. 420 pp.

U.S.D.A. Forest Service. 1989. *RPA Assessment of the Forest and Rangeland Situation in the United States,* 1989. Forest Resource Report No. 26, Washington, DC. 72 pp.

U.S.D.A. Forest Service. 1990. *An Analysis of the Timber Situation in the United States: 1989–2020.* Draft. Rocky Mountain Forest and Range Experiment Station, Ft. Collins, CO.

U.S.D.A. Forest Service. 1988. *The South's Fourth Forest: Alternatives for the Future.* Forest Resource Report No. 24, Washington, DC. 512 pp.

U.S. Department of Labor. 1988. The consumer price index. Chapter 19 in *BLS Handbook of Methods.* Bureau of Labor Statistics. Bulletin 2285. Washington, DC. Pp. 154–208.

U.S. Office of Management and Budget. 1972. Discount rates to be used in evaluating time-distributed costs and benefits. Circular A-94 (revised). Washington, DC. U.S. Govt. Printing Office.

Ulrich, A. H. 1990. *U.S. Timber Production, Trade, Consumption, and Price Statistics—1960–1988.* USDA Forest Service Misc. Publ. 1486, Washington, DC. 80 pp.

Washburn, C. L., and C. S. Binkley. 1993. Do forest assets hedge inflation? *Land Economics.* 69(3): 215–224.

Yohe, W. P., and D. S. Karnosky. 1969. Interest Rates and Price Level Changes, 1953–69. Reprint Series No. 49, Research Department, Federal Reserve Bank of St. Louis, MO. 38 pp.

Zinkhan, F. C., W. R. Sizemore, G. H. Mason, and T. J. Ebner. 1992. *Timberland Investments—A Portfolio Perspective.* Timber Press, Portland, OR. 208 pp.

6

CAPITAL BUDGETING IN FORESTRY

An enormous number of forestry decisions involve capital management. This capital might be landowners' or stockholders' cash, loans, or public agencies' tax revenues to be invested in buying new forests or improving existing tracts. Other capital assets could be thousands or millions of dollars worth of existing timber, land, equipment, and other resources. In fact the biggest inputs in forestry are time and capital. Labor is a relatively small factor. And since capital is limited, we have an economic problem: how best to manage assets—forest and nonforest—to maximize satisfaction of the owners or, more broadly, of society. I'll consider natural resources, including land, as capital in the sense that funds are spent to obtain them, and they represent a store of value. These assets yield benefits and could be put to other uses yielding larger or smaller benefits. Even resources like fish and wildlife, which aren't easy to value, have a capital value that you can see in the marketplace. For example, forestlands yielding hunting lease income will tend to sell for more than those without such income, other things being equal. The wildlife income is **capitalized** into the value of the property, becoming part of the capital asset.

This chapter expands on the last two and covers financial tools for ranking forest investments from best to worst. Chapters 7 and 8 further apply these concepts to forestry questions like optimal rotations, stocking levels, and thinning regimes.

THE CAPITAL BUDGETING PROBLEM

The **capital budgeting** problem is to decide how to invest money—the "capital budget"—so that its value is maximized. The kinds of spending considered here will be on fairly durable assets or projects that require an initial cost followed by incomes, and usually further costs, over several years. Examples are expenditures on bare land, reforestation, fertilization, timber for future harvest, equipment, entire forested properties, and processing facilities. First let's assume that all outputs can be measured in dollars. Nondollar values will be considered later.

In the broadest sense, investors try to spend money in a way that maximizes satisfaction. In financial terms, this means maximizing the value of assets for the owners. Managers, for example, may wish to maximize the value of a firm for stockholders. To do this, they need criteria for ranking investments and deciding which is best. Chapters 4 and 5 explained **net present value (NPV)** as a maximum willingness to pay for assets, given expected costs and revenues and a desired earning rate. That's one capital budgeting criterion: choose first the projects with the highest net present value. Let's now look at the pros and cons of NPV and several other criteria for ranking investments. *Throughout the chapter, let interest rates and cash flows be real (in constant dollars—see Chapter 5).* Assume that all taxes have been paid, so that cash flows and rates of return are after taxes (or before taxes for government agencies not paying taxes). Until Chapter 6, on risk, let the amount added to the discount rate for risk—the **risk premium**—be equal for costs and revenues and for any group of investments being compared, unless stated otherwise. Notation for this chapter is in Table 6-1.

CRITERIA FOR ACCEPTING OR REJECTING INVESTMENTS

I'll first describe four criteria for accepting or rejecting investment projects: net present value, internal rate of return, benefit/cost ratio, and payback period. Then let's consider problems in ranking investments from best to worst.

Table 6-2 shows two investments: project D has more *D*istant income than project N with *N*earer income. These are simple examples; in reality, many more costs and revenues would usually occur. You could think of both projects as afforestation investments, with D yielding timber harvests at ages 15 and 30, and N yielding Christmas trees at age 8 and a residual timber stand at age 30. First let's look only at project D; the ranking discussion considers both.

TABLE 6-1
NOTATION FOR CHAPTER 6

ARR = alternative rate of return	n = project life, years
B = annual nonmarket value, dollars	NPV = net present value
B/C ratio = benefit/cost ratio	PV = present value
C_y = cost in year y	r = real interest rate, percent/100
EAA = equivalent annual annuity	R_y = revenue in year y
IRR = internal rate of return	y = an index for years
MAR = minimum acceptable rate of return	

TABLE 6-2
CASH FLOWS FOR PROJECTS WITH DISTANT INCOME (D) AND NEARER INCOME (N)

Year	Cash flows, project D $	Cash flows, project N $
0	−400	−400
5	−100	−100
8		+1,200
15	+200	
30	+6,600	+2,500

Net Present Value

As shown in equation (4-6), net present value of a project is the present value of its revenues minus the present value of its costs:

$$\text{NPV} = \sum_{y=0}^{n} \frac{R_y}{(1+r)^y} - \sum_{y=0}^{n} \frac{C_y}{(1+r)^y} \tag{6-1}$$

where R_y and C_y are revenues and costs in any year y. Or, more specifically, from equation (4-5):

$$\text{NPV} = R_0 + \frac{R_1}{(1+r)^1} + \frac{R_2}{(1+r)^2} + \frac{R_3}{(1+r)^3} + \ldots + \frac{R_n}{(1+r)^n}$$
$$- C_0 - \frac{C_1}{(1+r)^1} - \frac{C_2}{(1+r)^2} - \frac{C_3}{(1+r)^3} - \ldots - \frac{C_n}{(1+r)^n}$$

Note that R_0 and C_0 are not discounted, since they are already at the present. Applying equation (6-1) to project D of Table 6-2, and assuming the decision maker wishes to earn a 6 percent real rate of return (the **minimum acceptable rate,** MAR), net present value is:

$$\frac{6{,}600}{(1.06)^{30}} + \frac{200}{(1.06)^{15}} - \frac{100}{(1.06)^5} - 400 = \$758 \tag{6-2}$$

Given an initial project cost of $400, this means that the investor could pay up to $758 *more* and still earn a 6 percent rate of return.

For project D, Figure 6-1 graphs the **present value** (PV) of revenues and PV of costs, computed with different interest rates shown on the x axis. The PV of revenues is the first term of equation (6-1), and PV of costs is the second term without the negative sign. Why does the PV revenues curve fall faster than PV costs as the interest rate is increased? [It's because revenues are further in the

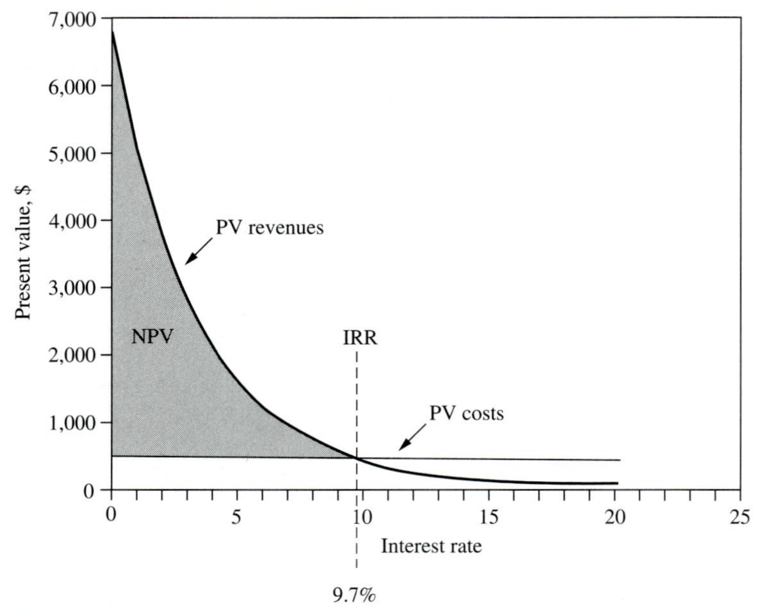

FIGURE 6-1
Present values, project D.

future than costs, which is typical of most investments, especially forestry.[1] Thus, the larger exponents (years) in the revenue denominators of equation (6-2) cause the present value of revenues to fall faster than PV costs as the interest rate is increased.] The shaded area, NPV in Figure 6-1, is PV revenues minus PV costs. You can see that for interest rates up to about 9.7 percent, NPV is greater than zero. If the investor computes NPV of project D with an interest rate greater than 9.7 percent, NPV will be negative (PV costs > PV revenues).

According to the NPV guideline, a project is acceptable if NPV is zero or greater. Projects with a negative NPV are unacceptable. In other words, the present value of revenues must be greater than or equal to the present value of costs, both computed with the investor's minimum acceptable rate of return (MAR).

Interpreting NPVs Net present values can easily cause confusion. For example, suppose you compute the net present value of project D for an investor with an MAR of 9.68 percent. Substituting 0.0968 in place of $i = 0.06$ in equation (6-2) will yield an NPV of about zero. According to the NPV criterion, the project is just barely acceptable. Incredulous, the investor exclaims, "You're asking me to spend $400 on my property, and it increases the value by *zero?*" "Well,

[1]Revenues aren't always more distant than costs. For example, borrowing money is a project where revenue (the loan amount) occurs first, and costs (loan payments) occur later. In that case the PV cost curve would be steeper than the PV revenue.

not exactly," you respond. "The value will increase by the initial expenditure of $400, but by no more than that." It's like putting $400 in a certificate of deposit earning 9.68 percent interest. Using 9.68 percent, the present value of the future receipts will be exactly $400, and subtracting the $400 deposit gives an NPV of $0. But immediately after deposit, the CD is worth $400.

If you didn't subtract the original cost when computing NPV, you couldn't tell from NPV alone whether a project was acceptable. For example, without subtracting the $400 initial cost, the present value of project D at 10 percent interest is $364. That figure alone, although positive, doesn't tell you much until you compare it with the $400 cost. That's why, by convention, for capital budgeting purposes, NPVs are net of original cost. But such values are easily misinterpreted. What adds to the confusion is that some types of present values do *not* have the initial cost subtracted; for example, when you compute a maximum bid price for an asset based on present value of future income, you should not subtract the seller's asking price. *When interpreting net present values, make sure you know whether original expenditure has been subtracted. Otherwise, serious misunderstanding can result.* In this chapter, original cost will always be subtracted in computing NPV. You'll see more on this issue in the chapter on valuation.

Internal Rate of Return

Another common criterion for evaluating investments is the **internal rate of return (IRR).** As given in equation (6-3), the IRR of a project is the discount rate at which the present value of revenues minus the present value of costs equals 0, or where NPV = 0:

$$\sum_{y=0}^{n} \frac{R_y}{(1 + \text{IRR})^y} - \sum_{y=0}^{n} \frac{C_y}{(1 + \text{IRR})^y} = 0 \qquad (6\text{-}3)$$

The IRR is the rate of return earned on funds invested in a project. The above equation also says that the *IRR is the interest rate at which the present value of revenues equals the present value of costs.* For project D of Table 6-2, Figure 6-1 shows that the IRR is about 9.7 percent, since at that interest rate the cost and revenue present value curves cross, and NPV = 0. Note that the IRR is unique (or "internal") to a project. The MAR is the minimum rate an investor wishes to earn; it is based on individual desires or the best earning rate widely available elsewhere and is external to a project being evaluated. In Chapter 4, equation (4-4) gives a simple IRR solution for investments with one cost and one revenue. But with more complex cash flow patterns, IRR usually must be found by trial and error. Many modern business calculators and spreadsheet programs can quickly find IRRs when you enter a project's positive and negative cash flows and the years when they occur. Without such tools, finding an IRR for a complex project can be tedious.

Table 6-3 shows a trial and error approach for finding the IRR of project D to the nearest hundredth of a percentage point, using equation (6-3). For the interest

TABLE 6-3
PROJECT D NET PRESENT VALUES AT DIFFERENT INTEREST RATES

(a) Interest rate %	(b) Present value, revenues, $	(c) Present value, costs, $	(d) Net present value $ (b − c)
0.00	6,800.00	500.00	6,300.00
4.00	2,145.96	482.19	1,663.76
9.00	552.36	464.99	87.36
9.66	465.20	463.06	2.14
9.67	464.00	463.03	0.97
9.68*	462.80	463.00	−0.20
9.69	461.60	462.97	−1.37
10.00	426.11	462.09	−35.98

*Internal rate of return to the nearest 1/100 percentage point. Trials first show that the IRR (or zero NPV) is between 9 and 10 percent. Then more detailed trials between these two rates reveal that 9.68 percent brings NPV closest to 0.

rates shown, columns (b) and (c) are the values plotted in Figure 6-1, and the last column gives NPVs. Since the net present value is \$0.97 at 9.67 percent interest and −\$0.20 at 9.68 percent, the IRR (where NPV = 0) must be between these two rates, and closer to 9.68 percent. Given the uncertainty in projecting most cash flows, such hairsplitting usually isn't too meaningful. In general, you can think of the IRR as an average rate of return on all capital invested in a project.

The IRR investment guideline says that a project is acceptable if its IRR is equal to or greater than the minimum acceptable rate of return. Projects with IRR < MAR are unacceptable. Looking at Figure 6-1, you can see that's just another way of expressing the net present value guideline: If the MAR is less than or equal to the IRR (here ~ 9.7 percent), the NPV will be ≥ 0, and the project will be acceptable. Otherwise it is rejected. An advantage of the IRR is that many investors with different MARs can look at one IRR and decide whether it's acceptable. But an NPV is meaningful only for an investor whose MAR equals the one discount rate used to compute that NPV.

Many firms use internal rate of return rather than net present value to decide whether to accept or reject projects. But it's important to note that for the accept/reject decision, the two guidelines are fundamentally the same (as long as there's a unique IRR solution). You can't compute an IRR without calculating net present values, and both guidelines essentially say, "Accept projects if NPV is ≥ 0, using the MAR" (see Figure 6-1). We'll see shortly, however, that for ranking several projects from best to worst, NPV and IRR don't always give the same ordering.

Benefit/Cost Ratio

A project's **benefit/cost ratio** (*B*/*C*) is the present value of benefits (revenues) divided by the present value of costs, using the investor's MAR. The *B*/*C* ratio, also called the **profitability index,** is:

$$B/C \text{ ratio} = \frac{\text{PV revenues}}{\text{PV costs}} = \frac{\displaystyle\sum_{y=0}^{n} \frac{R_y}{(1+r)^y}}{\displaystyle\sum_{y=0}^{n} \frac{C_y}{(1+r)^y}} \tag{6-4}$$

From this equation you can see that when PV revenues equals PV costs, the B/C ratio is 1, and NPV is 0. Also, if PV revenues exceeds PV costs, B/C must be greater than 1. And if PV costs exceeds PV revenues, $(B/C) < 1$. *Thus, according to the B/C ratio criterion, projects are acceptable when the B/C ratio is 1 or greater, and unacceptable if* $(B/C) < 1$. This gives the same accept/reject decisions as the NPV approach, because a negative NPV makes $(B/C) < 1$.

Using 6 percent interest, the benefit/cost ratio for project D in Table 6-2 is:

$$B/C = \frac{\dfrac{6{,}600}{(1.06)^{30}} + \dfrac{200}{(1.06)^{15}}}{\dfrac{100}{(1.06)^{5}} + 400} = \frac{1232.58}{474.73} = 2.60 \tag{6-5}$$

Since $(B/C) > 1$, the project is acceptable, as it was when a positive NPV was computed at 6 percent interest.

For project D, Figure 6-2 ties together the NPV, IRR, and B/C ratio criteria. Study this carefully, because it shows how the three criteria are very closely related. The vertical dashed line marks the IRR—where PV costs = PV revenues—and thus where $B/C = 1$. To the right of the dashed line, the interest rate exceeds the IRR: there NPV is negative, and therefore B/C must be < 1. So when the MAR exceeds a project's IRR, the project is unacceptable by any of the three foregoing criteria. To the left of the dashed line, where the IRR exceeds the MAR, NPV is positive, and $(B/C) > 1$. In that range, the project is acceptable by all three criteria. *If you can understand Figure 6-2 and sketch it from memory, with all its labels but without numerical values, you'll have a firm grasp of the three investment evaluation criteria discussed so far.*

Payback Period

Another criterion is to choose the project with the shortest **payback period,** which is the number of years it takes to recover the invested capital. For example, looking at Table 6-2, it takes 30 years for the net income from project D to offset the \$500 total investment; so the payback period is 30 years. (Some analysts will count only the initial investment, here \$400.) For project N, the payback period is 8 years. Based on payback period alone, project N would be preferred. Or a firm requiring payback periods of 8 years or less would accept project N and reject project D.

The major problem with payback period is that it says nothing about a project's present value or rate of return. For example, suppose project N yielded only \$500

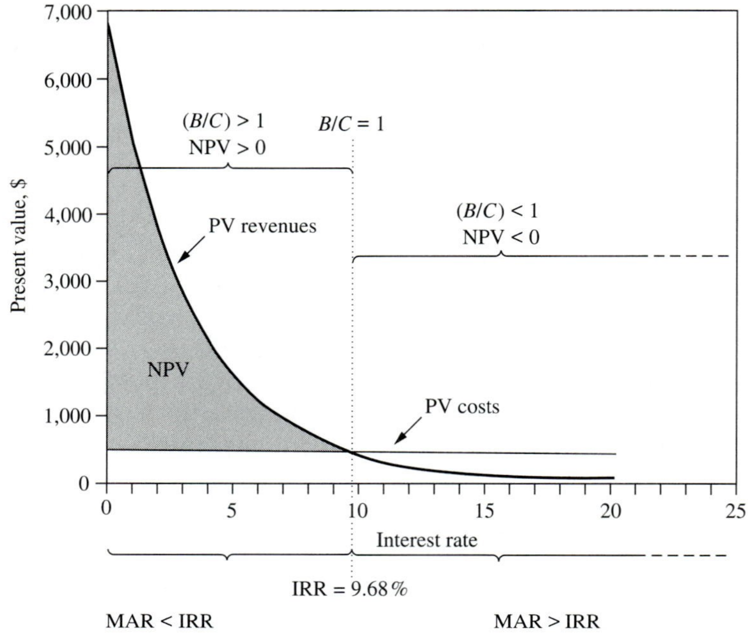

FIGURE 6-2
Capital budgeting criteria, project D.

in year 8 and nothing thereafter. Its payback period would still be 8 years, but the rate of return would be 0 percent (only a 0 percent discount rate would make the present value of costs and revenues be equal—at $500). Despite its shortcomings, payback period is often useful *if considered together with measures such as NPV and IRR.* For instance, investors with short time horizons or those projecting cash flow problems may reject ventures with long payback periods. Also, other things being equal, more distant income tends to be more uncertain; hence, long payback periods suggest greater risk. Banks can be wary of lending funds for projects with long payback periods.

THE REINVESTMENT RATE, ARR, AND MAR CONUNDRUM

A source of confusion in capital budgeting is the assumed rate of return that a project's capital and intermediate income could earn elsewhere. Theoretically, at a competitive equilibrium, we could expect the best rates of return on added capital to be the same in all areas, given equal risk (see Chapters 2 and 3). The real world isn't this neat. Investors often perceive several different available earning rates, even if risk premiums are equal, as assumed in this chapter. Suppose an investor's best alternatives yield 15 percent, 13 percent, 10 percent, and 8 percent, but projects earning the top rates aren't large enough to absorb all available capital. There's obviously no single **alternative rate of return** (ARR), which is why I avoid that term for the investor's discount rate.

Unless noted otherwise, this book will use the investor's minimum acceptable rate of return (MAR) to calculate present values or to judge IRRs. In the above example, that could be, say, 10 percent. Note that some people's MARs *exceed* available rates of return, in which case they don't invest at all. Also, some intermediate income from a project can often earn a rate of return—the **reinvestment rate**—which exceeds the MAR. Thus you need to clarify reinvestment rates and discount rates.

RANKING PROJECTS

The accept/reject decision is easier than ranking several projects from best to worst. The above criteria don't necessarily give the same rankings. As before, we'll assume that investors wish to maximize the value of their assets. Consider ranking the two projects in Table 6-2: project D with distant income and project N with nearer income. For comparability, both projects have the same capital requirements and the same 30-year life.

Mutually Exclusive, Independent, Divisible, and Indivisible

Certain features of projects are important in the ranking process. If several projects are **mutually exclusive,** only one can be chosen. An example is the choice between investments in planting pure loblolly pine timber, pure scotch pine Christmas trees, or only forage on a given acre. If projects are not mutually exclusive, they are **independent** and could all be adopted—for example, investments in fertilization, precommercial thinning, brush control, and purchase of a timber tract. If the two investments in Table 6-2 were reforestation choices for the same acre, they would be mutually exclusive; if for different acres, they'd be independent.

If you can invest in part of a project, it is **divisible**—as in the case of adding money to a savings account or acres to a fertilization project. If a project is all-or-nothing, as in buying a truck or a pulp mill, it is **indivisible.**

Possible Inconsistency between NPV and IRR Rankings

You can rank projects in order of decreasing net present value or internal rate of return. But one problem is that internal rate of return rankings aren't always consistent with NPV rankings. By plotting NPVs of projects D and N (from Table 6-2) over the interest rate, Figure 6-3 shows such a case. Notice that the curves for each project are the difference between PV revenues and PV costs: in the case of project D, that's the shaded area in Figure 6-1. Thus, in Figure 6-3, IRR is where the NPV curve meets the x axis (where NPV = 0). In the figure, the IRR for D is 9.7 percent compared with 14.5 percent for N.

Which project is better? To decide, we need more information. If the investor's hurdle rate is 9.7 percent or less, if the projects are independent, and if enough capital is available, both D and N should be accepted. But if they are mutually

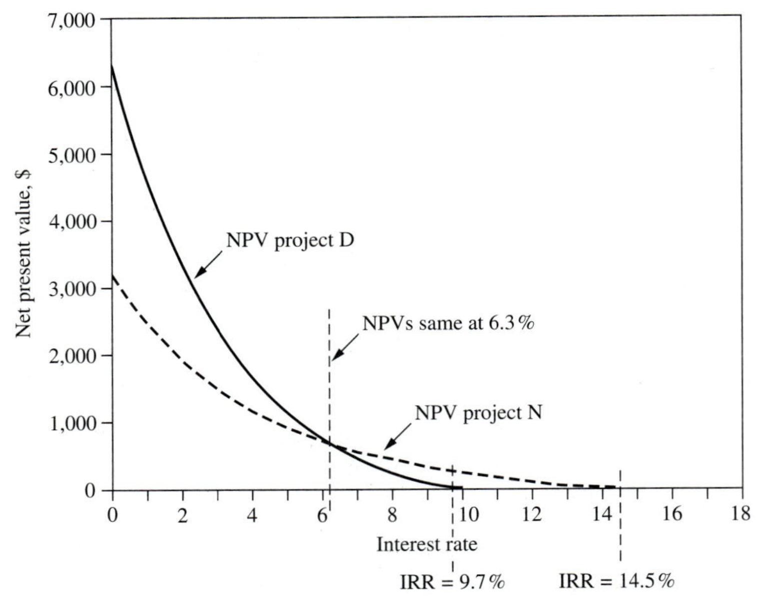

FIGURE 6-3
Net present values, projects D and N.

exclusive or if initially available capital is less than $800, we need to know which project to choose. Based on IRR, N wins at 14.5 percent. For minimum acceptable rates above 6.3 percent, project N also has the highest NPV, as shown in Figure 6-3. But at interest rates < 6.3 percent, project D has the highest NPV and is preferred according to the NPV criterion. With a 6.3 percent interest rate, you're indifferent between the two. If your MAR is less than 6.3 percent, say 5 percent, and the goal is to maximize NPV, project D is the correct choice in spite of its lower IRR. At 5 percent interest, D's NPV = $1,144.94, and N's NPV = $912.30.

This is confusing, because N's 14.5 percent rate of return sounds so much better than D's 9.7 percent. The above choice of D is based on the assumption that the *reinvestment rate* for intermediate income from either project is the 5 percent MAR. To see that D is best under this constraint, compare your future wealth in 30 years from investing in D or N, as shown below:

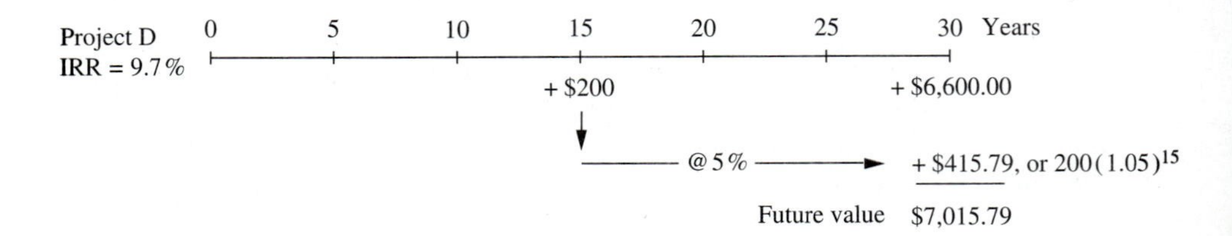

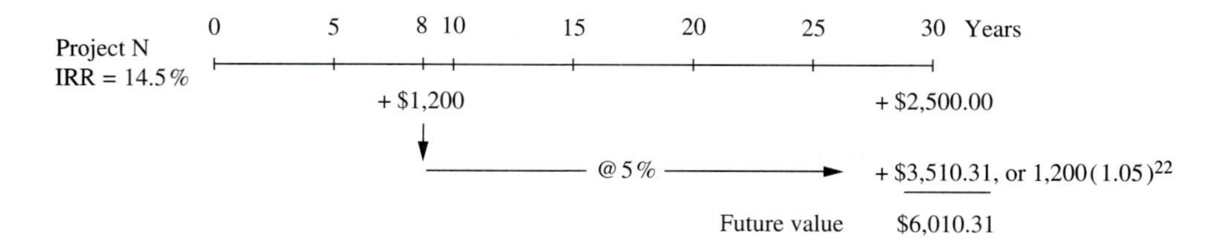

Since both projects have the same capital investment, we need only compare the future values of income accumulated at the 5 percent reinvestment rate. In each case, compound the first income at 5 percent interest for the number of years *remaining* in the period (15 years for D and 22 years for N), and add the final incomes. By investing in D, you'd have a higher future value ($7,015.79) than you would from N ($6,010.31), despite N's higher internal rate of return. Given the same investment lives, if the future value of D's income is greater, present value must also be greater. In part for this reason, net present value enjoys greater theoretical support than internal rate of return as a capital budgeting criterion. Note that projects with the same capital requirements and durations, as in Figure 6-3, will be ranked the same by benefit/cost ratio or NPV. If NPV > 0, $(B/C) > 1$. But if capital requirements differ, rankings by B/C ratio and NPV will *not* always be consistent.

Before rushing to condemn the IRR, note that in many cases, projects' NPV curves over interest never cross, in which case NPV and IRR will rank projects consistently. Also, even when NPV curves do cross, inconsistency is a problem only for a range of interest rates (e.g., under 6.3 percent in Figure 6-3).

Multiple IRRs Major costs during or at the end of a venture can sometimes cause more than one IRR to occur. In such cases the IRR criterion is ambiguous. For example, applying equation (6-3) to the hypothetical project below gives IRRs of 4.0 percent and 9.9 percent (both rates give NPVs of 0):

– $200 + $1,200 – $1,100

0 5 10 15 Years

Between 4.0 percent and 9.9 percent interest, NPV is positive, and at all other rates NPV is negative. This inconsistency in the NPV trend with increasing discount rate is just as ambiguous as multiple IRRs, since we normally expect NPV to steadily decrease as we increase the discount rate from zero. Thus, as Mills and Dixon (1982) point out, multiple IRRs occur when the investment itself is ambiguous. They note that the IRR criterion can display such ambiguity, while a single NPV calculation hides it. Thus, it is not at all clear that the potential for multiple IRRs is a disadvantage of the IRR criterion. In the above investment, you might wonder why one wouldn't stop the project after 10 years and avoid the $1,100 cost. But the cost might be a legal requirement such as restoring land after mining. Multiple IRRs can also occur when a venture starts with a positive cash flow from a loan, or no negative cash flow, followed by sufficiently large negative and positive flows. If a project has more than one IRR, you can compute a single NPV to evaluate the venture. But you must realize that the project will still have the above-mentioned ambiguity.[2]

NPV and Project Size

For consistency, capital requirements for projects D and N were kept equal. This is important, because looking at IRR or NPV alone, you can't tell how much capital a project requires. Below, you'll see how NPV can hide the amount of capital needed. Both investments I and II have one initial cost and one revenue a year later and are *indivisible* (all-or-nothing). Let the MAR be 6 percent.

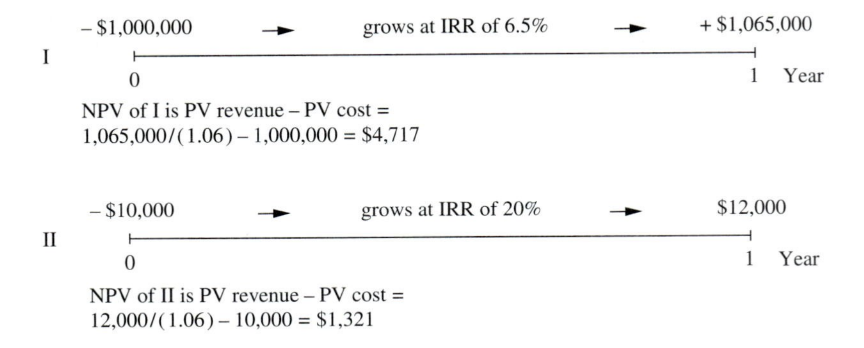

Which is best, I or II? Considering NPV alone, I is best, but II's IRR is an excellent 20 percent, compared with I's 6.5 percent. You really need more information. Are there more opportunities like II? Is there $1,000,000 available to invest in I? The key point of the example is that NPV alone doesn't really give enough information; nor does IRR. You cannot make valid comparisons or rankings of different-sized projects based on NPV or IRR alone.

Suppose no other projects beyond I and II can yield more than the 6 percent MAR, and projects I and II are *independent* (one doesn't preclude the other, say, fertilization and herbicide spraying). In that case, if capital is unlimited, both should be selected because they both earn more than the MAR.

Now suppose capital is limited to $1 million, or the projects are *mutually exclusive* (like homesite development versus reforestation on the same area). You can invest only in one. If 6 percent is the best you can do anywhere else, then project I

[2]Both the problems of NPV and IRR ranking-inconsistency and multiple IRRs can be solved by ranking projects with the *modified internal rate of return* (MIRR) (Brigham and Gapenski, 1991, p. 335), also called the *realizable rate of return* (Schallau and Wirth, 1980). To do this for a project, you first accumulate at the MAR each year's net revenues to a future value and discount each net cost at the MAR to year zero. The MIRR is the interest rate that discounts the future value of revenues to exactly equal the present value of costs. In comparing projects with the MIRR, you need to correct for unequal lives and to be aware that, for projects with intermediate incomes, the MIRR tends to get lower as project life lengthens (Klemperer, 1981).

alone will maximize present value at the $4,717 shown above. Suppose $10,000 of the $1 million is spent on project II and the remaining $990,000 must be invested at the 6 percent MAR, yielding 990,000(1.06) = $1,049,400 after one year. This would give no increase in net present value beyond the $1,321 shown above for II. The $990,000 compounded at 6 percent interest and discounted at 6 percent for the same time gives $990,000—just enough to offset the added $990,000 cost—and net present value is still $1,321:

NPV of $10,000 spent on II and $990,000 invested at the 6% MAR =

$$12{,}000/(1.06) - 10{,}000 + 1{,}049{,}400/(1.06) - 990{,}000 = \$1{,}321$$

The above procedure of adding other projects to a smaller project so that, in total, they equal the size of a larger one is called **normalizing** for size. But as just shown, if the rate of return on added capital equals the MAR, then net present value is unchanged. So in that case, normalizing for size isn't necessary with NPV calculations.[3] But in calculating IRRs and *B/C* ratios, you should always normalize for size, regardless of the MAR.

I first said that projects I and II were indivisible. If they were *divisible,* and you had $1,000,000 to invest, you'd maximize NPV by investing first in project II, which has the highest NPV per dollar of cost, and putting the remaining $990,000 in I.

Unequal project lives also can cause problems in comparing investments.

Correcting for Unequal Investment Lives

When two equal-sized projects have different durations, it may or may not be valid to compare their NPVs. If the shorter project could be repeated and earns more than the MAR, we need to *normalize for unequal lives* by repeating the project. If repetition isn't possible, and the reinvestment rate is the MAR, no correction is needed for NPVs. Consider the following two *mutually exclusive* reforestation projects S (short duration) and L (long duration) for the same area. Assume an 8 percent MAR.

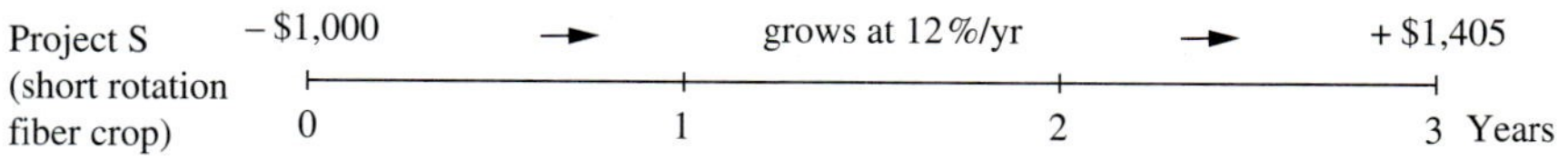

Project S (short rotation fiber crop): – $1,000 → grows at 12%/yr → + $1,405

Timeline: 0, 1, 2, 3 Years

$$\text{NPV} = 1{,}405/(1.08)^3 - 1{,}000 = \$115$$

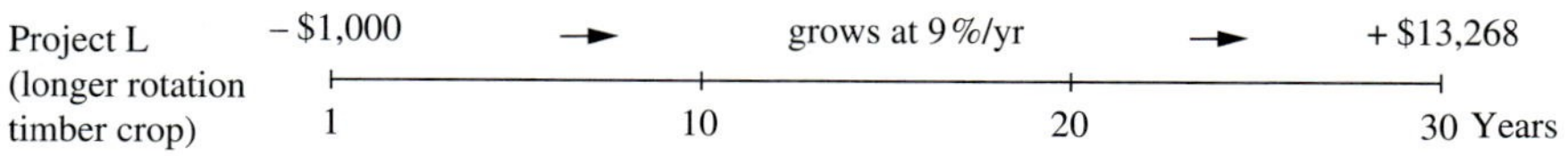

Project L (longer rotation timber crop): – $1,000 → grows at 9%/yr → + $13,268

Timeline: 1, 10, 20, 30 Years

$$\text{NPV} = 13{,}268/(1.08)^{30} - 1{,}000 = \$319$$

[3] If the rate of return on added capital *exceeds* the MAR, normalizing for size is necessary because it increases present value. Although an available rate of return greater than the minimum acceptable rate suggests that the MAR is too low, limited amounts of capital can sometimes earn more than the MAR.

Without capital limits, if the projects were independent, you'd accept them both, since the IRRs exceed the 8 percent MAR. But they're mutually exclusive, so which one should be accepted?

Let's evaluate the projects under two scenarios:

1 Repeating the shorter project is not possible, and the only opportunities for reinvestment of proceeds are at the 8 percent MAR.

In this case it is valid to compare the above NPVs, and L is preferred. In year 3, reinvesting the $1,405 return from S for 27 more years at the 8 percent MAR would give both projects a 30-year life. But this would have no effect on S's NPV because compounding the reinvestment out for 27 years at 8 percent [multiplying by $(1.08)^{27}$] and discounting it back for 27 years at 8 percent [dividing by $(1.08)^{27}$] gives no added net present value. This also means that the *B/C* ratio would be unchanged. But the above normalizing for length will reduce the IRR on the shorter project, so unnormalized IRRs should not be compared when the reinvestment rate equals the MAR.

This and the previous section show further advantages of the NPV criterion: it requires no normalizing for size or length of mutually exclusive projects when the reinvestment rate equals the MAR.

Now consider the second scenario:

2 The shorter project could be repeated 10 times until its duration equaled that of the 30-year project (or both could be repeated in perpetuity). This means that the reinvestment rate exceeds the MAR for at least some capital.

By repeating project S, $1,000 of the $1,405 harvest could be reinvested in planting at 12 percent, and the rest invested at the MAR of 8 percent. This may or may not offer a present value advantage over project L. A way to find out is to compute for each project an **equivalent annual annuity** (EAA): an equal annual real income with the same present value, over the project life, as the project's NPV, all computed at the same *real* MAR. (An **annuity** is a series of equal payments at regular intervals; here a one-year interval is specified.) The project with the highest EAA would be preferred. The equation for EAA is the same as the capital recovery formula [equation (4-12)]:[4]

$$\text{EAA} = \text{NPV}\,\frac{r}{1-(1+r)^{-n}} \tag{6-6}$$

Since you could conceive of these annuities continuing forever, the EAA implicitly gives both projects a common life of infinity, thus correcting for unequal

[4]Here an EAA is assumed to be an equal annual amount in real terms, or rising at the inflation rate. Thus, in computing an EAA, it's important to use a real interest rate r. Otherwise, the EAA increases as more inflation is added to the interest rate, making a project appear better. This result is inconsistent with the fact that inflation should not affect before-tax present values. As more inflation is added to the discount rate, the EAA rankings could incorrectly change. Using a real interest rate avoids this problem.

lives. Below are the EAAs for both projects, computed at the 8 percent discount rate:

$$\text{EAA, project S} = 115\frac{0.08}{1-(1.08)^{-3}} = \$44.62 \tag{6-7}$$

$$\text{EAA, project L} = 319\frac{0.08}{1-(1.08)^{-30}} = \$28.34 \tag{6-8}$$

Despite the fact that project S's NPV is lower than L's, the EAAs show that project S is preferred to L, *if you assume that the shorter project could be repeated at its IRR (here, 12 percent) for the life of the longer project.* If the shorter investment cannot be repeated, EAA is not a valid ranking criterion. In general, if the shorter project can be repeated, a correction for unequal lives should be made. The correction may not always change NPV rankings, but it could.[5]

Capital Rationing

If capital were unlimited, project scale or precise rankings wouldn't be relevant; managers could maximize asset values by investing in all projects with NPV $\geq$ 0, IRR $\geq$ MAR, or $(B/C) \geq 1$. But there's almost never enough capital to invest in all acceptable projects (unless the MAR is very high). Thus, the problem is to find the combination of independent projects that maximizes NPV without exceeding a limited investment budget.

Borrowing more money doesn't solve the problem. Even when lenders have abundant funds, they resist lending to firms with too much debt. There's a limit to borrowing, especially at reasonable interest rates. As a debt-ridden firm tries to borrow more, the risk of loan default increases, and lenders will charge higher interest rates (if they're willing to lend at all). Governments have similar problems: as they borrow more—by selling bonds—beyond a point, they must pay higher interest rates to get people to buy more bonds.

Let's consider only cases where *projects are nonrepeating,* and the reinvestment rate on all incomes equals the MAR (here, 10 percent). Assume that the major cost is in the first year and that there's enough income to cover later costs. As before, *the objective is to maximize wealth (NPV).* The optimum allocation of limited capital will depend on whether projects are divisible or indivisible.

Divisible Projects Suppose you're an analyst for a forest products firm that wants you to recommend which of the eight nonrepeating projects in Table 6-4 to invest in, given a \$100,000 budget. For most of the projects, initial costs C_0, lives, and IRRs differ. For simplicity, all projects yield equal annual income, and all costs are in year 0. Assume that output values rise at the inflation rate and are fixed in real terms. Thus, the interest rate and the IRRs are real. A forestry

[5]Given the assumed reinvestment of only \$1,000 of project S's payoff every 3 years, repeating S will reduce its 12 percent IRR. Thus, to compare IRRs, you should compute S's IRR based on its repetition for 30 years. But that still wouldn't assure consistency of IRR and NPV rankings, because income timing varies (see Figure 6-3).

TABLE 6-4
EIGHT HYPOTHETICAL, INDEPENDENT, NONREPEATING PROJECTS
(Reinvestment Rate = MAR = 10%)

(a) Project	(b) Initial cost, C_0 (dollars)	(c) Life (years)	(d) Annual revenue (dollars)	(e) NPV at 10% MAR (dollars)	(f) IRR (%)	(g) NPV per $1 cost (dollars) (e) : (b)[†]	(h) Accumulated initial cost, all projects (dollars)
1	20,000	14	3,396.50	5,020.95*	14.4	0.251	20,000
2	20,000	6	5,615.92	4,458.80	17.3	0.223	40,000
3	20,000	14	3,236.54	3,842.58	13.4	0.192	60,000
4	100,000	10	18,947.41	16,423.63	13.7	0.164	160,000
5	50,000	20	6,319.89	3,804.79	11.1	0.076	210,000
6	40,000	12	6,015.07	984.83	10.5	0.025	250,000
7	30,000	6	6,293.87	−2,588.55	7.0	−0.086	280,000
8	35,000	10	4,755.38	−5,780.25	6.0	−0.165	315,000

*For project 1, NPV $= 3{,}396.50\left[\frac{1-(1.10)^{-14}}{0.10}\right] - 20{,}000 = \$5{,}020.95.$

This is the present value (PV) of revenues minus PV costs. Discounting with the 14.4 percent IRR would give a zero NPV.

[†]Because this example has no intermediate costs, the *B/C* ratio would also give the correct ranking. But in general, when intermediate costs occur, the *B/C* ratio would *not* always give the correct ranking. For example, a project with a highly ranked NPV/C_0 could have a lower ranked *B/C* ratio if intermediate costs were high enough.

example would be the purchase of timber cutting rights for a price of C_0 on tracts where future harvests are in equal annual amounts for a contract period. Assume tract owners are willing to sell cutting rights on portions of the tracts. That makes the projects "divisible." Column (e) shows NPVs calculated at the 10 percent MAR.[6] You could immediately rule out all projects with NPV < 0 (projects 7 and 8). Total cost of the remaining projects is \$250,000 in column (h), but your budget for the year is only \$100,000. Which projects should be selected? We've already noted problems with ranking projects by IRR. So why not just take those with the largest NPVs, since we want to maximize NPV anyway? One problem lies in the capital constraint: the project with maximum NPV may be too large for your budget. That's not the case here; project 4 has the largest NPV and just exhausts the budget. But project 4 has a lower NPV than a combination of the smaller projects.

A solution is to rank projects by decreasing *NPV per dollar of initial cost* (NPV/C_0):

$$\frac{\text{Net present value}}{\text{Initial cost}} = \text{NPV}/C_0 = \frac{\sum_{y=0}^{n} \frac{R_y}{(1+r)^y} - \sum_{y=0}^{n} \frac{C_y}{(1+r)^y}}{C_0} \tag{6-9}$$

[6]Column (f) shows the problem with calling the investor's minimum acceptable rate the "alternative rate of return." For example, consider project 2, which earns 17.3 percent. Your next best alternative (project 1) earns 14.4 percent. In addition to project 2, five projects earn alternative rates of return higher than the MAR.

Since the numerator is NPV, this is a modified NPV criterion: projects with $(\text{NPV}/C_0) < 0$ are unacceptable. By choosing first those projects with the greatest NPV per dollar cost, you'll automatically maximize the net present value of your limited budget. Projects in Table 6-4 have already been ranked by decreasing NPV/C_0 [see column (g)]. Thus, simply take the first three projects and 40 percent of project 4. Table 6-5 shows that this combination exhausts the $100,000 budget and yields a total NPV of $19,892. This is the optimum selection, since no other combination of projects within the budget could yield a higher NPV. From the sections on project size and life, recall that no NPV correction is needed for unequal sizes or lives, given the above assumption that the reinvestment rate equals the MAR.

Figure 6-4 shows the selection process graphically. The y axis is NPV/C_0. Since the x axis is dollars of original cost, the width of each project's bar is its year-0 cost. The hatched area shows each project's positive NPV (original cost times NPV per dollar of cost). Because of the budget constraint, not all projects with positive NPV were accepted. But the graph shows why this ranking process assures NPV maximization. Therefore, the project combination must be optimal. Things are not so simple with indivisible projects.

Notice in Table 6-4 that ranking the projects by decreasing IRR is not optimal: following the IRR, you'd choose first project 2, then project 1, and 60 percent of project 4, yielding a total NPV of $19,334. Percentagewise, that's not much below the optimum, but in some cases, the difference could be substantial. Note also that ranking by NPV per $1 of initial cost addresses only the first year's budget constraint (usually the most important) but ignores the impact of budget constraints in later years.

Indivisible Projects If projects in Table 6-4 were not divisible, you could no longer use the above ranking by NPV per $1 cost and just skip projects pushing you over the budget. Such a procedure could sometimes lead to rejecting a project that would have fit in the budget under another ordering and that could have yielded a higher total NPV. For example, suppose projects in Table 6-4 were *indivisible,* and you ranked them by decreasing NPV/C_0. With a $100,000 budget, you'd pick numbers 1, 2, and 3; and with $40,000 left, you'd skip 4 and 5 and

TABLE 6-5
OPTIMAL SET OF PROJECTS FROM TABLE 6-3 GIVEN A $100,000 BUDGET

Project	Initial cost	Net present value
1	$20,000	$5,020.92
2	20,000	4,458.80
3	20,000	3,842.60
40% of 4	40,000	6,569.46
Totals	$100,000	$19,891.78 (maximum)

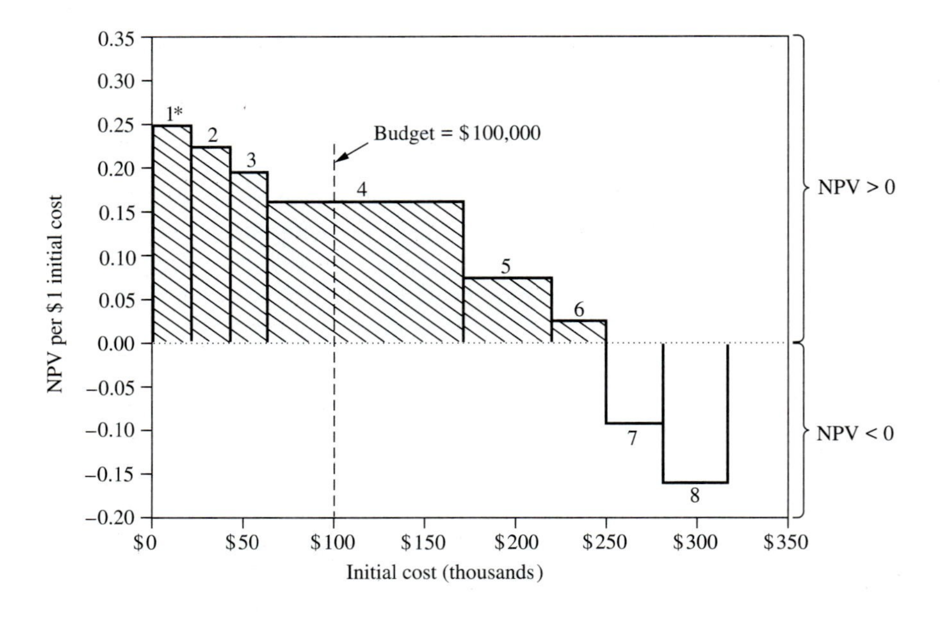

*Project numbers from Table 6-4.

FIGURE 6-4
Ranking divisible projects, given $100,000 budget.

would accept 6. Total NPV from column (e) would be $5,021 + $4,459 + $3,843 + $985 = $14,308. But that's suboptimal, because spending the full budget on project 4 gives an even higher NPV: $16,424.

When capital is limited and projects are indivisible, there is no simple decision rule that always guarantees a project ranking that will maximize net present value. By trial and error you could find the optimal combination of indivisible projects that maximized the present value of a limited budget. But that's very time-consuming when the number of possible projects is large. Although beyond the scope of this book, another approach is to use linear programming to find the optimum project combination. Such "activity scheduling" models with capital constraints are usually covered in forest management courses.

A SUMMARY OF CAPITAL BUDGETING

Given a wealth maximizing goal, the flowchart in Figure 6-5 summarizes the capital budgeting guidelines discussed so far, under limiting assumptions. Although it helps sort out the options, it's still very simplified. For example, the chart assumes that the reinvestment rate is always the MAR and the MAR is fixed over time, that projects are either mutually exclusive or independent, that projects are non-repeating, and that projects are either divisible or indivisible. Imagine the worst possible decision morass: The reinvestment rate for limited amounts of capital each year exceeds the MAR, the MAR changes over time, the group of possible

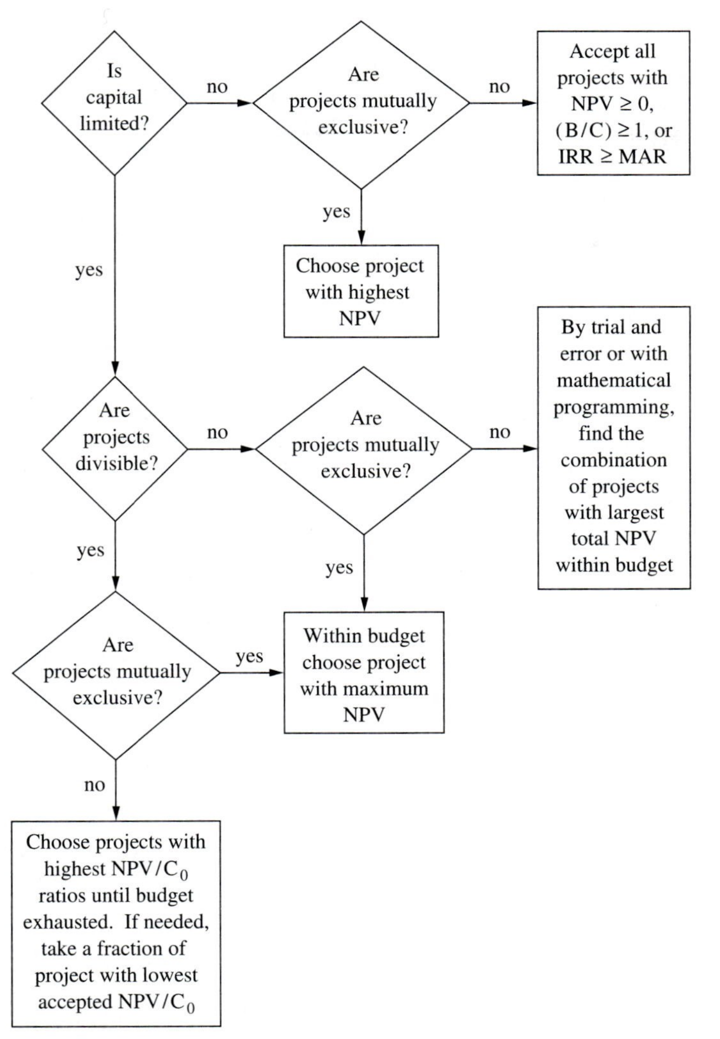

FIGURE 6-5
Capital budgeting flowchart—reinvestment rate equals minimum acceptable rate of return.

projects includes *both* mutually exclusive and independent projects, *both* repeating and nonrepeating, as well as *both* divisible and indivisible projects,[7] budget limitations change from year to year so that you must consider the level of net negative cash flows in future years as well as today, debt is incurred to extend

[7]When ranking projects by NPV/C_0 under capital rationing, by first selecting from mutually exclusive projects alone and then from independent ones, you might reject a project that would have been in a larger optimal set of indivisible, independent projects. One solution is to first consider all mutually exclusive and independent projects together. Then, as soon as you select a mutually exclusive project, discard all other mutually exclusive projects from that bundle.

budget limits (but how far?), and the degree of risk (expected variation in cash flows) differs among projects. When simplifying assumptions are relaxed and the number of options is large, the capital budgeting decision is tremendously complex.

Large simulation models and mathematical programming approaches are helpful but not always available for solving the above problems. In practice, options are usually limited, so the worst possible decision nightmares aren't common. Even so, choices aren't usually as fine-tuned as theoreticians might like. Several surveys have found that United States businesses, including the forest products industry, use internal rate of return most often to rank investments (Brigham and Gapenski, 1991; Redmond and Cubbage, 1985). While the IRR has theoretical shortcomings, it's easily understood, and for a wide range of forestry investments, Mills and Dixon (1982) and Haley (1969) find that NPV, IRR, and *B*/*C* ratio tend to yield similar forestry investment rankings under constrained budgets.

Ideal would be some foolproof guideline like "Choose projects in order of decreasing NPV or IRR," but no single approach applies to all situations. Even when conditions make the NPV guideline or NPV/C_0 seem appropriate, analysts should always give the investor other project measures such as IRR, payback period, and capital requirements over time. These are all relevant. To some extent, capital budgeting is an art that can't always be boiled down to a simple decision rule.

Several computer programs are available for conducting forestry investment analyses and applying many of the capital budgeting guidelines covered here. For lists of these, see Nodine et al. (1994) and Bullard and Straka (1993). When using such software, be sure you understand proper ways to include inflation (Chapter 5), taxes (Chapter 9), and risk (Chapter 10). And to make sure you know what's going on inside a program, always try to have it match some simple hand-calculated examples with a range of inputs. Sometimes these programs don't calculate results in the way that you expect, or they may have errors.

APPLICATIONS INCLUDING NONMONETARY VALUES

Although later we'll look at techniques for valuing nonmonetary outputs, capital budgeting tools allow us to consider such outputs without actually assigning dollar values to them. For example, suppose a public project is unacceptable by one of the financial criteria, but it will yield unmarketed recreation benefits. Below are examples of determining the minimum annual value of these benefits needed in order for the project to become acceptable. This doesn't actually place a value on the benefit, but the procedure often lets us see whether the minimum required value could reasonably be expected.

Perpetual Benefits

Suppose a State Parks Department considers spending $5,500 on a campsite with annual maintenance at $125. Given no camping fees, if the department needs a 7 percent real rate of return (MAR), what must be the minimum value of annual benefits B? *Assuming a perpetual life,* the present value of costs, and hence the present value of required benefits, is:

$$\text{PV costs} = 5{,}500 + 125/0.07 = \$7{,}286$$

where the second term is the present value of perpetual annual costs (equation 4-8). In order to be barely acceptable, NPV (or PV benefits of $\$B$/year minus PV costs) must be 0:

$$\text{PV benefits} - \text{PV costs} = B/0.07 - 7{,}286 = 0$$

Solving for the required perpetual annual benefit B,

$$B = 0.07(7{,}286) = \$510$$

Or, in general,

$$\begin{aligned}\text{Required perpetual real annual benefits} &= r|\text{negative NPV}| \\ &= \text{“annualized cost”}\end{aligned} \tag{6-10}$$

You can think of the NPV in equation (6-10) as the difference in NPVs of a project with and without investment for a nonmonetary good. For this example, to provide free camping, a park's financial NPV dropped by \$7,286. *For perpetual projects, the above equation shows that the required annual nondollar benefit is the interest rate times the loss in NPV incurred to provide the benefit.* While this doesn't place a value on the recreation, it helps analysts to determine whether the benefits are high enough in this example for the State Parks Department to earn its required 7 percent return. For instance, if projected use of the above campsite is 200 visits per year, benefits need to be \$510/200 = \$2.55 per visit. If people readily pay that much, or more, at fee areas (which they do), you could assume the site was worth at least that much, and the project would be justified. If you had simply asked, "Are the benefits worth at least as much as the \$7,286 present value of costs?" the answer wouldn't be as obvious.

The above \$510 is an example of **annualized costs** (equal annual costs) with the same present value as costs that are not annual. The time horizon above was perpetual; the finite case follows. In both equations (6-10) and (6-11) below, *make sure you use a real interest rate, since the equations assume that the annualized cost is fixed in real terms.* You'd use a nominal discount rate only if you thought the annual value would be fixed in current dollars, regardless of the inflation rate, and declining in real terms.

Terminating Benefits

Suppose investment project 8 in Table 6-4 was a proposed boat launching and picnic site with an expected 10-year life. Annual income would come from projected launching fees charged by the Park Agency. Nonboaters are expected to use the picnic area at no charge. The agency requires a 10 percent rate of return, but the project's internal rate of return is only 6 percent. What must the minimum value of picnicking be in order to raise the rate of return to 10 percent? Recall that the project's NPV of −\$5,780 was calculated with a 10 percent discount rate. Thus, still using 10 percent interest, if the present value of picnicking benefits would exactly offset that amount, making NPV = 0, the project would earn a 10 percent rate of return. (Remember that IRR is the discount rate at which NPV = 0.)

So all we have to do is find the annual picnicking benefit B with a present value of \$5,780 over a period of 10 years. The procedure is like finding the equivalent annual annuity (EAA) in equation (6-6):

$$\text{Required terminating real annual benefit} = B = |\text{negative NPV}|\frac{r}{1-(1+r)^{-n}} = \text{“annualized cost”} \quad (6\text{-}11)$$

As in equation (6-10), the negative NPV is in absolute value terms. Substituting the appropriate values for the example:

$$B = 5{,}780\frac{0.10}{1-(1.10)^{-10}} = \$941 \text{ per year}$$

Thus, if the annual picnicking benefits were worth at least \$941, the project would be acceptable: it would earn at least a 10 percent rate of return and have an NPV of 0 or greater. As above, the Park Agency could estimate projected use and judge whether the benefits were reasonable to expect. Suppose anticipated use was an average of 10 picnic groups per week, or 520 annually. Then required benefits would be 941/520 = \$1.81 per group. As above, you'd normally use a real discount rate in equation (6-11), assuming the offsetting benefits would be fixed in real terms and rise at the inflation rate in current dollars.

Although the above examples deal with costs, you can use equations (6-10) and (6-11) to annualize either costs or revenues for perpetual or terminating cases: For revenues, just enter a positive rather than a negative NPV.

Nonmonetary Summary

Remember that the above two sections don't find a value for nonmonetary benefits. The approaches simply organize information in a way that makes it easier to see whether benefits are worth the costs—just another example of economic thinking and asking the right questions.

You can generalize the procedure for finding the needed nonmonetary benefits to make any financial measure become acceptable: Calculate the measure, for example, NPV, B/C ratio, IRR, or NPV/C_0. *If unacceptable, calculate the needed addition to the present value of revenues in order to make the criterion become acceptable* (the above examples aimed for NPV = 0, but one could seek an NPV > 0, a B/C ratio > 1, or an IRR $>$ MAR). Then, use equation (6-10) or

(6-11) if you want to calculate the required annual benefit equivalent to the needed increase in present value. Questions at the chapter's end give more examples of such analyses.

The above types of calculations are relevant for public agencies (but that's not to say that they all necessarily make them). Individuals may also find cases where projects, say forestland purchases, may not yield an acceptable NPV strictly on cash flow projections. The above procedures could help them decide whether or not the annual nonmonetary benefits they expected from the property would be worth the annualized costs. Businesses primarily interested in financial rates of return and in income reports to stockholders aren't likely to consider nonmonetary outputs in the above manner. But these procedures are still useful in calculating annualized costs of "goodwill" practices like free forest recreation sites and scenic landscape management zones.

APPLICATIONS IN FOREST DAMAGE CONTROL

Another area for applying capital budgeting is in deciding how much to spend on controlling fire, insects, and diseases. You spend money on forest damage prevention and control (the cost) to prevent losing trees (the benefit). So you can rank damage control expenditures with measures like net present value, internal rate of return, or benefit/cost ratio to see where opportunities are best for improving net benefits. Cost estimates are often the easiest; for example, with bark beetle control, costs might be for spraying, doing precommercial thinning to maintain stand vigor, and removing beetle-damaged trees and other material providing breeding grounds. But benefits are often harder to estimate.

Let's start with a case of all-or-nothing damage control, considering only timber harvest revenues. On a given area, the damage control benefit is the net present value (NPV) of expected timber harvests without damage minus NPV with uncontrolled damage. If you've not subtracted control costs, this NPV difference is the maximum amount you can afford to spend on fully controlling damage, in present value terms. You wouldn't want to spend more preventing damage than its prevention was worth. Things get more complex when you consider damage such as reduced quality of scenery or water, where dollar values are hard to estimate, but the principle is the same.

In present value terms, the maximum you can spend on fully controlling damage and still earn the interest rate used in the present value calculations is:

$$\begin{aligned}\text{Benefits of damage control} &= \text{NPV of the forest with no damage}\\ &\quad - \text{NPV of the forest with uncontrolled damage}\\ &= \text{damage (loss)}\end{aligned} \tag{6-12}$$

Dealing only with the extremes of no damage versus uncontrolled damage, the equation tells you two things: (1) the benefit of completely controlling the damage and (2) the loss if you let damage occur. Note that *control costs have not been subtracted.* Figure 6-6 graphs the scenarios. The point N is NPV with no damage,

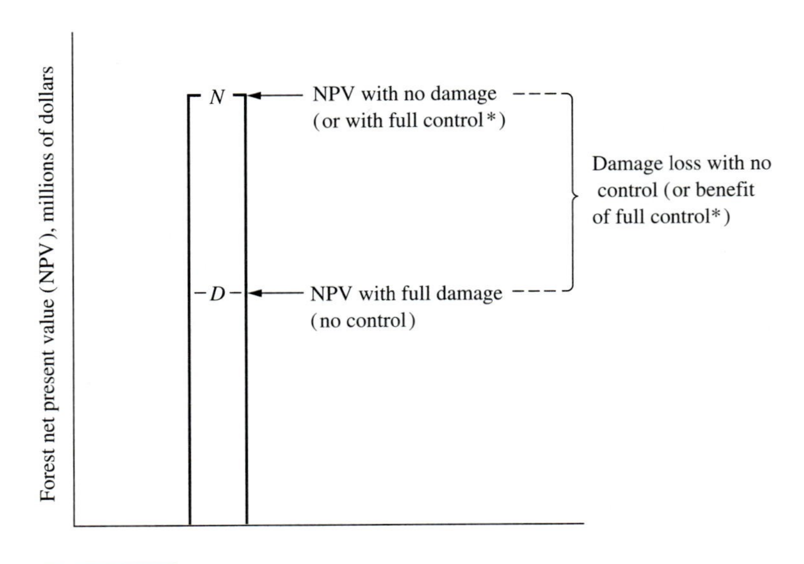

*Control costs have not been subtracted.

FIGURE 6-6
Damage loss and benefits of complete damage control.

and D is NPV with damage. Appendix 6A covers cases with various levels of damage control.

Using a simple example of all-or-nothing damage control in one timber stand, and assuming that all costs and benefits can be measured in dollars, let

PV_{Bn} = present value of all *b*enefits expected to occur with *n*o damage—e.g., harvest revenues and values of recreation, water, wildlife, etc.

PV_{Cn} = present value of all *c*osts with *n*o damage—e.g., costs of management and future reforestation, *excluding* damage control and prevention.

PV_{Bd} = present value of all *b*enefits with uncontrolled *d*amage—this would include any salvage cut after damage, plus future cuts and all other future benefits (some of which might be reduced) like recreation, wildlife, and water quality.

PV_{Cd} = present value of *c*osts with uncontrolled *d*amage—e.g., immediate and all later reforestation, erosion control, and management costs.

Remembering that net present value is the present value of benefits minus the present value of costs, equation (6-12) is:

$$\text{Benefits of damage control} = (\text{PV}_{Bn} - \text{PV}_{Cn}) - (\text{PV}_{Bd} - \text{PV}_{Cd}) \qquad (6\text{-}13)$$

Thus, the present value of damage control costs shouldn't exceed the benefits from equation (6-13). In measuring damage control costs, you should consider not only the first year's cost but the present value of all current and future damage control costs for the planning period. You can use this general framework for gauging the loss from any kind of forest damage and thus for finding your maximum willingness to pay, in present value terms, for preventing the loss.

A key point is that a timber loss isn't necessarily how much a logger would have paid for the damaged trees. Damaged trees might be too small to interest a logger, yet they still have value to the owner for future cutting purposes. That's why damage analyses look at present values of future benefits. Also, nontimber values need to be considered.

A Hypothetical Insect Problem—All-or-Nothing Control

Suppose you have a 19-year-old, even-aged forest scheduled for harvest in 11 years at age 30. Given progression of a bark beetle infestation, entomologists predict that, without an insecticide spray schedule, the timber would be so damaged by beetles next year that the wisest choice would be to do a salvage clear-cut yielding \$500 after attack and to replant at \$100 per acre. With a spray schedule starting next year, no damage would occur, and a harvest 10 years later would yield \$1,800/acre minus \$100 for reforestation. With or without spraying, after the first cut, later rotations would yield the same \$1,800/acre minus \$100 every 30 years. To correct for unequal lives, analyses continue in perpetuity for both cases. The present value of nontimber benefits would be \$50/acre with damage and \$65/acre without.[8] Using equation (6-13), next year's NPV of benefits from insect control, calculated with 6 percent interest, would be:

Benefits of damage control

$$
\begin{aligned}
&= \left[\frac{1{,}800 - 100}{(1.06)^{10}} + \frac{(1{,}800 - 100)/[(1.06)^{30} - 1]}{(1.06)^{10}} + 65\right] \\
&\quad - \left[500 - 100 + \frac{1{,}800 - 100}{(1.06)^{30} - 1} + 50\right] \qquad (6\text{-}14) \\
&= 1{,}214 - 808 \\
&= \$406 \text{ per acre}
\end{aligned}
$$

In the first set of brackets is NPV with no damage: the first term is the present value of the undamaged clear-cut in 10 years minus reforestation cost; the second term is the present value of subsequent harvests minus reforestation occurring every 30 years and discounted 10 years, since the series starts in 10 years. And the last term is the present value of nontimber benefits. In the second set of brackets is NPV with damage: the first two terms are the value of the salvage cut minus reforestation cost, and the third term is the present value of subsequent cuts minus

---

[8] See Chapter 14 for ways to value nonmarket outputs. Leuschner and Young (1978) have applied one of these methods to estimate losses in recreation values due to southern pine beetle (SPB) damage. See Leuschner et al. (1978) for an actual and more detailed estimate of (SPB) damages excluding nontimber damages.

reforestation every 30 years thereafter. The fourth term is the present value of non-timber benefits. In present value terms, \$406/acre is the maximum you can spend on insect control programs and still earn a 6 percent return on your investment, *if* you can attain the complete control assumed in equation (6-14). You needn't include costs or benefits that are the same with or without damage, since they'll cancel out of the equation. Examples could be annual overhead costs or hunting lease income.

If entomologists recommended spraying for 5 years, you could figure an equivalent annual annuity of the above \$406, using equation (6-6) at 6 percent interest:

$$406\left[\frac{0.06}{1-(1.06)^{-5}}\right] = \$96 \qquad (6\text{-}15)$$

Thus, you could spend up to \$96/acre/year for 5 years on insect control.

This is a highly simplified example. In reality, possible damage levels as well as control results are highly variable, so you'd need to deal with average values or **expected values** (see Chapter 10). Also, the above example covers only one maximum damage level and its complete control. Rather than the *single* level of complete damage control assumed above, with most kinds of damage there exists a range of control intensities (see Appendix 6A). Let's look at the theory of finding the optimum control expenditure.

Least Cost Plus Loss

The example will be fire control. Many activities can lessen the annual acreage burned on a forest. Total fire control costs include (1) *prevention,* e.g., public education, enforcing fire laws and closures, patrolling; (2) *presuppression,* e.g., maintaining equipment and supply inventories, training and maintaining crews, detection programs, building fire breaks; and (3) *suppression,* i.e., actual fire fighting costs. For simplicity, I'll combine these costs into one total, assuming a least-cost combination for any level of fire control achieved (realizing there are trade-offs—for example, if you spend more on prevention, suppression costs are lower).

For a hypothetical forest, Figure 6-7 plots total fire control costs per year as a 45-degree line from the origin, since the horizontal axis is dollars of control cost per year and the vertical axis is also dollars per year. The dashed line is total annual damage at different levels of control expenditure, starting at the indicated maximum annual damage with no control. The more you spend on control per year, the less the annual loss. The dashed damage curve convex to the origin shows typical diminishing returns to control efforts: eventually you reach a point where adding more to annual fire control cost doesn't decrease the loss much more. The dashed damage curve is average annual forest net revenue with no damage minus average annual net revenue with partially controlled damage—the amount by which net income is reduced below the no damage case. This differs from equation (6-12), which deals only with complete damage control. The solid curve of cost plus loss is the vertical sum of the control cost and loss curves, and

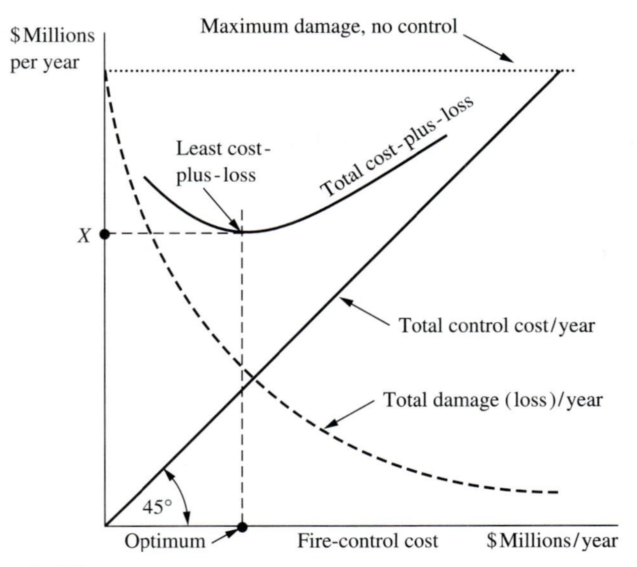

FIGURE 6-7
Model for optimizing annual fire-control expenditures on a hypothetical forest.
Gorte and Gorte (1979) review the history of such models. Marty and Barney (1981) discuss applications.

cost plus loss is minimized at $ X million per year. Thus, the optimum annual fire control expenditure is as shown on the horizontal axis in Figure 6-7.

The cost-plus-loss curve is a form of "total damage" with different control levels. Thus, the vertical distance between the dotted maximum damage line and the cost-plus-loss curve is the net benefit of control expenditures. So if you want to bend your mind a bit, the cost-plus-loss curve viewed *upside down* is a net benefit curve that is maximized at the optimum fire control cost. This is simple marginal analysis again: Keep adding to control cost until the added cost just equals the added benefit of loss reduction. At that point you've maximized the net benefit (or minimized the cost plus loss—see Appendix 6A for more detail).

Ideally, such analyses should be in present value terms, but in the case of fire, with a large forest area, we can expect that, on average, a given portion would be damaged annually with a certain equal annual control expenditure. Thus, the ratios of costs to benefits would be about the same in terms of average annual cash flows or present values. If you think of Figure 6-6 in present value terms, the model shows the optimum size of a project, where size is expressed as the present value of costs. Appendix 6A looks in more detail at the present value analysis of optimal damage control levels.

KEY POINTS

- For a given amount of capital, the capital budgeting problem is to find the pattern of investment spending that maximizes the resulting asset values.

- For the decision to accept or reject investment projects, four major criteria are net present value (NPV), internal rate of return (IRR), benefit/cost ratio (*B/C* ratio), and payback period.
- NPV is present value of revenues minus present value of costs, calculating with the investor's minimum acceptable rate of return (MAR).
- As calculated here, NPV is net of initial project cost, so that an NPV of zero does not mean the project is worthless. Some discounted cash flows have the initial costs subtracted; some do not. In order to correctly interpret NPVs, it's vital to know how initial costs have been treated.
- The IRR is the discount rate that equates the present value of revenues with the present value of costs.
- The *B/C* ratio is the present value of revenues divided by the present value of costs, calculated with the MAR.
- Payback period is the number of years it takes for project revenues to offset costs. This criterion gives no information about IRR or NPV. Thus, astute investors will rarely accept a project just because the payback period meets their approval.
- To compare and rank projects, determine first whether they are mutually exclusive, independent, divisible, or indivisible.
- If capital is unlimited, accept all projects with NPV ≥ 0, IRR $\geq$ MAR, or $(B/C) \geq 1$.
- If two projects are of unequal size, adding other projects to the smaller one until it equals the capital requirement of the larger one is called "normalizing for size." When comparing NPVs, such normalizing is not necessary if IRRs on the added projects equal the MAR but *is* necessary if IRRs exceed the MAR. Such normalizing is always necessary when comparing IRRs and *B/C* ratios.
- If two projects have unequal lives, NPV comparisons are valid if the shorter project cannot be repeated. If repetition is possible, equivalent annual annuities rather than NPVs should be compared.
- With a limited investment budget and divisible, independent projects, opportunities for any year should be ranked by decreasing NPV per dollar of initial cost. To maximize total NPV, invest in the best projects until the year's budget is exhausted.
- Under the above capital rationing, if projects are not divisible, ranking by NPV/C_0 will not necessarily maximize NPV. In that case trial and error or mathematical programming should be used to find the optimal set of projects.
- If dollar values aren't available for nonmonetary outputs from a project, pose the question, "What would the annual dollar value of nonmonetary benefits need to be to increase some financial criterion to a required level?" After computing these needed benefits, decision makers would be in a better position to gauge users' willingness to pay for them.
- The value of preventing a certain forest damage is the NPV of the forest without the damage minus the NPV of the forest with damage. In order for a damage control investment to earn at least the interest rate used in the calculations, the present value of control costs shouldn't exceed the value of damage prevented.

QUESTIONS AND PROBLEMS

6-1 A precommercial thinning investment of $190 now (year 0) is expected to increase timber harvest yields by $300, 15 years from now. On the same timber stand, a $100 fertilization investment in year 5 is expected to increase timber harvest value by an added $195 in year 15.

a Viewing these practices as a total package, what is the internal rate of return on this investment, to the nearest one percentage point?

b Why not figure rates of return for fertilization and thinning separately and average them to get IRR on the combination?

6-2 The benefit/cost ratio for an irrigation development project was computed at 1.7, using 6 percent interest. How will this *B/C* ratio change if 10 percent interest is used? (Costs occur earlier than revenues.)

6-3 Using 7 percent interest, the benefit/cost ratio for a federal reforestation project was 1. What was the internal rate of return? Explain your answer.

6-4 Using 10 percent interest, the U.S. Army Corps of Engineers calculated the benefit/cost ratio of a water impoundment project at 0.8, considering only dollar costs and revenues. They also expect nonmonetary recreation benefits from use of the reservoir. What must the present value of these recreation benefits be to obtain a *B/C* ratio of 1, given that the present value of project costs is $5 million?

6-5 The present value of costs was $10,000 for a new public campsite with an expected 17-year life. No camping fees are charged. What is the minimum fixed annual dollar value we must assume the recreation will yield in order for society to receive an 8 percent rate of return on the investment?

6-6 Suppose you're considering the following two mutually exclusive investments of different durations. Let your minimum acceptable rate of return be 7 percent. Note the different lives.

a Which has the greatest net present value?

b Which project would you choose, assuming the Christmas tree investment could be repeated?

c Which would you choose if Christmas trees could not be repeated?

Investment	Cost	Return
Christmas trees	$500 in year 0	1,080 in year *10*
Timber	$500 in year 0	4,400 in year *30*

6-7 Suppose you must choose between the two projects below, which are *independent* and *indivisible.* The only other available alternatives will earn a 7 percent rate of return, and your investment budget is limited to $5,000. Assume your MAR is 7 percent.

a Cost = $200 (year 0) — IRR is 15% — Revenue = $809 (10 Years)

b $5,000 (year 0) — IRR is 7.5% — $10,305 (10 Years)

How should you invest your money?

6-8 Suppose, in the previous question, the projects are *divisible,* and all other factors are the same. How should you invest?

6-9 A logging firm is evaluating logging equipment that costs $70,000 and is projected to yield a net after-tax income of $13,000 per year in constant dollars (of year 0) over a 7-year period. Assume that incomes occur at the end of each year and that the equipment will be sold for $5,000 (in constant dollars) at the end of year 7. At an 8 percent real minimum acceptable rate of return, what would be the net present value of such an equipment investment to the firm? Interpret this value: what does it mean?

This is a simplified example. What other types of factors might be considered in a "real-world" situation?

6-10 A park agency estimates the following campground development costs per unit:

Property purchase and construction	$3,500
Maintenance, in perpetuity	$300/year
Reconstruction, in perpetuity	$2,000 every 20 years

a Assuming the agency wants to earn a 5 percent real rate of return, what is the minimum real annual recreation value this campground must yield in perpetuity to justify the above costs?

b Given 5 percent interest and assuming that the campground will be used 125 nights each year, what will be the average cost per group night?

6-11 A park agency estimates the following costs for a new campground unit. Estimated life is 25 years, after which land use is expected to change.

Purchase and construction	$5,000
Annual maintenance	$250

a Assuming a 9 percent real MAR, what is the real minimum annual recreation value this campground unit must yield to justify the above costs?

b Based on a recreation user survey, the park agency feels that the cost per group night should not exceed $4. What is the minimum number of nights per year the campground must be used to justify the costs? Use 9 percent interest.

6-12 Assume $1,000 income today from salvaging a 20-year old timber stand damaged by fire this year. After a $500 reforestation cost, the area is expected to yield harvest income of $6,300 in 34 years. If the fire had not occurred, expected harvest income would have been $6,300 in 14 years. In either case, the $6,300 harvests minus $500 reforestation costs would continue in perpetuity every 34 years. Annual costs would be $20 and annual nontimber benefits would be $15 in both cases. Let the owner's discount rate be 6 percent, and assume no other costs or revenues. What is the dollar value of the fire damage (or the maximum amount the owner would have been willing to pay to prevent the fire)?

APPENDIX 6A: Marginal Analysis in Maximizing Benefits of Damage Control

Figure 6-6 and equation (6-12) equated loss from uncontrolled forest damage with the benefit of damage control because only full damage control was considered. In Figure 6-8, *excluding damage control costs,* the distance *DP* shows the benefit of partial damage control in present value terms: the NPV with partial control minus the NPV with

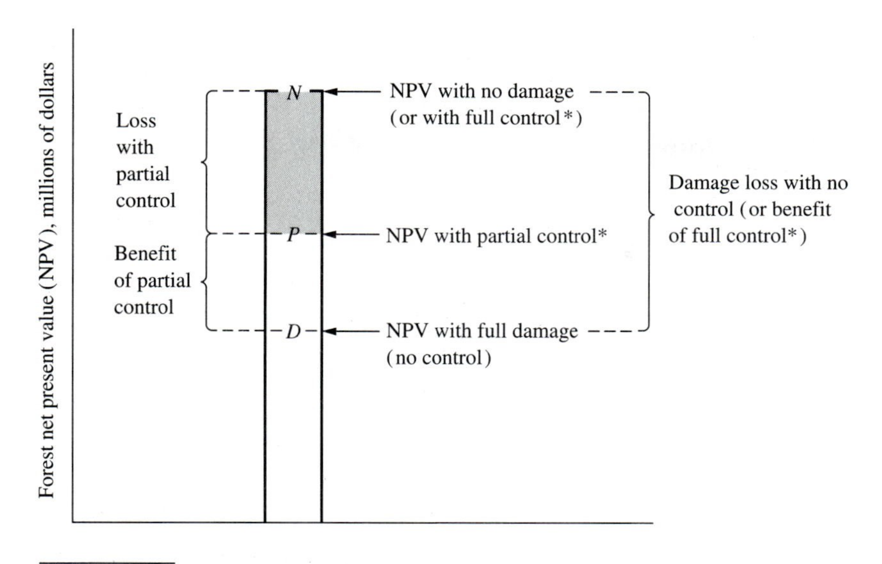

*Control costs have not been subtracted.

FIGURE 6-8
Damage loss and benefits under different levels of control.

full damage. You can imagine *P* in Figure 6-8 moving up or down depending on the amount spent on control. The shaded area in the figure is the loss with partial control:

$$\text{Damage loss with partial control} = \text{forest NPV with no damage} - \text{NPV with partial control} \tag{6A-1}$$

Figure 6-9a generalizes the Figure 6-7 fire control model to deal with any type of forest damage. The curves in the upper panel show the same relationships, but now in present value terms. The figure's dashed total damage loss curve is from equation (6A-1), which is the shaded portion of Figure 6-8 with varying damage control expenditure. The NPV of control cost is a 45-degree line from the origin in Figure 6-9 because the x axis is control cost and the y axis is also dollars.

In panel (b) of Figure 6-9, the dashed total benefit curve plots the difference between the dotted maximum damage line in panel (a) and the total damage curve (this difference is also the distance *DP* in Figure 6-8). The effect is to invert the dashed curve from the upper panel so that it comes from the origin in the lower panel. In Figure 6-9b you can see that net benefit is maximized where the slopes of the two curves are equal: where added benefit per dollar of control cost equals the added cost ($1). This is the same type of optimum where marginal revenue product equals marginal resource cost in Figure 2-7, except here the resource input is dollars of control cost, and values are present values.[9]

The key point of Figure 6-9 is that you can generalize the marginal analysis to optimize expenditures on controlling any type of damage like insects, fire, and diseases. The major problem is in estimating costs and benefits of different control levels. Abstracting from the forest damage problem, Figure 6-9b also shows the optimum scale of any divisible

[9]Rideout and Omi (1990) and de Steiguer (1991) stress the importance of including all benefits in forest damage analyses. Holmes (1991) shows how to include benefits and costs of price fluctuations caused by salvage of beetle-damaged timber.

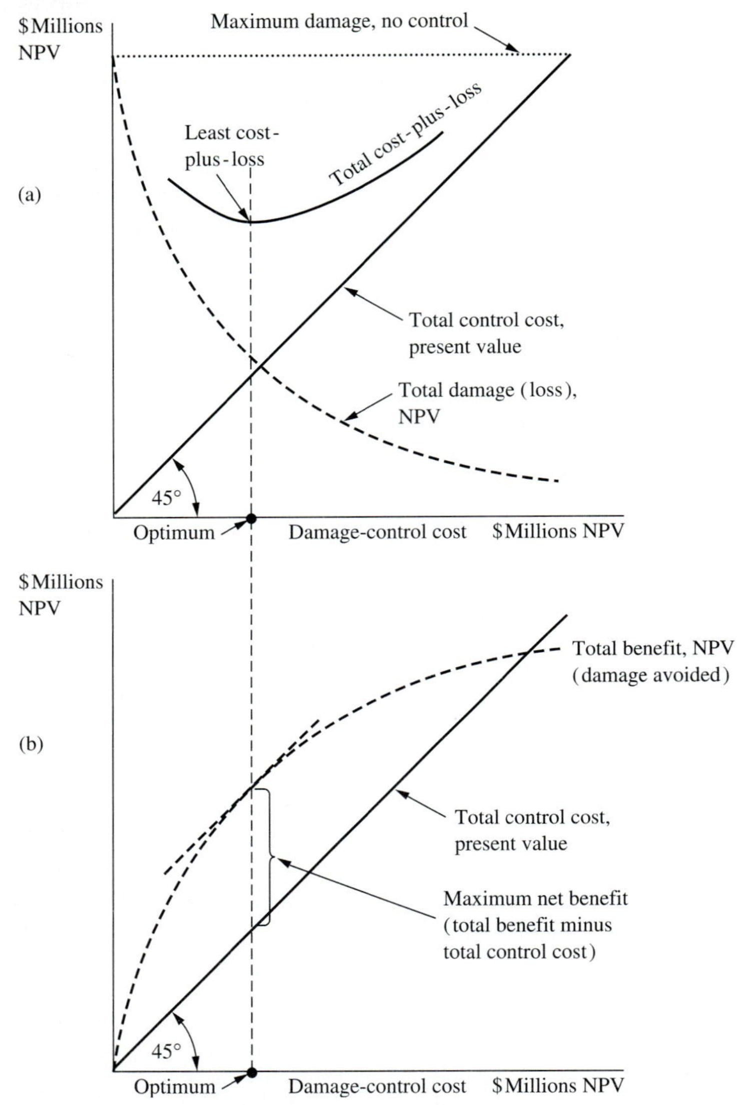

FIGURE 6-9
Model for optimizing present value of damage-control expenditures on a hypothetical forest.

project. Just let the *x* axis read "Present value of costs," and the solid and dashed curves read "Present value of costs" and "Present value of benefits." The optimal scale is where net present value is maximized.

REFERENCES

Baumol, W. J. 1977. *Economic Theory and Operations Analysis.* Prentice Hall, Englewood Cliffs, NJ. 695 pp.

Bierman, H., Jr., and S. Smidt. 1988. *The Capital Budgeting Decision.* Macmillan, New York.

Blank, L. T., and A. J. Tarquin. 1989. *Engineering Economy* (3rd edition). McGraw-Hill, New York. 531 pp.

Brigham, E. F., and L. C. Gapenski. 1991. *Financial Management—Theory and Practice* (6th edition). The Dryden Press, Chicago. 995 pp. plus app.

Bullard, S. H., and T. J. Straka. 1993. *Forest Valuation and Investment Analysis.* GTR Printing, Starkville, MS. 70 pp.

Collier, C. A., and W. B. Ledbetter. 1988. *Engineering Economic and Cost Analysis.* Harper and Row, New York. 635 pp.

Davis, L. S., and K. N. Johnson. 1987. *Forest Management* (3rd edition). McGraw-Hill, New York. 790 pp.

de Steiguer, J. E. 1991. *Comparison of Economic Criteria for Optimal Forest Damage Control.* U.S. Forest Service Southeastern Forest Exp. Sta. Research Note SE-362. Asheville, NC. 3 pp.

Gorte, J. K., and R. W. Gorte. 1979. *Application of Economic Techniques to Fire Management—A Status Review and Evaluation.* Intermountain Forest and Range Exp. Sta. Gen. Tech. Rept. INT-53. U.S. Forest Service, Ogden, UT. 26 pp.

Haley, D. 1969. A Comparison of Alternative Criteria for the Evaluation of Investment Projects in Forestry. Faculty of Forestry, University of British Columbia, Vancouver. 93 pp.

Holmes, T. P. 1991. Price and welfare effects of catastrophic forest damage from southern pine beetle epidemics. *Forest Science.* 37(2):500–516.

Klemperer, W. D. 1981. Interpreting the realizable rate of return. *Journal of Forestry.* 79(9):616–617.

Leuschner, W. A., T. A. Max, G. D. Spittle, and H. W. Wisdom. 1978. Estimating southern pine beetle timber damages. *Bulletin of the Entomological Society of America.* 24(1):29–34.

Leuschner, W. A., and R. L. Young. 1978. Estimating the southern pine beetle's impact on reservoir campsites. *Forest Science.* 24(4):527–537.

Marty, R. J., and R. J. Barney. 1981. *Fire Costs, Losses, and Benefits: An Economic Valuation Procedure.* U.S. Forest Service Intermountain Forest and Range Experiment Station Gen. Tech. Rept. INT-108, Ogden, UT. 11 pp.

Mills, T. J., and G. E. Dixon. 1982. *Ranking Independent Timber Investments by Alternative Investment Criteria.* U.S.D.A. Forest Service Pacific Southwest Forest and Range Experiment Station Research Paper PSW-166, Berkeley, CA. 8 pp.

Nodine, S. K., S. H. Bullard, T. J. Straka, and D. Gilluly. 1994. FORS complete guide to utilizing computer technology for forestry investment analysis. *The Compiler.* 12(3): 3–30.

Redmond, C. H., and F. W. Cubbage. 1985. *Capital Budgeting in the Forest Products Industry: A Survey and Analysis.* University of Georgia College of Agriculture Research Bulletin 333, Athens. 39 pp.

Rideout, D. B., and P. N. Omi. 1990. Alternate expressions for the economic theory of forest fire management. *Forest Science.* 36(3):614–624.

Schallau, C. H., and M. E. Wirth. 1980. Reinvestment rate and the analysis of forestry enterprises. *Journal of Forestry.* 78(12):740–742.

Van Horne, J. C. 1986. *Financial Management and Policy* (7th edition). Prentice Hall, Englewood Cliffs, NJ. 858 pp.

7

ECONOMICS OF FORESTLAND USE AND EVEN-AGED ROTATIONS

In an unregulated market, how would land use be determined? And once the use is resolved, what's the best management plan? Some lands are obviously unsuited for certain uses like agriculture: they may be too dry, too wet, too steep. And after the general type of use is set, for example, forestry, land features could preclude some species or management practices. So first you have to know what's physically possible. In forestry, much of your education centers on such knowledge: what species do well in certain environments and how they respond to given management practices. Once you eliminate the impossible or impractical options, economics can help in deciding which of the remaining ones to choose.

In a market economy, the principle of present value maximization is the driving force that determines land use. Land tends to be used for the activity that generates the greatest net present value of future satisfaction to the owner. And once the use is determined, this same principle defines how that use will be carried out. For example, if timber is the most valuable use, the best management regime is that which maximizes present value. In theory, this basic principle can help to determine things such as how many trees to plant per acre, whether or not to thin, when and how much to thin, how much to fertilize, and when to clear-cut (if at all) and replant.

This chapter applies some of the economic theory discussed so far to forest "stand level" decisions, which concern a fairly small, homogeneous timber type. "Forest level" decisions involve larger collections of stands. Sometimes what seems the best decision for one stand may be tempered in view of forestwide concerns. These are introduced in later chapters and treated extensively in forest management courses. Assume competitive conditions where landowners are price takers. Nonmonetary values are initially ignored but included later. All cash flows and interest rates will be real (excluding inflation), unless otherwise noted. Although taxes are not specifically considered, you could assume cash flows are after taxes, in which case the interest rates would represent after-tax rates of return. Or you could assume public agencies paying no taxes and earning before-tax rates of return. Let cash flows be **expected values;** for example, if 80 percent of the time a given investment is expected to yield $1,000 and 20 percent of the time to yield $500, the expected value is 0.8(1,000) + 0.2(500) = $900. An increase in the discount rate above the risk-free rate will reflect the reduced value to investors due to variance (risk) in the expected revenues. Chapters 9 and 10 cover taxes and risk in more detail. Notation for this chapter is in Table 7-1.

We'll start with **even-aged** silviculture, which means that the trees in any stand have been established at about the same time and are thus roughly the same age. If harvested, stands will be clear-cut, followed by reforestation and another even-aged stand. I discuss this method first only because the economic analysis is simplest, not because it's "best" in any sense. In many cases, partial cutting and developing uneven-aged, multistoried stands will yield greater satisfaction and social value—points that are covered in the next chapter.

LAND USE DETERMINATION

In an unregulated economy, land is bought by the highest bidder. Consider a rural tract of cutover timberland of all one site quality being offered for sale by the Woods Timber Company. Assume the land could be used for grazing, even-aged pulpwood, or even-aged sawtimber production. Soil type, slope, and other

TABLE 7-1
NOTATION FOR CHAPTER 7

Symbol	Definition	Symbol	Definition
a =	equal annual revenue, \$/acre	PV =	present value
B =	dollar value of nonmonetary benefit per year	q =	an index from 0 to $t - y$ for years between year y and clear-cutting age t
c =	equal annual cost, \$/acre	r =	real interest rate, percent/100
$(a - c)$ =	net annual cash flow, \$/acre (may be positive, zero, or negative)	R_y =	revenue in year y, \$/acre
C_y =	cost in year y, \$/acre	t =	rotation age (clear-cutting), also number of years between periodic occurrence of p
Δ =	change in		
fvg =	forest value growth percent/100		
H_y =	potential net clear-cutting revenue in year y, \$/acre	V_0 =	present value
L_y =	market value of land in year y, excluding crops, \$/acre	WPL =	willingness to pay for land, in present value terms, \$/acre
MAR =	decision maker's minimum acceptable rate of return, percent/100	WPL_1 =	willingness to pay for land, assuming land sale at end of one rotation, \$/acre
NPV =	net present value	WPL_∞ =	willingness to pay for land, assuming perpetual rotations, \$/acre
p =	fixed payment occurring every t years	y =	an index for years from 0 to t

TABLE 7-2
THREE BIDDERS FOR THE WOODS TIMBER COMPANY TRACT

Bidder	Prospective use	Costs $/acre	Revenues $/acre	Maximum bid price $/acre
1	Even-aged sawtimber production	\$150 reforestation now and every $t = 40$ years. \$50 thinning in $y = 10$ years and every 40 years thereafter. \$4 annual costs.	\$3,500 clear-cut income every $t = 40$ years. \$300 thinning income in $y = 20$ years and every 40 years after.	\$421 [equation (7-5)]
2	Even-aged pulpwood production	\$150 reforestation now and every $t = 20$ years. \$4 annual costs.	\$1,000 clear-cut income every $t = 20$ years.	\$284 [equation (7-6)]
3	Grazing	\$175 planting now.	\$20/year net revenue $(a - c)$	\$225 [equation (7-7)]

conditions rule out more profitable agriculture. For simplicity, assume that the land is isolated, and there is no demand for residential, recreational, or manufacturing uses.

Suppose three potential buyers projected different costs and revenues from the Woods tract, as shown in Table 7-2. Let these be typical land buyers from each user group. Many more cash flows would normally be considered. This is just a hypothetical case and doesn't reflect any region or actual situation. From these cash flows, financially astute investors would calculate net present values, given as land bid prices per acre in the last column of the table and explained below. Assume a 5 percent real **minimum acceptable rate of return (MAR)** for buyers.

Willingness to Pay for Bare Land—Perpetual Analysis

Initially, assume that potential buyers project land income in perpetuity (the next section allows the option of land sale at some future date). For even-aged timber production, the major income is a perpetual periodic series of clear-cutting revenues p per acre, at the end of every **rotation** of t years. At afforestation date, the present value V_0 of such income, using a discount rate of r, is given by equation (4-9) from Chapter 4:

$$V_0 = \frac{p}{(1 + r)^t - 1} \tag{7-1}$$

The problem with this equation is that not all the forestry net revenue p occurs every t years. By compounding each rotation's cash flows to rotation-end into one net value p, you can make forestry cases fit equation (7-1). At 5 percent interest, on the time line below, this compounding is shown per acre of the Woods tract, for the sawtimber cash flows in Table 7-2 (*temporarily excluding annual costs*). The compounded value p in equation (7-1) is the sum of the values encircled at the end of each 40-year rotation. In the example below, p is the sum of $796 + 3{,}500 - 216 - 1{,}056 = \$3{,}024$. The $150 reforestation cost in year 40 is not included in the year 40 sum, since that $150 belongs to the next rotation. Detailed calculations are shown at the end of the next paragraph.

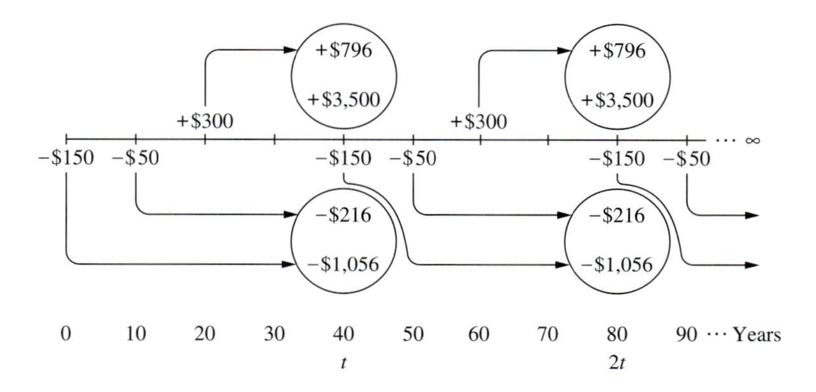

The above time line shows the repeating series of cash flows every $t = 40$ years. For each rotation, values are compounded at 5 percent interest for $t - y$ years to arrive at a net value of p every t years, as shown below for the first rotation of the sawtimber case:

Cash flow	**Value compounded for $t - y$ years to year $t = 40$, \$/acre**
Thinning income (year $y = 20$)	$+300(1.05)^{20} = +\$796$
Clear-cut income (year $y = t = 40$)	+3,500
Reforestation cost (year $y = 0$)	$-150(1.05)^{40} = -1{,}056$
Precommercial thinning cost (year $y = 10$)	$-50(1.05)^{30} = -216$
	Year 40 net compounded value $= p = \$3{,}024$

In equation form, this compounded value occurring every t years is:

$$\text{Net compounded value} = p = \sum_{y=0}^{t} R_y(1 + r)^{(t-y)} - \sum_{y=0}^{t} C_y(1 + r)^{(t-y)} \quad (7\text{-}2)$$

where R_y and C_y are costs and revenues. Since p occurs every t years in perpetuity, equation (7-2) can be substituted into equation (7-1), giving the formula below for **WPL$_\infty$**—the ***w*illingness to *p*ay for *l*and** per unit area, assuming perpetual rotations. Here the unit will be an acre, but any area could be used. Annual revenues and costs are cumbersome to accumulate to rotation-end and are added as a perpetual series $(a - c)/r$:

$$\text{WPL}_\infty = \frac{\sum_{y=0}^{t} R_y(1+r)^{(t-y)} - \sum_{y=0}^{t} C_y(1+r)^{(t-y)}}{(1+r)^t - 1} + \frac{a-c}{r} \tag{7-3}$$

Equation (7-3) is also known as the *Faustmann formula*, after Martin Faustmann (1849). The forestry literature calls this the **land expectation value** or **soil expectation value** (LEV or SEV), a translation from the German *Bodenerwartungswerte*. The notation WPL_∞ is more descriptive and also allows distinction between an infinite series of rotations and 1 rotation followed by land sale, WPL_1, discussed next. The WPL is a **net present value** designed for bare forestland—the present value of future revenues minus the present value of future costs, calculating just before reforestation. *Assuming only the cash flows in the equation,* WPL_∞ *is the maximum an investor could pay for bare land and still earn the minimum acceptable rate of return r.* In other words, someone paying $\$\text{WPL}_\infty$ per acre of bare land will earn a rate of return of r, if the assumed cash flows occur.[1]

Equation (7-3) can be made more specific for the sawtimber example:

$$\text{Sawtimber WPL}_\infty = \frac{H_t + R_y(1+r)^{t-y} - C_0(1+r)^t - C_y(1+r)^{t-y}}{(1+r)^t - 1} + \frac{a-c}{r} \tag{7-4}$$

where H_t is the clear-cut revenue in year 40, R_y is the thinning revenue in year 20, C_0 is the reforestation cost, and C_y is the thinning cost in year 10. You could expand equation (7-4) to include any number of revenues and costs R_y and C_y in any year y. Substituting the Table 7-2 sawtimber cash flows into equation (7-4), at 5 percent real interest, the maximum bid price per acre offered by sawtimber growers for the Woods tract is:

$$\text{Sawtimber WPL}_\infty = \frac{3{,}500 + 300(1.05)^{20} - 150(1.05)^{40} - 50(1.05)^{30}}{(1.05)^{40} - 1} - \frac{4}{0.05} = \frac{3{,}024}{6.04} - 80 = \$421/\text{acre} \tag{7-5}$$

[1] You can refine WPL equations to account for rising real prices (see Appendix 4B) and sequences of rotations with different lengths. However, with rotation lengths common in United States forestry, most of the land value comes from the first rotation, so WPL_∞, with its fixed rotation, gives a reasonable estimate, given the uncertainty about distant-future cash flows. Note that if you assume real timber price-increases relative to other prices, equation (7-3) stops the increase after one rotation (contrary to the approach in Appendix 4B). But relative price-increases of one good are unlikely to continue forever. The higher prices would stimulate greater wood supplies and encourage use of wood substitutes, both of which have a depressing effect on timber prices.

Equation (7-5) means that a buyer paying \$421/acre for the Woods tract would earn a 5 percent rate of return, if the indicated cash flows occurred. Recall from Chapter 6 that this 5 percent is the **internal rate of return (IRR)**, or the interest rate at which NPV = 0. Imagine that the above acre has been bought for \$421. Then the \$421 becomes a cost that you can move to the other side of the equal sign in equation (7-5) as a negative value. In that case, the equation value, which is an NPV, becomes 0, and the IRR is 5 percent (the r used in the equation). *If the projected cash flows occur, a land buyer paying more* (less) *than the computed WPL will earn a rate of return less* (more) *than r.* This same logic applies to any of the WPL equations.

Entering the Table 7-2 values for bidder 2 into equation (7-3), the average pulpwood grower's maximum bid per acre is:

$$\text{Pulpwood WPL}_{\infty} = \frac{1{,}000 - 150(1.05)^{20}}{(1.05)^{20} - 1} - \frac{4}{0.05} = \$284/\text{acre} \qquad (7\text{-}6)$$

A potential buyer of the Woods tract interested in grazing predicts that site preparation and grass seeding without removing stumps would cost \$175/acre, and annual revenue minus annual cost ($a - c$) would be \$20/acre, as shown in Table 7-2. Willingness to pay for the bare land is the present value of this income in perpetuity [equation (4-8)] minus the initial \$175:

$$\text{Grazing bid price for land} = \frac{a - c}{r} - C_0 = \frac{20}{0.05} - 175 = \$225/\text{acre} \qquad (7\text{-}7)$$

Figure 7-1 graphs the three bid prices per acre, under the given assumptions. Since a landowner generally sells to the highest bidder, sawtimber interests would tend to win in the bidding for the Woods tract, thus defining the land use. If \$421/acre were the average price paid for other similar tracts, it would be the **market value**, but you'd expect variation in prices from sale to sale. Market value is the most likely price that would be paid for an asset, if sold, assuming willing buyers and sellers.

In any of the land bid price equations, you can see that higher revenues would increase willingness to pay for land, and higher costs would reduce it. And that's reasonable, given the assumptions about competition and landowners being price takers: if expected costs increase, the landowner can't just increase output price (e.g., price of wood or farm crops). Owners have to take what comes, and bid prices adjust accordingly.

It is often unrealistic to assume perpetual identical rotations as in equation (7-3). A more flexible method of finding WPL includes only one rotation and adds the market value of land at rotation-end, as shown next.

Willingness to Pay for Land, Considering Future Land Value

In computing willingness to pay for bare forestland, the last section assumed that forestry would continue forever. This is realistic for some public agencies, but

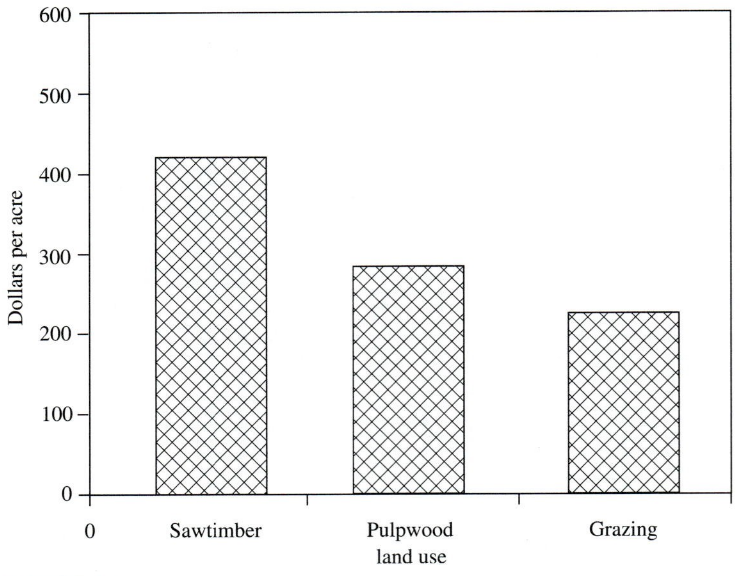

FIGURE 7-1
Hypothetical bare land bid prices for Woods tract.

many forest owners change land use or sell land before or after harvest, or at least consider the option. This applies even to large, stable forest products firms, which often sell forestland. To adapt equation (7-3) for land sale after harvest, simply calculate for one rotation, for each year y, the present value of all revenues minus the present value of all costs, making sure to include the market value of bare land (L_t) as a revenue after clear-cutting in year t. The result is WPL_1, a *w*illingness to *p*ay for bare *l*and, assuming land sale after *1* rotation:

$$\text{WPL}_1 = \sum_{y=0}^{t} \frac{R_y}{(1 + r)^y} - \sum_{y=0}^{t} \frac{C_y}{(1 + r)^y} + \frac{L_t}{(1 + r)^t} \tag{7-8}$$

To stress the importance of including the projected land value, L_t is shown separately rather than assuming it is part of the summed **present value** of revenues. Since equation (7-8) is not a perpetual series of net revenues, there is no need to accumulate cash flows to rotation-end (year t) as in equation (7-3). Note that, under competitive conditions, the projected (and current) market value of land is beyond a landowner's control and is independent of the other inputs into equation (7-8). Here, WPL equations represent net present values for one investor. The collective bidding behavior of all buyers sets the market value of land. This is an extremely important distinction: *Individual landowners' unique forestry costs or revenues do not affect what they can receive for their bare land on the open market* (assuming they aren't changing the character of the land by, for example, road building or drainage). Yet we noted that, *collectively,* the average net income from land will determine what buyers, on average, will bid for different types of land.

Chapter 11 gives more detail on the difference between one investor's willingness to pay for an asset versus the market value or the most likely selling price.

More specifically, equation (7-8) is:

$$\mathrm{WPL}_1 = \frac{H_t + L_t}{(1+r)^t} + \frac{R_y}{(1+r)^y} - C_0 - \frac{C_y}{(1+r)^y} + \frac{(a-c)[1-(1+r)^{-t}]}{r} \qquad (7\text{-}9)$$

The present value of annual revenues minus costs $(a - c)$ is separated out in the last term, as are revenues in year t (harvest and land value H_t and L_t) and the reforestation cost C_0. The remaining costs and revenues in any year y are shown as C_y and R_y. You can apply equation (7-9) to the Table 7-2 sawtimber case, assuming the land is expected to be sold for \$600 after clear-cutting in year 40 (constant dollars):

$$\mathrm{WPL}_1 = \frac{3{,}500 + 600}{(1.05)^{40}} + \frac{300}{(1.05)^{20}} - 150 - \frac{50}{(1.05)^{10}} - \frac{4[1-(1.05)^{-40}]}{0.05} = \$446$$

The land bid is slightly higher than the \$421 WPL_∞ computed for the perpetual case, because the \$600 land sale price exceeds the WPL_∞. If the value of L_t above exactly equaled the land bid price computed, equation (7-9) would be equivalent to WPL_∞ [equation (7-4)].[2] But the land sale case of equation (7-9) is important, because projected land value can differ greatly from WPL_∞.

Another reason for simulating land sale is that the perpetual analysis leaves out the income tax due when land is sold. This overestimates after-tax income for private landowners who plan future land sale. State and federal income taxes on land sales can be significant even when real land values are constant, because inflation generally causes nominal land values to increase (see Chapter 5). After-tax analyses will be covered in Chapter 9.

Interpreting the Land Bid Prices

You should think of the computed WPLs as maximum bid prices, considering financial factors only. Nonmonetary benefits to landowners can be added to the above framework. Sometimes, based purely on income, a nontimber use may yield the greatest present value for a tract of bare land; however, someone may buy the

---

[2]Take a simple WPL_∞ equation based only on perpetual harvest incomes and no other costs and revenues:

$$\mathrm{WPL}_\infty = \frac{H_t}{(1+r)^t - 1} \qquad (1)$$

That's equivalent to the present value of clear-cutting at age t and selling the land for WPL_∞:

$$\mathrm{WPL}_\infty = \frac{H_t + \mathrm{WPL}_\infty}{(1+r)^t} \qquad (2)$$

If you multiply both sides of (2) by $(1 + r)^t$, bring WPL_∞ to the left side, combine terms, and solve for WPL_∞, you'll get equation (1). The same equivalence holds for cases with many costs and revenues, but the proof is messier.

land for the nontimber price and plant trees because of their nonmonetary benefits of beauty and open space. Or we can note some tracts held in timber but yielding virtually no income. In many of these cases, aesthetic and recreational benefits make the owner willing to pay more for the land than the present value of timber income alone. So, in the broadest sense, bid prices for land are based on the present value of the utility it yields to the bidder. But buyers may not make their highest estimated bid if they feel they could purchase the property for less. Chapter 14 discusses problems in measuring nonmonetary benefits.

Financial analysts computing bid prices for any property should calculate a range of NPVs using different interest rates and cash flow scenarios. This is called **sensitivity analysis**: it shows how sensitive the result (here, NPV) is to changes in selected inputs. Decision makers can then judge which scenario is most reasonable and make a bid. For one tract, bidders considering the same use will probably have different inputs and generate different bids, the highest bidder becoming the buyer.

In many cases, discounted cash flow equations aren't used to find bid or asking prices for land. For example, in the above case, if the Woods Timber Company saw that similar tracts in the area had recently sold for about $400 per acre, that would likely be the company's minimum acceptable price, unless the firm was desperate for a quick sale. But many of the tracts in the $400 average could have had sale prices based on NPVs. Also note that past sale prices may not necessarily be a guide to current ones. For instance, in parts of the Southeast and some other locations in the United States, average real forestland sale prices declined during the mid-1980s, due to a large amount of land for sale and lower estimates of future timber prices. During other periods, the reverse has occurred.

Whether or not land buyers actually use NPV equations, some form of discounted cash flow determines the sale price, even if it's the seat-of-the-pants discounting mentioned earlier: commercial buyers pay less for land than the sum of all future net incomes. And the higher the potential income, the more they pay for the property. Often variations of the above equations are actually used in calculating land bid prices; in other cases the formulas are "models" of buying behavior. Usually forestland is bought together with timber of several ages, so that net present values are computed for an entire property, rather than for land alone. The above topics will be treated further in Chapter 11.

Looking at Figure 7-1, you could imagine other types of land where features like soil, slope, and location could lead to agriculture or other uses generating higher bid prices than forestry. And even for the Woods tract, other input assumptions could change the ranking of bids. If prices of beef relative to wood became high enough, grazing could be the highest value use on the tract. Or if expected pulpwood prices increased relative to sawtimber, pulpwood WPL could exceed that for sawtimber. Also, if you use a higher real interest rate (say, 7 percent), the ranking of bid prices can be reversed. Why?[3] As you'll see in the chapter

[3]The further in the future an income is, the more its present value is depressed by an increase in the interest rate. Forestry's WPL is thus reduced by a greater percentage than agriculture's when *r* is increased.

on timber supply, many areas in the United States have switched back and forth between agriculture and forestry, often as a result of changing expectations about prices of farm crops relative to wood. In Figure 7-1, you can think of this as changes in the relative height of land bid price bars.

All this explains what's meant by markets allocating land to the "highest and best use"—the use bringing the greatest present value of satisfaction *to the owner.* As you saw in Chapter 3, as a result of the **market failures** of unpriced positive and negative side effects in land use, the greatest utility to the landowner isn't necessarily maximum benefit to society. *Thus, the market theoretically yields a private optimum land use pattern—by the Figure 7-1 mechanism—but the social optimum may differ.*

Factors Affecting Forestland Values Let's consider factors that affect buyers' willingness to pay for forestland—and hence the land's market value—strictly from a wood production view. In equations (7-3) and (7-8), whatever increases the average owner's revenue R_y increases the land value. And the reverse is true for costs C_y, under competitive conditions where landowners are price takers. With each of the examples below, be sure to add *"other things being equal."*

- Higher site quality boosts the volume and value of harvests and thus increases land values.
- Land values tend to be higher where timber markets are strong—where mills must compete for wood and thus pay higher prices.
- Within any market area where mills pay a given price for delivered logs, lands farther from mills tend to have lower values because log hauling costs are higher. That reduces loggers' willingness to pay for stumpage to be harvested—more on this in Chapter 11. **Stumpage** is live timber, "standing on the stump."
- Factors that increase logging costs will tend to reduce land values, for example, steep slopes, high water table, unstable soils, or restrictive logging regulations (following the reasoning above).
- Conditions leading to high site preparation costs—for example, a tendency for heavy brush competition following afforestation—will decrease land value.
- Higher taxes tend to reduce land values (see the chapter on taxation).
- If timberland owners or buyers project a higher valued land use after clearcutting [increasing L_t in equation (7-8)], land value increases. That's why forestland values tend to increase as you approach population centers.

Remember, these cases relate to timber production. Certain factors could have opposite effects for other land uses. For example, some steepness could be an advantage for recreation lands. Also, the above points are often hard to document because so many things occur at once.

Now let's see how WPL equations can be used to find the most efficient ways to manage forests.

OPTIMAL ROTATION AGE FOR EVEN-AGED TIMBER

For one type of land use like even-aged loblolly pine pulpwood on a given site, you could calculate separate WPLs for different ways to manage such timber. Management variables include plant spacing, site preparation choices, thinning frequency and intensity, fertilization options, and clear-cutting age (rotation). *The economically optimum stand management plan for an owner is that which maximizes the net present value or willingness to pay for bare land just before afforestation.* Let's now consider just one management variable: the optimum age to clear-cut an even-aged timber stand.

Annual Forest Value Growth Percent

Think about two extreme cutting ages for a stand of timber: (1) cut the trees too early, and they're so small that they yield no revenue; (2) wait too long, and potential clear-cutting revenue from year to year stagnates or even decreases due to mortality. In both cases, WPL will be less than maximum. If you calculate WPL for different rotation ages between these extremes, the WPL rises, reaches a maximum at the economically optimum clear-cut age and then falls again. This occurs because, beyond the optimum rotation age, the value growth rate of the forest falls below the owner's minimum acceptable rate of return (MAR).

As shown by the tree cross sections in Figure 7-2, this decline in value growth rate is a result of the way trees grow. The hatched areas represent a year's growth ring of the same thickness. You can see that one ring on the larger tree is a much smaller percent of the cross-section area than one ring on the smaller tree. For one year, a percent increase in cross-section area is roughly the same percent increase in merchantable tree volume. Thus, *the same thickness growth ring becomes a smaller percent of tree volume as trees get larger.* Translating to value, a tree's percentage value growth rate also eventually declines with age. True, value per unit volume increases as a tree grows from pulpwood to sawtimber size and sometimes to valuable veneer logs. But eventually a tree's product mix stabilizes, and in addition to the mechanics of Figure 7-2, the growth rings

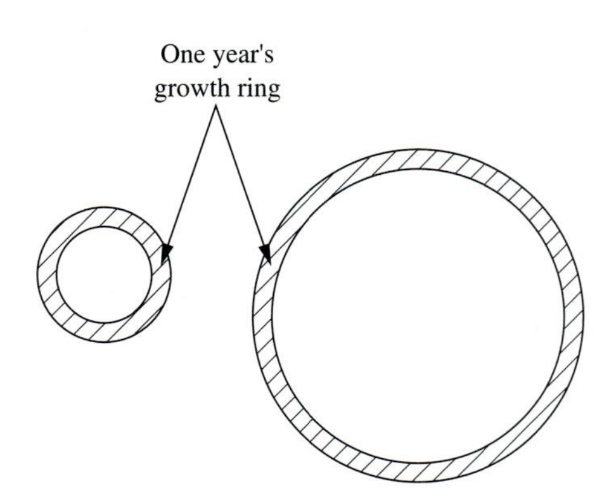

FIGURE 7-2
Declining annual volume growth percent with increasing tree age. Since tree volume is proportional to cross-section area, one year's growth ring of the same width becomes a smaller percent of tree volume as tree diameter (and age) increases.

also become smaller. So, unless timber prices are rising sharply, percentage value growth rates of timber stands ultimately decline.

Since forests are capital assets, their percentage value growth rates are crucial to investors. It's just like the discussion of profits: profits must be measured as a percent of capital value in order to see how efficiently capital is working. The same goes for forests: your annual growth per acre in volume or value is much less important than knowing this value as a percent of the total forest value per acre. One hundred dollars' worth of growth in one year on an acre of timber could be impressive if it was 15 percent of the total forest value per acre: if it was only 1 percent, the owner could do better by investing the forest capital elsewhere.

How do we measure the annual percentage value growth rate of a forest? Just divide one year's increase in value by the previous year's forest value. For example, in Figure 7-3, ignoring annual costs and revenues, the **forest value growth percent** between years y and $y + 1$ is $\Delta H/(H_y + L)$. If you could have cut the timber and sold the land for $1,000 last year, and it increased in value by $100 over the year, the value growth percent would be $100/1{,}000 = 0.10$, or 10 percent for last year. But a few clarifications are vital in calculating forest value growth percent:

1 *Make sure to include the market value of land in the forest value.* You need to earn interest on all the capital, which includes timber value plus land value.

FIGURE 7-3
Forest value growth.

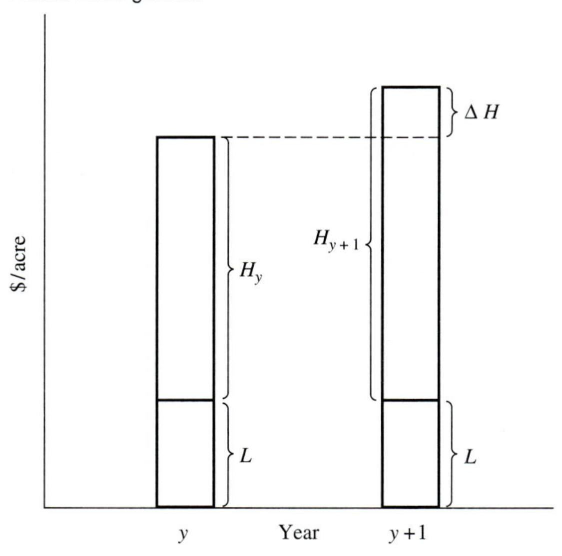

L = market value of land (here, constant)
H = stumpage value or potential harvest value

$$\left[\begin{array}{c}\textit{forest value growth percent}\\ \textit{between years } y \textit{ and } y+1\end{array}\right] = \left[\begin{array}{c}\Delta H/(H_y + L),\\ \textit{ignoring other revenues and costs}\end{array}\right]$$

Initially, we'll assume the land could be sold at any time for a given market value L_y, which the owner sees as independent of costs and income from ownership.

2 Since we are deciding whether to cut trees now or later, in this case, *the timber "value" should be the potential clear-cutting revenue (stumpage value),* net of harvesting costs. This **stumpage value** is the amount a buyer would be willing to pay for standing timber (stumpage) for immediate cutting. Potential timber clear-cutting revenue plus land value will be called the forest **liquidation value.** It's important to note that someone may be willing to pay *more* than the liquidation value if the forest is growing rapidly and has a high future harvest value; in that case it's worth more to hold the forest than to clear-cut it now.

3 *The computed increase in forest value must incorporate any annual cost or revenue incurred as a result of owning the forest.* For example, if forest liquidation value increases by $100 over a year, and costs for that year are $5, the true net value increase is $95. Or if annual nontimber revenues exceed costs by $5, the value increase is $105. In general, just add the sum of annual revenues minus costs $(a - c)$ to the year's increase in forest value.

The forest value growth percent between years y and $y + 1$ is the increase in forest liquidation value divided by the liquidation value in year y:[4]

$$\text{Annual forest value growth \%} = \frac{(H_{y+1} + L_{y+1}) - (H_y + L_y) + (a - c)}{H_y + L_y} \tag{7-10}$$

For example, suppose at age 24 a stand's stumpage value H is $500 and at age 25 it is $560; annual revenues minus costs are $(a - c) = -\$5$, and land value is fixed at $L = \$250$. Using equation (7-10), if land value is constant, it cancels out of the numerator, and forest value growth percent between ages 24 and 25 is:

$$(560 - 500 - 5)/(500 + 250) = 0.073, \text{ or } 7.3\%$$

An owner with an MAR of 8 percent wouldn't be satisfied with this value growth rate, assuming no other values were received. The decision would be to either clear-cut or accelerate the growth by thinning or fertilization if economically feasible. If the above $5 cost had been omitted, value growth percent would be 8 percent; ignoring land value would give an 11 percent value growth; forgetting both costs and land value would yield a 12 percent value growth. All three would be incorrect measures of forest value growth, if you annually consider the option of cutting the timber and selling the land. Note that equation (7-10) gives a *current* value growth percent from one year to the next. To obtain a projected annual value growth percent over *several years,* you'd need to compute forest

[4] A simpler equation, but perhaps not as intuitively appealing, is:

$$\text{Annual forest value growth \%} = \frac{H_{y+1} + L_{y+1} + (a - c)}{H_y + L_y} - 1$$

values at the beginning and end of the period and calculate the annual percentage value change over the period.[5]

If annual costs, revenues, and land value were negligible, the forest value growth percent in equation (7-10) would simply be the percent change in potential timber clear-cutting revenue $(H_{y+1} - H_y)/(H_y)$, the **timber value growth percent**, but this is not in general a suitable estimate of forest value growth percent. Note the terms: *forest* will mean trees plus land; *timber* will mean trees alone. Looking back to Figure 7-3, you can see that, *for any given change in potential clear-cutting revenue, the higher the land value, the lower the forest value growth percent will be,* given a fixed real land value. That's also evident in equation (7-10), where increasing the land value in the denominator decreases the value growth percent, if the land value is fixed. These relationships explain why it's hard to justify growing timber on high valued land. It would be valid to omit land value from equation (7-10) if the landowner received enough nonmonetary benefits from landownership alone to justify holding the land regardless of timber benefits. In that case, the owner would hold timber as long as its value growth percent was at least as great as the MAR, ignoring land value. But the initial assumptions exclude nonmonetary benefits: these are included later and tend to lengthen optimal rotations.

Make Sure to Correct for Inflation Although this chapter started by assuming all values were in constant dollars, the values collected for figuring forest value growth percent with equation (7-10) might be in current dollars. In that case, the value growth percent would be in nominal terms, which wouldn't mean much until you knew the inflation rate over the relevant period. The real forest value growth percent/100 (fvg) would be $[(1 + \text{nominal fvg})/(1 + f)] - 1$, where f is the inflation rate/100 (see Chapter 5). You then compare this with your real MAR. You can also compare nominal forest value growth percent with a nominal

---

[5]Approximations of forest value growth percent can be computed for multiyear periods, but it's difficult to include annual costs and revenues. Using equation (4-4) and ignoring $(a - c)$, the projected forest value growth percent between age y and $y + n$ is:

$$\text{Forest value growth \%} = \sqrt[n]{\frac{H_{y+n} + L_{y+n}}{H_y + L_y}} - 1$$

For example, if today, at age 25, stumpage value is \$900 and land value is \$200, and at age 30, these projected values are \$1,600 and \$225, the next 5 years' annual forest value growth percent is:

$$\text{Forest value growth \%} = \sqrt[5]{\frac{1{,}825}{1{,}100}} - 1 = 0.107, \text{ or } 10.75\%$$

However, if $(a - c)$ is negative (positive), the above value growth percent is overstated (understated). Incorporating $(a - c)$ over the n-year period would require using the trial and error internal rate of return method of equation (6-3).

MAR, but you must be sure the same amount of inflation is in both rates. The rest of the chapter continues to assume all values are real.[6]

Equation (7-10) is a capital performance indicator for the forest. If the forest value growth percent equals or exceeds the MAR, let the timber grow; if forest value growth percent < MAR, clear-cut and invest the revenue where it can earn more. (For the moment we're ignoring thinning and nonmonetary values.) The effect of such a rotation guideline is to maximize the willingness to pay for land or the net present value of the forest at rotation-start. Let's now look at WPL maximization.

Maximizing Willingness to Pay for Land

The following example applies the above rotation guideline to loblolly pine on a good site in the southeastern United States. The general principles and procedures will apply to any tree species. Assuming no thinning, we'll simulate different clear-cutting ages and find the one that gives the maximum WPL or willingness to pay for bare land, given a 5 percent MAR. For now, let's consider only monetary values. These are carried to the nearest cent only to show some of the marginal conditions: in practice, such precision isn't realistic.

Columns (a) and (b) in Table 7-3 give possible clear-cutting ages and harvested timber volumes in cubic feet per acre for a pine pulpwood example. The market value for bare land is held at $300/acre in constant dollars, planting and site preparation are $120/acre, annual revenues minus costs ($a - c$) are $-$4/acre, and, assuming pulpwood objectives only, the real stumpage price for harvested timber, net of harvesting costs, remains fixed at $0.25 per cubic foot [column (c)] or about $22/cord, with no increase to allow for higher valued sawtimber as tree diameters increase.[7] **Stumpage price** is stumpage value expressed in dollars per unit volume of timber. Column (d) is potential clear-cutting revenue, or column (b) times (c). Column (e) shows timber value growth percent, or the annual percent change in potential harvest value from column (d).

In column (g) is the forest liquidation value or what the landowner could receive in any year by clear-cutting the trees and selling the land [the market value of land in column (f) plus the stumpage value in column (d)]. Column (h) shows

[6]Note that if forest values and cash flows exactly share in the inflation rate, you'd multiply (H_{y+1} + L_{y+1}) and ($a - c$) by ($1 + f$) in equation (7-10) to get nominal growth percent. In general, forests tend to be affected by inflation in this way. In that case, the above inflation correction in the nominal forest value growth percent would leave the *real* fvg percent unchanged. *That means the optimal rotation would also tend to be unchanged by inflation.*

[7]Examples in this chapter are simplified to show basic principles and do not model representative loblolly pine management. More realistic examples would include other costs and revenues, taxes, and alternative assumptions about stumpage prices, risk, interest rates, and management methods. Market values for land are highly variable; in the South, during the late 1980s and early 1990s, $250 to $300 per acre of bare forestland has been common for large rural tracts away from development pressure, but some values have been outside this range. Bare forestland values can be much lower in regions with longer rotations or slower timber growth, for example, drier ponderosa pine sites in the Pacific Northwest and in zones with poor timber markets.

TABLE 7-3
LOBLOLLY PINE PULPWOOD CLEAR-CUT OPTIONS*

(a) Age, yr	(b) Vol, cu ft per acre	(c) $/cu ft	(d) Harvest $/acre (b)(c)	(e) tvg % (%ch.d)	(f) Mkt. val land $/acre	(g) Liq. val $/acre (d)+(f)	(h) fvg % eq. (7-10)	(i) WPL_1 $/acre eq. (7-9)	(j) MAI cu ft/yr (b)/(a)	(k) Holding value $/acre (NPV) eq. (7-13)
0					300.00	300.00				422.53
1					300.00	300.00				447.65
2					300.00	300.00				474.04
3					300.00	300.00				501.74
4					300.00	300.00				530.83
5					300.00	300.00				561.37
6					300.00	300.00				593.44
7					300.00	300.00				627.11
8					300.00	300.00				662.46
9					300.00	300.00				699.59
10	634.93	0.250	158.73		300.00	458.73		130.73	63.49	738.57
11	997.02	0.250	249.25	57.0	300.00	549.25	18.86	167.91	90.64	779.49
12	1349.28	0.250	337.32	35.3	300.00	637.32	15.31	199.43	112.44	822.47
13	1691.71	0.250	422.93	25.4	300.00	722.93	12.80	225.81	130.13	867.59
14	2024.31	0.250	506.08	19.7	300.00	806.08	10.95	247.53	144.59	914.97
15	2347.07	0.250	586.77	15.9	300.00	886.77	9.51	265.03	156.47	964.72
16	2660.01	0.250	665.00	13.3	300.00	965.00	8.37	278.73	166.25	1016.96
17	2963.11	0.250	740.78	11.4	300.00	1040.78	7.44	288.99	174.30	1071.80
18	3256.38	0.250	814.10	9.9	300.00	1114.10	6.66	296.17	180.91	1129.39
19	3539.82	0.250	884.96	8.7	300.00	1184.96	6.00	300.59	186.31	1189.86
†20	3813.43	0.250	953.36	7.7	300.00	1253.36	†5.43	‡302.53	190.67	1253.36
21	4077.21	0.250	1019.30	6.9	300.00	1319.30	4.94	302.27	194.15	
22	4331.15	0.250	1082.79	6.2	300.00	1382.79	4.51	300.05	196.87	
23	4575.27	0.250	1143.82	5.6	300.00	1443.82	4.12	296.11	198.92	
24	4809.55	0.250	1202.39	5.1	300.00	1502.39	3.78	290.65	200.40	
25	5034.00	0.250	1258.50	4.7	300.00	1558.50	3.47	283.85	201.36	
26	5248.62	0.250	1312.16	4.3	300.00	1612.16	3.19	275.90	201.87	
27	5453.41	0.250	1363.35	3.9	300.00	1663.35	2.93	266.95	201.98	
28	5648.37	0.250	1412.09	3.6	300.00	1712.09	2.69	257.15	201.73	
29	5833.49	0.250	1458.37	3.3	300.00	1758.37	2.47	246.63	201.15	
30	6008.79	0.250	1502.20	3.0	300.00	1802.20	2.26	235.50	200.29	

*Assumes land sale at rotation-end at $300/acre. All values in constant dollars. Volumes are from Weih et al. (1990).

†Optimum rotation age.

‡Maximum willingness to pay for land (WPL_1).

Abbreviations: tvg % = timber value growth percent
%ch.d = annual percent change in column (d)
Liq.val = liquidation value
fvg % = forest value growth percent

C_0 = $120.00 — 20.00 = optimum rotation age
$a - c$ = $-4.00 — $1,253.36 = liquidation value at optimum rotation age
$/cu ft = $0.25 — $300.00 = market value of land
$r = 0.05$ — $302.53 = maximum WPL_1
0.000 = annual rate of increase in stumpage price/cu ft

the forest value growth percent calculated with equation (7-10). For example, at ages 14 and 15, land and timber liquidation values are \$806.08 and \$886.77, with $a + c = -4$, and forest value growth percent according to equation (7-10) is:

$$(886.77 - 806.08 - 4)/806.08 = 0.095,$$

or 9.5 percent, as shown in column (h) of Table 7-3 at age 15.

For each age of our pine pulpwood stand, Figure 7-4a plots the 5 percent MAR and the forest value growth percent from column (h). The value growth percent declines for the reasons discussed under Figure 7-2. The graph indicates that if the trees are left to grow beyond age 20.7, the forest value growth percent drops below the owner's MAR. Where these two curves cross is the economically optimal rotation: from a financial view, owners would not wish to hold a forest after its net value growth percent drops below their minimum acceptable rate of return. In percent terms, as you lengthen the rotation year by year, you can think of the forest value growth percent curve as the marginal revenue, and the MAR as the marginal cost (the cost of earning opportunities forgone). So it's marginal revenue = marginal cost all over again. For example, column (h) of Table 7-3 shows that holding the trees for the twentieth year will bring a net value increase of 5.43 percent—still better than the alternative of 5 percent. But if you grow the trees from age 20 to 21, the value growth percent drops to 4.94 percent—no longer acceptable, given the 5 percent MAR. So, on the one-year grid of the table, the optimal rotation is 20 years, not as precise as the graphic solution but more exact than one can usually be in actual management situations.

Comparing columns (e) and (h) of Table 7-3, the annual timber value growth percent at any age exceeds the forest value growth percent. For example, at the optimal rotation of age 20, the timber value growth is 7.7 percent—more than 2 percentage points above the forest value growth of 5.43 percent. In order to provide a net 5.43 percent growth on the value of the entire forest (trees and land together), and to cover annual costs, the percent increase in timber value (stumpage value) alone must exceed the forest value growth percent. The extent of this excess depends on the levels of land value and $(a - c)$. Timber value growth percent is therefore not a useful guideline when finding optimal rotations for forest owners desiring a financial return on land value. For example, the timber value growth in column (e) of Table 7-3 stays above the 5 percent MAR through age 24, which is longer than the economic rotation. In Table 7-3, the same holds for the volume growth percent, which is the percent annual change in the column (b) volume. Although not shown, volume growth percent in this table equals value growth percent because the price per cubic foot is constant.

In column (i) of Table 7-3 is the WPL_1 from equation (7-9), assuming clearcutting and the possible option of land sale at age t:

$$WPL_1 = \frac{H_t + L_t}{(1 + r)^t} - C_0 + \frac{(a - c)[1 - (1 + r)^{-t}]}{r} \qquad (7\text{-}11)$$

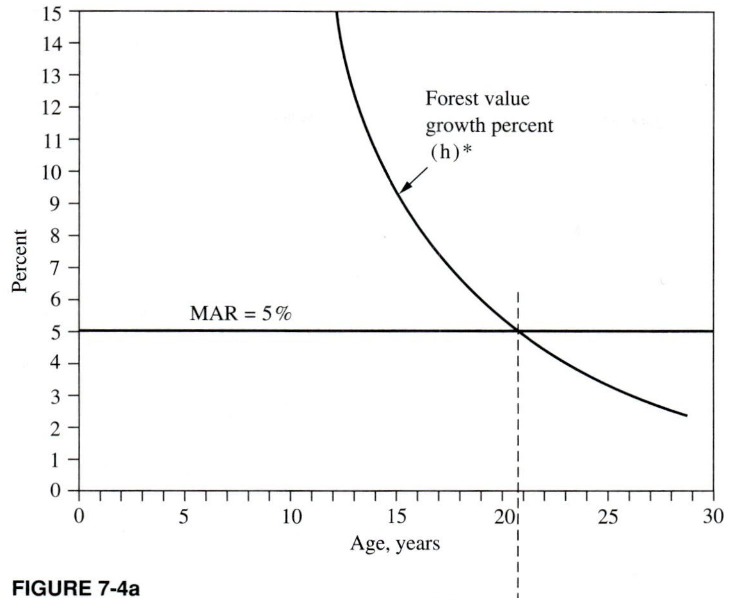

FIGURE 7-4a
Optimal pulpwood rotation, 5% interest rate ($300 land value).

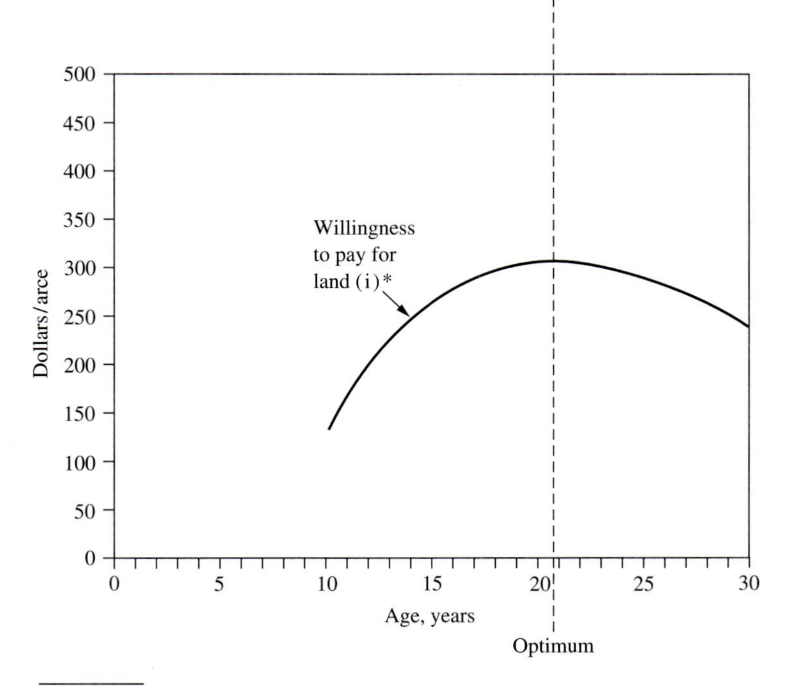

*Letters in parentheses are the columns in Table 7-3 which are graphed.

FIGURE 7-4b
Willingness to pay for land, pulpwood, 5 percent interest.

The sign of $(a - c)$ will be negative if annual costs exceed annual revenues, and positive if $a > c$. Entering the values for a rotation age of 20,

$$\mathrm{WPL}_1 = \frac{953.36 + 300}{(1.05)^{20}} - 120 - \frac{4[1 - (1.05)^{-20}]}{0.05} = 302.53 \qquad (7\text{-}12)$$

as shown in column (i) at age 20. This also is the highest possible WPL in column (f), verifying that 20 years is the optimal rotation.

Figure 7-4b plots the WPL_1 values from column (i) of Table 7-3, showing the more precise optimal rotation of about 20.7 years, as it ties into Figure 7-4a. So if you follow the rotation guideline of Figure 7-4a, you'll also maximize the WPL_1 (a net present value). Think of the process this way: as long as your net forest value grows into the future faster than the rate at which you discount that value to the present, you'll increase present value by growing the forest one more year. But the moment the rate of forest value increase is less than your discount rate, postponing the clear-cut one more year will decrease the NPV. Note the incremental thinking. The same reasoning applies in absolute terms: holding the forest one more year brings a net increase in potential liquidation value of $[(L_{y+1} + H_{y+1}) - (L_y + H_y) + (a - c)]$, the marginal revenue. During that year, you forgo the chance to earn interest on the liquidation value at your MAR, or $r(L_y + H_y)$, the marginal cost. You hold the forest until marginal revenue drops below marginal cost.

Note that the level of annual costs and revenues [or net level of $(a - c)$] will affect the optimal rotation when maximizing WPL_1. The WPL_1 equation realistically keeps the *market value* of land L_t independent of the owner's $(a - c)$. Thus, you can see from equation (7-10) that an increase in the owner's projected annual costs c will decrease the forest value growth percent at any age. The reverse is true for annual revenues a. In the framework of Figure 7-4, this means that higher c shifts the forest value growth percent downward and shortens the optimal rotation. Higher a has the opposite effect.[8]

[8]Many analyses conclude that fixed annual cash flows don't affect the optimal rotation under the assumption that higher (lower) costs will reduce (increase) land value. In that case, the negative value of perpetual annual c reduces land value by c/r. So, while the presence of c decreases the annual forest value growth by c, it also decreases the opportunity cost of holding land by $r(c/r) = c$. Since c reduces the added value of growing trees one more year by the same amount that c reduces the added cost, the optimal rotation would be unchanged. This is relevant in a general equilibrium sense, where changes in income to land will affect the market value of land. It also applies where an infinite time horizon must be used, and WPL_∞ is a surrogate for the market value of land. But *individual* landowners, as price takers, are likely to see the market value of land as fixed and independent of their projected cash flows and management plans. From this static partial equilibrium view, the level of $(a - c)$ *does* affect optimal rotations. Analogous conclusions hold for future stumpage price levels. A higher fixed stumpage price, say \$0.50/ cubic feet in Table 7-3, lengthens the optimal rotation slightly with the WPL_1 formulation because the land value L_t is fixed. But in a WPL_∞ or general equilibrium approach, doubling price also doubles land value, so the optimal rotation would be unchanged. These points are unlikely to be significant, since, over time, market values of forestland tend to rise and fall with changes in the average levels of $(a - c)$ and stumpage prices. But the WPL_1 approach is relevant for the landowner doing sensitivity analyses of alternative cash flow scenarios.

Maximizing Mean Annual Increment

For each clear-cut age, column (j) of Table 7-3 gives the **mean annual increment** (MAI) in cubic feet (harvest volume divided by harvest age). The MAI is the average annual timber volume growth and is maximized at 201.98 cubic feet per acre per year by harvesting at age 27. No other rotation could yield more volume over time. The rotation that maximizes MAI is sometimes called a **biological rotation**, in contrast to the economic rotation discussed above. The U.S. Forest Service, on its commercial timber production areas, chooses the rotation age that maximizes MAI. However, this approach doesn't assure efficient capital use that maximizes the present value of society's capital. Although here the MAI-maximizing rotation exceeds the economically optimal rotation, that won't necessarily always be the case (as you'll see in Table 7-4). For example, if the interest rate is low enough and stumpage prices are increasing, or if nontimber values are sufficiently high, economically optimal rotations can *exceed* the MAI-maximizing rotation. Moreover, the measure of volume (e.g., cubic feet, or various board foot measures) for computing MAI can sharply affect the biological rotation age. *In general, biological rotations take no account of costs and benefits and are unlikely to maximize satisfactions from forest management.*

Keep an Eye on the Market Value of Land

Calculating willingness to pay for land as WPL_∞ [equation (7-3)] assumes that the rotation-end land value is the calculated WPL_∞ (see footnote 2). Since this is usually not the case, it's more realistic to calculate WPL_1 and enter the best estimate of rotation-end land value L_t.

Since the maximum WPL_1 of \$302.53 in Table 7-3 is so close to the assumed \$300 market value of land, the value for WPL_∞ using equation (7-3) would have been essentially the same as WPL_1. Thus the optimal rotation that maximizes WPL_∞ for our example is also the same 20 years. However, for the usual case where $WPL_1 \neq WPL_\infty$, the two approaches can yield different optimal rotations. For one thing, consider the annual dollar opportunity cost of holding land: r (land value), or what you could earn elsewhere by investing the land value in an alternative investment earning r. Maximizing WPL_1 automatically considers the correct land holding cost by including the estimated potential land sale price. Maximizing WPL_∞ considers land holding cost based on the computed WPL_∞, which can be much lower than the market value of land.

In calculating optimal rotations or bid prices for land, should one use WPL_1 or WPL_∞? The infinite time horizon is relevant for cases where land sale is absolutely impossible or highly improbable, for example, on certain public lands. The WPL_∞ is easily used in the Table 7-3 format by calculating column (i) with equation (7-3) and entering the maximum WPL_∞ in column (f) instead of the market value of land. However, when the calculated WPL is much lower than market value, assuming all values have been included, one could raise the question that perhaps another management plan or a different land use should be adopted. So with either approach, you should keep an eye on the market value of land. That's especially important for private firms: if they manage so that the present value of net revenues from their forests is much less than market value, they're ripe for a takeover.

Due to high costs, low revenues, or a high MAR, a calculated WPL can often be far below the market value of bare forestland. In that case a prospective land buyer wouldn't purchase the land, ignoring other values. But suppose you're a private landowner with the option of selling land or changing land use, and you own immature timber and compute a WPL significantly below the market value for that land. In that case, in the optimal rotation framework of Figure 7-4, you should use WPL_1, which considers the market value of land at rotation-end. The WPL_∞ formulation would underestimate the opportunity cost of holding land and would therefore give too long a rotation.

What if your WPL, computed by either method, is much *higher* than the market value of land? This is not a likely situation because it suggests you know something that the market doesn't, or that you may be using a discount rate lower than the market rate: economists would call both cases "market imperfections." But in cases where such imperfections did occur, landowners would find the infinite WPL_∞ the logical approach to use in the Figure 7-4 rotation framework if they were not planning to sell land at rotation-end.

Holding Value versus Liquidation Value

In Table 7-3, in contrast to the column (g) liquidation value from clear-cutting and selling land, column (k) shows the **holding value**: the net present value of holding the forest and selling land and timber *at the optimal rotation age t.* At any age y, if we let q be an index for the number of years $(t - y)$ between y and the optimal rotation age t, holding value is:

$$\text{Holding value at age } y = \sum_{q=0}^{t-y}\left[\frac{R_q}{(1+r)^q} - \frac{C_q}{(1+r)^q}\right] \tag{7-13}$$

At any age y, equation (7-13) is the present value of all revenues R_q minus the present value of all costs C_q occurring from age y through t. At age t the clear-cut income is included, and the market value of land remains either as an asset or as a sale income (you need to recognize the land value but not necessarily sell it). Using the loblolly pine pulpwood example with a 20-year rotation and the 5 percent discount rate, you can calculate the holding value for the forest at age $y = 12$, in which case $t - y = 8$:

$$\text{Holding value at age } y = \frac{H_t + L_t}{(1+r)^{(t-y)}} + \frac{(a-c)[1-(1+r)^{-(t-y)}]}{r} \tag{7-14}$$

$$= \frac{953.36 + 300}{(1.05)^8} - \frac{4[1-(1.05)^{-8}]}{0.05} = \$822.47$$

as shown in column (k) at age 12. The second term above is the present value of equal annual cash flows between ages 12 and 20.[9] You would need to include in equation (7-14) other revenues or costs occurring between ages y and t. For example, a $200 thinning revenue at age 15 (or three years in the future) would be entered as $+200/(1.05)^3$.

Figure 7-5 graphs the market value of land, the **liquidation value**, and the holding value from columns (f), (g), and (k) in Table 7-3. As long as the owner's holding value exceeds the liquidation value, it is worthwhile to hold the forest and not clear-cut. When the forest holding value exceeds the liquidation value, the timber is said to be *financially immature*. The optimal rotation age is where holding value equals liquidation value, or slightly over 20 years in this case, consistent with the previous analyses. Beyond that age, the timber is **financially mature**, and the owner could accumulate greater wealth by investing the age 20 clear-cutting income plus the $300 land sale at 5 percent interest. Or a new rotation could be started, assuming the WPL_1 was equal to or greater than the $300 land value.

As defined above, holding value discounts the optimal rotation age revenue. But if you're at or beyond the rotation age, you could also compute the net present value of a still later liquidation. For example, starting at age 20 in Figure 7-5 and looking ahead to a projected clear-cut at age 30, such a present value at 5 percent interest is $1,075.51 at age 20, using the Table 7-3 values in equation (7-14) and letting t be 30 years, rather than the optimal rotation. (Can you calculate that value?) Since the age 20 liquidation value is greater, at $1,253.36 in column (g) of Table 7-3, you shouldn't wait until age 30 to harvest, based on financial criteria alone.

In general, to decide whether to continue holding any asset: *If net present value of future cash flows exceeds current liquidation value, hold the asset. If net present value of future cash flows is less than current liquidation value, sell.* These general guidelines apply to any asset held strictly for investment purposes, for example, land, art, casks of wine or whiskey, stocks, individual trees, or entire forests. Using equation (7-13), you can compute today's holding value for any asset, assuming liquidation or sale in some year t, for example, the year 2010. You may or may not have intermediate costs and revenues. Just let the future revenues be R_q and future costs be C_q, and let y be the current year. But it's important to simulate several liquidation dates t to make sure you don't overlook the one yielding the highest net present value.

Although Figure 7-5 relates to one hypothetical landowner, average holding values for similar stands would tend to define a market value. Where competitive markets exist, buyers planning to hold timber tend to pay more for immature timber than its stumpage value for immediate harvesting purposes. I've seen this in sales of immature timber in the Douglas-fir and southern pine regions. It's generally a mistake to assume that the market value of financially immature timber equals the stumpage value per unit volume times the units of volume. But exceptions exist, and these will be discussed in Chapter 11.

[9]Equation (7-14) is analogous to equation (7-11) for WPL_1, except that C_0 doesn't appear in equation (7-14), and discounting is only for the number of years until rotation end ($t - y$ years).

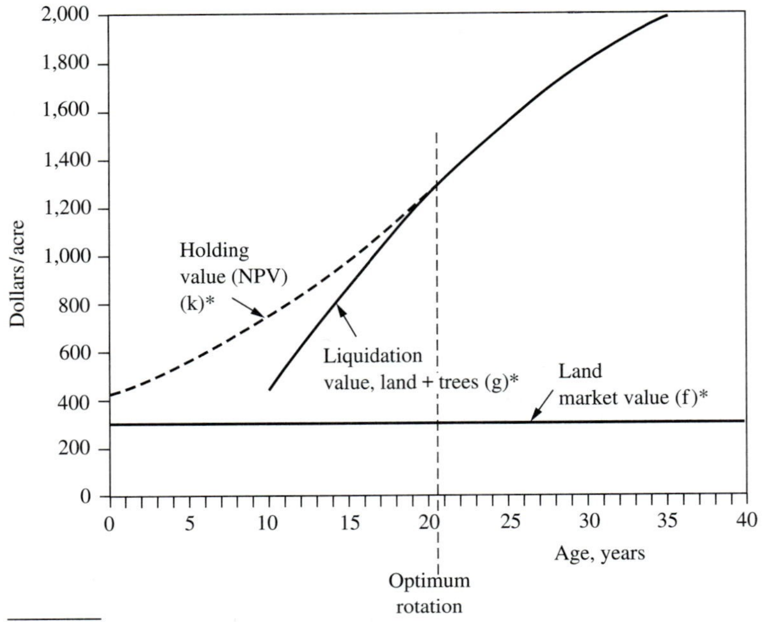

* Letters in parentheses are the columns in Table 7-3 which are graphed.

FIGURE 7-5
Loblolly pine pulpwood forest values.

The constant real land value in Figure 7-5 models an equilibrium case where real stumpage prices are constant. It's important to note that land value doesn't have to increase in real terms for forestry to be economically attractive. In current dollars, forestland values tend to increase when inflation is greater than zero, since prices of forest outputs usually rise with inflation. And where output values increase in real terms, real forestland value will rise over time, because the WPL after each rotation will be higher due to larger expected incomes.

Optimal Rotations and the Interest Rate

So far our example has used a 5 percent real interest rate. Still assuming the $300 per acre land value, suppose Table 7-3 were recomputed to simulate owners using discount rates of 3 and 7 percent. The resulting optimal rotations are shown using growth percent analysis in Figure 7-6a and maximum WPL_1 in Figure 7-6b. This approach assumes that the market value of land remains at $300/acre. We're only simulating views of landowners or potential buyers with different discount rates. In this example, regardless of the value someone ascribes to land (WPL), the land tends to sell for the market value of $300/acre. In that case the forest value growth percent curve remains the same as before.[10]

[10] A general equilibrium analysis would have the three interest rates be market rates, in which case the WPLs would be different market values of land. For such an analysis, each interest rate would generate a different forest value growth percent curve due to the changing land value.

The logic of Figure 7-6 is the same as that in Figure 7-5: forest owners will clear-cut timber when the forest value growth percent drops below their MAR. Therefore, the higher the discount rate, the shorter the economically optimal rotation. And in Figure 7-6b you can see that the higher the discount rate, the lower the maximum WPL will be. In Figure 7-6b the optimal rotation for someone with a 7 percent discount rate is about 17 years, which generates a WPL of $170 per acre. Such a person would not be willing to pay the full $300 market value for bare land if no other income or benefits were expected.

Looking at Figure 7-6, it's tempting to think that lower interest rates would favor the environment. True, lower rates lengthen rotations and increase the net present value of forestry investments. Since forestry's benefits are in the distant future, from a bare land view, lower interest rates would tend to boost forestry's WPL more than other land uses: that would put forestry in a better bidding position for bare land, resulting in more forests. But lower discount rates also favor development projects such as dams, highways, and factories. These can damage the environment and cause more pollution and energy consumption. So, although lower interest rates would favor forestry, it's not at all clear that they'd improve the environment overall.

Impact of Stumpage Prices Increasing with Timber Age

As average stand diameter increases, gradually a larger percentage of the volume will be higher valued sawtimber and veneer logs. The effect is to boost the stumpage price per cubic foot as the stand ages. Part of this price rise could also be due to real increases in stumpage value of any product type, a trend that analysts project in many regions. You could model these price trends in many ways, for example, by calculating average stand diameter at each age and including an equation making price per cubic foot an increasing function of average diameter, based on market projections (see Chapter 13). Or you could predict separate product volumes, such as cords of pulpwood and board feet of sawtimber at each age, along with prices for each product. For long-term projections, the latter approach may be a bit inflexible, since distant-future products are so hard to predict.

To model stumpage prices increasing with time and timber size, column (c) of Table 7-4 allows the real stumpage price per cubic foot to increase at 3.5 percent annually after age 19. The table is similar to the pulpwood case of Table 7-3, but here the percentage of more valuable sawtimber volume increases with stand age; at the same time, prices of sawtimber and pulpwood could also be increasing. The loblolly pine yield function, the $300/acre land value, $120/acre afforestation cost, $4/acre annual cost, and 5 percent real discount rate are the same as in Table 7-3. The calculation procedures are identical with those of the pulpwood case of Table 7-3. You'll see in column (h) of Table 7-4 that, after age 19, the forest value growth percent at any age is higher than in the pulpwood case. This is due to the rising price per cubic foot. In the context of Figure 7-4a, *compared to a constant stumpage price per unit volume, an increasing stumpage price over time will raise the forest value growth percent curve and lengthen the optimal rotation.* Now the forest value growth percent in column (h) doesn't fall below

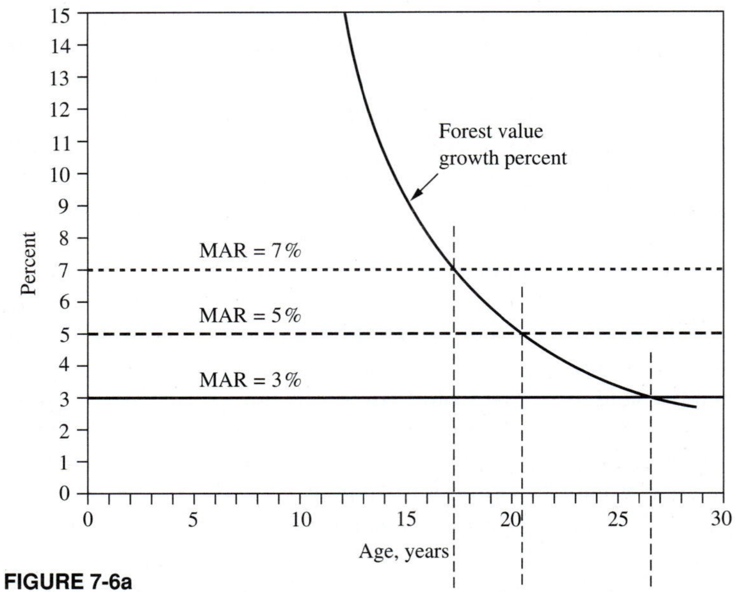

FIGURE 7-6a
Optimal pulpwood rotations, different interest rates ($300 land value).

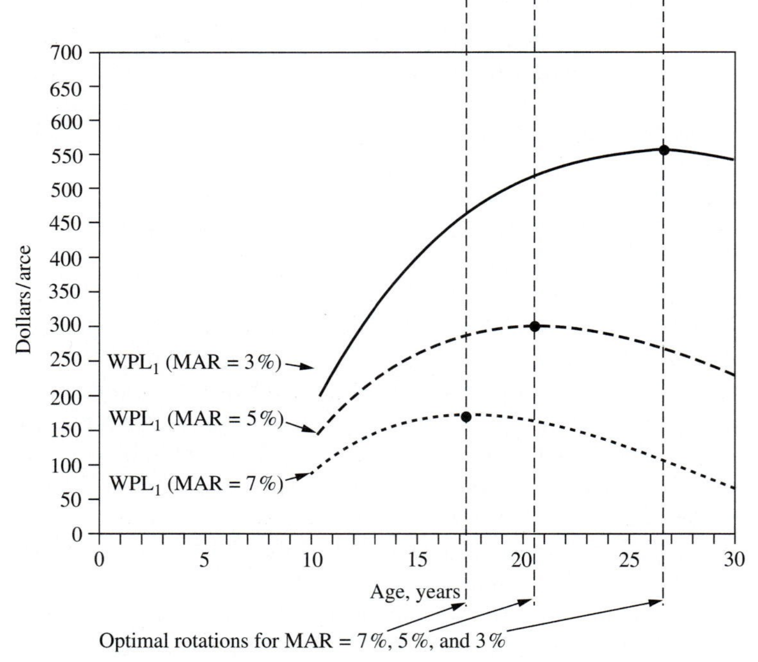

FIGURE 7-6b
Willingness to pay for land—pulpwood.

TABLE 7-4
LOBLOLLY PINE SAWTIMBER CLEAR-CUT OPTIONS*

(a) Age, yr	(b) Vol, cu ft per acre	(c) $/cu ft	(d) Harvest $/acre (b)(c)	(e) tvg % (%ch.d)	(f) Mkt. val land $/acre	(g) Liq. val $/acre (d) + (f)	(h) fvg % eq. (7-10)	(i) WPL_1 $/acre eq. (7-9)	(j) MAI cu ft/yr (b)/(a)	(k) Holding value $/acre (NPV) eq. (7-13)
0					300.00	300.00				519.72
1					300.00	300.00				549.71
2					300.00	300.00				581.20
3					300.00	300.00				614.26
4					300.00	300.00				648.97
5					300.00	300.00				685.42
6					300.00	300.00				723.69
7					300.00	300.00				763.87
8					300.00	300.00				806.07
9					300.00	300.00				850.37
10	634.93	0.250	158.73		300.00	458.73		130.73	63.49	896.89
11	997.02	0.250	249.25	57.0	300.00	549.25	18.86	167.91	90.64	945.73
12	1349.28	0.250	337.32	35.3	300.00	637.32	15.31	199.43	112.44	997.02
13	1691.71	0.250	422.93	25.4	300.00	722.93	12.80	225.81	130.13	1050.87
14	2024.31	0.250	506.08	19.7	300.00	806.08	10.95	247.53	144.59	1107.41
15	2347.07	0.250	586.77	15.9	300.00	886.77	9.51	265.03	156.47	1166.79
16	2660.01	0.250	665.00	13.3	300.00	965.00	8.37	278.73	166.25	1229.12
17	2963.11	0.250	740.78	11.4	300.00	1040.78	7.44	288.99	174.30	1294.58
18	3256.38	0.250	814.10	9.9	300.00	1114.10	6.66	296.17	180.91	1363.31
19	3539.82	0.250	884.96	8.7	300.00	1184.96	6.00	300.59	186.31	1435.48
20	3813.43	0.259	986.72	11.5	300.00	1286.72	8.25	315.10	190.67	1511.25
21	4077.21	0.268	1091.90	10.7	300.00	1391.90	7.86	328.33	194.15	1590.81
22	4331.15	0.277	1200.51	9.9	300.00	1500.51	7.52	340.30	196.87	1674.35
23	4575.27	0.287	1312.56	9.3	300.00	1612.56	7.20	351.05	198.92	1762.07
24	4809.55	0.297	1428.06	8.8	300.00	1728.06	6.91	360.62	200.40	1854.17
25	5034.00	0.307	1547.02	8.3	300.00	1847.02	6.65	369.05	201.36	1950.88
26	5248.62	0.318	1669.43	7.9	300.00	1969.43	6.41	376.38	201.87	2052.43
27	5453.41	0.329	1795.27	7.5	300.00	2095.27	6.19	382.64	201.98	2159.05
28	5648.37	0.341	1924.54	7.2	300.00	2224.54	5.98	387.87	201.73	2271.00
29	5833.49	0.353	2057.18	6.9	300.00	2357.18	5.78	392.10	201.15	2388.55
30	6008.79	0.365	2193.16	6.6	300.00	2493.16	5.60	395.37	200.29	2511.98
31	6174.25	0.378	2332.43	6.4	300.00	2632.43	5.43	397.71	199.17	2641.58
32	6329.88	0.391	2474.91	6.1	300.00	2774.91	5.26	399.15	197.81	2777.65
†33	6475.68	0.405	2620.54	5.9	300.00	2920.54	†5.10	‡399.72	196.23	2920.54
34	6611.65	0.419	2769.20	5.7	300.00	3069.20	4.95	399.47	194.46	
35	6737.79	0.433	2920.81	5.5	300.00	3220.81	4.81	398.40	192.51	
36	6854.09	0.449	3075.22	5.3	300.00	3375.22	4.67	396.57	190.39	
37	6960.57	0.464	3232.29	5.1	300.00	3532.29	4.54	393.99	188.12	

Abbreviations defined in Table 7-3.

*Assumes rising price/cu ft after age 19 and land sale at rotation-end at $300/acre. All values in constant dollars. Volumes are from Weih et al. (1990).

†Optimum rotation age.

‡Maximum willingness to pay for land (WPL_1).

$C_0 = 120.00$	33.00 = optimum rotation
$a - c = -4.00$	2,920.54 = liquidation value at optimum rotation age
Initial $/cu ft = 0.25	300.00 = market value of land
$r = 0.05$	399.72 = maximum WPL_1
	0.035 = annual rate of increase in stumpage price/cu ft, starting $t = 19$

the 5 percent MAR until age 34, which means the optimal rotation is 33 years. You can verify this by seeing that the maximum WPL of $399.72 occurs in column (i) at age 33.[11] Note that this rotation is longer than the 27 years that maximizes mean annual increment in column (j) (in Table 7-3, the optimum was shorter than 27 years).[12]

In projections like the above, it's vital to try many stumpage price scenarios: for example, different fixed prices for all average stand diameters; prices that rise over time but don't change as a result of increasing diameter; prices that increase only because of larger diameter but at different rates. The possibilities are almost endless, but we know so little about what will happen several decades into the future that it's best to simulate many price scenarios, attach probabilities to them, and present results to decision makers. You should also note that a distant-future rotation based on present value maximization is just a preliminary estimate, which is likely to change as conditions change (see "Adaptive Management" below).

Published Stumpage Price Projections Whatever method you use to model stumpage price-changes as a stand grows, you should check whether the projected average stumpage price for mature timber is in some sense reasonable. For many regions, experienced researchers have made stumpage price forecasts based on projected supply and demand interactions. For example, in the mid-1980s, for southern pine stumpage prices, several studies projected real rates of increase ranging from 0.4 to 2.4 percent per year between 1990 and 2010. The median forecast was a real 1.9 percent per year (Binkley and Vincent, 1988). Remember to project from trend-line prices, rather than actual current prices, which can be quite volatile (see Figure 5-2).

Looking Ahead to Alternative Futures

Table 7-3 assumed that only pulpwood markets were available or that pulpwood was needed to supply a mill. In that case, the stand reached financial maturity earlier than for the case of Table 7-4 where we predicted a sawtimber harvest. This shows the importance of not looking only at the current annual forest value growth percent to make rotation decisions. Dramatic examples of this point exist in Japan where I have seen some stands with branches pruned very high to obtain

---

[11] A WPL above the market value of land, as shown here, is not necessarily typical. Adding taxes and risk could markedly reduce the expected revenues to the owner and lower the willingness to pay for land.

[12] In Table 7-4, at the optimal rotation, the timber *value* growth percent in column (e) exceeds the forest value growth percent, as it does in Table 7-3. But the *volume* growth percent [annual percent change in column (b)] is now below the value growth percent because price is rising. The percent change in cubic foot volume between ages 32 and 33 is now 2.3 percent (not shown in the table). Again, this emphasizes the point that *timber* volume or value growth percents are not suitable guidelines for determining economically optimum rotations when a financial return on land value is a concern. You need to look at *forest* value growth percent, including land and timber together.

knot-free lumber. I asked a Japanese colleague whether the small crowns caused slow volume growth rates. He said, "That's what we want in this stand; for some decorative natural wood surfaces, consumers pay exceptionally high prices for clear lumber with very close growth rings." I was thinking in terms of a very low volume growth percent that would also translate to a small value growth percent at young ages before trees were large enough for sawlogs. But starting with small trees of a very low current value growth percent, looking ahead to a potential future value as high as $5,000 per 1,000 board feet, suddenly the silviculture made sense. The projected average value growth percent became much higher but would be realized only if growth rings were very close together. The economically optimal rotation in this case extended well beyond an age where the current annual forest value growth percent at first seemed very low.

The above case shows how important it is to maximize present values based on a number of projected scenarios rather than just measuring the annual forest value growth percent at selected ages and harvesting when this percent drops below the MAR. This is vital when evaluating thinning alternatives, as discussed in the next chapter.

Adaptive Management

You should view the WPL-maximizing rotation as an initial best estimate. Years later, the rotation based on value growth analysis is likely to be very different. Final harvest decisions are often tempered by conditions like weather, log shortages and surpluses, stumpage price peaks and troughs, new wood utilization technology, or land-use changes. For example, prolonged wet weather can halt logging on many areas, thus making some mills temporarily bid high prices for accessible timber rather than suffer a costly mill closure. That could make it optimal to cut some timber at younger ages than originally calculated. Conversely, harvests are sometimes postponed during price slumps. This harvest flexibility is often called *adaptive management*. So the optimum rotations based on maximum WPL aren't nearly as precise as they appear.

Using Spreadsheet Programs

Tables 7-3 and 7-4 were constructed on a personal computer with a spreadsheet program—a powerful analytical tool with which foresters should be acquainted. The equation computing cubic foot volume per acre as a function of age is stored in the cells of column (b) in the spreadsheet.[13] The other equations indicated in the

[13]Volumes are to a 4-inch top, assuming 90 ft^3 per cord, with cord volumes from Weih et al. (1990). The volume equation in column (b) of Tables 7-3 and 7-4 is:

$$\text{cubic feet/acre} = -3526.7143 + 465.32143A - 4.9157143A^2$$

where A is age in years. This is a regression equation suitable only for ages 10 through 40.

column headings are stored in those columns. When an equation is in a cell, the spreadsheet automatically computes its value. Variables used in these equations are picked out of other columns, for example, age in column (a), or from elsewhere in the spreadsheet. Values for variables listed below the table are stored in separate cells so that you can easily change them and recalculate the entire spreadsheet in an instant. For example, you could enter different reforestation costs, annual costs or revenues, interest rates, stumpage prices, or rates of stumpage price growth and immediately see the effect on WPL and the optimal rotation. Other scenarios, like tax alternatives, can be readily built into spreadsheets.

Once you've constructed a spreadsheet, you can easily enter new equations, for example, for a different species in column (b) of Table 7-4. Also, a spreadsheet program can create graphs, such as those in this chapter, by slicing through different columns or rows in a spreadsheet. If you haven't already become familiar with such programs, I urge you to do so; they're extremely useful in forestry investment analyses.

Optimal Rotations and the Regulated Forest

Suppose you have a "fully regulated" forest with equal areas in each one-year age class. Then you could annually cut an equal area of rotation-aged timber forever. Some analysts claim that the economically optimal rotation for such a case is shorter than that outlined in Figure 7-4, if the owner is constrained to develop and maintain the regulated forest. But several questions can be raised about such a shortening. Picture a regulated forest on a rotation shorter than that shown in Figure 7-4. That would require annually liquidating assets growing faster than the MAR. Owners of such forests could increase their wealth by postponing harvests until the Figure 7-4 optimum was reached. They would no longer have full regulation on their original land, but most owners aren't under such constraints, especially if they can purchase forested land on the open market. Moreover, in a competitive market, land goes to the highest bidder, and the buyer who plans rotations as outlined in Figure 7-4 will be able to outbid owners planning shorter rotations. Thus, in an unconstrained market, it's difficult to build a theoretical defense for rotations shorter than those that maximize the willingness to pay for land.[14]

Nonmonetary Values and the Optimal Rotation

When standing trees yield nonmonetary values like aesthetic benefits to landowners or recreationists, economically optimal rotations can be longer than the foregoing analyses suggest. In fact, if nonmonetary benefits are large enough, the optimal rotation is infinitely long, as in wilderness areas. Figure 7-7 shows forest liquidation values (market value of land and clear-cut value of timber) in years y and $y + 1$. Let the hatched extension on the year $y + 1$ bar be the dollar value

[14]For a discussion of these issues, see Oderwald and Duerr (1990) and Hultkrantz (1991).

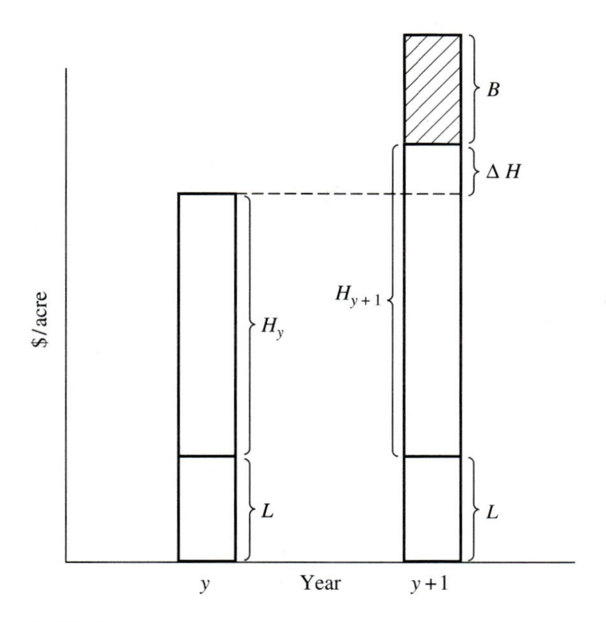

B = nonmarket benefit from holding the trees one year (from year y to $y + 1$)
L = market value of land (here, constant)
H = stumpage value or potential harvest value
ΔH = increase in stumpage value from year y to $y + 1$
Note: By year's end, B is "consumed," so it isn't a permanent addition to liquidation value $(H_{y+1} + L)$. For the next year $(y + 2)$, a new B is added to the new ΔH.

FIGURE 7-7
Adding nonmarket benefit to forest value growth.

of aesthetic benefits B received from the standing timber between the two years. Recipients could be individual landowners or users of public lands. You could think of this dollar value as a willingness to pay for the benefit, or a compensation required should it be lost (sometimes called **willingness to sell**). Excluding the nonmonetary benefit and value of $(a - c)$, the forest value growth percent is $\Delta H/(L + H_y)$ between years y and $y + 1$. But if you include the nonmonetary benefit, the forest value growth percent increases to $(\Delta H + B)/(L + H_y)$. This raises the forest value growth percent curve in the Figure 7-4a framework and lengthens the optimal rotation. In Figure 7-7, the aesthetic benefit B is "consumed" every year, so it's not a cumulative addition to stand value the way timber growth is.[15]

[15]Contrary to the general equilibrium reasoning about $(a - c)$ in footnote 8, the presence of B will always increase the optimal rotation even if B increases the land value. This is because B arises only when mature timber exists, so the aesthetic value has a substantially lower present value than a perpetual series B/r. Thus the value of B will always exceed the opportunity cost of any property value-increase B may cause, where this cost is $r \cdot$ (property value-increase).

Suppose the sawtimber forest in Table 7-4 at age 33 is owned by a nonindustrial private landowner who is willing to pay at least \$0.20/acre/day for aesthetic benefits received from the forest. This is 360(0.20), or \$72/year. Given the \$4 annual cost, $(a - c)$ is \$68/year. While the pure financial calculation with equation (7-10) gave a forest value growth percent of 5.10 percent [column (h)] at age 33, this percent jumps to 7.7 if the value of +\$68 is entered for $(a - c)$ in equation (7-10) [try it, working with column (g) of Table 7-4]. In Figure 7-8, the solid forest value growth percent curve is from column (h) of Table 7-4 without aesthetic benefit and shows a financially optimal rotation of about 33 years at point *F*. The dashed curve in Figure 7-8 plots a recomputed value growth percent including the above aesthetic benefit. With the latter, the forest value growth percent doesn't drop below the 5 percent MAR until after age 45, which is the optimal rotation considering the above nonmonetary value (point *N* in Figure 7-8). Except for the added aesthetic value, the same yield function and other inputs from Table 7-4 were used, although the table shows yields only to age 37.

Note that the above aesthetic benefit *B* was constant at \$72 per acre per year. Often such values increase with timber age, so the forest value growth percent curve is raised more than shown in Figure 7-8. The effect would be to lengthen the rotation even more.

Woodland owners are certainly not consciously making rotation decisions by using graphs like Figure 7-8. But the diagram shows that economic thinking is adaptable for nonmonetary values as well as financial calculations. And it shows how forest owners can be economically rational—in the sense of maximizing

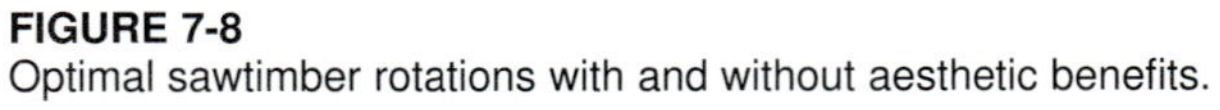

FIGURE 7-8
Optimal sawtimber rotations with and without aesthetic benefits.

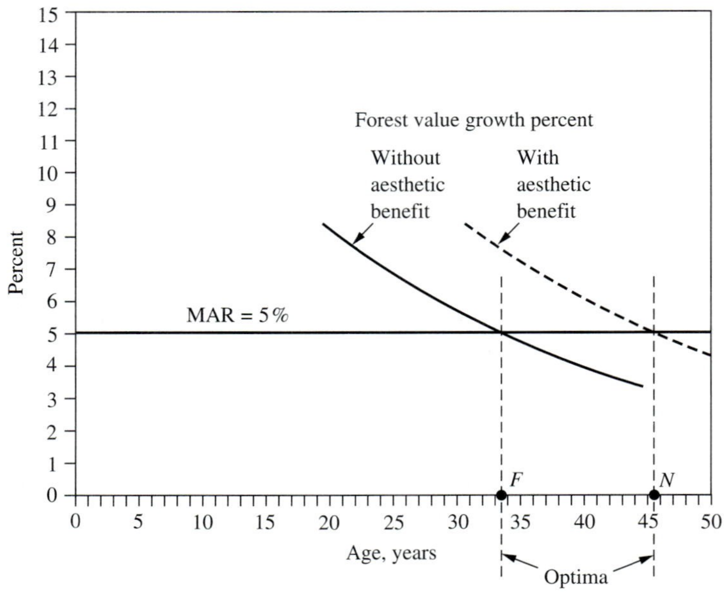

utility—when choosing rotations far longer than those based strictly on financial returns. As shown below, in many cases it's economically logical not to harvest at all. Note that I'm viewing economics broadly as a way of thinking and not purely as financial analysis.

The above effects of nonmonetary benefits on landowners' rotation decisions don't generally apply to forests owned by industry or absentee investor groups because such owners do not receive the aesthetic benefits. If such owners use financial rotations and the rest of society receives the nonmonetary benefits, we have a socially optimal rotation (N in Figure 7-8) longer than the private optimum (F). This is another example of the market failure, "unpriced positive side effects," discussed in Chapter 3. But this divergence between social and private rotations applies only if someone actually receives the nonmonetary benefit. If the forest is being used only for timber production, and no one else sees or uses it, the financially optimal rotation is also the social optimum.

"Perpetual" Rotations Where values of land and stumpage are low, the above rotation-lengthening effect can be significant. For example, in remote areas such as public lands in parts of the Rocky Mountains, it can cost nearly as much to log and haul timber to the mill as the mill is willing to pay for the delivered logs. And sometimes delivered mill prices wouldn't even cover delivery costs. Thus, stumpage value of standing timber can be very low and sometimes even negative. For those cases, in Figure 7-7, the bar for $(L + H_y)$ will be fairly short relative to $(\Delta H + B)$, and annual forest value growth percent including nonmonetary benefits can be above an acceptable rate of return indefinitely. Such a case is shown in Figure 7-9, where the dashed forest value growth percent curve doesn't cross the MAR curve, and it would be optimal to never harvest. This is very likely to happen in remote and high mountain areas where stumpage values are typically low and nonmonetary values high. That explains the location of many wilderness areas and why they often make good economic sense.

The idea of infinite rotations (or at least as long as the trees live) runs into trouble when liquidation values of timber rise. In Figure 7-7, that means H_y gets larger relative to $B + \Delta H$, thus lowering the forest value growth percent curve to where it might cross the MAR curve in Figure 7-9. Such an intersection point suggests at least a partial cut. On the other hand, very high aesthetic benefits can justify holding even valuable timber. Think of an extreme case: Suppose your front yard is graced by an enormous and beautiful shade tree planted by your great granddaddy. You think, "There's no way I'll ever cut that tree; it brings me such pleasure every day." But what if a specialty hardwood sawmill owner tells you the tree is a black walnut with unusual grain patterns and high volume for which he's willing to pay $30,000? This isn't just hyperbole: such cases exist, and the trees are often cut. To the owner, the annual aesthetic benefit sometimes becomes too small as a *percent* of the tree's very high liquidation value, which, invested elsewhere, will bring greater satisfaction.

Suppose the community viewed the above walnut tree as an asset of unusual beauty and historic significance, and 30,000 people were willing to pay $1 each to save the tree. Should the owner cut it? Now we're in the realm of property

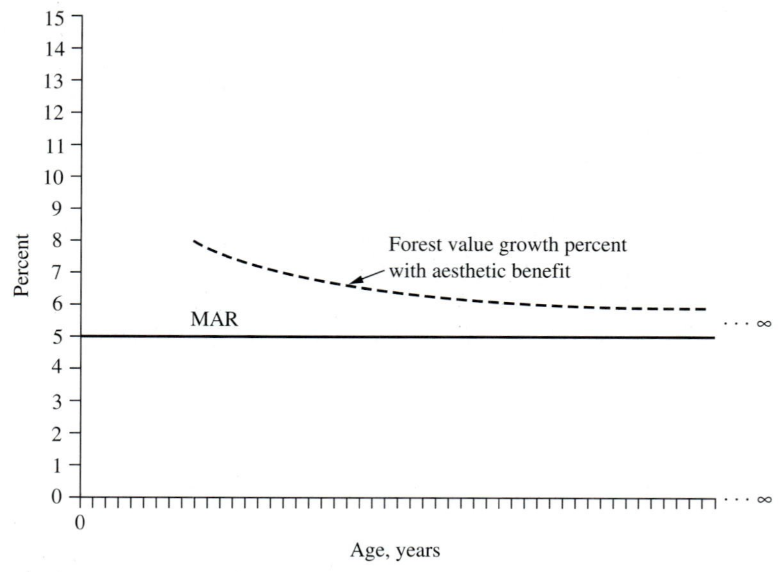

FIGURE 7-9
Perpetual rotation with aesthetic benefits.

rights and liability rules discussed in Chapter 3. As a heavy equipment manufacturer has stated in its advertisements, "There are no easy answers, just difficult questions"—but questions we can ill afford to ignore in forestry, as community values and environmental concerns become more important.

KEY POINTS

- Markets tend to allocate land to uses generating the highest bid prices for bare land. While this is a private optimum, the socially optimum pattern may differ due to unpriced positive and negative side effects (see Chapter 3).
- In theory, the maximum willingness to pay for bare land (WPL) in a given use is the present value of revenues minus the present value of costs, computed just before land use establishment.
- For even-aged forestry, a WPL calculation can assume a perpetual series of rotations (WPL_∞) or can have cash flows terminate with final harvest plus market value of land. The latter usually provides the most realistic estimate of value in forestry.
- As trees become larger, eventually their annual percentage increase in potential harvest value will decline. When this percent increase includes land value and incorporates annual costs and revenues, it is called "forest value growth percent." The potential harvest income from standing timber is called "stumpage value."
- The economically optimal rotation is at the age where the forest value growth percent equals the owner's minimum acceptable rate of return (MAR). This is also the clear-cutting age yielding the highest willingness to pay for land (WPL).

- Timber value growth percent (annual percent increase in potential clear-cutting revenue) will exceed forest value growth percent.
- The biological rotation that maximizes mean annual increment is usually not economically optimal.
- Forest liquidation value is the potential income from clear-cutting and selling the land. Holding value of an even-aged forest is the present value of future liquidation at the optimal rotation age.
- When forests are financially immature, holding value exceeds liquidation value. For financially mature forests, holding value is less than or equal to liquidation value.
- The higher the MAR, the lower the WPL and the shorter the optimal rotation.
- The greater the rate of increase in stumpage price per unit volume, the longer the economically optimal rotation.
- Optimal rotations that maximize WPL are initial best estimates, which usually change as the future unfolds.
- The rotation that maximizes WPL is still optimal even in a fully regulated forest, as long as markets exist for forestland and owners have the flexibility to buy and sell forested acres.
- Adding nonmonetary benefits derived from standing timber will lengthen the optimal rotation, sometimes infinitely.

QUESTIONS AND PROBLEMS

7-1 Suppose you want to buy bare land and estimate that reforestation costs will be $100 per acre after land purchase and after each harvest. Projected harvest income is $3,000/acre in 40 years and every 40 years thereafter. Annual management costs will be $3/acre at the end of each year. Based only on these real cash flows in perpetuity, what is the maximum you can pay per acre for the land and still earn a 5 percent real rate of return?

7-2 In Chapter 4, question 4-12, use the Faustmann formula to compute the maximum willingness to pay for bare land (WPL_∞). [See equation (7-4).]

7-3 Suppose you own an acre of forest. One year ago the land was worth $400/acre and the stumpage value of timber was worth $1,500. Now the stumpage value is $1,670, with land still at $400/acre. Annual costs are $5/acre. What was your forest value growth percent (fvg %) for the past year?

7-4 **a** Suppose in the previous question your minimum acceptable rate of return (MAR) is 10 percent, and there are no chances to increase the forest value growth percent. Given no other forest benefits, what should you do?

b Suppose you receive nonmonetary benefits from the standing timber. How large would such annual benefits need to be to justify holding the timber—in other words, to give you an annual value growth rate of 10 percent?

7-5 Suppose in question 7-3 you received $50 worth of aesthetic benefit per acre of the forest over the past year. What was the forest value growth percent for the past year, including the aesthetic benefit? How does this affect the optimal rotation? Why?

7-6 In 1980, the stumpage value of an acre of my even-aged timber was worth $2,000 with land at $100/acre. In 1983, the stumpage was worth $2,693 with land at $100/acre. What was the annual forest value growth percent (land and timber) over this period?

7-7 Suppose, other things being equal, that land values were $500/acre in the previous question. How would forest value growth percent be affected?

7-8 Under even-aged management, let the economically optimal rotation be where the annual forest value growth percent equals the minimum acceptable rate of return (MAR).

a If the MAR increases, will optimal rotation increase or decrease?

b Suppose you compute an optimal rotation at 50 years for one set of stumpage prices. Now suppose prices per MBF will rise more rapidly over time and will rise more rapidly with tree size. What happens to the optimal rotation (given the same MAR and land value)?

c Suppose the forest owner receives nonmonetary benefits from the standing timber, not considered above. What happens to the optimal rotation? Why?

7-9 **a** Given the following yield table for even-aged Bogaloosa Blue Beech, which of the indicated rotation ages would maximize willingness to pay for bare land? Assume 7 percent interest, $100/acre reforestation cost, harvest values as shown, and no other costs or revenues. Assume a perpetual time horizon without land sale.

b Which rotation would maximize the mean annual increment?

Age, years	Volume per acre, MBF	Stumpage value, $/MBF
20	15	$30
30	25	50
40	34	66
50	41	80

7-10 Suppose Ms. Barker plans to spend $110 on planting loblolly pine on an acre of bare land. She projects a pulpwood clear-cut yielding $1,100 in 22 years. Her projected annual costs are $3/acre, and she expects a brush control cost of $30/acre at age 7. She plans to sell the land for $500 after harvesting in 22 years. If she wants to earn a real 5 percent rate of return, what is her willingness to pay for the land? (All values are real, after taxes.)

7-11 **a** In the previous question, what would be Ms. Barker's "holding value" for the forest at age 15?

b How would this value compare with the liquidation value at age 15, assuming 22 years was her economically optimal rotation?

Appendix 7A: Rates of Return on Afforestation

There's much confusion about how to treat land value in calculating rates of return on afforestation investments. The decision depends on the situation, as outlined below. All values will be in real terms.

If Timber Revenue Is Land's Only Benefit

Suppose the Woods Timber Company buys forestland at $\$L_0$ per acre and spends C_0 on planting, for the sole purpose of gaining future timber income H_t and land value L_t in year

t. Assuming no other costs and revenues, Woods' rate of return on the investment in land and planting is the value for r below, which equates the present value of revenues with costs:[16]

$$L_0 + C_0 = \frac{H_t + L_t}{(1+r)^t} \tag{7A-1}$$

Solving equation (7A-1) for r:

$$(1+r)^n = \frac{H_t + L_t}{L_0 + C_0}$$

$$\text{Internal rate of return on } L_0 + C_0 = r = \sqrt[n]{\frac{H_t + L_t}{L_0 + C_0}} - 1 \tag{7A-2}$$

Letting C_0 be fixed at some optimum level, if landowners complain that rates of return on $C_0 + L_0$ are too low, they have only themselves to blame: they paid too much for land. In equation (7A-1) you can see that if less were paid for land L_0, the value of r would have to be higher in order to maintain the equality. But if perceived r is above the market rate, buyers will bid up the price of land until r declines to the market rate. Thus, economists say that future benefits (and costs) are "capitalized" into land values.

Equation (7A-1) is a simple model of land use equilibrium: on average, investors, through competitive bidding for land, will earn the market rate of return on investments in land and land use. But this is true only in an average sense, especially with forestry's long time horizons. After the fact (*ex post*), some investors will earn more and some less. As we'll see in the chapter on risk, forestry rates of return that sometimes seem low may often be quite acceptable, given the risks involved.

Another important lesson from equation (7A-2) is that, if land sale is a viable option, one should *not* calculate a rate of return on afforestation cost alone without including the land value. For example, suppose $t = 20$, $C = \$100$, $H_t = \$900$, and $L_0 = L_t = \$384$, assuming constant real land value. Substituting these values into equation (7A-2) gives a 5 percent rate of return on the *total* investment of $L_0 + C = \$484$. If $L_t > L_0$, the rate of return would exceed 5 percent. It would be misleading to calculate a rate of return on only the afforestation cost by leaving out the land value in equation (7A-1) and solving for r, as follows:

$$C_0 = \frac{H_t}{(1+r)^t}$$

$$C_0(1+r)^t = H_t \tag{7A-3}$$

$$(1+r)^t = H_t/C_0$$

$$r = \sqrt[t]{H_t/C_0} - 1$$

[16]If you simplify WPL_1 equation (7-9) to include only H_t and L_t and exclude all costs except C_0, you have:

$$WPL_1 = \frac{H_t + L_t}{(1+r)^t} - C_0$$

Let $WPL_1 = L_0$. Moving C_0 to the left, you get equation (7A-1) above.

Using the numerical example,

$$r = \sqrt[20]{900/100} - 1 = 0.116 \tag{7A-4}$$

or 11.6 percent. This isn't a valid rate of return because it ascribes all income to the reforestation expense and ignores the capital in land that must also earn interest. The correct return in this case is 5 percent on C_0 and L_0 combined.

If Land Yields Other Benefits

The present value of other benefits (or costs) such as hunting revenues could be added to the right side of equation (7A-1), thus boosting (or decreasing) the rate of return on $L_0 + C$. In the right side of equation (7A-1), suppose r is the landowner's MAR, and the present value of future benefits is *less than* $L_0 + C$ by an amount NPV. That would mean the project earns less than the MAR. By substituting the negative NPV into equation (6-11), you could find the annual nonmonetary benefit required for the rate of return to equal the MAR.

If a landowner receives enough nontimber benefits, priced or unpriced, to make land ownership worthwhile without any reforestation, then land value needn't be considered, and equation (7A-3) would yield a valid rate of return for reforestation. Some argue that equation (7A-3) is also relevant for owners legally unable to sell their forestland. Others suggest that all capital values should still be considered.

REFERENCES

Binkley, C. S., and J. R. Vincent. 1988. Timber prices in the U.S. South: Past trends and outlook for the future. *Southern Journal of Applied Forestry.* 12(1):15–18.

Duerr, W. A. 1960. *Fundamentals of Forestry Economics.* McGraw-Hill, New York. 579 pp.

Faustmann, M. 1849. Berechnung des Werthes welchen Waldboden sowie noch nicht haubare Holzbestände für die Waldwirtschaft besitzen. *Allgemeine Forst und Jagd-Zeitung.* 25:441–455. (English translation in Linnard, W., and M. Gane. 1968. Martin Faustmann and the evolution of discounted cash flow. *Commonwealth Forestry Institute Paper 42.*)

Gaffney, M. 1957. Concepts of financial maturity of timber and other assets. Agricultural Economics Info. Ser. No. 62. North Carolina State College, Raleigh. 105 pp.

Gregory, G. R. 1987. *Resource Economics for Foresters.* John Wiley & Sons, New York. 477 pp.

Hultkrantz, L. 1991. A note on the optimal rotation period in a synchronized normal forest. *Forest Science.* 37(4):1201–1206.

Newman, D. H. 1988. The optimal forest rotation: A discussion and annotated bibliography. Gen. Tech. Rept. SE-48. USDA Forest Service Southeastern Forest Experiment Station, Asheville, NC. 47 pp.

Oderwald, R. G. and W. A. Duerr. 1990. König-Faustmannism: A critique. *Forest Science.* 36(1):169–174.

Samuelson, P. A. 1976. Economics of forestry in an evolving society. *Economic Inquiry.* 14:466–492.

Weih, R. C., J. E. Scrivani, and H. E. Burkhart. 1990. PCWTHIN version 2.0 user's manual. School of Forestry and Wildlife Resources. Virginia Polytechnic Institute and State Univ., Blacksburg. 31 pp.

8

OPTIMAL TIMBER STOCKING

The last chapter assumed a given number of trees per acre and considered when to clear-cut an even-aged stand and whether timber production was the most profitable land use. Now let's assume that timber growing is financially the best use of a given area, but you want to know how many trees to plant or seed per acre and how much volume to maintain at any age. After afforestation, volume can be varied by **thinning**—cutting some trees and leaving the rest to grow until the next thinning or a clear-cut. Or perhaps with some species and in some areas, you'd never clear-cut and would continue partial cutting, thus developing an **uneven-aged** or multi-aged stand. The latter option is increasingly attractive in scenic areas and where soils are unstable. These are all questions of the **optimum** stocking, or number of trees per acre to carry over time—the topic of this chapter. And as before, the optimum is the plan that maximizes **net present value (NPV)** of *all* outputs, using the owner's **minimum acceptable rate of return (MAR).** To encompass the entire stand life, NPV should be calculated just before afforestation. Such an NPV is the **willingness to pay for land (WPL),** which, for **even-aged** stands, is equation (7-4) or (7-9). With all the variables involved, the optimal stocking problem is tremendously complex. Let's simplify it by considering one facet at a time.

Here, all values and interest rates are **real.** Cash flows are considered as expected or average values, and risk is assumed to be correctly accounted for in the discount rate. You could assume that taxes have already been subtracted from cash flows, in which case the MAR would be after taxes. Or analyses could be for tax-exempt investors, where the MAR would be the desired earning rate before taxes. Taxes are covered in the next chapter. First we'll consider only timber

TABLE 8-1
NOTATION FOR CHAPTER 8

a =	annual revenue, \$/acre	R_t =	added revenue at rotation age t, \$/acre
c =	annual cost, \$/acre	t =	rotation age, years
C_0 =	afforestation cost, \$/acre	u =	cutting cycle, years between harvests in an uneven-aged stand
C_y =	cost in year y, \$/acre	v_a =	real annual *value* growth rate of a tree or group of trees, percent/100
d =	development period, years, between establishment and first harvest for an uneven-aged stand	v_o =	annual *volume* growth rate of a tree or group of trees, percent/100
g_r =	real annual rate of stumpage price growth (percent/100)	WPL =	willingness to pay for bare land, NPV, \$/acre (may assume land sale after clear-cut or assume perpetual harvesting with ∞ subscript)
h =	revenue from partial cutting, \$/acre	y =	an index for years, from 0 to t, or from 0 to n
H =	clear-cutting revenue, \$/acre		
MAR =	minimum acceptable rate of return		
n =	number of years in an investment period		
NPV =	net present value, dollars		
r =	real interest rate, percent/100		
RPI =	number of rings per inch of radial tree growth		

values and then incorporate nontimber outputs. Table 8-1 shows notation used in this chapter. Let's begin with even-aged stands and the question of how many trees to plant per acre.

OPTIMAL PLANTING DENSITY

To understand the idea of optimal planting density, it helps to consider extremes: If you plant one tree per acre, it will grow fast, and planting costs will be low, but total volume per acre will be very small, and the tree will be limby and perhaps not too straight. Plant 20,000 trees per acre, and you'll have very high planting costs as well as a "dog hair" stand with low **merchantable,** or salable, volume growth and yield. Somewhere in between is the optimum that maximizes net present value. The trick is to simulate yields and WPLs, for example, in spreadsheets like Table 7-3—constructing a different table for each planting density. By entering new costs for each planting density, you could select the optimum density and rotation combination with the greatest WPL.[1]

Unfortunately, for most species, we aren't able to predict very well the future merchantable yields for a wide range of planting densities. Existing computer-based growth models can make such predictions for major species like Douglas-fir and loblolly pine (see Curtis et al., 1981; and Weih et al., 1990), but they may not always be reliable at low planting densities. Thus, Table 8-2 gives hypothetical WPL values for three planting densities. For each planting density, the maximum

[1]Planting densities are given in trees per acre or as a spacing, for example, 6 feet × 8 feet. To find the number of trees per acre for any planting spacing, say 8 feet × 8 feet, divide the number of square feet per acre (43,560) by the product of the spacing. For example, the 8 × 8 spacing yields 43,560/(8 × 8) = 681 trees per acre. The appropriate square spacing for any desired number of trees per acre (TPA) is thus the square root of (43,560/TPA).

TABLE 8-2
HYPOTHETICAL DATA AT DIFFERENT PLANTING DENSITIES FOR A GIVEN SPECIES AND SITE QUALITY

Planting density (trees per acre)	Planting cost, $/acre	Optimal rotation, years*	WPL†, $/acre
300	60	34	230
400	70	35	250
500	80	36	200

*The optimal rotation (which maximizes WPL at any planting spacing) often lengthens with increasing planting density, as shown, but that may not always be the case.
†Willingness to pay for land [for example, using equation (7-4)].

WPL value and accompanying rotation could be calculated with spreadsheets like Tables 7-3 and 7-4 and recorded in columns (3) and (4) of Table 8-2. For example, with even-aged management of a given species and site quality, for each planting density, equations based on yield simulations could give harvest volumes at different ages. These volumes would be multiplied by projected stumpage values to get harvest values for each age. Other inputs to enter in the WPL equation would be planting cost for each density (for example, $30/acre fixed cost plus $0.10/seedling, including labor), real interest rate, and other costs and revenues. For the case of clear-cutting, WPL values could be calculated with equations such as (7-4) and (7-9) as shown in Chapter 7. Non-clear-cutting systems will be covered shortly.

Figure 8-1 graphs WPL over planting density from Table 8-2 and shows that WPL is maximized at an optimal density of 400 trees per acre for this hypothetical case.[2] Optimal planting density can be affected by things like site quality, interest rate, cutting regime, relationship between stumpage price and tree diameter, taxes, planting costs, and tree species. Thus, you need separate analyses for different cases.

Think about the necessary marginal conditions for the optimal planting spacing: Starting with a light planting density, as you add one more seedling per acre, the extra cost is, say, $0.10, which is the added cost per seedling in present value terms. At rotation-end, the extra seedling gives added harvest income per acre, and discounting that to planting date gives the seedling's contribution to present value—the extra revenue per unit of input. Eventually, because of crowding, this added revenue declines. Present value is maximized in Figure 8-1 when the added cost of one more seedling exactly equals its contribution to NPV (marginal resource cost = marginal revenue product in present value terms). If we add one more seedling beyond the optimum number, its addition to present value is less than its cost, and present value declines.

[2]Several loblolly pine simulations have found increasing WPL with decreasing planting density down to 400 or 300 trees per acre when stumpage value per unit volume increases with tree diameter (see Broderick et al., 1982; and Caulfield et al., 1992). However, because of data limitations at low densities, these studies haven't always shown the peaking of WPL that should theoretically occur.

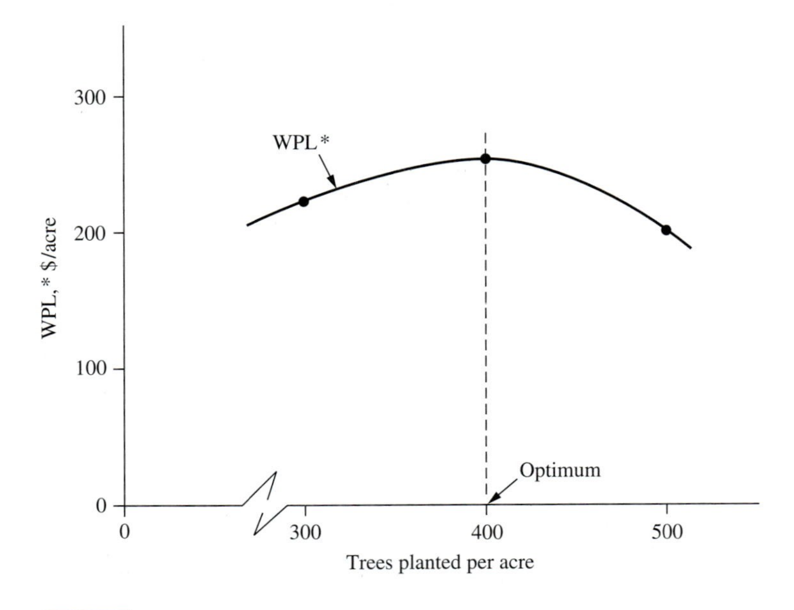

*Willingness to pay for land.

FIGURE 8-1
Optimum planting density from Table 8-2.

Actual applications are much less precise than the theory suggests, because outcomes are hard to predict. A special problem is to predict how wider planting spacing will affect stem quality and future harvest value. Nevertheless, you should know about optimizing planting density, since our ability to predict outcomes of different densities is improving, and such applications will become more common. In general, many forest managers have moved to lighter planting densities since the 1960s, in part because of the above economic factors. On the other hand, you might aim for a somewhat closer planting spacing than suggested by Figure 8-1 if you want flexibility in making early partial cuts to capture good stumpage markets without sacrificing stand health.

OPTIMAL EVEN-AGED THINNING REGIMES

Thinning removes some trees to give remaining trees more space, moisture, nutrients, and sunlight for growth. Within a fairly wide range of intensities, thinning doesn't significantly increase total volume growth over one rotation. But thinning can increase merchantable volume growth and can provide the following advantages:

- By removing poorly formed or suppressed trees, growth can be concentrated on the more valuable trees, for example, those producing high-grade sawlogs, veneer logs, or poles, as opposed to lower valued pulpwood.

- By cutting slowly growing trees, forest value growth percent of the remaining stand can be improved [see equation (7-10)].
- Some thinnings yield income before rotation-end. Other things being equal, earlier income boosts present value more than late income.
- Aesthetically, some people prefer a thinned stand compared to a dense forest with closely spaced trees.

In a **precommercial thinning,** felled trees are too small to be salable and are left lying in the woods. *Commercial thinnings* yield logs large enough to be sold.

Like the optimal planting density problem, which is a form of thinning at age 0, the optimal thinning regime is that which maximizes net present value or the willingness to pay for land. The problem is that thinning doesn't always increase present value.

Figure 8-2 diagrams the optimal thinning problem in even-aged stands. Simultaneously we need to find the optimal planting density, the type of thinning, the thinning age or ages, the volume to remove in each thinning, the number of species, and the optimal clear-cutting age. The decision becomes even more complex if you consider different types of thinning: *thinning from above* (removing mainly the tallest or "dominant" and "codominant" trees), *thinning from below* (removing trees from the understory), or *row thinnings* (removing selected rows in plantations). In general, the question is what trees of which size to remove when. If you consider mixed species stands, you can literally have trillions of

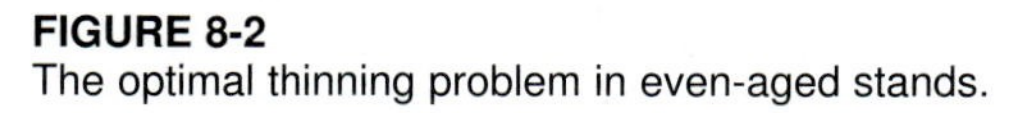

FIGURE 8-2
The optimal thinning problem in even-aged stands.

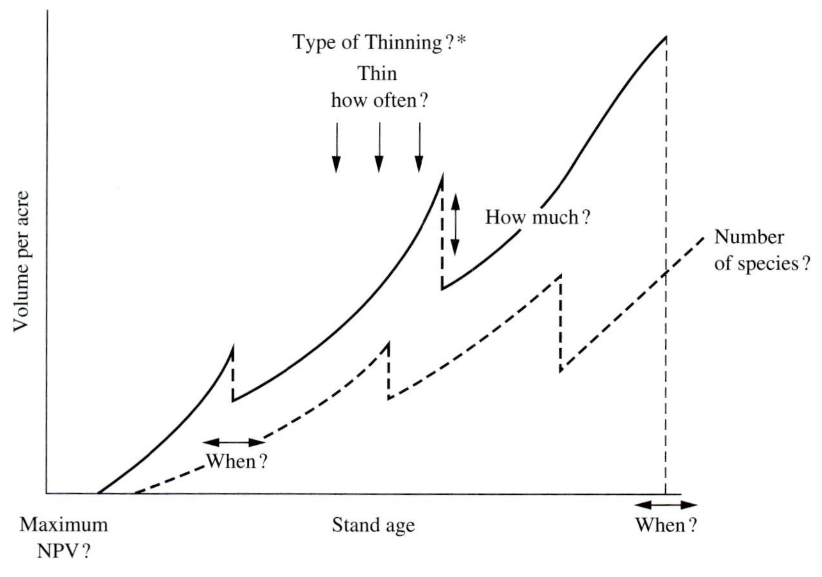

*Thinning from above, from below, row thinning?

thinning alternatives within one rotation. Each alternative is called a **thinning regime.** For example, one regime might be to thin from above at ages 15 and 25, removing 30 percent of the cubic foot volume each time, and to clear-cut at age 35. In Figure 8-2, the solid line shows one thinning regime.

Computerized mathematical programming approaches are sometimes used to solve thinning problems but are beyond the scope of this book.[3] A simpler way to approximate a solution is to simulate a limited number of thinning alternatives with a computer-based growth model.[4] You can then choose the regime generating the maximum WPL, using modifications of equation (7-4) or (7-9). In the WPL equation, you'd enter thinning revenues or costs as R_y or C_y values for each regime. Starting with bare land, equation (7-5) gave a numerical example of WPL_∞ for the case of one thinning followed by a clear-cut. If you're starting with an established stand, choose the thinning regime that maximizes the net present value of future cash flows from the stand, using the "holding value" equation (7-13). Let's now consider separate facets of thinning decisions.

Optimal Thinning Intensity

In finding the optimal intensity for any thinning, we decide how much volume to remove (see Figure 8-2). To maximize WPL, we should remove from the forest any trees with expected value growth rates less than the owner's minimum acceptable rate of return (MAR). The relevant value is the trees' stumpage value or liquidation value explained in the last chapter. As you'll see, this thinning guideline is difficult to apply because when you remove one tree, the expected value growth rates of neighboring trees will increase.

Figure 8-3 diagrams the theory of optimal thinning intensity for the case of one thinning. The vertical axis shows dollars per acre of land value plus potential harvest value or stumpage value. Assume that, without thinning, the economically optimal rotation (which maximizes WPL as in the last chapter) is at age t. At age y, consider removing volume increments with values of b', c', and d'. In a thinning, you'd remove trees evenly spaced throughout the densest parts of the stand and having the slowest value growth rates. If left in the forest, increment b' would grow to a value of e' in $t - y$ years. Over the next $t - y$ years, its projected value growth rate v_a [using equation (4-4)] is therefore:

$$v_a = \sqrt[t-y]{\frac{e'}{b'}} - 1 \qquad (8\text{-}1)$$

---

[3]See Davis and Johnson, 1987, Chapter 13; Dykstra, 1984, Chapter 10.

[4]Examples of the many models that simulate thinnings are PCWTHIN for loblolly pine (Weih et al., 1990) and DFSIM for Douglas-fir (Curtis et al., 1981). Several models for the southeastern United States are incorporated in YIELDplus (Hepp, 1990). New-growth models are constantly being developed. One way to find the latest simulators is to contact forest biometricians at a university doing research on the species of interest. With any model, you should compare outputs with field studies in your area. Such validation is extremely important if you're simulating scenarios beyond the model's database or if the model has flaws.

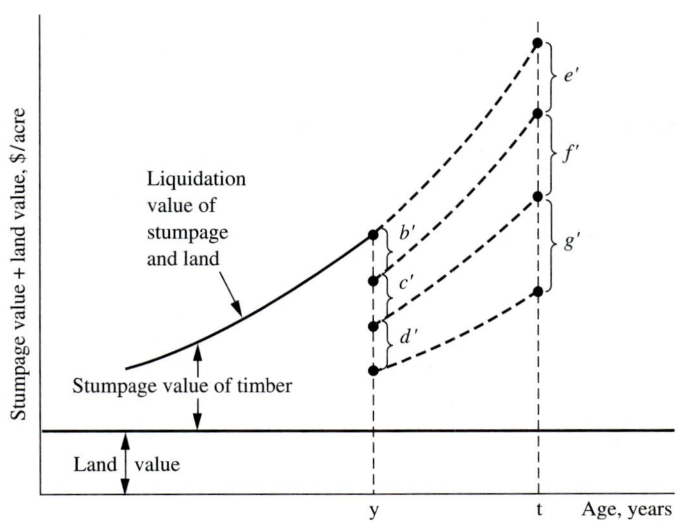

FIGURE 8-3
Even-aged thinning intensities.

where e' and b' are stumpage value increments in Figure 8-3. *If* $v_a \geq$ *MAR, don't thin,* because the trees are growing faster than your MAR. Since the first increment b' contains the slowest growing trees, you know that the remaining increments c' and d' are growing even faster and should therefore also remain in the stand. *If* v_a *in equation (8-1) is less than the MAR, remove increment* b', for it could earn more elsewhere. For example, suppose the potential harvest value of b' in Figure 8-3 is $100, and it could grow to a value of $e' = \$180$ over a period of $(t - y) = 10$ years. Substituting these values into equation (8-1), $v_a = 0.06$, or 6 percent. If MAR $\leq$ 6 percent, don't thin; if MAR $>$ 6 percent, the stocking value b' should be removed. Thinning decisions are "marginal" analyses looking at incremental units of timber volume, so you can ignore annual costs and land value since they're already considered in WPL.[5] If the stand is thinned, the growth on residual trees will increase, and a new optimal rotation would need to be selected as done in the last chapter.

If increment b' is removed, consider increment c', which if left in the forest would grow to a value of f' at the following rate:

$$v_a = \sqrt[t-y]{\frac{f'}{c'}} - 1 \qquad (8\text{-}2)$$

[5]In the previous chapter's discussion of rotation ages, recall that the forest value growth percent had to include both timber and land, because we were considering the growth rate of the entire capital investment [see equation (7-10)]. However, in making marginal decisions about cutting part of a stand before rotation-end, you need only consider whether or not an extra unit of capital—an individual tree or group of trees—will earn an acceptable rate of return. In choosing the rotation that maximizes WPL, the land value will be automatically considered by including it in equation (7-8) or, for the perpetual case, by including an infinite series of rotations in equation (7-3).

where f' and c' are stumpage value increments in Figure 8-3. Using the above reasoning, remove unit c' if its expected value growth rate is less than the MAR. But note that removing the first increment (b') will allow the next one (c') to grow faster. Additional increments of volume are evaluated in the same way. In this example you wouldn't remove all volume, since the optimal clear-cutting age was already determined to be longer than age y. *To roughly estimate the optimal thinning intensity at age y, remove all volume increments with expected value growth rates less than the MAR.* This is only an approximation because you can sometimes remove trees growing faster than the MAR and still increase the stand's present value if the thinning sufficiently boosts residual stand growth. Thus, present value–maximizing simulations with a reliable growth model are likely to produce the best thinning prescriptions.

Figure 8-4 shows one simulation study's curve of WPLs for different percent volume removals in a single loblolly pine thinning from below at age 20 on an average site quality (Broderick et al., 1982). In this case the present value was based on an infinite series of 30-year rotations [an adaptation of equation (7-4)], using 6 percent real interest and assuming production of sawlogs, peelers, veneer logs, and pulpwood. The curve shows a maximum WPL when about 33 percent of the volume was removed. Note that the curve is fairly flat over a wide range of thinning intensities, showing that, in this case, the exact percentage volume

FIGURE 8-4
Effect of thinning intensity on net present value of loblolly pine.

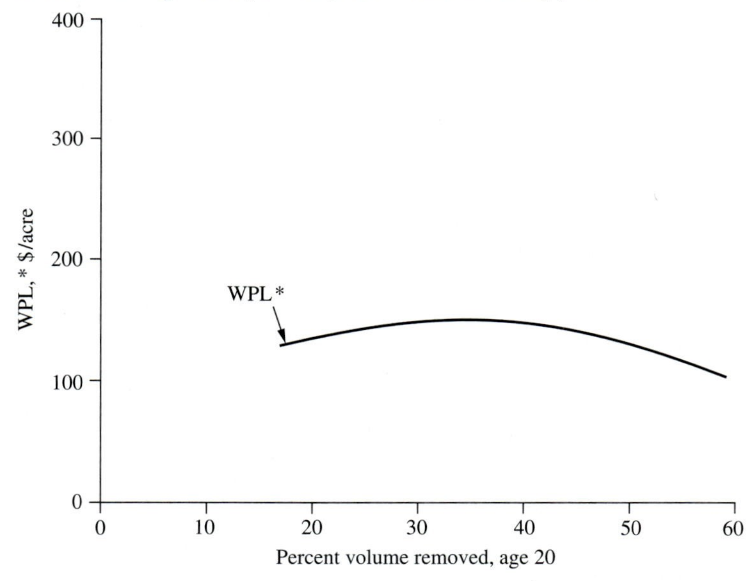

*Willingness to pay for land (net present value). Adapted from Broderick et al. (1982), by permission.

removed wasn't crucial within that range. The same study found that the optimal planting density and clear-cutting age were 436 trees per acre and 30 years.

The above study is based on one set of physical and economic variables; changing those factors will generally change the optimum thinning regime. Examples of such *physical* factors are:

- Species and the type of growth model used
- Site quality
- Original stand density
- Type of thinning (from below, from above, or row thinning)

Examples of *economic* variables affecting the optimal thinning regime are:

- Interest rate used to compute present value
- Current and projected stumpage values and the degree to which stumpage value increases with tree diameter
- The degree to which higher logging costs reduce stumpage values (In a thinning, loggers must work around the remaining trees and prevent damage to them. Thus, logging costs per unit volume harvested will be higher than in clear-cutting, and stumpage values will be lower.)

Optimal Thinning Timing and Frequency

The optimal thinning timing and frequency is that which maximizes net present value or WPL. From strictly a timber growth view, you might be tempted to do frequent light thinnings. But the lighter the thinning, the higher the logging costs are likely to be per unit volume cut. Thus, less frequent and heavier thinnings tend to yield greater net harvest income per unit volume. Consider the timing of a single thinning: If too early, trees are so small that stumpage values are low or negative, and the stand can become overstocked again before final harvest. But if too late, the stand can stagnate early in the rotation. Somewhere in between lies the optimum. And perhaps more than one thinning would increase present value.

As before, the simplest approach to finding an optimum is to use a growth simulator to compute WPL values for different thinning timings and frequencies. You can then choose the combination with the maximum WPL for the most reasonable cost and price scenarios. As with optimal rotations, optimal thinning regimes are much more exact on paper than they are in the field. By repeated simulations, you can reach precise conclusions about the optimal thinning regime for one set of assumptions about prices, initial stand density, site quality, logging costs, price/diameter relationships, etc. But often you'll find that changing the thinning age and the number and intensity of thinnings, within some range, will have little effect on present value. For example, in Figure 8-4, note how flat the WPL curve is over a wide range of thinning intensities. Thus, you often arrive at flexible thinning guidelines.

As a simulation example, the above-cited study reached the following conclusions for average loblolly pine sites under one set of prices and costs, starting with about 450 trees per acre: When growing pulpwood, sawlogs, and veneer logs, one thinning (from below) removing about 30 percent of the volume at about age 18 provided the maximum present value at a 6 percent real discount rate. When growing pulpwood only, thinning offered no financial advantage—not a surprising result because the thinning-induced increase in tree diameters brought no higher value per cubic foot. And the advantage of earlier income was offset by higher logging costs for thinning than for clear-cutting. Results will change with the physical and economic variables mentioned above. For quite a range of conditions, it didn't make much difference whether somewhat more or less volume was removed or whether thinning occurred a bit earlier or later. However, for exceptionally dense stands, different growth models, or other price situations, prescriptions could differ. And it's a good idea to revise simulations: new growth models become available, and there are changes in markets, prices, utilization, and harvesting methods.

Type of Thinning

Earlier, I mentioned thinning from above, thinning from below, and row thinning in plantations. As before, you can simulate thinning alternatives and choose the one with the highest present value. But growth models may not offer choices of thinning types. As often as not, thinning is a combination of thinning from above and below. In removing trees of low projected value growth and poor quality, stems are often taken from both the upper and the lower crown canopy. Also, the shade tolerance of a species will often dictate the type of thinning. For example, with an intolerant species like Douglas-fir, natural mortality is high in the understory, and thinnings tend to be from above. On the other hand, in loblolly pine, thinnings have been predominantly from below. Regardless of traditions, it pays to simulate alternatives and see if another approach may be financially better.

As shown next, marking guides for thinning can sometimes solve the problem of deciding whether to thin from above or below. You simply remove trees with projected value growth percents lower than the MAR. Row thinnings make no use of such guidelines, but the method—for example, removing every second or third row in a plantation—is sometimes preferred because costs of timber marking and logging are usually lower than with other types of thinning.

Marking Guides for Thinning

Suppose simulations suggest that the economically optimal harvest plan for a given species and site is one thinning removing 30 percent of the volume at age 20, followed by a clear-cut 10 years later. Which trees should be removed? In addition to removing poorly formed and suppressed trees, marking guides can help choose which trees to cut in a thinning. Based on the principles discussed under "Optimal Thinning Intensity," you should remove trees with projected annual value growth percents below the owner's minimum acceptable rate of return. These trees are considered **financially mature.** But this guide needs some qualification, as mentioned shortly. Rather than projecting the value growth percent of a volume amount such as b' in Figure 8-3, you can predict the value growth percent for individual trees. It's best to start with volume growth percent and then adjust for value growth.

Volume growth percent tables are available for some species, as shown in Table 8-3 for Douglas-fir on average sites in the Pacific Northwest. The table shows projected annual percentage rates of Scribner board foot volume growth over the next five years for trees of a given diameter at breast height (DBH) and projected number of growth rings per inch of diameter. You can find the latter for the last few years with an increment hammer and then project the future rings per inch. Such projections can be only rough estimates, since the future growth rate depends in part on the amount of release given by thinning.

Table 8-3 shows that for a given number of rings per inch, the volume growth percent declines with increasing tree diameter, as explained in the previous chapter. If you're growing only one product like pulpwood and prices are constant, volume growth percent equals value growth percent. But if stumpage prices are rising and if the tree develops over time into higher grade logs, its value per unit volume will increase. In that case, you need to adjust the Table 8-3 values upward to get a tree's value growth percent.

Suppose you expect the real stumpage price for a given tree diameter to increase at an annual rate of g_r over the next n years. This increase could be from a general rise in stumpage prices or from a tree's expected change to a higher percentage of sawlogs and peeler logs, or from a combination of both. If projected annual volume growth percent is v_o, the volume will increase by a factor of $(1+v_o)^n$ over n years. Combining the increases in price and volume, the annual value growth percent (expressed as percent/100) is:

$$\text{Value growth rate} = v_a = (1 + v_o)(1 + g_r) - 1 \tag{8-3}$$

TABLE 8-3
AVERAGE ANNUAL VOLUME GROWTH PERCENTS FOR DOUGLAS-FIR TREES FOR NEXT FIVE YEARS*

Diameter breast high, inches	Projected rings per inch†						
	4	6	8	10	12	14	16
9	39.4	30.9	25.7	22.2	19.6	17.6	16.0
12	13.6	9.8	7.8	6.5	5.6	5.0	4.5
15	9.0	6.4	5.1	4.3	3.7	3.3	3.0
18	7.1	5.1	4.1	3.4	3.0	2.7	2.5
20	6.2	4.5	3.6	3.1	2.7	2.5	2.3
22	5.5	4.1	3.3	2.9	2.5	2.3	2.2

*From Flora and Fedkiw (1964). Based on board feet, Scribner rule, to a 6-inch top, site index 140 (base age 100).

†Based on sample of recent growth.

For example, consider a 12-inch DBH Douglas-fir expected to grow at 12 rings per inch. Table 8-3 shows an annual **volume growth percent** of 5.6 percent for the next five years ($v_o = 0.056$). Suppose the real stumpage value per cubic foot over that period is expected to increase at 2 percent per year for such a tree. Substituting into equation (8-3), the annual **value growth percent** will be 0.077, or 7.7 percent.

Unfortunately, for many species, volume growth percent tables aren't available. The Douglas-fir volume growth rates would probably not apply well to, say, loblolly pine, since the latter usually has a different taper, and other log rules might be used. Projected volume growth rates will vary with species, form class, expected rates of height growth (site quality), merchantability limits, and the volume measure used. Also, for thinning guidelines, be sure that volume growth percent tables give *projected* estimates, not current volume growth percents, say, for the last year. For any given number of rings per inch, projected rates of volume growth will be less than current rates, since trees increase in diameter. If you're interested in volume growth percent tables for certain species, check with biometricians in the relevant region. Or you could construct your own with volume tables for trees of different diameters and heights. On a given site, for a tree of d-inch DBH, with a given projected number of rings per inch, you could calculate the DBH and height in, say, 10 years, look up the tree's future volume, and calculate the annual volume growth rate over the period, adapting equation (4-4):

$$\text{Volume growth \% over next 10 years} = \left(\sqrt[10]{\frac{\text{future tree volume}}{\text{today's tree volume at } d\text{-inch DBH}}} - 1 \right) 100 \qquad (8\text{-}4)$$

The simplest financial maturity marking guide says, "If the *projected* value growth percent equals or exceeds the owner's MAR, the tree should be retained, otherwise it should be removed." But it's not always easy to project value growth percent, especially in dense stands where current value growth percents of nearly all trees may be below the MAR. In such cases, if simulations show that thinning rather than clear-cutting is optimal, you should constrain the above marking guideline so that you don't cut more than the prescribed volume removal. The "leave" trees will be released enough that their value growth percents will become acceptable.

In some stands, if you leave only trees with value growth rates above the MAR, it may be possible to remove even more trees without sacrificing annual growth. In such cases, the loss of the cut tree's growth is more than offset by the gain in growth of surrounding trees. Thus, you shouldn't look at a tree's projected value growth rate in a vacuum: financial maturity marking guides give only rough approximations of optimal stocking. Ideally, stand-level simulations should indicate the optimal volume to remove, and financial maturity marking guides can show which trees to cut, *within this constraint.* As in the "Optimal Thinning Intensity" section, the land value is considered in WPL maximization and can be ignored in the marginal decision to cut one tree.

For marking large thinnings, checking each tree for rings per inch and value growth percent is much too time-consuming, so you could develop marking guides based on the financial maturity concept but translated into tree vigor classes and showing which trees to cut and leave. Appraising foliage appearance, crown ratio, and position in the stand, you could estimate, say, three recognizable vigor classes, with a range of projected rings per inch (RPI), based on sample measurements. For example, 7 RPI and less could be high vigor; 8 to 12 RPI, medium vigor; and over 12 RPI, low vigor. Table 8-4 gives a hypothetical example of such a marking guide showing the diameters above which trees should be removed, given three vigor classes and different real MARs. In this case, subject to a previously determined optimal thinning intensity, if the owner's MAR is 6 percent, cut trees of high vigor with diameters over 14 inches, trees of medium vigor with diameters over 9 inches, and trees of low vigor with diameters over 8 inches.

UNEVEN-AGED MANAGEMENT

Even-aged management with eventual clear-cutting often yields the greatest timber volume, lowest logging costs, and highest willingness to pay for land (WPL) based strictly on financial values. But uneven-aged management, without clear-cutting, looks better and can sometimes create less soil disturbance. On the other hand, compared to clear-cutting, to cut the same volume, partial cutting requires roading and harvesting a larger area. Absence of clear-cutting can benefit some types of wildlife, but other animals like deer and many game birds thrive on a combination of forested and open areas. With some species like ponderosa pine, in drier and warmer parts of its range, regeneration succeeds better under an existing stand because seedlings often can't survive the intense heat and dryness after clear-cutting. So clear-cutting systems aren't always the most profitable, especially if aesthetic values are included.

Figure 8-5 shows an uneven-aged stand with three age classes. In such a stand, you could cut the oldest trees every $t/3$ years, where t is the rotation age.

Starting with bare land in year 0, Figure 8-6 graphs the land and stumpage values (liquidation value) of a 1-acre stand to be under uneven-aged management. After an even-aged **development period** of d years, partial cuts worth h dollars

TABLE 8-4
HYPOTHETICAL TREE MARKING GUIDE—CUT TREES LARGER THAN INDICATED DBHs

Real MAR, %	High vigor, 7 RPI* and less	Medium vigor, 8–12 RPI	Low vigor, over 12 RPI
4	22-inch DBH	16-inch DBH	14-inch DBH
6	14-inch DBH	13-inch DBH	11-inch DBH
8	12-inch DBH	11-inch DBH	10-inch DBH

*Projected rings per inch. User develops description of tree characteristics for each vigor class, based on past observations of trees in each RPI range.

FIGURE 8-5
Three age-class uneven-aged stand.

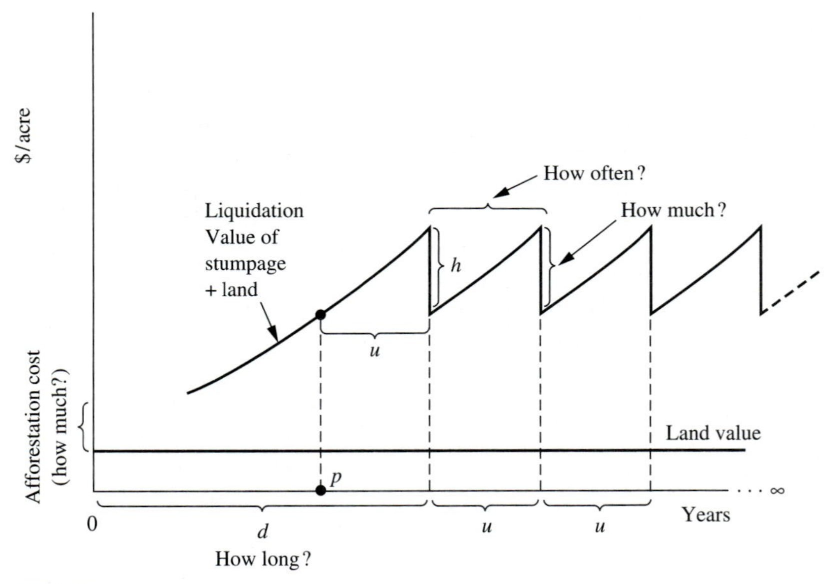

FIGURE 8-6
Liquidation values when developing an uneven-aged stand.

occur every u years, which is the **cutting cycle.** The stand is gradually converted to an uneven-aged structure after d years. For many species, natural regeneration will occur beneath the remaining trees if cuttings are heavy enough during good seed years and soil is exposed. Another approach is to plant or seed between existing trees after each harvest. Some species respond better than others to such management. Generally, intolerant or sun-loving species don't reproduce well in the shade of an overstory. Thus, uneven-aged management of intolerant species often means conversion to shade tolerant species, unless measures are taken to assure intolerant tree regeneration.

Silviculturists and biometricians can provide data on regeneration and growth under partial cutting; here we're only considering ways to evaluate alternatives. With uneven-aged management, eventually harvest would equal growth; a multi-aged forest would develop with u years between age classes, and cutting, mainly from the older trees, could continue forever without clear-cutting. For example, suppose after a development period of 30 years, partial cuts occurred every 10 years and the manager aimed for a maximum age of 30 years. Eventually the forest would have three age classes as in Figure 8-5: 10, 20, and 30 years just before harvest, and 0, 10, and 20 just after harvest.

Willingness to Pay for Bare Land

Cash flows from developing an uneven-aged stand from bare land, as in Figure 8-6, are shown on the time line below, ignoring annual revenues and costs $(a - c)$:

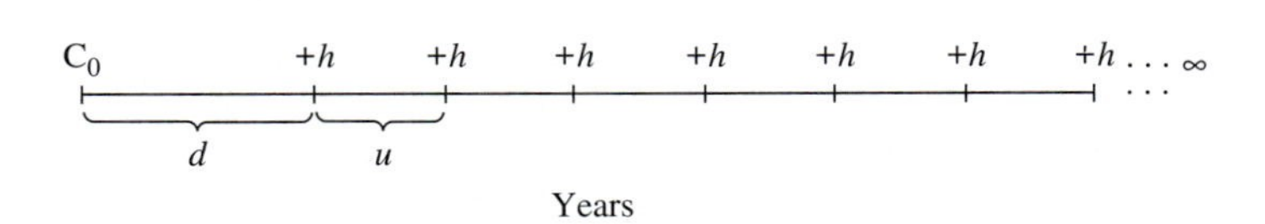

Starting with bare land, WPL_∞ for uneven-aged management is:

$$\text{Uneven-aged } WPL_\infty = \frac{h + \dfrac{h}{(1+r)^u - 1}}{(1+r)^d} + \frac{(a-c)}{r} - C_0 \tag{8-5}$$

In the first term, the numerator is the first harvest occurring in d years plus the year d present value of all future harvests every u years thereafter. The denominator discounts this value to year 0. The other terms are the present value of net annual cash flows minus original afforestation cost. This approach assumes equal real harvest incomes in perpetuity.

As an example, assume a 30-year development period, the first harvest of \$800 occurring in $d = 30$ years and every $u = 10$ years thereafter, $(a - c) = -\$2$, and afforestation cost $=$ \$70, all per acre values. Using equation (8-5) with a 6 percent real MAR, WPL is \$212/acre, assuming we start with bare land. Assume a public agency without taxes.

In the above examples, the stand is even-aged for the first d years, after which natural regeneration occurs beneath the existing stand every u years and the stand becomes uneven-aged. This is a very simple example, since harvests may not always be equal, especially when starting with an even-aged stand.

If you ignore most of the development period and compute the present value of harvests occurring every u years, you'll get the present value of timber and land in a fully stocked uneven-aged forest ready to yield its first harvest in u years. For example, in Figure 8-6, the NPV in year p, u years before the first harvest, is:

$$\text{Uneven-aged NPV, } u \text{ years before harvest} = \frac{h}{(1 + r)^u - 1} + \frac{a + c}{r} \tag{8-6}$$

This NPV 10 years before the first harvest in the above numerical example would be $800/[(1.06)^{10} - 1] - 2/0.06 = \$978$/acre. This is far above the \$212 WPL because it includes the present value of timber, which is $d - u$ years old, or age 20 in this example. The NPV in equation (8-6) generally won't lie exactly on the curve of land plus stumpage value, as explained shortly. Note that the last two equations assume costless natural regeneration after harvests. Regeneration costs would be subtracted from h in equations (8-5) and (8-6).

Optimizing Uneven-Aged Management

In uneven-aged management, we need to simultaneously find the optimal combination of planting spacing, development period, cutting cycle, and type and level of harvest (see questions in Figure 8-6). As with thinning, the optimal plan is that which maximizes the net present value. This NPV could be WPL or could be the NPV of future cash flows from an existing stand. You could find the optimum by simulating alternative harvest and income patterns and choosing the plan with the highest WPL. But most computer-based uneven-aged growth models don't permit starting with alternative planting spacings. If they did, by repeated simulations, you could estimate the optimal combination of planting spacing, development period, cutting cycle, and cutting intensity that maximized WPL, using equation (8-5). Again, this problem lends itself well to mathematical programming solutions.[6] As uneven-aged growth models improve, such analyses could be readily done.

Some uneven-aged growth models exist for mixed-species and single-species stands already established.[7] With such simulators you could place yourself, for example, at point p in Figure 8-6 and project alternative partial harvest levels and intervals. The optimal harvest pattern would be that which maximized the NPV in equation (8-6). Or for an understocked stand, an initial development period d could be longer than the cutting cycle, in which case, the NPV to maximize would be equation (8-5) without the afforestation cost C_0. You could use the financial maturity marking guides discussed under thinning to decide which trees to remove, within the constraints of a predetermined optimal harvest level and timing.

---

[6]See Haight (1990) for problem formulations and Hotvedt and Ward (1990) for applications to established stands.

[7]For example, TWIGS (Miner et al., 1988). For stands of uneven-aged loblolly-shortleaf pine managed under the selection system, Farrar et al. (1984) give equations and tables to predict growth and potential harvest after different cutting cycle lengths, for stands of given initial basal area and merchantable volume.

In the foregoing cases of an existing stand, what if the highest calculated NPV was less than the liquidation value of land and trees? For example, in Figure 8-6, suppose the NPV in year *p* with equation (8-6) was less than the upper curve of potential revenue from clear-cutting the timber and selling the land. This would suggest, strictly from a financial view, that the owner would be better off by clear-cutting and either selling the land or managing it for some other use yielding a WPL at least as high as the land value, for example, grazing or even-aged forestry. But nonmonetary factors could change that decision, as shown in the last section. In any case, with the analyses of thinning and uneven-aged management, decision makers should compare liquidation values with NPVs.

TIMBER STAND IMPROVEMENT

Timber stand improvement (TSI) is an investment in a stand that increases its future harvest revenue. Examples are fertilization to yield more timber growth, precommercial thinning and improvement cutting to distribute future growth on the most valuable trees, and pruning to provide more valuable knot-free wood. These are all variants of the optimal stocking problem: how much of each type of tree should we have? For the simplest case of even-aged forests, assuming only clear-cutting, with each TSI you have an investment of C_y in year y that yields an *added* real harvest revenue of R_t at harvest age, $t - y$ years later. The TSI investment is acceptable if the rate of return on the investment of C_y is at least equal to the MAR:

$$\text{Rate of return on TSI} = \sqrt[t-y]{\frac{R_t}{C_y}} - 1 \qquad \text{(8-7)}$$

The principle is perfectly general for any TSI, say, in Monterey pine in Chile or Douglas-fir in Oregon. Since the TSI is a marginal investment, you needn't include land value. For example, suppose, in an immature stand, you're considering spending $C_y = \$50$ per acre on fertilization that will yield an expected value of *increased* real after-tax harvest revenue of $R_t = \$98$ in 10 years. Substituting into equation (8-7), the real rate of return is 0.07, or 7 percent, which is acceptable as long as it exceeds your MAR. To include TSI into your willingness to pay for land, just add the TSI cost in an even-aged WPL equation as a C_y, and increase harvest incomes appropriately. If the rate of return on TSI exceeds the MAR, the TSI will increase WPL.

INCLUDING NONMONETARY VALUES

Several trends are kindling more interest in uneven-aged forestry: some localities have passed ordinances against clear-cutting; the U.S. Forest Service's "new perspectives" program of the early 1990s promised more sensitivity to protecting scenic values and ecosystems; many nonindustrial private forest owners are interested in timber harvest income but don't wish to clear-cut; in visually sensitive areas on all forest ownerships, the public is increasingly upset about clear-cutting. In 1994, Representative John Briant of Texas introduced a bill before Congress that would have banned clear-cutting and all even-aged management on federal lands, but the bill died in committee.

Where partial cutting is most profitable, there's little conflict; examples are some uneven-aged northern hardwood stands and parts of the ponderosa pine region. But thorny trade-offs occur in cases when clear-cutting yields the most profit, as is usually true for two of our commercially most important species, Douglas-fir and loblolly pine. Both of these species reproduce best in full sunlight and grow fastest in even-aged stands. But in some understocked loblolly pine stands, and for some interest rate and price combinations, uneven-aged systems may be more profitable (see Chang, 1990; and Redmond and Greenhalgh, 1990). Although rarely practiced, *uneven-aged management is possible in Douglas-fir and southern pines,* but volume growth is slower and species composition may change over time [see Foss (1990) and Franklin and DeBell (1972)]. Also, logging costs per unit volume with partial cuts are higher than with clear-cuts. Let's look at a way to compare higher income from clear-cutting with greater nonmonetary values from uneven-aged management.

Figure 8-7 shows two cutting alternatives for an existing acre of mature timber. The solid lines indicate clear-cutting now ($\$H$) and every 40 years thereafter. The dashed lines assume cutting part of the volume ($\$h$) every 10 years and converting the stand to a four-age-class uneven-aged forest. To correctly compare the different income intervals, assume a perpetual time horizon. Also, note that we're starting with the same stocking level for both alternatives, for valid comparisons. As a simple example, let $H = \$2{,}000$ and $h = \$700$, assume costless natural regeneration, and assume no other costs and revenues.[8] With clear-cutting, you get $2,000 now and every 40 years thereafter; with conversion to uneven-aged management, it's $700 now and every 10 years thereafter. Computing the difference in today's NPVs per acre at 6 percent real interest:

$$
\begin{aligned}
\text{Clear-cutting NPV} &= 2{,}000 + \frac{2{,}000}{[(1.06)^{40} - 1]} = \$2{,}215/\text{acre} \\
\text{Uneven-aged NPV} &= 700 + \frac{700}{[(1.06)^{10} - 1]} = \$1{,}585/\text{acre} \qquad (8\text{-}8) \\
\text{NPV difference} &= \$630
\end{aligned}
$$

Think of these NPVs as a maximum willingness to pay for the existing stand and land under each management system, before any cutting. For this hypothetical example, what would the extra *annual* nonmonetary benefit (for example,

[8]Since we'll compute the difference in NPVs of the two cutting alternatives, other costs and revenues could be omitted if they were the same for each option—present values of other costs and revenues would cancel out. If other costs and revenues differed between alternatives, you'd have to include them in the present values in equation (8-8).

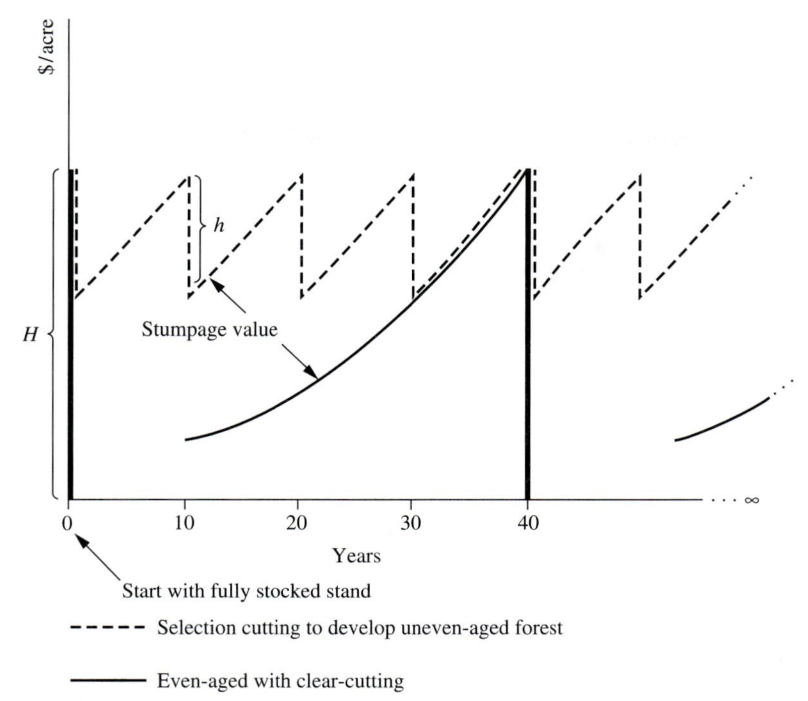

FIGURE 8-7
Stumpage values of an existing stand under clear-cutting vs. selection cutting.

aesthetics) from the uneven-aged forest have to be, in dollars, to make up the \$630 per acre NPV loss? From equation (6-10), for the perpetual case, multiply the NPV difference by the interest rate to annualize the perpetual loss of income: $630(0.06) = \$38$/acre/year.[9] You could then ask whether the added annual aesthetic benefit from uneven-aged management was worth the annualized loss in harvest income. This example is highly simplified. In reality, the aesthetic difference between the options becomes less per year over the first 40 years; starting with a mature even-aged stand, the uneven-aged harvests wouldn't necessarily be equal; and many other costs and revenues could be included. But the key point is to compare present value differences between choices and gauge whether gains in nonmonetary benefits are worth the value forgone, if any. Or thinking back to Chapter 3, under the amenity rights framework, ask whether the extra financial gains from clear-cutting are enough to fully compensate those damaged thereby.

[9]Recall the formula for the present value (PV) of a perpetual annual benefit B : PV $= B/r$. Solving for B, $B =$ (PV)r. If, in both cases, after 40 years, you assume selling land and timber (land at \$400/acre), the PV difference is \$569. Assuming both alternatives have the same aesthetic appeal for the last 10 years, the annualized PV difference, or required annual benefit B for the first 30 years, using equation (6-11), is:

$$B = 569 \frac{0.06}{1 - (1.06)^{-30}} = \$41$$

Although my example of an alternative to clear-cutting was uneven-aged forestry, other harvesting approaches exist. Examples are:

- *Group selection,* sometimes known as patch cutting in small clear-cuts with diameters about 2.5 times the height of edge trees
- *Seed-tree harvest,* where mature trees, usually less than 10 percent of the original volume, remain as scattered trees or groups of trees for varying lengths of time
- *Shelterwood cutting,* which is similar to the seed-tree approach but leaves more trees (seed-tree and shelterwood methods usually result in even-aged stands)
- *Two-aged* forest, starting with a shelterwood cut (for example, on a 40-year rotation, cut about half the trees every 20 years)
- *Old-growth snag retention,* which leaves, say, two trees per acre for wildlife enhancement (rather close to clear-cutting)
- *No harvesting at all*

You could make net present value comparisons of these and other alternatives[10] and evaluate benefits and costs of attaining a wide range of outputs including aesthetics, water quality, timber, fish and wildlife, and soil protection (more on this in Chapter 15). Refinements of the analyses in this section can be very helpful in designing environmentally sensitive management plans.

Many foresters have remained wedded to the most profitable harvesting method, which usually means clear-cutting. This may not be a problem on many flat areas, out of the public view. But clear-cutting in easily visible, mountainous forests has caused major conflicts. For example, the U.S. Forest Service clear-cut several areas close to well-traveled highways in West Virginia in the late 1960s and before. This led to the Monongahela flap over scenic damage and the resulting (although temporary) ban on harvesting in many National Forest areas in the early 1970s. Such problems could have been avoided had the U.S. Forest Service been willing to do partial cutting in visually sensitive areas.

Greater trauma from cutting restrictions occurred in the Pacific Northwest in the early 1990s. For years, U.S. Forest Service clear-cutting had been insensitive to scenic and other values in the region. Thus the public and the courts mistakenly believed the choice was between clear-cutting and no harvesting at all. They chose the latter on many public forest areas in the Northwest, with the result of sudden job losses. Had more partial cutting been done in recent decades, we might never have had the economic and ecological disasters that occurred in the Northwest.

The above reasoning does not imply that we'd have some degree of harvesting on all areas and no untouched forests. Think back to Chapter 3 where you saw that zero environmental damage is often optimal; also, remember from Chapter 7 the cases when perpetual rotations can be optimal. You can incrementally analyze increasingly smaller partial cuts. So an infinitely small partial cut implies

[10]For example, see Sedjo and Bowes' (1991) analysis of Douglas-fir and western hemlock management alternatives on Washington's Olympic Peninsula.

no harvesting at all, which is what many believe is socially optimal in wilderness areas and in some of the few remaining stands of public old-growth timber not reserved as wilderness. Let's now look at other ways to evaluate partial cutting methods where nontimber values are at stake.

Simple Betterness Method

The **simple betterness method** can include nonmonetary values when comparing harvesting alternatives. Rickard et al. (1967) illustrated the method with old-growth Douglas-fir stands for which they computed NPVs of several harvesting alternatives such as clear-cut and plant Douglas-fir, shelterwood cut and regenerate to Douglas-fir, hold for 40 years with mortality harvest and then shelterwood cut, and do selection cutting and convert to an uneven-aged forest of tolerant species. To illustrate the approach, Figure 8-8 displays dollar and nondollar outcomes of five such harvesting alternatives, including a no cutting option, in order of decreasing net present value of timber income on a forested area. The utility bars are broken to suggest preference ranking by some group without any cardinal measure of satisfaction. Utilities from alternatives could be ranked on an arbitrary scale of, say, 1 to 10, based on photographs or visual computer simulations over time; descriptions of soil and water quality; and opportunities for hunting, fishing, and other recreation. If recreation could be marketed, its yields would be included in the dollar NPV bars.

The simple betterness method allows decision makers to eliminate certain *dominated* alternatives that, compared to another plan, provide either less aesthetic benefit and the same NPV, less NPV and equal nondollar benefit, or less

FIGURE 8-8
Dollar and nondollar outcomes of five harvesting alternatives. (Adapted from Rickard et al., 1967)

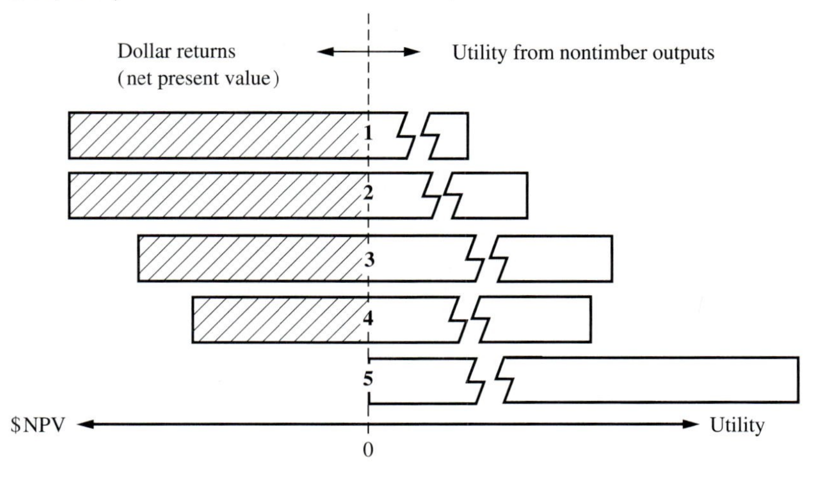

Alternative 2 dominates 1; 3 dominates 4.

of both dollar and nondollar outputs. For example, in Figure 8-8, alternative 1 is dominated by 2, and 3 dominates 4. So we're left with *nondominated* alternatives 2, 3, and 5. How would you choose between them?

Choosing from Nondominated Alternatives

Figure 8-9 shows several nondominated timber harvesting alternatives like 2, 3, and 5 from Figure 8-8, ranked in a hypothetical order of decreasing NPV and increasing nondollar benefits. Starting with 1 in Figure 8-9, you could ask whether the added or "marginal" nondollar benefit (MB) from cutting restrictions moving us to alternative 2 is worth the marginal cost of NPV forgone (MC). If so, choose 2; if not, stay with 1. Moving down the diagram, you could decide, as above, how far to go in the direction of increased nondollar benefits.

The framework of ranked nondominated harvesting alternatives in Figure 8-9 is also useful for seeking the optimal environmental damage imposed by a landowner on the rest of society, using the two liability rules in Chapter 3. In Figure 8-9, suppose that the NPV bars are the landowner's total benefits, and the nondollar bars accrue to the rest of society. Using the **victim liability** approach, you could start with alternative 1, allowing profit-maximizing forest management, and require the landowner to manage for increasingly higher environmental benefits at the expense of lower NPV, as long as the public's willingness to pay for added benefits (MB) was enough to compensate landowners for the loss of income (MC). Under **damager liability,** you'd start with no damage allowed (alternative 4) and relax environmental restrictions as long as the landowner's marginal benefit (MB) exceeded the public's marginal cost or compensation required (MC) to willingly endure the resulting environmental damage. For reasons given in Chapter 3, the damager liability framework is likely to yield less environmental damage than victim liability.

FIGURE 8-9
Four nondominated harvesting alternatives.

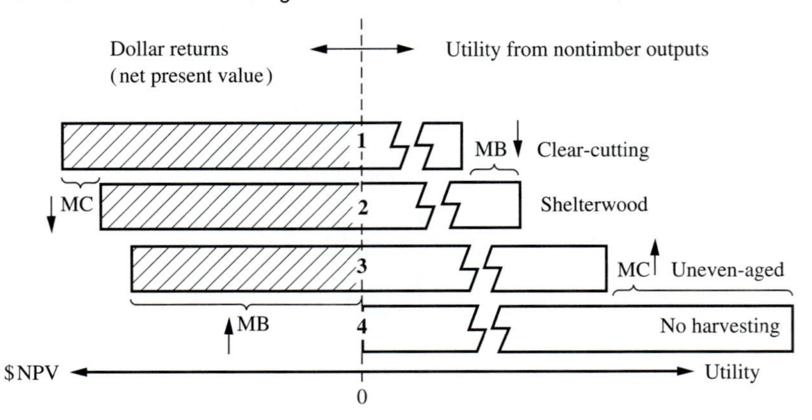

Remember from Chapter 3 that if temporary employment and income losses result from regulations that move us from alternative 1 downward in Figure 8-9, the marginal benefit (MB) to society becomes less. Conversely, starting at alternative 4, if employment benefits result from relaxing cutting restrictions and moving upward in Figure 8-9, the public's marginal compensation (MC) becomes less. Granted, none of the measures in Figures 8-8 and 8-9 can be precise, but the framework defines the nature of environmental problems in harvesting and could help in designing optimal management plans.

Heavier Partial Cuts to Get More Beauty

Looking only at financial returns, the optimal even-aged thinning intensity and regime is that which maximizes the willingness to pay for land (WPL), adapting either equation (7-4) or (7-9). People often prefer a more open stand into which they can see farther, which they'd get if thinning were heavier and possibly earlier than the financial optimum. For an even-aged stand, Figure 8-10 shows this by comparing visual quality (in terms of viewer satisfaction) for (1) a financially optimal thinning and clear-cut (solid line) and (2) a heavier and earlier thinning followed by a later clear-cut (dashed line). Although thinnings can increase visual quality, I assume a temporary decrease in viewer satisfaction immediately following thinning (Hull and Buhyoff, 1986).

Suppose you had bare forestland, and you calculated WPL_∞ for the two options in Figure 8-10: say the financial optimum yielded $WPL_\infty = \$300$ per acre, and

FIGURE 8-10
Visual quality rankings for two different thinning regimes in an even-aged stand.

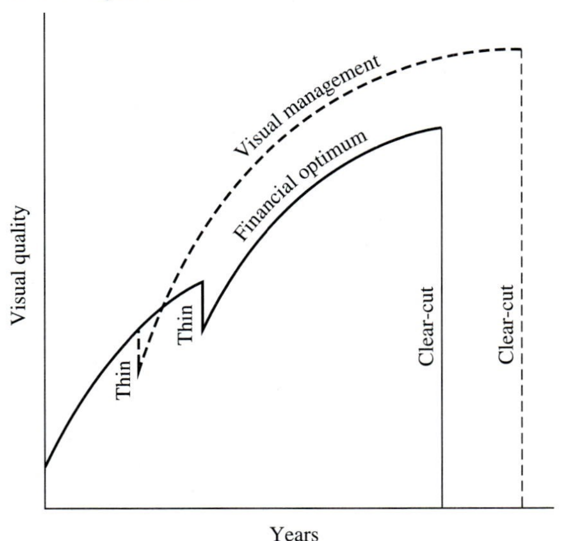

the visual management regime, $250, both at your MAR of 7 percent. In the same way you compared present value differences in equation (6-10), the perpetual annualized $50 present value difference is 0.07(50) = $3.50 per acre per year. If you felt that aesthetic benefits of visual management averaged at least that much, you'd choose the latter. This is a very crude averaging, since aesthetic benefits are the same under each system until the first thinning. Photographs or computer visualizations of different thinning intensities over time would help in such decisions.

KEY POINTS

- Planting the optimal number of trees per acre will maximize the net present value (NPV) of a particular forest management regime.
- Precommercial and commercial thinning can often increase the NPV of forest management plans by concentrating growth on the most valuable trees and bringing earlier income.
- Thinning types include thinning from above, thinning from below, and row thinning.
- In seeking the thinning regime that maximizes NPV of even-aged stands, we need to simultaneously find the optimal initial stocking, thinning ages, thinning intensity, type of thinning, and clear-cut age.
- Computer-based growth models are useful for simulating thinning alternatives and choosing the optimum thinning regime.
- After finding the optimal volume to remove in a thinning, financial maturity guides can identify which trees to cut: those with projected value growth percents less than the owner's minimum acceptable rate of return.
- Uneven-aged management involves partial cuts, usually at regular intervals without ever clear-cutting, and some regeneration occurs after each cut. The result is a multi-aged stand.
- To find the optimal uneven-aged management regime that maximizes stand NPV, you must simultaneously find the optimal initial stocking, development period, cutting cycle, harvest type, and harvest intensity.
- When NPV for a clear-cutting regime is greater than that for uneven-aged management, you can determine whether the extra annual nondollar benefits from uneven-aged management are great enough to offset the "annualized" difference in NPVs.
- By ranking stand management alternatives according to NPV and relative nondollar benefits, you can use the *simple betterness method* to eliminate options that have both lower NPV and lower nondollar benefits. From the remaining alternatives, you can use marginal analysis to choose the best option.
- For years, many foresters have stubbornly clung to clear-cutting in visually and ecologically sensitive areas. Other harvesting systems, easier on the eyes and on the environment, do exist, but they're usually not as profitable. Simple techniques can help us decide how much to temper profit-maximizing harvest methods to accommodate nontimber values.

QUESTIONS AND PROBLEMS

8-1 Explain verbally, without any equations, the concept of an economically optimal planting density. Think in terms of planting increasingly larger numbers of trees per acre. Show this graphically.

8-2 What are the major factors to vary in determining optimal thinning regimes in even-aged stands?

8-3 Explain the concept of an optimal thinning regime.

8-4 What are the major economic variables affecting the optimal thinning regime?

8-5 **a** Explain the concept of an economically optimal thinning intensity (percent volume removed) given one thinning age y and one rotation age t.

b Graphically, show the general idea of an economically optimum thinning intensity.

c Suppose the next 100 ft^3 of volume considered for thinning in your timber stand would grow in value from $20 in year $y = 15$ to $30 in year $t = 35$. If your minimum acceptable rate of return is 6 percent, should this volume be removed?

8-6 **a** Starting with newly regenerated land, suppose you are planning to develop an uneven-aged selection forest yielding net harvest revenues of $600/acre in constant dollars every 10 years forever, after a development period of 30 years. You expect natural regeneration after each cut. Net annual revenues minus costs are −$2/acre. If you wish to earn a real 5 percent rate of return, what is your maximum willingness to pay for such land?

b Suppose the land were bare and you needed to spend $100/acre on reforestation initially and $25/acre for underplanting after each cut. All other assumptions are the same. What is your maximum willingness to pay for the bare land?

8-7 In question 8-6a, suppose you are at the end of the development period and your stand is 30 years old. Rather than developing the uneven-aged forest, you now have the option of clear-cutting the stand for a net income of $1,500/acre now and every 30 years thereafter. Let reforestation costs be $100/acre after each clear-cut; natural regeneration occurs with uneven-aged management. Show how you could help decide which management system to use. (Annualize the loss in present value when using uneven-aged management, and ask the "right question.")

8-8 Suppose the cutting cycle in an uneven-aged selection forest is 10 years. You estimate that if you increase your stocking by 1 MBF/acre, then your harvest 10 years later (and every 10 years thereafter) would be worth $116/acre *more* than it would have been without the addition to stocking. The value of the added 1 MBF is $100/acre. What is the percent rate of return earned by the 1 MBF increase in stocking? (This rate of return is the marginal annual value growth percent of the addition to stocking.)

8-9 Explain how you could use the simple betterness method to compare several forest harvesting alternatives with different aesthetic values.

REFERENCES

Broderick, S. H., J. F. Thurmes, and W. D. Klemperer. 1982. Economic evaluation of old-field loblolly pine management alternatives. *Southern Journal of Applied Forestry.* 6(1):9–15.

Caulfield, J. P., D. B. South, and G. L. Somers. 1992. The price-size curve and planting density decisions. *Southern Journal of Applied Forestry.* 16(2):24–30.

Chang, S. J. 1990. An economic comparison of even-aged and uneven-aged management of southern pines in the mid-south. In Hickman (1990), pp. 45–52.

Curtis, R. O., G. W. Clendenon, and D. J. DeMars. 1981. *A New Stand Simulator for Coast Douglas-Fir: DFSIM User's Guide.* U.S.D.A. Forest Service. Gen, Tech Rept. PNW-128. Portland, OR. 79 pp.

Davis, L. S., and K. N. Johnson. 1987. *Forest Management* (3rd edition). McGraw-Hill, New York. 790 pp.

Duerr, W. A., J. Fedkiw, and S. Guttenberg. 1956. *Financial Maturity: A Guide to Profitable Timber Growing.* U.S. Department of Agriculture. Technical Bulletin 1146. Washington, DC.

Dykstra, D. P. 1984. *Mathematical Programming for Natural Resource Management.* McGraw-Hill, New York. 318 pp.

Farrar, R. M., Jr., P. A. Murphy, and R. L. Willett. 1984. *Tables for Estimating Growth and Yield of Uneven-Aged Stands of Loblolly-Shortleaf Pine on Average Sites in the West Gulf Area.* Arkansas Agric. Exper. Sta. Bulletin 874. Fayetteville. 21 pp.

Fedkiw, J., and J. G. Yoho. 1960. Economic models for thinning and reproducing even-aged stands. *Journal of Forestry.* 58:26–34.

Flora, D., and J. Fedkiw. 1964. *Volume Growth Percent Tables for Douglas-Fir Trees. USDA Forest Service.* PNW Forest and Range Experiment Station, Portland, OR. 145 pp.

Foss, T. 1990. New forestry—A state of mind. *Inner Voice.* 2(1):4–6.

Franklin, J. F., and D. S. DeBell. 1972. Effects of various harvesting methods on forest regeneration. In *Even-Age Management.* Proceedings of a symposium. Oregon State University School of Forestry. Paper 848. Corvallis. Pp. 29–57.

Haight, R. G. 1990. Economic models for evaluating plantation and uneven-aged forestry. In Hickman (1990), pp. 25–34.

Hepp, T. E. 1990. *YIELDplus User Manual—Timber Yield Forecasting and Planning Tool* (version 2.1). Tennessee Valley Authority, Norris. 85 pp.

Hickman, C. A. (Ed.). 1990. *Evaluating Even and All-Aged Timber Management Options for Southern Forest Lands.* Proceedings, Southern Forest Economics Workshop. Monroe, LA. March 1990. U.S.D.A. Forest Service. Gen. Tech. Rept. SO-79. Southern Forest Experiment Station, New Orleans, LA. 149 pp.

Hicks, L. L. 1991. Plum Creek Timber Company's new forestry experiments: Lessons learned and future directions. *Forest Perspectives.* 1(3):9–10.

Hotvedt, J. E., and K. B. Ward. 1990. A dynamic programming optimization model for uneven-aged loblolly-shortleaf pine stands in the mid-South. In Hickman (1990), pp. 35–43.

Hull, R. B., and G. J. Buhyoff. 1986. The scenic beauty temporal distribution method: An attempt to make scenic beauty assessments compatible with forest planning efforts. *Forest Science.* 32(2):271–286.

Miner, C. L., N. R. Walters, and M. L. Belli. 1988. *A Guide to the TWIGS Program for the North Central United States.* U.S.D.A. Forest Service North Central Experiment Station. Gen. Tech. Rept. NC-125. St. Paul, MN. 105 pp.

Redmond, C. H., and R. Greenhalgh. 1990. An economic analysis of even-aged and uneven-aged management on nonindustrial private land in southern Arkansas. In Hickman (1990), pp. 69–79.

Rickard, W. M., J. M. Hughes and C. A. Newport. 1967. *Economic Evaluation and Choice in Old-Growth Douglas-Fir Landscape Management.* U.S.D.A. Forest Service Pacific Northwest Forest and Range Exp. Sta. Research paper PNW 49. 33 pp.

Sedjo, R. A., and M. D. Bowes. 1991. *Managing the Forest for Timber and Ecological Outputs on the Olympic Peninsula.* Resources for the Future. Discussion Paper ENR92-02. Washington, DC. 70 pp.

Tesch, S. D., and J. W. Mann. 1991. *Clearcut and Shelterwood Reproduction Methods for Regenerating Southwest Oregon Forests.* Oregon State University. Forest Research Lab. Research Bull. 72. Corvallis, OR. 43 pp.

Weigand, J. F., and R. W. Haynes. 1991. Economic considerations for green tree retention. *Forest Perspectives.* 1(3):11–12.

Weigand, J. F., and R. W. Haynes. 1991. *"High Quality" Forestry Alternatives for Western Washington—A Problem Analysis.* Pacific Northwest Research Station. U.S.D.A. Forest Service. Portland, OR. Draft Version.

Weih, R. C., J. E. Scrivani, and H. E. Burkhart. 1990. *PCWTHIN Version 2.0 Users Manual.* Virginia Polytechnic Institute and State University, Department of Forestry, Blacksburg. 31 pp.

9

FOREST TAXATION

It seems like taxes are constantly nibbling away our income—in direct and painful ways, as with taxes on income, sales, property, and inheritance; or more subtly, as with taxes on gasoline, restaurant meals, hotel rooms, and imports. Thinking back to Chapter 3, you could imagine an ideal market that would supply nearly all our needs in optimal quantities. In such a utopia, few taxes would be needed. But you'll also recall the **market failures** that keep a free market from maximizing social welfare: property rights not enforced, imperfect competition, imperfect information, immobility of labor or capital, existence of public goods that are not easily sold in units to individuals, unpriced negative and positive side effects, unsatisfactory income distribution, and economic instability. So, to solve these problems, governments regulate some enterprises and provide education, national defense, public health, police, parks, recreation, welfare payments, research, subsidies, transportation, and other services. We fund these activities with user fees and taxes on individuals and businesses. Public agencies are tax-exempt, but many agencies such as the U.S. Forest Service and the National Park Service make payments to counties in place of taxes.

In this chapter you'll learn about several taxes and how they affect forestry. For private forest owners, all taxes combined are one of the largest expense categories; thus, they are vitally important in investment analysis. I'll emphasize basic principles and methods of analysis rather than specific laws. *Remember that the laws mentioned are constantly changing, so for any actual analysis, you'd need to get up-to-date information about things like tax types, tax rates, definitions of taxable income and value, exemptions, relevant dates, allowable deductions, and administrative details.* As you go through this chapter, try to think beyond tax analysis:

the material gives major insights into capital allocation in market economies, impacts of government policies on resource allocation, earning rate measurement, the importance of cash flow timing, and your own personal finances. Let's first consider general taxation concepts.

GENERAL FOREST TAXATION CONCEPTS

The major taxes on forests are *income taxes, property taxes,* and *death taxes.* The federal government and most states levy income taxes as a given percentage of net forestry income. Local governments impose property taxes as a percentage of real estate values, including forested property, or they substitute another tax for the property tax. State and federal death taxes and gift taxes apply to noncorporate forests.

Below are some of the factors that affect the way landowners and policy makers analyze tax impacts. Notation for this chapter is in Table 9-1.

Tax Shifting

While not all forest owners sell stumpage in perfectly competitive markets, most are primarily **price takers.** They take the output price given by the market and

TABLE 9-1
NOTATION FOR CHAPTER 9

a	equal annual revenue, \$/acre
b_t	taxing district budget that will be financed by the property tax
c	equal annual cost, \$/acre
$(a - c)$	net annual cash flow, \$/acre (may be positive, zero, or negative)
C_y	cost in year y, \$/acre
G	average annual harvest revenue, \$/acre
H_t	clear-cutting revenue at rotation age t, \$/acre
I_n	inflated or nominal income in year n
i_a	after-tax nominal rate of return or discount rate
MAR	minimum acceptable rate of return (must specify whether before or after taxes)
MBF	thousand board feet
n	number of years of compounding or discounting
NPV	net present value or present value of revenues minus present value of costs
p	amount of fixed payment occurring regularly in a series
PV	present value
r	real before-tax discount rate, percent/100
r_a	real after-tax discount rate
R_y	revenue in year y
T_i	income tax rate, percent/100
T_a	property tax rate, percent/100—called an *ad valorem* tax (from Latin meaning "based on value")
T_y	yield tax rate, percent/100
t	number of years between periodic occurrences of a payment p (for timber examples, t is the rotation age)
V_0	value at the start of an investment period
WPL	willingness to pay for bare land, \$/acre
WPL_a	willingness to pay for bare land after *ad valorem* taxes, \$/acre
WPL_p	willingness to pay for bare land after productivity taxes, \$/acre
WPL_y	willingness to pay for bare land after a yield tax, \$/acre
y	an index for years

can't simply raise their stumpage price to "pass forward" or "shift" a tax-increase to consumers. If one forest owner tries such a price-increase, there are generally enough other suppliers that stumpage buyers will go elsewhere. This reasoning also applies to firms cutting their own timber to produce, say, paper or lumber. In the short run, a higher forest tax cannot be passed forward into higher prices of such products, because they are sold in national and international markets. A price-increase by one firm will cause buyers to shop elsewhere.

Given the fairly competitive timber markets in the United States, we can expect that, in the short run, higher forest taxes will be shifted mainly into lower land values. We'll view the mechanism shortly. For now, remember that your **willingness to pay for bare land (WPL)** is the **net present value** of future cash flows therefrom. If taxes reduce net income, buyers tend to bid less for land, other things being equal, and land values decline.[1] This process is called **tax capitalization.** Recall that to *capitalize* can mean to compute present value; in this case, we say that the tax is capitalized into lower land value. The combination of competition and tax capitalization means that various states can and do have different forest tax levels and still sell forest products in national and international markets at competitive prices. Other things being equal, regions with higher taxes will have lower land values.

While **short-run** tax capitalization is reasonable to assume, in the **long run** (a time long enough for capital and land devoted to forestry to change), taxes can be partially shifted into higher forest products prices. For example, suppose a new tax is imposed only on forestry, so that on some lands, tree farmers' bid-prices for land are depressed below what ranchers could bid for the same land. Such a tax could cause "marginal timberlands"—those just barely more profitable than the next-best use—to shift from timber production to grazing. As shown in Figure 9-1, one rotation later (for example, in 40 years), the tax-induced change in land use could shift the timber supply curve leftward, because at any given price, less would be supplied. This shift brings a lower harvest and higher stumpage price (Q_1, P_1) than would have occurred without the tax (Q, P). But such land-use changes will occur only if bid prices for the competing land uses are originally fairly close and if new tax differences are large enough. Moderate tax-changes are unlikely to cause major shifts in land use. And today's landowners won't normally predict tax-induced changes in prices when they make decisions. Thus, *tax analyses in this book assume that* (1) *decision makers' price projections will be unaffected by most tax alternatives*[2] and (2) *other things being equal, the more taxes reduce income from an asset, the less investors will pay for it.*

Elasticity and Forward Shifting of Taxes In Figure 9-1, you can sketch a *more* **elastic** *demand* curve (flatter, given the same scales) and see that the same

[1]For empirical evidence of property taxes being shifted into lower values of land in forestry and agriculture, see Turner et al. (1991) and Pasour (1973).

[2]The most notable exception is a **value added tax (VAT)**, which is levied as a percentage of value added to products at each stage of production. Generally a VAT increases prices immediately, because sellers are legally allowed to add the tax to their quoted prices [Klemperer and O'Neil (1982)]. Although the U.S. government doesn't levy a VAT, most of the major industrial nations do, and Congress has often discussed the idea.

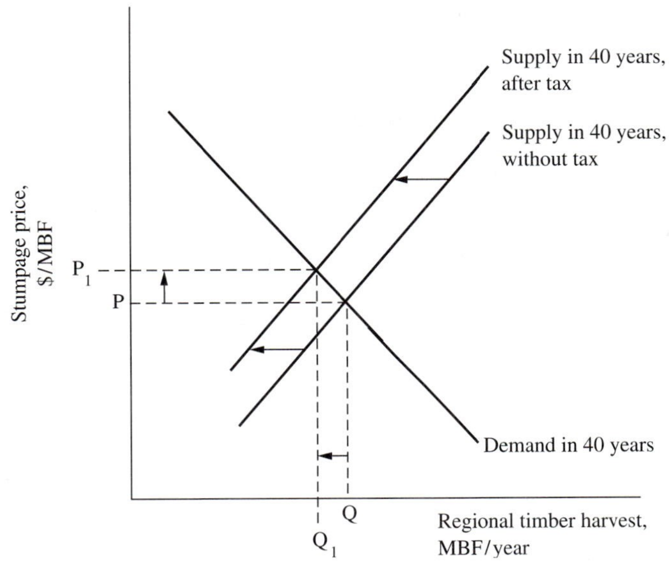

FIGURE 9-1
Long-run impacts of a tax causing land to shift out of timber production.

tax-induced supply shift *causes a smaller price-increase* than shown. Likewise, sketch a more **inelastic** demand, and the price increase is *greater* than shown. In both cases use the same supply curves.

In Figure 9-1, suppose you sketch *more elastic supply* curves, given the same MBF tax-induced leftward supply shift and using the same demand curve. Then the *price-increase is less* than shown. And if supply curves are more *inelastic,* the tax-induced price-increase is *greater.*

So in general, for any tax that shifts any product supply curve leftward, forward shifting of taxes into higher prices will be greater if demand and supply are more inelastic.

Tax Neutrality with Respect to Land Use

When we like the way a market allocates resources, it's most efficient to levy a tax that won't change that allocation. Such a tax is **neutral.** If, before a tax, welfare was being maximized, and a tax changes the allocation of resources, then welfare is reduced (ignoring for the moment the net benefits of the tax-funded program). Examples of changes in resource allocation could be a shift in land use patterns, a change in optimal holding periods for assets, or, in general, changes in investment and consumption patterns. Taxes can change *total* spending on consumption and investment, but here we're concerned with changing the *relative* expenditures in different categories.

Figure 9-2 shows the principle of a tax that is neutral with respect to land use. The open bars depict hypothetical bid prices of three user groups for one

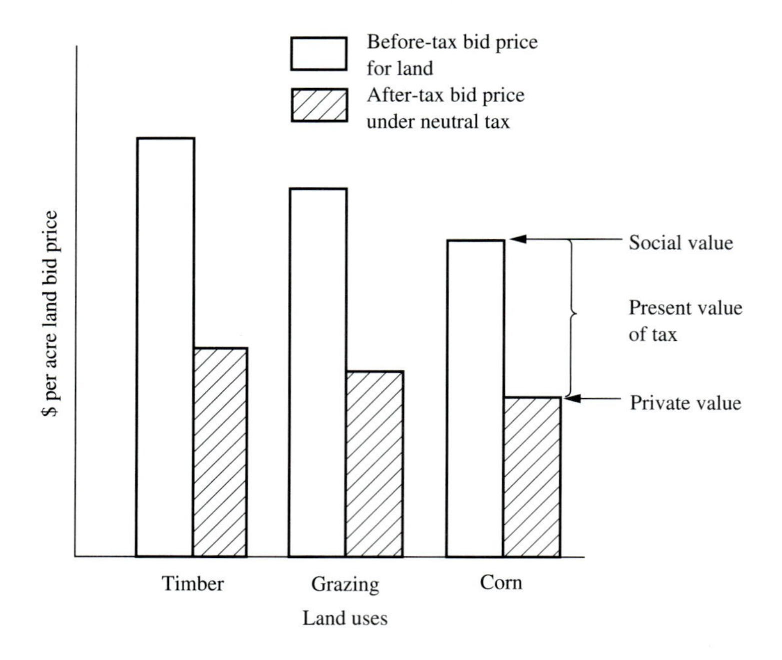

After taxes, forestry still wins in the bidding for land.

FIGURE 9-2
Example of neutral tax where forestry is financially the best use (bids which three uses might generate for 1 acre of bare land). (Adapted from Klemperer [1980], by permission.)

type of land, without taxes. Think of these bids as WPLs, examples of which are equations (7-4) and (7-7). Assuming no market failures, sawtimber is socially the best use for such land because it generates the highest net present value. Since sawtimber growers are the highest bidders, the market would allocate such land to sawtimber production. Now, without worrying about the type of tax, suppose a capitalized tax reduces by 50 percent the amount these potential buyers can bid for land. The taxes would become a negative entry in the WPL equations for each bidder, and the after-tax bids would be the hatched bars in Figure 9-2. Because all bids are reduced by the same percentage, the best use before taxes is also best after taxes, and the market allocation of land is unchanged by the tax. We therefore say that the tax is neutral with respect to land use. This tax is efficient because it hasn't changed the market's efficient determination of land use.

With the same land use example, Figure 9-3 shows a *nonneutral* tax that reduces by *different* percentages the bids these potential buyers can make for the land. Again, don't worry about the type of tax. The hatched bars show the after-tax bids. Since the after-tax bid by ranchers is now the highest, over time the use for this type of land would tend to change to grazing. This result is inefficient because sawtimber production is still socially the best use for the land. Note that the difference between before-tax and after-tax land values is the present value of the tax revenues received by the government. After taxes, the land's value

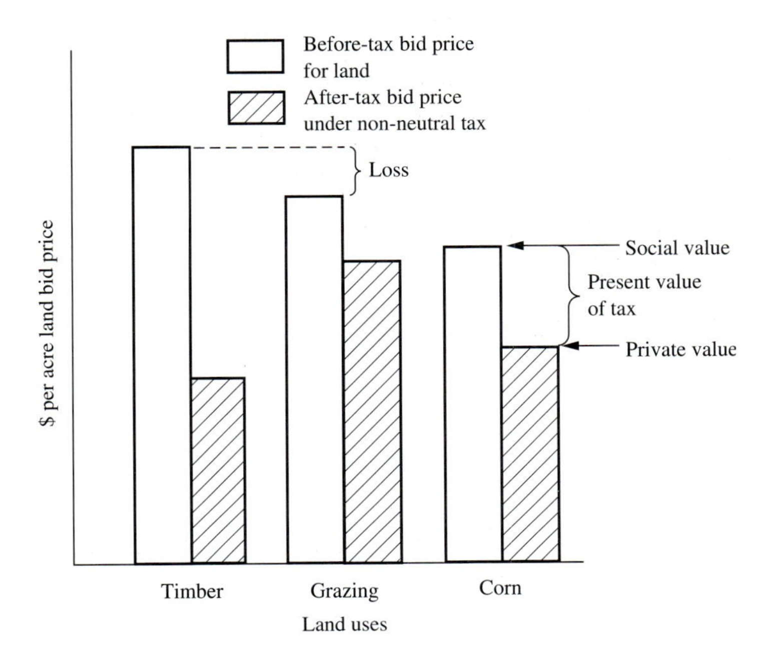

After tax, grazing wins in the bidding for land, while forestry is still socially the best use. In what sense is forestry "best"?

FIGURE 9-3
Example of nonneutral tax where forestry is financially the best use (bids which three uses might generate for 1 acre of bare land). (Adapted from Klemperer [1980], by permission.)

to society is the value to the landowner (the hatched bar) *plus* the present value of taxes to the government. Thus, the true social value is the height of the open bar or the before-tax value. The tax in Figure 9-3 is inefficient or nonneutral because it stimulates a second-best land use. Such a tax prevents the market from generating land uses with the highest social values. In Figure 9-3, the present value of losses caused by the nonneutral tax is the vertical distance between the open bars for sawtimber and grazing, labeled “loss.”

The foregoing covers tax neutrality with regard to land use, one of the most important types of neutrality in forestry. But we can also have neutrality with respect to rotation length, land use establishment costs, use of labor, capital, and many other factors. Few taxes are neutral in all respects, except for a tax on land value alone, which is mentioned again later.

INCOME TAXES

Federal Income Taxes

Just as you pay federal taxes on your wage or salary income, private forest income is also taxed. Taxable income is total revenue minus *deductible* costs, and your tax for any year is the tax rate times the taxable income. Tax law defines which costs are "deducted" (subtracted) and when. Taxable income is of two types: (1) **capital gains** income, or the difference between the sale price and purchase price of capital assets held longer than one year, and (2) **ordinary income,** or other income like wages, rents, interest, and income from producing goods and services. Most income from timber sales is a capital gain. Before 1987, federal income tax rates on gains from capital held longer than a specified period were substantially lower than on ordinary income. Thereafter, corporations have paid the same tax rate on ordinary income and capital gains. Between 1987 and 1991, individuals paid nearly the same rates on ordinary income and capital gains. But as of 1993, for qualifying individuals, the maximum tax rate on capital gains was 28 percent, while on ordinary income it was 39.6 (see Table 9-2).

Chapter 5 explained that income taxes are paid on the difference between asset sale price and original purchase cost, which is called the **basis** in tax lingo (see Figure 5-4). For timber established by the seller, the basis is the afforestation cost, including costs like site preparation and vegetation control. Forest owners need to keep good records of purchase costs or establishment costs, since these are usually deducted many years later for tax purposes, when trees are harvested. Federal tax law—the *Internal Revenue Code*—allows gradually deducting from current income up to $10,000 of qualifying afforestation costs each year, as shown in Appendix 9A.

As of 1995, federal income tax rates on personal and corporate incomes were as shown in Table 9-2. But federal income tax rates have changed often, and the taxable income brackets are indexed for inflation, so it's best to check with the Internal Revenue Service (IRS) about current details when doing financial analyses. Likewise you'll always want to learn of the latest regulations about deductions, tax credits, and other rules that are often revised.

TABLE 9-2
FEDERAL INCOME TAX RATES, 1995

Single noncorporate taxpayer*		Corporate taxpayer	
Ordinary taxable income/yr†	**Tax rate**	**Taxable income/yr‡**	**Tax rate**
0–$23,350	15%	0–$50,000	15%
$23,351–$56,550	28%	$50,001–$75,000	25%
$56,551–$117,950	31%	$75,001–$100,000	34%
$117,951–$256,500	36%	$100,001–$335,000	39%
Over $256,500	39.6%	$335,001–$10,000,000	34%
		$10,000,001–$15,000,000	35%
		$15,000,001–$18,333,333	38%
		Over $18,333,333	35%

*Other categories, subject to different tax rates, are married taxpayers filing jointly or separately, and heads of households.

† *Brackets, which are annually indexed to inflation, apply only to 1995. Initial capital gains income is combined with ordinary income and taxed at the same rates up to 28 percent, which remains the maximum rate for capital gains.*

‡ Applies to both ordinary and capital gains income.

In order to get benefits of lower capital gains tax rates, noncorporate forest owners must have held harvested timber longer than one year before cutting and must meet specific tests about method of sale, regularity of income, and taxpayer status (see Siegel et al., 1995).

Effects of Deducting Costs

In calculating taxable forest income, you can subtract or *deduct* from gross income any expenses incurred in earning that income. This includes expenses like management, labor, supplies, equipment, timber sale preparation, and logging, as well as other taxes paid and interest on debt. But the timing of deductions is important. For nondepreciating assets like timber, land, and other natural resources, any basis or capital expenditure like reforestation-costs over $10,000 or purchase of timber and land can be deducted only when the asset is sold. This type of deducting at asset sale date is called **capitalizing** (an unfortunate word choice, since to capitalize also means to find the present value, but you can usually catch the meaning by context). Some costs like fertilization and temporary roads are considered capital expenditures and can be gradually deducted ("amortized") over their useful lives.

The law allows forest owners to deduct all annual maintenance costs in the year they occur. Such deducting is called **expensing.** Any deduction saves taxes, and the earlier you can get the savings from a deduction, the better, in present value terms (remember from Chapter 4 that the further in the future any benefit is, the lower its present value). Suppose there's no forest income against which to expense forestry maintenance costs in the year they occur. Then such costs can be deducted from nonforest income, for example, from other business income or from personal income for individuals, if the owner meets certain tests about participation and regularity of income (see Haney and Siegel, 1995; and Siegel et al., 1995).

Consider $1,000 of a firm's nonforest income taxed at 34 percent ($T_i = 0.34$). The tax is:

$$T_i(\text{taxable income}) = 0.34(1{,}000) = \$340$$

Now, imagine $100 of forest maintenance costs that are deducted from the $1,000. The tax is now:

$$0.34(1{,}000 - 100) = 0.34(1{,}000) - 0.34(100) = 340 - 34 = \$306$$

The tax savings as a result of deducting the $100 are $34, or T_i(cost). So if you're calculating the "effective" level of a forestry cost that can be expensed against income earned elsewhere, subtract the tax savings from the cost. The effective cost above is $100 - 34 = \$66$. In general:

$$\text{Effective cost} = \text{cost} - T_i(\text{cost}) = \text{cost}(1 - T_i) \qquad (9\text{-}1)$$

If you have annual taxable *income* of $a - c$ (or gross revenue minus deductible costs), the after-tax net income is $(a - c)$ minus the tax, or taxable income multiplied by $(1 - T_i)$:

$$\begin{aligned}\text{After-tax income} = \text{taxable income} - \text{income tax} &= (a - c) - T_i(a - c) \\ &= (a - c)(1 - T_i)\end{aligned} \tag{9-2}$$

If annual forest income $(a - c)$ is negative, equation (9-2) still looks the same but is interpreted like equation (9-1), because the cost can be deducted against income earned elsewhere.

The above concepts are very important for personal finance: Any time you incur a deductible expense, like interest on a home loan, multiply it by $(1 - T_i)$ to reflect your tax savings. For example, suppose \$900 of your \$1,000 monthly mortgage payment is tax-deductible interest and property taxes, and you pay a 30 percent income tax rate. You'll save 0.30(900) = \$270 in taxes each month, so your *effective* interest and property taxes are 900(1 − 0.30) = \$630. Your effective total mortgage payment is 1,000 − 270 = \$730. And that's still better than paying \$730 of nondeductible rent, because your mortgage payment is also slowly building equity in a home.

State Income Taxes

Most states with income taxes levy rates between 5 and 10 percent for both individuals and corporations, although both don't necessarily pay the same rate. States usually define taxable income in nearly the same way as the federal government (see Bettinger et al., 1989 and 1991).

When computing after-tax cash flows, just add the *effective* state tax rate to the federal tax rate. *But remember that the effective state tax rate is lower than the actual rate, because state taxes are deductible for federal income tax purposes.* Thus, as in equation (9-1), the effective state tax rate is:

$$\text{Effective state tax rate} = \text{state tax rate }(1 - \text{federal tax rate}) \tag{9-3}$$

For example, if the state income tax rate is 6 percent and the federal rate is 34 percent, the effective state tax rate is 0.06(1 − 0.34) = 0.0396, or about 4 percent, to account for federal income tax savings from deducting state taxes. In that case, the combined effective state and federal income tax rate T_i would be[3] 0.34 + 0.04 = 0.38.

Before including taxes into discounted cash flow analyses, let's first consider the effect of taxes on appropriate discount rates.

[3] In almost all states, federal income taxes cannot be deducted in figuring state taxes, so you don't have to adjust the federal tax rate. In the rare cases where that's needed, just multiply the federal tax rate by (1 minus the state tax rate) to get the effective federal tax rate.

Rates of Return before and after Taxes

Chapters 4 through 6 showed ways to calculate internal rates of return on investments, either without taxes or assuming you'd already subtracted taxes. Now, with an income tax example, let's calculate before-tax and after-tax rates of return. Let's assume competitive conditions where investors are price takers and can't raise their output prices to pass taxes on to consumers. In that case, a tax lowers an investor's net income and reduces the rate of return on original costs. Thus it's important for private investors to know their true earning rates after paying taxes. First consider an *annual yield investment,* for example, a $1,000 savings account earning 8 percent interest, or $80 annually before taxes. Given a 30 percent income tax rate, the tax is 0.30(80) = $24, so after-tax income is 80 − 24 = $56, which implies a 5.6 percent rate of return after paying income taxes. Another way of computing this is 8(1 − 0.30) = 5.6 percent. So, for annual yield investments, a 30 percent income tax reduces both income and the rate of return by 30 percent, and the nominal rate of return after income taxes is:

$$\text{After-tax rate of return on annual yield asset} = i_a = i_b(1 - T_i) \qquad (9\text{-}4)$$

The above after-tax rate of return equation applies only to annual yield investments, not to delayed-yield assets. When cash flows are irregular, you need to calculate after-tax rates of return based on actual cash flows; no simple rule of thumb like equation (9-4) applies for reducing the before-tax rate of return. For example, suppose you buy an asset for $1,000, it grows at a before-tax rate of 8 percent annually, and you sell it after 10 years for $1{,}000(1.08)^{10}$ = $2,158.92. The 30 percent income tax on the capital gain is 0.30(2,158.92 − 1,000) = $347.68, so your return from the sale is 2,158.92 − 347.68 = $1,811.24, after taxes. Using equation (5-9), the nominal after-tax rate of return is:

$$\text{After-tax rate of return} = i_a = \sqrt[10]{\frac{1{,}811.24}{1{,}000}} - 1 = 0.061 \qquad (9\text{-}5)$$

or 6.1 percent, which is only 24 percent lower than the 8 percent before-tax rate, not 30 percent lower as in the annual yield case. *The longer the payoff delay, the less an income tax will reduce the rate of return.* For the above kind of single-input (V_0), single-output (I_n) investment, a general expression for the nominal rate of return after income taxes is:

$$i_a = \sqrt[n]{\frac{I_n - T_i(I_n - V_0)}{V_0}} - 1 \qquad (9\text{-}6)$$

Since all the above values are in current dollars, this section's rates of return are **nominal;** to find real rates, use equation (5-8). For investments with many cash flows, there's no simple equation for an after-tax rate of return; you'd need to use the trial-and-error approach of equation (6-3), entering all projected taxes as costs. While the above deals only with income taxes, in general, rates of return after any tax are computed in a similar way, by simply entering the taxes as negative amounts in the years they occur.

Appropriate Discount Rates To find the correct net present values of future after-tax cash flows for private investors, you need to use their **minimum acceptable rates of return** (**MAR**) *after taxes.* Theoretically, you could get the same answer by discounting before-tax cash flows with a before-tax interest rate. However, for a given risk level and tax rate, if an investor wishes to earn, say, a 6 percent return after taxes, *there is no unique before-tax rate of return that will always guarantee 6 percent after taxes* [as shown in equations (9-4) through (9-6)]. That's why it's so important for private investors to discount cash flows after taxes, using an after-tax MAR. Examples are in the next section.

Now consider public agencies investing tax money collected from the private sector. It's often argued that public investments should earn at least as high a return as the private sector earns. Following that argument, the government's discount rate should be the average rate of return on private investments *before taxes,* since the government doesn't pay taxes. Some analysts have estimated this rate to be 10 percent in real terms, for investments of average risk (see the chapter on inflation). As noted earlier, others have argued that the government should use a "social discount rate" lower than the private rate. In addition, the chapter on risk notes problems with using one rate for all investments. Other nontaxable entities, like pension funds, should also use before-tax MARs to evaluate investments.

The next section shows how to use after-tax discount rates in net present value calculations after income taxes.

Willingness to Pay for Land after Income Taxes

As an example of computing net present value after income taxes, consider the willingness to pay for land. Equation (7-4) gave the willingness to pay for bare forestland (WPL_∞), ignoring taxes, assuming perpetual rotations. Now let's assume private landowners paying a combined effective federal and state income tax rate of T_i. If a land buyer wishes to earn an after-tax real rate of return of r_a, the WPL, *computing in real terms after income taxes,* is:

$$\begin{aligned}\text{After-tax WPL}_\infty &= \frac{H_t - T_i\left[H_t - \dfrac{C_0}{(1+f)^t}\right] - C_0(1+r_a)^t - C_y(1-T_i)(1+r_a)^{t-y}}{(1+r_a)^t - 1} \\ &\quad + \frac{(a-c)(1-T_i)}{r_a}\end{aligned} \tag{9-7}$$

As in equation (7-4), the numerator of the first term is the rotation-end accumulated net income occurring every t years. In this numerator, the second term is the tax on the capital gain from harvesting; since the discount rate and all values are in real terms, the basis (C_0) is divided by $(1 + f)^t$, because at deduction date, its value is in current dollars (see the chapter on inflation). The cost in year y is multiplied by $(1 - T_i)$ to reflect its deductibility against other income.[4] Note that the equation's last term includes tax treatment of the perpetual annual $(a - c)$.

In equation (9-7), suppose, for a given acre of bare forestland, a buyer's expected harvest income is $H_t = \$4{,}000$, $t = 35$, afforestation cost is $C_0 = 120$, $T_i = 0.34$, projected inflation is $f = 0.05$, precommercial thinning cost is $C_y = \$75$ at age $y = 15$, $a = \$3$, $c = \$2$, and the real, after-tax MAR is $r = 0.06$. The WPL is:

$$\text{WPL}_\infty = \frac{4{,}000 - 0.34\left[4{,}000 - \dfrac{120}{(1.05)^{35}}\right] - 120(1.06)^{35} - 75(1 - 0.34)(1.06)^{35-15}}{(1.06)^{35} - 1} + \frac{(3-2)(1-0.34)}{0.06} = \$245/\text{acre} \qquad (9\text{-}8)$$

The above result means that if the indicated revenues, inflation, taxes, and other costs actually occur, buyers who pay $245 for such acres will earn a 6 percent real rate of return after taxes. In the previous two equations, you can see how taxes can be capitalized into lower land values. Given the same MAR, if the tax rate became higher, the negative tax components in the equations would be larger, and willingness to pay for land would decrease. The opposite is true if taxes decline. In general, to calculate the after-tax **net present value (NPV)** of income from any asset, follow the normal NPV procedure, including all taxes in the costs and discounting with the investor's after-tax MAR.

For simplicity, thinning revenues in the last two equations are omitted. In figuring the tax on thinning income, part of the basis C_0 can be deducted. With one approach, if you thinned 25 percent of the volume, you could deduct 25 percent of the basis and the rest at rotation-end. The following section covers basis calculations for forests with several age classes.

If you expected to sell land at rotation-end, you'd adapt equation (7-9) for WPL_1 to include income taxes on harvests and tax savings on costs, as done in equation (9-7) above. But you'd also have to subtract a capital gains tax on the difference between land purchase price and projected sale price.

Depletion

For multi-age-class forests, it's usually simpler not to separately keep track of original purchase or establishment cost (the basis) on each stand for income tax purposes. If you buy timberland, you need to record in separate *basis accounts* the price paid for land, premerchantable trees, and merchantable timber. As you

---

[4]The tax savings $T_i(C_y)$ occurring elsewhere accrue to the forest, since they wouldn't occur without the forest. However, in computing how much to pay for assets from whose taxable income C_y is deducted, do *not* count the tax savings again.

TABLE 9-3
ORIGINAL BASIS OF TIMBERLAND PURCHASE FOR INCOME TAX RECORDS

Category	Basis (cost)	Units	Depletion unit (basis/units)
Land	$15,000	50 acres	$300/acre
Premerchantable timber	5,000	25 acres	200/acre
Merchantable timber	20,000	200 MBF	100/MBF
Total	$40,000		

buy more timber and land, you add the costs to the proper basis accounts. Any reforestation cost is added to the premerchantable timber account. You also need to record the total units in each basis account: acres of land and premerchantable timber and volume units of merchantable timber [say, cords, cubic feet, or thousand board feet (MBF)]. Then you express each basis type in dollars per unit, which is called the **depletion unit,** for example, dollars per MBF (or total merchantable timber basis divided by MBF of timber).

From the revenue per unit of timber cut, say $/MBF, deduct the basis per MBF (the depletion unit) to determine taxable income. To illustrate, suppose you bought 50 acres of forested land in 1988 for $40,000, which was allocated[5] to the basis accounts shown in Table 9-3.

Over time, you'd "adjust" the basis for each category by adding dollars and units if you bought more of each or subtracting if you sold some. You'd add merchantable volume growth under units and reduce acres of premerchantable timber as it became merchantable, shifting its basis and volume into the merchantable row. Any changes in basis or units would require recalculating the depletion unit in the last column. Let's ignore timber growth and further purchases and say you sold 30 MBF of merchantable timber in 1994 for $130/MBF, net of all costs, yielding 30(130) = $3,900. To find the *deductible basis,* multiply the above $100 depletion unit by the 30 MBF cut: 100(30) = $3,000. Your taxable income would be 3,900 − 3,000 = $900. Your tax, say, at 28 percent, would be 0.28(900) = $252.

Now you'd adjust the merchantable timber basis in Table 9-3 to reflect the harvest and basis deducted: The new timber basis would be 20,000 − 3,000 = $17,000, and the volume would be 200 − 30 = 170/MBF, leaving the depletion unit unchanged. In this way you could never deduct more than the original basis. Sales of land with or without premerchantable timber would be handled as the

[5]At purchase date, you'd have to divide a single property price into the three categories according to proportions given by regional market values of land, premerchantable timber, and merchantable timber. For example, based on similar sales in the area, suppose 1988 market values of the three categories on your $40,000 purchase were $18,750, $6,250, and $25,000 for a total of $50,000 (you got a good deal!). This means that 37.5 percent of the value was in land [100(18,750/50,000)], 12.5 percent in premerchantable timber, and 50 percent in merchantable timber. Applying those percentages to the $40,000 purchase price gives the basis allocation shown in Table 9-3 (see Siegel et al., 1995).

timber sale above. You'd figure the deductible basis and adjust the basis records in Table 9-3.

The above is just one way to deduct a timber sale basis. The law allows different approaches, which are important to analyze because they can affect net present values and optimization decisions. Also remember from Chapter 5 that inflation can decrease the tax benefit of basis deductions under current laws in the United States (as of 1995).

These are the most basic elements of forest income taxation. Excellent references for more detail are Siegel et al. (1995), Haney and Siegel (1995), and CCH (1995). Brief timber tax updates are sometimes available in Bishop (1993). Tax laws and their interpretation change often, and for answers to detailed income tax questions, forest owners may need to contact tax specialists. The most authoritative details on federal income taxes are found in the U.S. Internal Revenue Code and in the U.S. Treasury Regulations and Revenue Rulings.

PROPERTY TAXES

Property taxes are usually levied by counties and smaller government units to provide revenues for local services, mainly primary and secondary education. Originally, property taxes were an annual percentage of wealth, including financial assets and household belongings. This tax is often called the **ad valorem tax,** from Latin meaning "based on value." Today, most localities levy property taxes only on real estate values—land, buildings, and other improvements, including business property, although some areas still include cars, boats, and certain investments. Timber is theoretically a part of real estate, but for reasons covered below, property taxes on timber are usually modified or replaced by other taxes. But forestland is often subject to unmodified property taxes, as are log inventories, equipment, and timber processing facilities.

To determine taxable value, or **assessed value,** of property, each county has an assessor. Increasingly, assessed value is market value: the most likely sale price agreed upon between willing buyers and sellers. But some districts let assessed value be a certain percent of market value (this percent is called the "assessment ratio"). Here I'll assume that assessed value is 100 percent of market value, since the assessment ratio doesn't affect taxes collected (see footnote 6). Taxing districts, for example, counties and cities, are usually limited in the amount they can increase property tax revenues without voter approval. Yearly, the district sets the property tax rate to collect just enough revenue to equal the approved budget for education, police, roads, and other public services. The district tax rate for any year is:

$$\text{Tax rate} = \frac{\text{district budget}}{\text{assessed property value in district}} = T_a = \frac{b_t}{A} \tag{9-9}$$

For example, if the next year's district budget is $33 million, and the assessed value of all the district's taxable property is $2,538.46 million, next year's tax rate will be:

$$\text{Tax rate} = 33/2{,}538.46 = 0.013, \text{ or } 1.3\%^{6}$$

Property tax rates are often between 1 and 2 percent of market value. In the above district, if you owned property assessed at $200,000, say, a home or taxable forestland, your tax for next year would be 0.013(200,000) = $2,600. I'll call the foregoing an **unmodified property tax**—the current property tax rate applied to full market value of property annually—and will now show how it can cause problems when applied to forests.

The Unmodified Property Tax on Forests

Assuming an even-aged stand, Figure 9-4 is based on Figure 7-5, with the "holding value" relabeled "market value." Remember that holding value is the present value of future harvest and land value expected at rotation age [see equation (7-13)]. Averaging over many forest purchases, holding value tends to be the amount paid for immature forests, hence the label "market value," which is the most likely selling price. If an unmodified property tax were applied to this stand over time, the annual tax would be the local tax rate times the forest's market value. Until harvest, both market value and the tax would increase over time.[7]

If you owned only one age of immature timber, you'd pay gradually increasing property taxes over time, with no timber income until harvest date, perhaps decades later. In such a case the tax might stimulate premature harvesting to reduce taxes—no timber volume, no timber tax. Another disadvantage is that forest property taxes are expensive to administer: assessors must keep track of changing timber volumes and values on all properties. And a further problem is that the property tax is nonneutral; it bears more heavily on land uses requiring more capital (*capital-intensive* uses). Below are two explanations of this tax bias.

Property Tax Bias against Long Payoff Periods Figure 9-5 shows market values of two land uses over time in the absence of taxes: (1) short rotation pulpwood and (2) longer rotation timber on the same type of land. In this example,

[6]In some areas this would be expressed as 13 "mills," or $13 per $1,000 dollars of assessed value (10 mills is 1 percent). Rates are also given as dollars per $100 of assessed value or, in this case, $1.30. In this equation and in (9-9), you can see that it doesn't matter whether assessed value is 100 percent of market value or not, as long as it's a uniform percentage for all property. For example, if assessed value were 50 percent of market value, the tax rate in equation (9-9) would be twice as high in order to raise the required budget, and your tax in the following example would be the same.

[7]Figure 9-4 is based on before-tax values, so in equilibrium, after the property tax, market value would be somewhat lower, since taxes tend to be capitalized into lower land values. Theoretically, the unmodified property tax would also shorten the optimal rotation slightly because this tax's effect is to boost the interest rate the forest has to earn. Not only does forest capital have to earn the owner's MAR, but it now must pay a percentage of its value in property taxes. Remember from Chapter 7 that raising the interest rate decreases the economically optimal rotation.

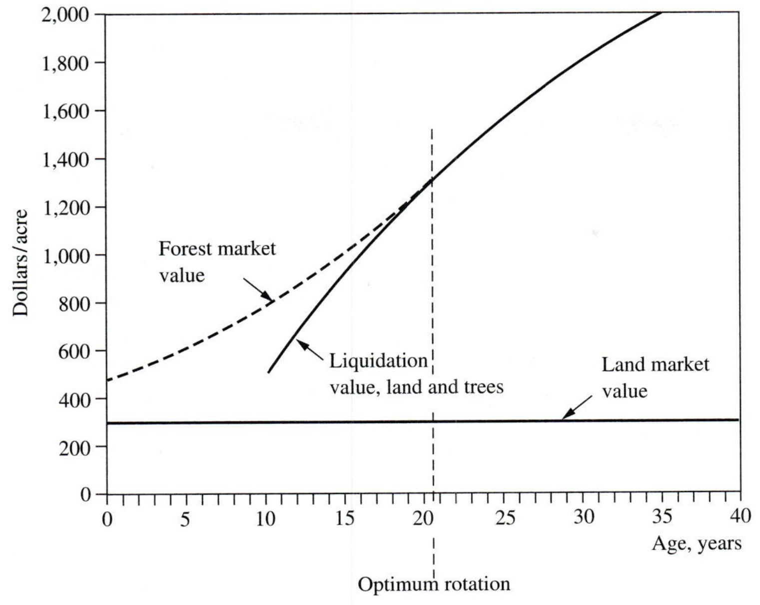

FIGURE 9-4
An unmodified property tax would be applied annually to even-aged forest market value.

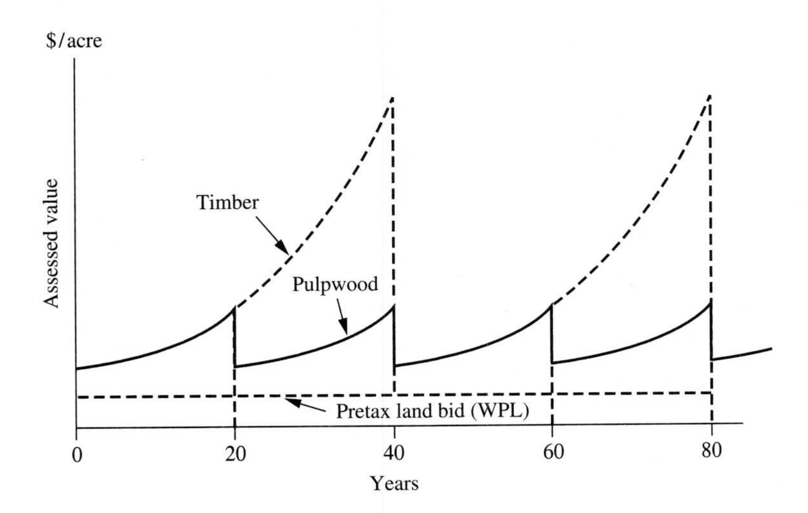

Which use can bid more for land after an unmodified property tax?

FIGURE 9-5
Two uses with same pretax bid for land, same establishment costs, different payoff periods. (Adapted from Klemperer, 1982.)

both uses have the same establishment cost (C) and before-tax willingness to pay for land (WPL), and that can occur only if timber's more distant-future income is much higher than pulpwood's. Now, imagine an annual property tax on full market value of each use. Which pays more tax over time? Timber, because taxable value over time will be greater; of the two uses, timber is more capital-intensive.

Starting with bare land, the present value of taxes will be greater for the timber example. Thinking back to Figure 9-2, the willingness to pay for bare land after taxes is the before-tax WPL minus the present value of taxes. Thus, the after-tax WPL for timber growers will be less than the after-tax WPL for pulpwood growers in this example. Since the WPLs in Figure 9-5 were equal before taxes, the property tax is biased against the use with a longer payoff period. The tax is nonneutral, because compared to the pretax case, pulpwood growers will be more likely to win the bidding war for the land in question after the property tax. But in the Figure 9-5 example, the land uses are equally desirable from society's view, because the pretax WPLs are equal. Society's present value from each land use is the present value of private income plus the present value of the taxes.

The foregoing means that *land uses with long payoff periods will tend to be more heavily burdened by the property tax.* Thus, timber growers, due to their delayed incomes, are likely to find that the property tax will reduce their bids for land by a larger percentage than is the case for other competing users of land. But the case isn't watertight, because the property tax is also biased against land uses with high establishment costs.

Property Tax Bias against High Establishment Costs Figure 9-6 shows hypothetical market values of two land uses with the same payoff periods and pretax WPLs but with different establishment costs: a short rotation fiber crop and Christmas tree farming. To maintain the same WPL while having to pay higher planting costs, the Christmas tree farm must yield higher harvest values. The result is that the land use with the highest establishment cost has the highest taxable value over time (it's the most capital-intensive) and pays the highest property taxes, other things being equal. Since in Figure 9-6 the pretax WPLs are equal, and the present value of tax costs is highest for Christmas tree farming, its after-tax WPL must be lowest. *This demonstrates a bias of the property tax against land uses with high establishment costs;* after the tax, such uses are less likely to win in the bidding process for land, other things being equal. This again illustrates how the property tax is nonneutral; it could stimulate market choices of socially second-best land uses.

Modifying the Property Tax So the property tax has several strikes against it in forestry: It is expensive to administer, may stimulate premature cutting, and is biased against capital-intensive land uses like forestry. As a result, the property tax is not generally levied in its unmodified form on forests in the United States. Either assessed values of forests are reduced, or the tax is replaced by another. Then why spend time studying the unmodified property tax on forests? The reason is that modified forest taxes cause nonforest property owners and legislators

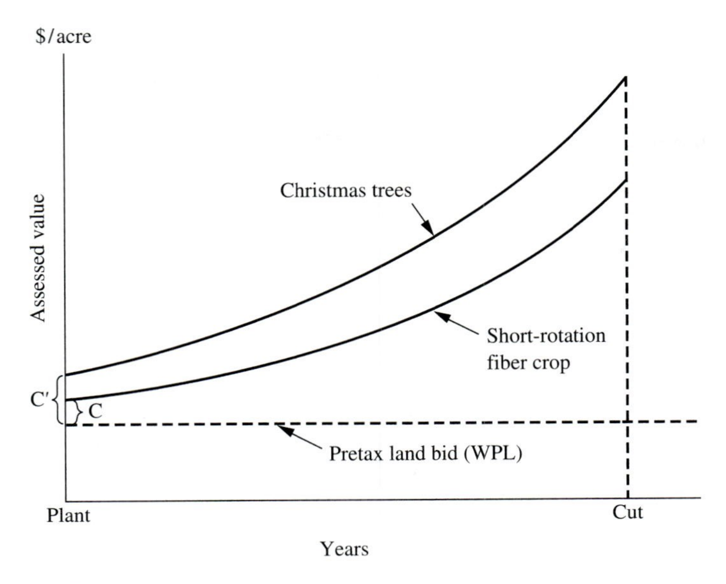

Which use can bid more for land after an unmodified property tax?

FIGURE 9-6
Two land uses with same pretax bid for land, same payoff period, different establishment costs. (Adapted from Klemperer, 1982.)

to ask whether forests are paying their fair tax share. The district property tax rate equation (9-9) can help explain why. If forest property taxes are reduced by exempting a part of forestland or timber value, the equation's denominator decreases, and in order to collect enough taxes to meet the budget, the district tax rate will increase. Thus, any forest tax reduction will increase taxes for nontimber properties. Conversely, if forest taxes are increased, nonforest owners will pay lower taxes. Thus, foresters are often asked to defend special forest taxes in place of the unmodified property tax.

In some states, the property tax is levied on forestland only, excluding timber values. An economywide tax on land value alone, exempting all improvements, has great theoretical appeal. This tax, known as a **site value tax,** avoids the non-neutrality of the unmodified property tax. A capitalized site value tax doesn't tend to distort land use decisions, since it reduces everyone's bid price for land by the same percentage. But such a radical change is hard to implement, and offering it to forestry without including other land uses would certainly be nonneutral and biased in favor of forestry.

When nearby development boosts forestland prices, some localities allow the land to be assessed, for property tax purposes, at lower forestland values typical of undeveloped areas. This is called "current use" taxation and usually requires that approved forestland later developed for nonforestry uses would have to pay part of the taxes that would have been due without current use valuation. Such programs also prevent assessors from assigning high development values to large forest areas that couldn't all realistically be developed. As for the property tax

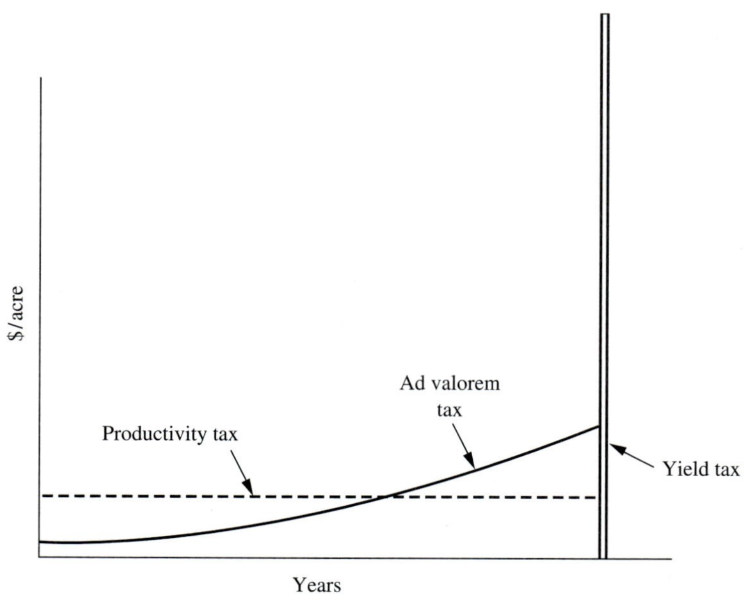

FIGURE 9-7
Three types of taxes on an even-aged stand. (From Klemperer, 1976.)

on timber alone, perhaps the most common practice is to replace the property tax with a completely different tax.

TAXES TO REPLACE THE PROPERTY TAX

Figure 9-7 shows the three major types of forest taxes as they'd occur on an even-aged acre of forest: the property tax, starting low and gradually increasing over time; the yield tax paid at harvest date; and the productivity tax at a given amount per acre, depending on site quality but independent of timber stocking. In several states, either yield or productivity taxes have replaced the property tax on forests.

Forest Yield Taxes

A yield tax is levied as a percentage of harvested stumpage value. Rates are usually around 5 to 10 percent of stumpage receipts, with no deductions allowed. Remember that stumpage value is the amount the landowner could receive for standing timber, just before harvest. In figuring federal income taxes, all other taxes are deductible, so you'd multiply your yield tax by $(1 - T_i)$ to account for the federal tax savings, according to equation (9-1). On a harvest of $\$H$ per acre, the effective yield tax would be $T_y(1 - T_i)H$. When applicable, this would appear as a negative amount in any forest cash flow calculation, say, in the numerator of a WPL equation like equation (9-7). For example, given a 6 percent yield tax on \$1,000 of harvested timber (stumpage value), a corporation paying a 38 percent income tax would pay an effective yield tax of $0.06(1 - 0.38)1{,}000 = \$37.20$.

Forest Productivity Taxes

Productivity taxes are roughly a fixed amount annually, in real terms, regardless of the amount of standing timber. In theory, better quality sites pay higher productivity taxes than poor land. Because such taxes don't increase with timber growth, they may be higher than the property tax on land alone, although this isn't always the case. The tax is the local property tax rate applied annually to a computed "productivity value," which is based on the present value of future income. For different forest sites and types, and assuming average stocking levels, the taxing authority usually calculates typical annual per acre forest harvest revenues (G) minus average costs. Assuming this amount could occur annually, it is capitalized with equation (4-8) as a perpetual annual series, as follows, using an interest rate determined by legislative guidelines:

$$\text{Taxable productivity value per acre} = \frac{G - c}{r} \tag{9-10}$$

The value for G is an estimated average annual growth per acre per year times the current stumpage price for the species in question. For even-aged timber, average annual growth is the rotation-aged yield per acre divided by the number of years in the rotation. Since better site qualities grow more timber, they'll have higher taxable productivity values. Note that equation (9-10) assumes the harvest income actually occurs every year. But only well-stocked forests can yield annual harvests, for example, an uneven-aged forest or, as diagrammed in Figure 9-8, a **regulated forest** of even-aged stands. For a 9-year rotation, this simple 9-acre "regulated model" has 9 age classes and could yield $\$H$ annually, in perpetuity. Given this model, the value for G in equation (9-10) would be $H/9$.

Because equation (9-10) assumes annual income, which bare land couldn't yield, the productivity value, in principle, is the average per acre value of land and timber in an annual sustained yield forest. Such a forest could be either uneven-aged or multi-aged as in Figure 9-8. However, in applying productivity taxes, the following errors usually keep values modest: In equation (9-10), often nominal interest rates are used in the denominator and real cash flows in the numerator (Chapter 5 explains how this undervalues assets). Also, estimated annual volume growth is often conservative.

The productivity tax T_a(productivity value) is entered in any discounted cash flow analysis as a negative amount, accounting for deductibility for income tax purposes, using equation (9-1). For example, in the WPL equation (9-7), the annual tax per acre would be entered as part of the annual cost c.

Tax Equity

If you modify or replace property taxes, choosing the tax level is controversial. For example, how high should the yield tax rate be? What should be the modified assessed value under the property tax or productivity tax? Most would agree that taxes should be fair. The problem is that equity goals conflict. Consider the following equity guides:

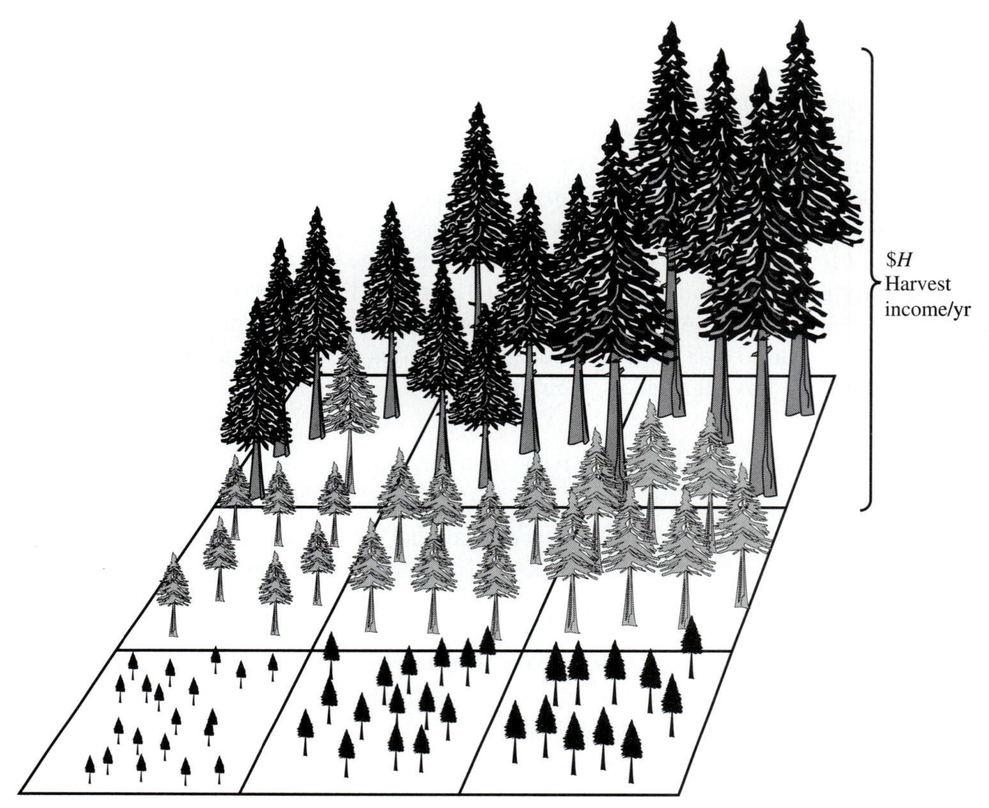

FIGURE 9-8
9-acre, 9-age-class, fully regulated forest model.

- **a** *Tax the same percentage of market value of all property annually* (this is a property tax equity guide, but as shown, it leads to a nonneutral tax).
- **b** *Tax the same percentage of net income annually from all enterprises.* This is an income tax equity guide and can conflict with guide (a) above. For example, if a tax under guide (a) annually takes 1 percent of all firms' asset values, the tax is a very large percentage of income for firms earning a low percent of asset value (100 percent of income for a year when a firm's income is only 1 percent of asset value). But when income is a high percent of asset value, the tax is a low percentage of income.
- **c** When changing forest tax systems—say, from a property tax to a yield tax—*let a new tax raise as much regional tax revenue as the old one.* But if no one has determined whether the old system was in some sense fair, the new one is unlikely to be fair. Furthermore, changing tax systems usually causes shifts in tax burdens between taxpayers.
- **d** *Keep forest tax burdens in one state similar to those in other states.* Firms often argue that such equity is needed for states to compete with one another. But some states already have much higher forest taxes than others, and they all compete effectively in national markets. This occurs because

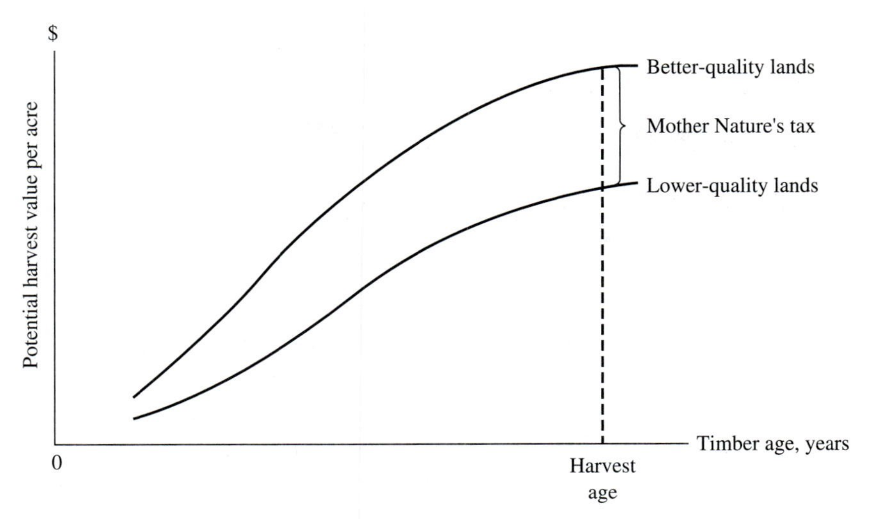

FIGURE 9-9
Mother Nature's "tax."

a forest tax is much like the "tax" that mother nature extracts as forest site quality declines, as shown in Figure 9-9. Low site lands produce less timber income than high site lands, but owners of different site quality lands can still compete with one another. Buyers simply pay less for poorer lands, other things being equal, and forestry continues. Similarly, taxes reduce land values more in some regions than in others, and forestry continues. This is the tax capitalization argument made earlier. Such reasoning shouldn't suggest that legislators tighten the tax screws as much as they dare. But it does show that higher forest taxes in one state wouldn't necessarily keep that state's timber growers from competing with those in other states.

e *Allow an after-tax forestry rate of return equal to that in other ventures.* This equity guide can lead to inefficient resource allocation, because it suggests that enterprises earning low rates of return should pay, say, lower income tax rates than those with higher rates of return. Low rates of return in one sector should be the market's signal to produce less in that sector, not a signal to reduce taxes. Lowering taxes for inefficient sectors would inhibit a desirable movement of capital from low rate of return activities to those with higher rates of return. Go back to Figure 4-4, and note that the efficient allocation of capital is where the before-tax rate of return on the last dollars invested in each sector is equal. The tax system shouldn't distort this process. This is another form of tax neutrality argument and is discussed in more detail under "Tax Policy Issues."

Given all the conflicts between equity guides, how can policy makers decide on a fair forest tax level? Perhaps one of the best approaches is to try taxing about the same percentage of forest net income as is taxed in other enterprises [guide (b)], although this is hard to do with some types of taxes when harvests are infrequent. At the same time, in the interest of efficient resource allocation, legislators should also try to avoid grossly nonneutral taxes.

THE IMPORTANCE OF TAX TIMING

Let's now consider the effects that different taxes have on forestland use decisions. In evaluating new tax options, legislators often seek some minimum acceptable level of taxes; for example, they may feel that, statewide, a new tax should raise at least as much revenue as a previous system. Since most states have a fairly even distribution of timber age classes, let's use the 9-acre 9-age-class model of Figure 9-8 to represent a region from which we wish to collect $90 of tax revenue annually. Though the scale is small, conclusions apply to large regions. The three tax options are, as shown in Figure 9-10, (1) collect all $90 as a yield tax from the harvested acre only, (2) collect $10 per acre per year as a productivity tax, and (3) collect an annual ad valorem tax starting at $2 for the first age class and increasing by $2 per age class up to a maximum of $18 for the ninth age class. For now, read the *x* axis of Figure 9-10 as "age class acres."

From the entire model, each tax raises exactly $90 per year. Thus, if one person owned the model forest, you might wrongly conclude that the type of tax didn't matter. True enough, the present value of taxes on the *whole forest* in perpetuity is the same for each system; for example, assuming 8 percent interest,

FIGURE 9-10
Property tax, productivity tax, or yield tax on 9-age-class model.

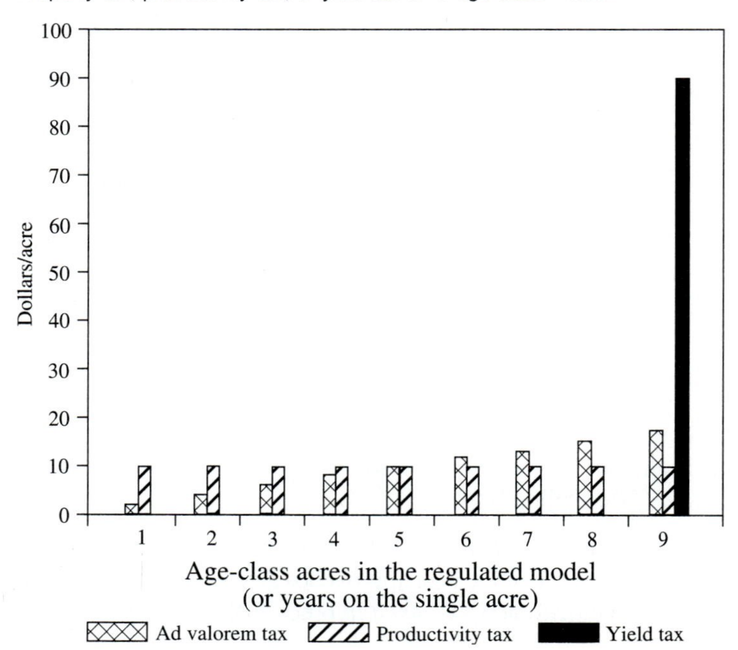

PV = 90/0.08 = \$1,125. But if you place yourself at rotation-start on individual acres, present value of the three taxes can vary greatly, and thus the willingness to pay for bare land would differ. When switching from the regulated forest to 1 acre, imagine that, in real terms, the acre of timber growing over time is just like progressing through the age-class acres on the regulated model. In that case, tax payments on progressively older age-class acres in the regulated model are the same as taxes for each year on a single acre of timber as it gets older. Thus, for each acre in the model, total taxes paid over one 9-year rotation are \$90 for each of the above three taxes, as shown in Figure 9-10, now reading the x axis as "years on a single acre."

Remember that the closer to the present a cash flow is, the greater its present value. In Figure 9-10, on 1 acre, productivity taxes are spread more toward the present, yield taxes are postponed the longest, and ad valorem tax timing is in between. Thus, you can rank these taxes by rotation-start present values, *given that each tax is the same number of dollars over one rotation:*

$$\text{PV productivity tax} > \text{PV ad valorem tax} > \text{PV yield tax} \qquad (9\text{-}11)$$

If you're purchasing bare land, think of it this way: What would you rather pay over the first rotation, \$90 in 9 years (the yield tax) or \$10/acre/year (the productivity tax)? The longer you can postpone a tax, the better off you'll be, financially, so at planting date, you'll prefer the yield tax. At rotation-start, using 8 percent interest, the per acre present value of one rotation's taxes for each tax in the above example will be:

$$\text{PV productivity tax} = 10\left[\frac{1-(1.08)^{-9}}{0.08}\right] = \$62.47 \qquad (9\text{-}12)$$

[using equation (4-10)]

$$\text{PV ad valorem tax} = \frac{2}{(1.08)} + \frac{4}{(1.08)^2} + \ldots + \frac{16}{(1.08)^8} + \frac{18}{(1.08)^9} = \$56.11 \qquad (9\text{-}13)$$

$$\text{PV yield tax} = \frac{90}{(1.08)^9} = \$45.02 \qquad (9\text{-}14)$$

With the long rotations typical of forestry, the present value differences will be much greater than in the above example. In an after-tax WPL calculation, the present value of taxes would be subtracted from the before-tax WPL (in the above example it would be WPL_1, since taxes were computed for only one rotation).[8]

[8]For the unmodified annual property tax at rate T_a on the previous year's full market value, the WPL_a can be calculated by adding T_a to the discount rate in a WPL equation. For example, decision makers desiring a 5 percent rate of return and expecting a 1 percent unmodified property tax should use a 6 percent discount rate to compute WPL_a without actually entering the taxes in the equation. If a property tax takes x percent of an asset's value yearly, the owner's rate of return is reduced by x percent. Thus, in the foregoing example, an apparent 6 percent return before taxes is actually 5 percent for the owner, after taxes.

Thus, given the PV ranking of taxes in equation (9-11), the after-tax WPLs will be ranked in reverse order, as follows, given that each tax is the same number of dollars over one rotation (subscripts are for yield, ad valorem, and property taxes):

$$\text{WPL}_y > \text{WPL}_a > \text{WPL}_p \tag{9-15}$$

The lower an area's WPL, the less likely it is to be used for forestry. Based on the discussion of switching back and forth between the regulated forest and single acres in the forest, we can conclude the following: In any region with a roughly even distribution of timber age classes, *if tax alternatives are to raise about the same annual tax revenue, the yield tax will be the least discouraging to forestland use, the productivity tax the most discouraging, and the ad valorem tax somewhere in between.*[9] This discussion ignores the effects of taxes on optimal rotation ages and impacts of real timber price-increases, but such factors, at reasonably expected levels, aren't enough to change the above general conclusions.

On the regulated forest model, the present value of the three taxes was the same. Thus, if land values differ under each tax, then timber values must differ in the opposite direction. For example, compared with the other taxes, the yield tax, being due at harvest date, has the most depressing effect on willingness to pay for timber as you approach rotation-end. But from a land use policy view, the WPL impacts are the most relevant. Table 9-4 briefly summarizes advantages and disadvantages of the three forest taxes.

Principles in this section help to explain why the timing of forestry cash flows is so important. *Other things being equal, in order to maximize WPL, try to place costs late in the rotation and revenues early.* In starting plantations, investors often seek ways to postpone expenses that can increase harvest income. For example, if you can gain the same improvement in final yield by spending \$50 more on site preparation now or \$50 on precommercial thinning in 12 years, go for the precommercial thinning; that will yield a higher WPL. Or again from a WPL view, if late and early fertilization give similar volume boosts, fertilize late in the rotation. Real-world decisions are rarely this simple, but the principle is important.

MEASURING TAX IMPACTS

Many analysts have shown theoretical impacts of different taxes on variables like optimal rotation age, planting spacing, or thinning regimes. Recall from Chapter 7 that *an economically optimum level of any variable is that which maximizes WPL.* Imagine the impact of an unmodified ad valorem tax on optimum rotation age: since the tax is x percent of market value each year, the effect is to reduce the rate of return by x percent. To simulate such a tax, timberland buyers, when

[9]One way to drive home this point is to think of a fictitious "reforestation tax" paid at planting date. To raise \$90/year in the regulated forest example, such a tax would be \$90 on the reforested acre. Having a present value of \$90 at rotation-start on 1 acre, this tax would be the most discouraging to forestland use, among the taxes considered, although it would raise the same annual regional revenue if all harvested acres were reforested.

TABLE 9-4
ADVANTAGES AND DISADVANTAGES OF FOREST TAX ALTERNATIVES

Tax type	Advantages	Disadvantages
Property tax (ad valorem)	Familiar system. Tax revenues predictable.	Biased against capital-intensive land uses. Expensive to administer. Can stimulate premature cutting.
Productivity tax	Low administrative cost after initial land classification. Tax revenues predictable. Cannot cause premature cutting. Prevents unreasonable higher use assessed values. Won't penalize intensive forestry.	Equity problems: forests with high and low income can pay same taxes per acre of a given site. Widespread disagreement on ways to find taxable value. Most discouraging to reforestation for a given level of regional taxes raised.
Yield tax	Tax paid when income received. Cannot stimulate premature cutting. Administrative costs lower than unmodified property tax. Least discouraging to reforestation for a given level of regional taxes raised.	Tax revenues may be unpredictable. Little agreement on the appropriate tax rate. Landowners fear rate changes.

discounting pretax incomes, would simply add x percent to their discount rates. Since a higher discount rate shortens the optimal rotation (see Figure 7-6a), the unmodified ad valorem tax has the same effect. One can show very slight rotation-lengthening effects of a yield tax at common rates. The productivity tax, since it's the same level regardless of timber stocking, has no appreciable effect on the economically optimal rotation.

From a practical policy standpoint, it's not especially useful to show that a given tax will theoretically increase or decrease the optimum level of some variable like rotation, thinning age, thinning intensity, or planting spacing. It's important to simulate, for example, with spreadsheets, exactly the degree to which a given tax will affect certain variables for selected species, prices, inflation rates, etc. For instance, with many species, a yield tax at typical rates will lengthen the optimal rotation by less than a year or two—usually less than the accuracy in determining such variables. A way to tell for sure is to construct a spreadsheet like Table 7-3, incorporate different taxes at various rates into the relevant equations, and observe the impacts on the management regime that maximizes WPL. Results can vary greatly, depending on species, sites, interest rates, price projections, tax levels, and other factors.

DEATH TAXES AND OTHER TAXES

In order to provide public revenue and lessen concentrations of wealth, federal and state governments impose taxes on inherited assets. These so-called death taxes apply only to noncorporate holdings and thus aren't a problem for most of the largest forest ownerships. But some worry that death taxes on personal estates can force harvest or sale of timberlands to pay the tax. Others argue that this is a problem only when it causes premature harvest or divides forests into uneconomic units.

The thought of passing wealth to heirs satisfies potential donors as well as receivers. But sometimes a donor or heir feels that, beyond some level of inheritance, the added utility from the remaining bequest would be less to the heir than to some other segments of society.[10] Death taxes could stem, in part, from such reasoning, although many would question whether the utility gained from the government's tax expenditures would exceed the utility that heirs or other groups would have received from that same revenue. Such attitudes lead many wealthy people to limit inheritance for their next of kin and to establish charitable tax-exempt trusts where heirs retain some control over expenditures.

Almost all states levy some form of death tax, but rates (usually well under 20 percent) and exemptions vary widely. Federal death taxes are reduced by the amount of state death taxes paid, up to certain limits that change periodically. Currently, all asset transfers between spouses are tax-free at the federal level and below a certain level in all states. As of 1995, the first $600,000 of an estate is exempt from federal death taxes, after which heirs pay a progressive tax rate rising from 37 percent of asset values to 55 percent for amounts over $3,000,000. An added 5 percent is levied on taxable amounts between $10,000,000 and $18,340,000. Exemptions at the state level are commonly lower, but tax rates are usually lower too. With a carefully worded will, a married couple can leave at least $1.2 million to heirs, free of federal death taxes.[11] Gifts of over $10,000 per year ($20,000 for a married couple) and not earmarked for tuition or medical expenses are taxable unless to spouses, governments, or nonprofit organizations. For death tax purposes, qualified forestlands can be appraised at typical forestland values ("current use value"), thus lowering taxes where lands have higher market value for nonforest uses.

If you're interested in legally minimizing death taxes, estate planning can be complex. You need to consider state and federal laws, estate valuation, gifts, exemptions, tax rates, possible forms of business organization and trusts, opportunities for tax deferral, loans, drafting a will, and life insurance to lessen tax burdens. So it's usually worthwhile to seek expert advice (see Haney and Siegel, 1993, for details).

Several states have minor taxes in addition to those mentioned above. Examples are low fees per acre of forest, or "severance taxes" at a low level per unit volume harvested. These taxes usually support services like fire protection, forestry research, and reforestation assistance.

---

[10] Again the specter of "interpersonal utility comparisons" raises its ugly head—one which economists are loath to deal with. On the other hand, policy makers can ill afford to ignore questions about the benefits of millions of dollars going to one group versus another.

[11] A trust account of up to $600,000 set aside before death can appreciate to any value and later pass free of federal death taxes to the heirs (although they will pay capital gains taxes, usually at levels below death taxes, on the amount above the original value).

TAX POLICY ISSUES

Forestry interests often seek tax reductions or so-called tax incentives in all areas of forest taxation. Forest owners maintain that such policies would allow more intensive forest management and stimulate greater timber output. We often argue that the tax system should encourage forestry, or at least not be discouraging. This goal isn't very useful, because all taxes are discouraging, and the least discouraging tax is no tax at all. The following is from a 1989 Society of American Foresters tax policy statement and is typical of forestry-oriented tax positions: "The Society of American Foresters favors a federal tax system that encourages opportunities for investing in private forestry ..." (S.A.F., 1991, p. 54).

The above policy sounds reasonable enough at first glance. But the same type of argument could be made for stimulating output of any good or service. How are legislators to know which outputs to encourage? With one sector taxed less than another, we'd have a nonneutral, inefficient tax, which results in lower financial output than a neutral tax, as noted in item (e) under "Tax Equity." In a market economy, increased consumer needs are expressed through higher prices, which in turn stimulate production. For this mechanism to work best, many economists suggest neutral taxes that don't distort price signals and don't favor one type of investment over another. If we like the way the market allocates resources, we should design taxes that are least likely to change that allocation.[12] This rationale was behind Congress' 1986 repeal of reduced income tax rates for long-term investments like forestry, compared to other income. The income that received lower tax rates was called **long-term capital gains** and was from the sale of assets held longer than one year. Part of this tax reduction for long-term capital gains was later reinstated for some individual tax brackets (see Table 9-2).

In defense of equal tax rates for all types of income, a U.S. Treasury Department report states that "any ... differential in tax rates among assets can reduce economic efficiency by causing capital to be reallocated to assets with lower before-tax returns" (OTA, 1985). Table 9-5 explains this quotation. On the left side of the table are "average rate of return investments" earning a hypothetical 8 percent real before-tax rate of return (ROR). The right side has "below-average rate of return investments" earning 5 percent before taxes, for example, certain forestry investments on low site lands. Suppose in row 2 we impose a 50 percent income tax. After-tax rates of return drop to 4 percent and 2.5 percent for average and below-average ventures, assuming annual income. Under a uniform tax in a freely competitive market, it's unlikely that much capital would be invested in below-average ventures. And that's as it should be. Society is better off if capital is shifted from 5 to 8 percent rates of return (assuming similar risk and that all benefits are reflected in rates of return).

[12]Short-term timber supply tends to be inelastic: a 1 percent increase in stumpage price brings less than a 1 percent increase in harvest. Thus, some analysts suggest that the government should give extra incentives to make private stumpage supply more responsive to price, for example, by providing timber tax reductions. The chapter on timber supply and the sections on market failure in Chapter 3 discuss weaknesses in this argument.

TABLE 9-5
HYPOTHETICAL REAL RATES OF RETURN AND INCOME TAX RATES

Average rate of return investments		Below-average rate of return investments	
Tax rate	**After-tax rate of return**	**Tax rate**	**After-tax rate of return**
0%	8% (social)	0%	5 % (social)
50%	4% (private)	50%	2.5% (private)
50%	4% (private)	20%	4 % (private)

Now suppose holders of below-average investments successfully convince Congress to decrease their tax rate by 60 percent (from 50 to 20 percent). Then in row 3 of Table 9-5, the average after-tax rates of return are 4 percent for both average and below-average investments. Everyone is happy, except for the efficiency-minded economist who sees the true "social" rates of return in row 1 as 8 and 5 percent: the private returns plus tax revenues. By equating after-tax returns, this inefficient, nonneutral tax has prevented a desirable shift of capital from low social rates of return to high rates of return. In fact, if after-tax returns on some below-average ventures were slightly above 4 percent, the "tax preference" or "tax subsidy" [in row 3, column (2)] could cause some capital to shift from high to low before-tax rates of return—the concern expressed in the above quote from the Treasury report. For every million dollars shifted from the right side of Table 9-5 (5 percent rate of return) to the left side (8 percent), society gains 3 percent, or $30,000 per year. Tax policy shouldn't impede this desirable shift. In Table 9-5, eliminating the tax preference in row 3 would make this beneficial reallocation more likely.

In the absence of market failures, the above arguments against tax preferences apply to any government assistance to businesses given only because rates of return are lower than average. Such subsidies foster inefficient use of capital.

Unequal Concentration of Losses and Gains

Even though Table 9-5 shows that potential gains from rescinding a tax preference can exceed losses, the problem is that losses will often be concentrated painfully in certain sectors. But gains from diverting capital to higher return ventures will be scattered throughout the economy and not so readily noticed. This is the problem of concentrated losses and diffuse gains, a common result of many policies, which are therefore hard to implement. An example is the 1986 repeal of the former capital gains tax preference for timber income; this brought concentrated losses to timber growers, but larger diffuse gains scattered across a greater number of short-term ventures receiving whatever investment might have been diverted from forestry. Before the repeal, Boyd and Hyde (1989) estimated that national efficiency-gains of this action would exceed losses by at least $240 million annually. For years this beneficial repeal had been strongly resisted by timber interests, where the losses would be concentrated.

Conversely, some proposed nonneutral taxes (like tax subsidies) or other policies could yield concentrated gains to a small group and even greater, but very diffuse, losses to large groups. Lobbying pressure for such policies can be intense on the part of gainers, while the individual losers may hardly feel the difference. One example is the effort by the timber industry to have Congress reinstate the capital gains tax preference for timber income.

Using Taxes to Gain Social Objectives

If a market is maximizing welfare, a nonneutral tax can change this allocation and reduce welfare. But if a market failure prevents welfare maximization, then a nonneutral tax that lessens the failure could *increase* welfare. For example, burning gasoline causes air pollution. A selective sales tax on gasoline reduces its consumption and lowers pollution.

Using the tax system to get desired results, as shown above, avoids more costly control measures like gasoline rationing and fosters individual freedom. But you have to be sure that the tax has the expected result and not also an unintended effect. For example, some argue that a capital gains tax preference will encourage reforestation, but there is no assurance that landowners will necessarily use the tax savings for reforestation. If more reforestation is a tax objective, a more effective approach is the current reforestation **tax credit,** which reduces the landowner's income taxes by a certain percentage of reforestation expenses (see Appendix 9A). That way the tax benefit could be gained only by reforesting. A more fundamental question in this case is whether it would be socially desirable to have more reforestation at the expense of less investment with perhaps higher returns elsewhere: if you haven't changed total investment, more spent on one activity means less spent elsewhere. Before using taxes to influence forestry, one should ask, "Is there a market failure?" And if so, what is the most efficient way to solve it?

A market failure sometimes attacked with tax policy is the case of insufficient forested open space in urban areas—a case of unpriced positive side effects in forestland use: if forest owners could sell open space benefits, they'd provide more. Instead, they sell their urban forests at high prices to developers, especially when pressured with exorbitant property taxes on their valuable land. Local governments wishing to preserve open space often allow such lands to be assessed at lower forestland values to reduce annual property taxes. This "current use" taxation does retard development in forested areas, but eventually, if forested land values skyrocket with enough nearby building, most landowners or their heirs will sell to developers. So in the long run, current use taxation will not guarantee open space in urban areas. Governments can secure open space by buying lands, although it's expensive, especially if they wait too long. Or lands can be *zoned* to stay in open space uses, but then zoned landowners complain that zoning reduces their property value. Can you explain how this occurs? (Open space uses

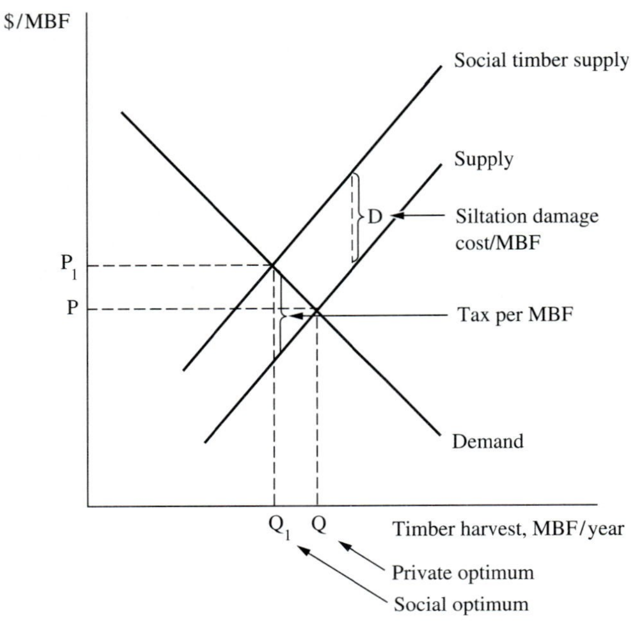

FIGURE 9-11
Regional timber supply and demand equilibria with and without a harvest tax equal to siltation damage per thousand board feet.

generally yield less income than development, so buyers pay less for the land if it is sold—think of the WPL equations in Chapter 7.)

A tax per unit volume harvested is sometimes proposed as a way to reach optimal levels of environmental damage from logging, but it's not likely to be very effective. Figure 9-11 shows the theory, using a logging-caused water siltation example. With the given regional timber demand and supply curves before the harvest tax, the unregulated, competitive market tends to yield a harvest of Q at price P. Without considering siltation damage, to the right of Q, the industry's cost of supplying one more MBF (read off supply curve) exceeds someone's willingness to pay for it (read off demand curve). If the siltation damage cost is $\$D$ per MBF in Figure 9-11, the correct supply curve from society's view is the "social timber supply" curve that shows the combined cost to industry and the rest of society of supplying one more MBF. In that case the socially optimal supply is at Q_1 and price P_1. Beyond output Q_1, the cost of supplying one more MBF (including siltation damage) exceeds the willingness to pay for it. If producers pay a tax equal to $\$D$/MBF, in theory, they automatically seek the optimal output of Q_1.

The problem with such tax proposals is that, even if you could estimate the siltation cost per MBF, the resulting output reduction might not be anything near an optimum. For example, if it's cheapest to drive equipment through streams, that's what loggers will do under cutthroat competition, even with the tax per MBF. The output tax theory incorrectly assumes that decreasing timber output will automatically reduce environmental damage to the optimal level. This tax also falls unfairly on types of harvests where little siltation occurs. Chapter 3 covers more effective environmental damage solutions; the section on pollution fines discusses a tax per unit of damage. But output taxes are likely to be far cheaper to administer than direct regulations or fines per unit of damage.

KEY POINTS

- Taxes are levied to finance programs that most of us feel are best provided by the public sector. Major taxes on forests are income taxes, property taxes, and death taxes.
- In competitive markets, much of a forest tax tends to be "capitalized" into lower land values, rather than being passed forward into higher wood prices. Thus timber growers in states with different tax levels can compete effectively.
- A neutral tax has the least likelihood of changing market allocation of resources. Thus, when we like the way the market functions, for example, in determining land use, neutral taxes are desirable. But few taxes are completely neutral.
- Federal and state income taxes are levied as a percentage of timber harvest income minus original purchase or establishment cost. This original cost is the "basis," which becomes a "depletion unit" when expressed as an amount per unit of timber or land sold.
- You can save taxes by deducting certain annual costs c from income taxed at rate T_i, so that the cost is effectively $c(1 - T_i)$.
- Private investors should discount after-tax cash flows with their after-tax minimum acceptable rate of return (MAR), which is less than their before-tax MAR. Untaxed entities should discount with before-tax MARs.
- For a fixed income tax rate, there is no simple rule of thumb that always reduces a before-tax rate of return to an after-tax rate of return.
- The willingness to pay for land (WPL) after taxes is the net present value of before-tax income from land minus the present value of the taxes, all discounted with an after-tax interest rate.
- Property taxes are biased against capital-intensive land uses like forestry.
- Forest taxes in place of the property tax are the yield tax as a percentage of harvest value and the equal annual productivity tax, regardless of timber stocking.
- For any amount of annual regional forest tax revenue raised, yield taxes tend to be the least discouraging to forestland use, and productivity taxes tend to be the most discouraging.
- In trying to design fair forest taxes, most equity guidelines are conflicting.
- In the absence of market failures, offering reduced tax rates to low rate of return enterprises can lead to inefficient resource allocation.
- Beneficial tax laws are hard to pass if gains are diffuse and smaller losses are concentrated. Welfare-reducing taxes are sometimes enacted when gains are concentrated and greater losses are diffuse.
- A selective tax can sometimes be used to attain forest policy goals, but this may not always be the most effective approach and can have unintended consequences.

QUESTIONS AND PROBLEMS

9-1 Consider a $150 precommercial thinning cost that is not deducted for forest capital gains tax purposes. This cost can be deducted against ordinary income, which is taxed at 34 percent. What is the effective level of the thinning cost after considering tax savings?

9-2 Compute the willingness to pay for bare land (WPL_∞) for the following forestry investment *after taxes,* given these assumptions:

Reforestation and site preparation cost at start of every rotation C = $100/acre
Harvest revenue (H) every 30 years t, in perpetuity = $2,500/acre
Annual costs of property taxes and administration c, in perpetuity = $2.50/acre/year
Annual net hunting lease revenue a, in perpetuity = $3.00/acre/year
Precommercial thinning cost P 15 years y after each planting = $75/acre
Income tax rate T = 34%
Real minimum acceptable rate of return r = 7%

Assume that the bidder has ordinary income in addition to timber revenues. All items are in constant dollars, where applicable. Ignore inflation.

9-3 Suppose, in the previous question, the average inflation rate was 5 percent per year. Recompute the WPL_∞ to include effects of inflation (compute in real terms, deflating any expense deducted later than the year it occurred).

9-4 A prospective buyer of 1 acre of bare forestland estimates that reforestation costs will be $100 per acre after land purchase and after each harvest. Projected harvest income is $3,000 per acre in 40 years and every 40 years thereafter. Annual costs would be $2, starting the year after land purchase and continuing in perpetuity.

Assume that the forestland buyer must pay a 28 percent capital gains tax on harvest income. Assume further that for income tax purposes, her annual forestry costs can be deducted from her ordinary, nonforestry income, which is subject to a 31 percent income tax.

Based only on these costs and revenues, what is the buyer's maximum bid price for the acre of bare land, after the above taxes? Use 6 percent interest, and assume that taxes do not affect future stumpage prices.

9-5 Suppose you purchased timberland 12 years ago for $10,000 and sold it today for $53,500. You had to pay a 28 percent capital gains tax on the sale. Compute your before-tax and after-tax rates of return. Ignore all other costs and revenues. Are these nominal or real rates of return?

9-6 **a** Consider two prospective land uses that both generate the same maximum pretax bid price for a given tract of bare land (assuming the same interest rate).

Land use I: Scotch pine Christmas tree farming. Land use establishment cost = $100/acre, income occurs every 10 years.

Land use II: Cottonwood pulpwood production. Establishment cost = $100/acre, income occurs every 18 years.

After a 2 percent annual property tax on market value of the property, how will the after-tax land bid prices generated by the two uses change, *relative to each other?* (Assume no other taxes.)

b After the above tax, which of the two land uses is likely to be chosen for the tract under free market conditions (assuming no more profitable use is possible)? Which land use is socially more desirable, from an income view only?

9-7 Assume that over the course of one 25-year rotation, a state wishes to collect exactly $500 of taxes from each acre of even-aged loblolly pine of a particular site quality. The state is considering collecting the $500 as either an annual ad valorem (property) tax, an equal annual productivity tax, or a yield tax. Comparing the three tax systems, how is the relative ranking of after-tax bid prices for bare forestland likely to appear? Assume the same stumpage prices in all three cases.

9-8 Suppose agricultural lands and forestlands near a large city are rapidly shifting to urban use. Assume further that assessed values under the property tax are based on "highest and best" use. What tax policies might be considered to help maintain open space land uses at the urban fringe? What would be the benefits and costs of such programs? Comment on how effective or ineffective such policies might be.

9-9 Suppose, in your local government district, the assessment ratio is reduced from 100 percent to 40 percent for property tax purposes. The original tax rate was 2.2 percent (or $22 per $1,000 of property value). If total market value of taxable property doesn't change, and if the tax revenue to be collected remains the same, what will the new tax rate be?

9-10 What is tax neutrality? Give examples of different kinds of neutrality. What is the advantage of tax neutrality? When might governments want to impose a nonneutral tax?

APPENDIX 9A: Reforestation Tax Credit and Amortization

Generally, reforestation costs must be capitalized for U.S. federal income tax purposes, that is, deducted many years later when the trees are harvested. However, for planted or seeded commercial timber crops of at least 1 acre, not used for Christmas trees or shelterbelts, forest owners may, over 7 years, gradually deduct (**amortize**) from current income up to $10,000 of reforestation and associated costs each year.

If trees are to be growing for 7 years or more, a landowner may claim as a *tax credit* 10 percent of the costs eligible for the above amortization. A tax credit is a direct tax reduction. Suppose Maria spent $10,000 on reforestation in 1995, all of which was eligible for amortization. The credit would reduce her income taxes by $1,000 for that year. But she must then reduce her amortization by half of the credit, or in this case by $500, so she can amortize $9,500 of her reforestation costs over the next 7 years.

The amortization schedule is that $\frac{1}{14}$ of the allowable expense is deducted from taxable income in the reforestation year, $\frac{1}{7}$ is deducted in each of the next 6 years, and $\frac{1}{14}$ in the seventh year. Maria can amortize $9,500, so her deduction for 1995 is $(\frac{1}{14})9{,}500 = \$679$. For the next 6 years, she deducts $(\frac{1}{7})9{,}500 = \$1{,}357$; and in the year 2002, she deducts $\frac{1}{14}$, or $697.

If Maria is in a 28 percent tax bracket, her tax savings are 28 percent of these deductions in the years they occur, and she also saves $1,000 from the credit in the first year. In addition to the major tax saving from the tax credit, landowners also benefit from being able to amortize an allowable reforestation cost rather than deduct it when the trees are cut ("capitalize" it), usually many decades later. As you know from Chapter 4, as well as from intuition, the sooner you can get any benefit, the greater its value is to you.

The reforestation amortization and tax credit were designed mainly to encourage reforestation by nonindustrial private forest owners. Major forest products firms have such large reforestation expenditures that the $10,000 limit makes the tax benefit a relatively small percentage of their total costs. But in the above example, Maria's tax savings were the $1,000 credit plus 28 percent of $9,500, or $2,660 over 7 years—a total of $3,660, or almost 37 percent of her 1995 reforestation cost.

REFERENCES

Bettinger, P., H. L. Haney, and W. C. Siegel. 1989. The impact of federal and state income taxes on timber income in the South following the 1986 Tax Reform Act. *Southern Journal of Applied Forestry.* 13(4):196–203.

Bettinger, P., H. L. Haney, and W. C. Siegel. 1991. The impact of federal and state income taxes on timber income in the West following the 1986 Tax Reform Act. *Western Journal of Applied Forestry.* 6(1):15–20.

Bishop, L. M. 1993. *Tax Tips for Forest Landowners for the 1993 Tax Year.* Management Bulletin R8-Mb 66. Published yearly. USDA Forest Service, Cooperative Forestry, Southern Region, Atlanta, GA. 2 pp.

Boyd, R. G., and W. F. Hyde. 1989. *Forestry Sector Intervention—The Impacts of Public Regulation on Social Welfare.* Iowa State Univ. Press, Ames. 295 pp.

Campbell, G. E. 1990. An investigation of the rule-of-thumb method of estimating after-tax rates of return. *Forest Science.* 36(4):878–893.

Commerce Clearing House. 1995. *U.S. Master Tax Guide.* CCH, Chicago. 672 pp.

Council of Economic Advisors. 1992. *Economic Report of the President.* U.S. Govt. Printing Office, Washington, DC. 423 pp.

Davis, L. S., and K. N. Johnson. 1987. *Forest Management* (3rd edition). McGraw-Hill, New York. 790 pp.

Haney, H. L., and W. C. Siegel. 1993. *Estate Planning for Forest Landowners—What Will Become of Your Timberland?* U.S. Forest Service Southern Forest Experiment Station. Gen. Tech. Rept. SO-97. New Orleans. 186 pp.

Haney, H. L., and W. C. Siegel. 1995. *Federal Income Tax Strategies for Timber Owners.* A video-seminar and work-reference book for 1994. The University of Georgia Center for Continuing Education, Athens.

Klemperer. W. D. 1976. Impacts of tax alternatives on forest values and investment. *Land Economics.* 52(2):135–157

Klemperer, W. D. 1977. Unmodified property tax—Is it fair? *Journal of Forestry.* 75(10):650–652.

Klemperer, W. D. 1980. *Equity Considerations in Designing Substitutes for Unmodified Forest Property Taxes.* Proceedings of Symposium on State Taxation of Forest and Land Resources. Lincoln Institute of Land Policy. Cambridge, MA. pp. 41–54.

Klemperer, W. D. 1982. An analysis of selected property tax exemptions for timber. *Land Economics.* 58(3):293–309.

Klemperer, W. D. 1983. Ambiguities and pitfalls in forest productivity taxation. *Journal of Forestry.* 81(1):16–19.

Klemperer, W. D. 1987. Income tax reform and the forest economy. *Tax Notes.* July 6, pp. 101–104.

Klemperer, W. D. 1989. An income tax wedge between buyers' and sellers' values of forests. *Land Economics.* 65(2):146–157.

Klemperer, W. D., and C. J. O'Neil. 1982. Potential implications of a value added tax for forestry. Proceedings of Forest Taxation Symposium in Williamsburg, VA. Virginia Polytechnic Institute and State University. Bulletin FWS-4-82. Blacksburg. pp. 146–157.

Leuschner, W. A. 1984. *Introduction to Forest Resource Management.* Wiley, New York. 298 pp.

Office of Tax Analysis. 1985. *Capital Gains Tax Reductions of 1978. A Report to Congress from the Secretary of the Treasury.* U.S. Govt. Printing Office, Washington, DC. 199 pp.

Pasour, E. C., Jr. 1973. *Real Property Taxes and Farm Real Estate Values in North Carolina.* Economic Res. Rept. 25. North State Univ., Economics Dept., Raleigh.

Siegel, W. C., W. L. Hoover, H. L. Haney, and K. Liu. 1995. *Forest Owners' Guide to the Federal Income Tax.* U.S.D.A. Forest Service. Agricultural Handbook No. 708. Washington, DC. In press.

Society of American Foresters. 1991. *SAF Forest Policies and Positions.* Society of American Foresters, Bethesda, MD. 65 pp.

Turner, R., C. M. Newton, and D. F. Dennis. 1991. Economic relationships between parcel characteristics and price in the market for Vermont forestland. *Forest Science.* 37(4):1150–1162.

10

RISK ANALYSIS

Earlier chapters projected values as if they were single quantities or averages like a $2,000 per acre white pine harvest in 30 years or an expected level of visits to a park. But we know that if we plant the same species many times on similar sites, future harvest revenue can vary widely due to weather, mortality, planting stock, insects, disease, fire, stumpage price fluctuations, and other factors. Or if we set aside a forest park, the number of visitors over time could vary, depending on population trends, income levels, people's preferences, weather, and other recreation opportunities. Projecting a single revenue or benefit implies an average of possible outcomes but says nothing about the amount of variation, which is called **risk**. The downside of revenue risk is the probability of earning less than the average return.

Here you'll learn about quantifying risk and how it can affect investment evaluations. Parts of this chapter may seem rather theoretical, and frustrating too, because you'll find that many financial calculations that may have seemed precise are really quite arguable. But bear with me: this is important material because future values and costs in forestry are so uncertain.[1] By looking at a few of the many ways to deal with investment risk, this chapter will help you come closer to defining the ever-elusive "optimum" allocation of investment funds. Unless

[1] *Uncertainty* indicates that an outcome is not known for sure, nor can one estimate its probability of occurrence. For example, we don't know the probability that a dramatic new wood processing technology will be developed next year. *Risk* indicates that based on past experience, we can estimate the probability of some outcome like a flood, during a given time. This chapter deals mainly with risk, although uncertainty is mentioned under "Simulation."

TABLE 10-1
NOTATION FOR CHAPTER 10

c_r =	certainty-equivalent ratio, or CE as a proportion of $E(R)$, for revenues [c_r = CE/$E(R)$]
CE =	certainty-equivalent of a risky revenue. CE is the dollar amount that, if received with certainty, would give the investor the same satisfaction or "utility" as the risky revenue with expected revenue, $E(R)$.
E(PV) =	expected present value
$E(R)$ =	expected value of a risky revenue at one time such as that shown in Figure 10-1 [see equation (10-1)]
k =	risk premium in the RADR (percent/100). RADR = $r_f + k$.
n =	payoff period or number of years between initial investment and final return
N =	number of values in a set of possible values
P_m =	probability of occurrence for the mth value in a distribution of values from 1 to N
PV_{CE} =	certainty-equivalent present value of a risky future revenue (i.e., a future CE discounted with the risk-free discount rate). This is considered the correct present value.
r_f =	real risk-free discount rate (percent/100), assumed here to be 0.03
RADR =	risk-adjusted discount rate for discounting single expected values (percent/100)
R_m =	a set of revenues in a probability distribution, where m is an index from 1 to N
σ =	standard deviation
σ^2 =	variance

otherwise mentioned, assume that all values and discount rates are real, after taxes, or that investors are tax-exempt. Table 10-1 lists all notation for the chapter.

RISK-FREE AND RISKY REVENUES

As an example of a risky revenue, Figure 10-1 shows possible yields from a hypothetical investment after a five-year period. The probability is 0.25 that revenue will be between \$0 and \$4,000, which is interpreted as a 25 percent probability of receiving the midpoint value of \$2,000. Other midpoints are similarly interpreted. If you made this investment hundreds of times, 25 percent of the time you'd get \$2,000, 50 percent of the time \$6,000, and 25 percent of the time \$10,000. Such a graph is called a **probability distribution.** These functions are often shown as continuous smooth curves, but initially I'll show only bar graphs or **probability histograms.** For a complete distribution, the probabilities must sum to 1 or 100 percent. Many repetitions of the above investment would yield an **expected value** for the revenue of $E(R) = 0.25(2{,}000) + 0.5(6{,}000) + 0.25(10{,}000) = \$6{,}000$. We also call this an **expected revenue.** So the expected value of any set of revenues is the sum of the possible revenues multiplied by their probabilities of occurrence.[2] In general terms:

[2]The probability distribution in Figure 10-1 is symmetrical, so that the $E(R)$ of \$6,000 is also the simple average of possible values in this special case [(2,000 + 6,000 + 10,000)/3 = 6,000]. But this isn't always so. If the distribution is skewed to the left or the right, the $E(R)$ is not the average.

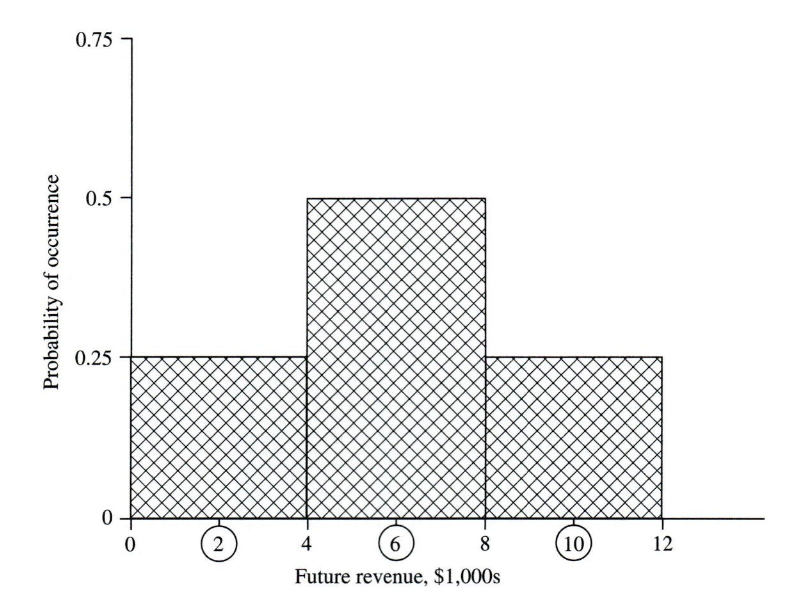

◯ = midpoint value for each outcome.

FIGURE 10-1
Risky revenue with a $6,000 expected value in five years.
(Three possible revenue outcomes: $2,000, $6,000, or $10,000, with probabilities of occurrence of 0.25, 0.50, and 0.25. For this example, the mean cash flow is E(R) = 0.25(2,000) + 0.50(6,000) + 0.25(10,000) = $6,000.)
Adapted from Klemperer et al., 1994, by permission.

$$\text{Expected value of revenues} = E(R) = \sum_{m=1}^{N} P_m R_m \tag{10-1}$$

Throughout the book, "expected value" will have this special meaning, whether applied to revenues, costs, net present values, rates of return, or utility. Here the emphasis will be on revenues. Earlier chapters expressed cash flows as expected values. But an expected value doesn't reflect the variation around the mean. The higher this variation, the greater the risk. For example, if the Figure 10-1 revenue could be expected to vary from, say, −$6,000 to +$18,000 and still average $6,000, its risk would be greater. Note that Figure 10-1 is a symmetrical probability distribution where the expected value equals the most frequently occurring value (the mode). Sometimes distributions are skewed so that the expected value is greater than or less than the mode.

As an example of risk-free income with no variation, consider the $6,000 yield from a five-year U.S. government bond pictured in Figure 10-2. Such bonds are purchased for a sum and yield a larger sum n years later, without intermediate dividends. Not all bonds are risk-free—some businesses and local governments have defaulted on bond issues—but the federal government

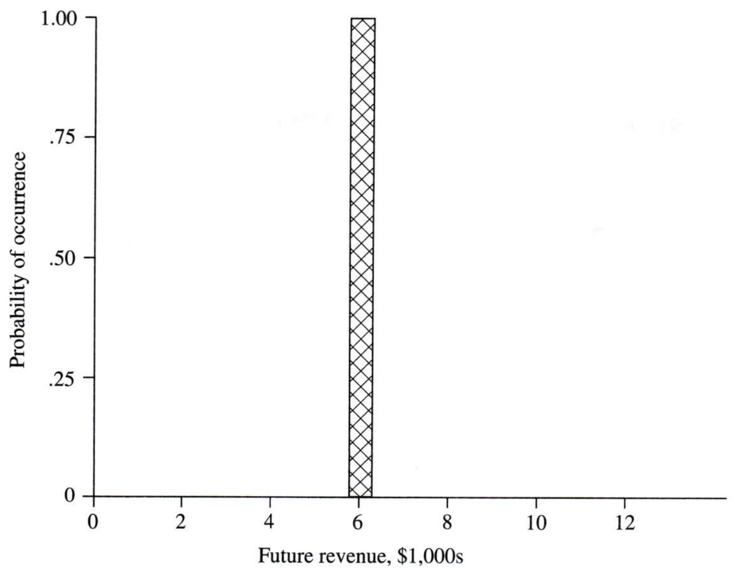

FIGURE 10-2
Risk-free revenue with a $6,000 expected value in five years.

never has, so its bonds are considered risk-free.[3] The Treasury guarantees this yield with 100 percent certainty to the purchaser (probability = 1). Thus, the expected value is $6,000 [or (1)6,000]—the same as the $E(R)$ of the Figure 10-1 revenue, except that the bond yield has no variation or risk. Now ask yourself, "Which $6,000 expected revenue is worth more to me, that in Figure 10-1 or 10-2?" Most people will answer "Figure 10-2" and are thus considered **risk-averse;** given the same expected value, they dislike greater variation in returns (See Appendix 10A).[4] Thus, for most investors, a risky revenue is worth less than a sure one of the same expected value. Those indifferent between the Figure 10-1 and 10-2 revenues are called **risk-neutral;** the amount of variation in returns makes no difference to them, as long as the expected value is the same. Those preferring greater variation are **risk seekers;** for them, the displeasure from possibly receiving less-than-average returns is more than offset by the chance of higher-than-average gains.

[3]There is, however, the risk of the bond losing real value due to unanticipated inflation. This is compensated for, although imperfectly, by higher fixed nominal interest rates when inflation is expected. The only truly risk-free bonds have yields continuously indexed to the inflation rate. But such inflation-indexed bonds are not issued by the U.S. government.

[4]Recent work shows that risk-seeking attitudes are common toward two-outcome gambles with low probability of gains (Tversky and Kahneman, 1992). But such outcome patterns aren't common in forestry. Moreover, Tversky and Kahneman's conclusion is not inconsistent with risk aversion on average, since they show close to 90 percent of subjects being risk-averse when the probability of the larger of the two gains exceeds 0.5.

Measuring Variation

In a set of revenues R_m, risk is often measured with the **variance,** which is the sum of squared deviations from the expected revenue, each squared deviation weighted by its probability of occurrence:

$$\text{Variance} = \sigma^2 = \sum_{m=1}^{N} [R_m - E(R)]^2 P_m \tag{10-2}$$

Variance (read as sigma squared) is in units squared; for example, with revenues, it's dollars squared. Because that's hard to interpret, we often measure dispersion about an expected value with the **standard deviation,** which is the square root of the variance:

$$\text{Standard deviation} = \sigma = \sqrt{\sum_{m=1}^{N} [R_m - E(R)]^2 P_m} \tag{10-3}$$

The variance and standard deviation for the Figure 10-1 revenue with an expected value of $6,000 are calculated below.

Deviation $[R_m - E(R)]$	**Deviation² × probability** $[R_m - E(R)]^2 P_m$
$2,000 − $6,000 = −$4,000	$16,000,000(0.25) = $4,000,000
$6,000 − $6,000 = 0	0(0.50) = 0
$10,000 − $6,000 = $4,000	$16,000,000(0.25) = $4,000,000
	Variance = σ^2 = $8,000,000
	Standard deviation = $\sigma = \sqrt{\sigma^2}$ = $2,828

You've probably learned or will learn in a statistics course that a "normal" probability distribution has 68.3 percent of its values within 1 standard deviation of the expected value, 95.5 percent within 2 standard deviations, and 99.7 percent within 3 standard deviations. *For any given expected value,* the higher the variance or standard deviation, the greater the risk. But if expected values of two distributions vary, you can't compare their standard deviations. For example, both may have the same percentage variation or risk, but the distribution with the higher expected value will have the higher standard deviation and variance. You solve this problem by calculating a measure of relative risk—the **coefficient of variation**—which is the standard deviation divided by the expected value. For the revenue case:

$$\text{Coefficient of variation} = \frac{\sigma}{E(R)} \tag{10-4}$$

For the above example, the coefficient of variation is \$2,828/\$6,000 = 0.47, which means that the standard deviation is 47 percent of the expected value. The higher the coefficient of variation, the greater the risk.

Now let's consider problems in estimating present values of expected revenues with different risk levels.

PRESENT VALUES OF RISKY REVENUES[5]

Risk-free assets earn a risk-free rate of return, and thus their present values should be computed with a risk-free discount rate. Table 5-4 showed risk-free rates of return to be about 1 to 3 percent in the United States, historically. Based on Simon's (1990) long-term projection, I'll assume that the real risk-free rate of return r_f is 3 percent. In that case, for an investor with a risk-free **minimum acceptable rate of return (MAR)** of 3 percent, the correct present value of the Figure 10-2 risk-free revenue is:

$$\text{Correct present value of risk-free revenue} = \frac{\text{risk-free revenue}}{(1 + r_f)^n} = \frac{6{,}000}{(1.03)^5} = \$5{,}176 \qquad (10\text{-}5)$$

Thus, assuming average investors with real risk-free MARs of 3 percent, the bond would sell for \$5,176 today if it were to yield \$6,000 in 5 years (constant dollars). That would guarantee the buyers 3 percent real interest. Anyone with a risk-free MAR greater than 3 percent wouldn't buy the bond. During several periods, including parts of the early 1990s, risk-free bonds have yielded less than 3 percent and at other times more than 3 percent. Remember that we're dealing in constant dollars; with inflation, nominal returns are higher.

In evaluating future returns of any given expected value, most investors, being risk-averse, assign a lower present value to returns with greater variance. Consider the risky and risk-free revenues in Figures 10-1 and 10-2, both occurring in 5 years and having the same \$6,000 expected value. Equation (10-5) shows that the correct present value of the Figure 10-2 risk-free \$6,000 is \$5,176 for an investor with a 3 percent risk-free MAR. If risk-averse, that same investor would discount the Figure 10-1 risky expected revenue of \$6,000 with a **risk-adjusted discount rate (RADR)** higher than the risk-free rate to reflect the reduced satisfaction, or lower present value, due to risk. Thus, with risk aversion,

$$\text{Present value of a given expected revenue} = \frac{\text{risky } E(R)}{(1 + \text{RADR})^n} < \frac{\text{risk-free } E(R)}{(1 + r_f)^n}$$

or for our example, (10-6)

$$\frac{\text{risky \$6,000}}{(1 + \text{RADR})^5} < \frac{\text{risk-free \$6,000}}{(1.03)^5}$$

[5]This section draws heavily from Klemperer et al. (1994), by permission.

Since the numerators are the same, the RADR must exceed the risk-free discount rate for the inequality to hold.

In the marketplace, we see that investors who are satisfied with, say, a 3 percent real rate of return on risk-free bonds generally demand higher risk-adjusted rates of return on riskier investments, *thus demonstrating risk aversion, on average, for all investors.* Therefore, analyses here will assume risk aversion, unless otherwise stated. The question now is how much higher than the risk-free rate should the RADR be?

Determining Risk-Adjusted Discount Rates

The U.S. Office of Management and Budget (OMB) recommends that expected values from government projects be evaluated with a 10 percent real, risk-adjusted discount rate (USOMB, 1972). This reflects an average before-tax rate of return on private capital in the United States at various levels of risk. Although not all agencies follow the guide (the U.S. Forest Service uses a 4 percent real discount rate), the OMB reasons that government projects should earn as much as private ones.[6] If that doesn't occur, OMB supporters argue that you could shift capital from lower to higher rates of return and improve total returns to all investment. The problem with the OMB approach is that the 10 percent real hurdle rate applies to investments of average risk and average duration. If investments of a given size and duration are to bring the same satisfaction to a risk-averse investor, low-risk investments don't need to earn as much as 10 percent, while some high-risk ones need to earn more. Thus, OMB's 10 percent hurdle rate is biased against low-risk investments and in favor of high-risk ventures. The same criticism applies to any use of a single RADR for discounting expected yields from investments of differing risk.

Let's review an approach to finding the appropriate RADR for discounting a risky expected revenue such as the $6,000 in Figure 10-1. We could ask the investor, "What sure income—the **certainty-equivalent**—would give you the same satisfaction as the risky revenue?" For a risk-averse investor, this certainty-equivalent would be less than the expected revenue [$CE < E(R)$], reflecting the reduced value due to risk. *Since a future certainty-equivalent is conceptually risk-free, it should be discounted with the risk-free rate* r_f, *resulting in the*

[6]This follows Hirshleifer's (1964) view that public and private investors are risk-averse and should use risk-adjusted discount rates exceeding the risk-free rate. But Samuelson (1964) suggests that the large numbers of government projects and taxpayers provide a risk pooling effect where extraordinary losses and gains offset one another in any year. That view supports a risk-free discount rate for risky government projects. But decision makers or groups within government agencies often behave like risk averters. Moreover, Hirshleifer (1966) notes that setting the public discount rate below the private rate is like charging different prices for the same commodity. In that case, shifting dollars from public to private investment could improve welfare. Arrow and Lind (1970) suggest that, if society's risk-free time preference rate is, say, 5 percent, the private sector may be underinvesting at, say, 10 percent because individuals and firms can't take advantage of risk pooling, which society could collectively do. This could be seen as a market failure, although not normally labeled as such. While the issue of a government risk premium is unresolved, this chapter makes no distinction between private and public risk premiums.

correct present value of a risky expected revenue: PV_{CE}. You could also get the same correct present value by discounting the expected revenue $[E(R)]$ with an RADR $> r_f$. The appropriate RADR results in the following equality:

$$\text{Correct present value} = \frac{\text{CE}}{(1+r_f)^n} = \text{PV}_{\text{CE}} = \frac{E(R)}{(1+\text{RADR})^n} \tag{10-7}$$

This says that an expected revenue's certainty-equivalent discounted with a risk-free rate should equal the expected revenue discounted with a risk-adjusted rate. Under risk aversion, $E(R) > \text{CE}$, and therefore RADR $> r_f$. Rearranging the last two terms of equation (10-7) and solving for the RADR:

$$(1+\text{RADR})^n = \frac{E(R)}{\text{PV}_{\text{CE}}}$$

$$\text{RADR} = \sqrt[n]{\frac{E(R)}{\text{PV}_{\text{CE}}}} - 1 \tag{10-8}$$

As a numerical example, suppose your certainty-equivalent of the \$6,000 risky expected revenue in Figure 10-1 was \$4,000. Substituting these values into equation (10-7), and remembering that $n = 5$ and $r_f = 0.03$, you can solve for the RADR:

$$\text{Correct present value} = \frac{4{,}000}{(1.03)^5} = 3{,}450 = \frac{6{,}000}{(1+\text{RADR})^5}.$$

Solving for RADR as in equation (10-8),

$$\text{RADR} = \sqrt[5]{\frac{6{,}000}{3{,}450}} - 1 = 0.11703, \text{ or about } 11.7\% \tag{10-9}$$

As a check, you'll see that discounting the risky \$6,000 at 11.703 percent for 5 years will give the correct present value of \$3,450.

The amount by which RADR exceeds r_f is the **risk premium** for revenues:[7]

$$k = \text{RADR} - r_f \tag{10-10}$$

In the above example, the risk premium is $11.7 - 3 = 8.7$ percentage points. Unfortunately, there's no universal risk premium or RADR: it depends on the payoff period, the amount of risk in the revenue, and the investor's degree of risk aversion, as explained below.

[7]Remember that here we're dealing only with risky *revenues.* At the chapter's end, you'll see that under certain conditions, the risk premium for risky *costs* is *negative,* in which case the risk-adjusted discount rate would be less than the risk-free rate.

So far I've simplified the problem by assuming a five-year payoff period and the one risk level (variance) shown in Figure 10-1. In the above example, the $4,000 certainty-equivalent was 67 percent of the expected revenue ($6,000). This percent/100 is called the **certainty-equivalent ratio:**

$$c_r = \frac{\text{certainty-equivalent}}{\text{expected revenue}} = \frac{\text{CE}}{E(R)} = \frac{4{,}000}{6{,}000} = 0.67 \qquad (10\text{-}11)$$

Revenue risk and risk aversion of the investor will cause CE to be less than $E(R)$. The value of c_r decreases with increasing revenue risk (variance) or with more risk aversion of the investor. Therefore, c_r is a measure of a revenue's risk as perceived by the investor. Appendix 10A covers risk aversion in more detail from a utility theory view.

Variation in the Risk Premium

Solving equation (10-11) for the certainty-equivalent, $\text{CE} = c_r[E(R)]$. From equation (10-10), $\text{RADR} = (r_f + k)$. Substituting these values for CE and RADR into equation (10-7), and solving for k, you have a general measure of the risk premium to add to any risk-free discount rate r_f, given a payoff period n and degree of perceived revenue risk c_r:

$$\frac{c_r[E(R)]}{(1 + r_f)^n} = \frac{E(R)}{(1 + r_f + k)^n}$$

$$(1 + r_f + k)^n = \frac{E(R)(1 + r_f)^n}{c_r[E(R)]} \qquad (10\text{-}12)$$

$$k = \frac{(1 + r_f)}{\sqrt[n]{c_r}} - (1 + r_f)$$

Using equation (10-12) and assuming $r_f = 0.03$, Table 10-2 shows how the appropriate risk premium, $k(100)$, changes for different c_r and n. For any given perceived risk level c_r, the table shows that the correct risk premium declines with increasing payoff period. The risk premiums in Table 10-2 are not highly sensitive to real risk-free discount rates ranging from 2 percent to 4 percent.

If you can approximate an investor's c_r for a given expected revenue, you can determine the appropriate risk premium from Table 10-2. After estimating and displaying a revenue probability distribution and its expected value, you can ask the investor, "What sure income would give you the same satisfaction as this expected value?" The answer is the certainty-equivalent, with which you can find c_r from equation (10-11). Use Table 10-2 to find the risk premium, or if the investor has a risk-free MAR substantially above the 3 percent assumed in the table, find the premium with equation (10-12).

TABLE 10-2
CORRECT RISK PREMIUMS [$k(100)$] IN PERCENTAGE POINTS*
[Risk-Free Discount Rate is 3 Percent; see Equation (10-12)]

Payoff period, n (yr)	c_r = CE as a proportion of $E(R)$†												
	1	0.9	0.8	0.7	0.6	0.5	0.4	0.3	0.2	0.1	0.05	0.01	0.001
1	0.0	11.4	25.7	44.1	68.7	103.0	154.5	240.3	412.0	927.0	1975.0	10197.0	102897.0
2	0.0	5.6	12.2	20.1	30.0	42.7	59.9	85.1	127.3	222.7	357.6	927.0	3154.1
5	0.0	2.2	4.7	7.6	11.1	15.3	20.7	28.0	39.1	60.2	84.5	155.7	307.1
10	0.0	1.1	2.3	3.7	5.4	7.4	9.9	13.2	18.0	26.7	36.0	60.2	102.5
20	0.0	0.5	1.2	1.9	2.7	3.6	4.8	6.4	8.6	12.6	16.6	26.7	42.5
30	0.0	0.4	0.8	1.2	1.8	2.4	3.2	4.2	5.7	8.2	10.8	17.1	26.7
40	0.0	0.3	0.6	0.9	1.3	1.8	2.4	3.1	4.2	6.1	8.0	12.6	19.4
50	0.0	0.2	0.5	0.7	1.1	1.4	1.9	2.5	3.4	4.9	6.4	9.9	15.3
100	0.0	0.1	0.2	0.4	0.5	0.7	0.9	1.2	1.7	2.4	3.1	4.9	7.4
150	0.0	0.1	0.2	0.2	0.4	0.5	0.6	0.8	1.1	1.6	2.1	3.2	4.9
200	0.0	0.1	0.1	0.2	0.3	0.4	0.5	0.6	0.8	1.2	1.6	2.4	3.6

*From Klemperer et al. (1994), by permission.
†Certainty-equivalent as a proportion of the expected revenue [c_r = CE/$E(R)$], or certainty-equivalent ratio. Investments have one revenue in n years. RADR = $r_f + k$, or here RADR = $0.03 + k$.

For example, suppose Marilyn estimates that the expected value of a white pine harvest to be received in 20 years is $25,000. She is indifferent between this risky income and a sure $15,000 in 20 years, which means that her certainty-equivalent is $15,000 and her CE ratio c_r is 15/25 = 0.60. Assuming her risk-free MAR is 3 percent, you can find her correct risk premium as 2.7 percent in Table 10-2 under $c_r = 0.6$, and $n = 20$. You could get this more precisely as 2.665 percent (or 0.02665) with equation (10-12). This means Marilyn's risk-adjusted discount rate should be 3 + 2.665 = 5.665 percent for discounting the risky $25,000. Her correct present value would be:

$$\$25{,}000/(1.05665)^{20} = \$8{,}305$$

Note that if you know the investor's certainty-equivalent, you don't need the RADR: just discount the CE with the risk-free rate. In Marilyn's case, that's $\$15{,}000/(1.03)^{20} = \$8{,}305$, or the same as the present value above. But since we usually don't have certainty-equivalents and must discount expected revenues with RADRs, it's important to understand which RADRs might be reasonable under different conditions.

A reasonable range of risk premiums isn't likely to include the high values in the upper right quadrant of Table 10-2, since shorter payoff periods usually have lower perceived risk (higher c_r values). The low-risk premiums in the lower left quadrant of the table are also not reasonable, since high c_r values (implying low risk) are not likely with long payoff periods. Thus, the most reasonable forestry risk premium values are likely to fall around a diagonal starting in the upper left corner of Table 10-2.

One objective of all this theorizing is to show how impossible it is to derive one suitable discount rate for expected values from all types of risky investments. In fact, the above examples deal only with a revenue at one date, not a bad model for some types of forestry investments where major income occurs at one time. But when incomes are scattered over time, in theory you'd need a different RADR for each revenue, considering its variance and timing. Since this isn't practical, we should at least try, for different classes of investments, to derive appropriate RADRs. One discount rate isn't usually suitable for a broad range of investments, but it's sometimes suggested. For example, given the assumed 3 percent risk-free discount rate, OMB's 10 percent RADR implies a risk premium of 7 percentage points for all government investments. In Table 10-2, at shorter payoff periods you can easily see how such a premium could be correct. For example, at a five-year payoff period, the premium is 7 when c_r is between 0.70 and 0.80. This suggests a certainty-equivalent at 70 or 80 percent of an expected revenue—a very plausible scenario. However, for long-term projects such as forestry, what does Table 10-2 imply?

FORESTRY AND THE RISK-ADJUSTED DISCOUNT RATE

Under even-aged softwood management, rotation lengths vary from 20 to 35 years in the South, 40 to 80 years in the Pacific Northwest, and can exceed 80 years if you consider nontimber values. For such long payoff periods, how great would the perceived risk need to be for the OMB risk premium of 7 percentage points to be correct? You can answer this by solving equation (10-7) for the certainty-equivalent:

$$\text{Correct present value} = \frac{\text{CE}}{(1 + r_f)^n} = \frac{E(R)}{(1 + \text{RADR})^n}$$

Solving for CE,

$$\text{CE} = \frac{E(R)(1 + r_f)^n}{(1 + \text{RADR})^n} \tag{10-13}$$

Using this equation, assuming a $5,000 risky revenue $E(R)$ received after different payoff periods, the last column of Table 10-3 shows future certainty-equivalents implied by using OMB's 10 percent RADR, if 3 percent is the risk-free rate, which means a 7 percentage-point risk premium. For example, in the first row, under these conditions, the certainty-equivalent of a $5,000 risky income expected from, say, a stock sale in 3 years is $4,105, using equation (10-13):

$$\text{CE} = \frac{5{,}000(1.03)^3}{(1.10)^3} = \$4,105$$

This means that, if a 10 percent RADR is correct, an investor must be indifferent between a $5,000 risky return in 3 years and a sure $4,105 received at the same date. This is an entirely plausible result implying that a sure $4,105 in

TABLE 10-3
REQUIRED FUTURE CERTAINTY-EQUIVALENTS FOR A 10 PERCENT RISK-ADJUSTED DISCOUNT RATE TO BE CORRECT (r_f = 0.03)

Payoff period (n, years)	Risky future income [$E(R)$, dollars]	Required future certainty-equivalent (CE, dollars)*
3	$5,000 (e.g., stock market)	$4,105
7	$5,000 (e.g., Christmas trees)	$3,156
30	$5,000 (e.g., loblolly pine)	$696
50	$5,000 (e.g., Douglas-fir)	$187
100	$5,000 (e.g., long rotation hardwood)	$7

*Equation (10-13). The certainty-equivalent is the dollar amount that, if received with certainty at the end of the payoff period, would give the investor the same satisfaction as the risky expected revenue $E(R)$ in column (2).

3 years discounted at the risk-free 3 percent gives the same present value as the risky $5,000 discounted for 3 years at the risk-adjusted 10 percent—$3,757 in either case. The second row of Table 10-3 is likewise reasonable: Someone could easily be indifferent between a risky $5,000 Christmas tree revenue expected in 7 years and the calculated sure $3,156 shown in the last column. Thus, in such ventures, OMB's 10 percent discount rate could be appropriate for many investors.

Now look at the longer payoff period of 30 years in row 3 of Table 10-3. It seems questionable for some investors to be indifferent between the risky $5,000 loblolly pine revenue in 30 years and the $696 computed certainty-equivalent on the same date. Rows 4 and 5 of Table 10-3 show that for OMB's 10 percent RADR to be appropriate, an investor would have to be indifferent between an expected $5,000 Douglas-fir harvest in 50 years and a guaranteed $187 on the same date, or a public agency with a long time horizon would be indifferent between an expected $5,000 hardwood yield in 100 years and a guaranteed $7 on the same date.[8] (Discounting $7 at a 3 percent risk-free rate for 100 years gives the same present value as the risky $5,000 discounted at 10 percent—about 36 cents in either case. Try it.)

There's little evidence that timber investors think most long-term forestry ventures are so risky that certainty-equivalents would be as low as Table 10-3 shows. To get higher certainty-equivalents that reflect less risk, you'd substitute

[8]A more general way to evaluate the perceived risk required for a given RADR is to solve for the certainty-equivalent ratio c_r in equation (10-12). For example, the Table 10-3 Douglas-fir investment with n = 50, r_f = 0.03, and OMB's 7 percentage-point risk premium yields c_r = 0.037. That means the investor's certainty-equivalent would need to fall to 3.7 percent of the expected value if OMB's 10 percent RADR were to be appropriate.

an RADR less than 10 percent in equation (10-13). A growing body of literature tells us that forestry in most regions of the United States is not viewed as a high-risk venture.[9] This work also shows that forest asset values are often less variable than common stock portfolios, and even less risky in a diversified portfolio, due to the zero or negative correlation of forest values with common stock indexes. Timber has the added advantage, unlike agricultural crops, that it stores well unharvested. If stumpage prices plummet, you can postpone harvest and cash in on later price spikes. So price variability can increase present values of forests (Haight, 1991). Although a good deal more research needs to be done, Table 10-2 suggests that appropriate risk-adjusted discount rates in forestry may often decline with increasing payoff period. Chen (1967) reaches the same conclusion for current valuation of future dividends from stocks.

Assuming present value-maximizing behavior, Berck (1979) found that private forest owners in the Douglas-fir region have chosen harvest rates as if they discounted future timber revenues at a 5 percent real before-tax interest rate. Using a 3 percent real risk-free interest rate, this implies a 2 percentage-point risk premium. One interpretation could be that holding Douglas-fir timber requires a risk premium lower than the industrial ventures on which OMB's 10 percent rate is based.

It's important to recognize that we cannot generalize forestry conclusions from Table 10-2 to all investments. The above inferences deal only with forestry investments in the United States. Very risky investments, say, buying lottery tickets, would have enormous revenue variance and low c_r values for risk-averse individuals. Therefore, regardless of lottery payoff periods, you'd discount expected yields with high RADRs.

It has long been known that, for any given revenue risk and real interest rate, the longer the payoff period, the lower the risk premium in the correct RADR. Simulations also show that, within limits, this relationship can hold even when perceived revenue risk *increases* over time, as is the case in forestry. But earlier analyses overlooked the potential for low-risk premiums with payoff periods of several decades because such time horizons are uncommon outside of forestry.

Conclusions about Forestry Risk-Adjusted Discount Rates

Foresters often make optimistic projections of volume yields and prices, ignoring the possibility of less desirable outcomes. For example, you might look up in a yield table the projected volume for a species and multiply that by some optimistic stumpage price prediction to get a future harvest income. That ignores much biological and economic variability and would exceed the true expected value. You could justifiably discount such an optimistic revenue with a rate higher than the RADRs discussed here. The RADRs used here (and those in the finance literature) are meant only for discounting *expected revenues,* not optimistic values. In

[9]See Fortson, 1986; Mills and Hoover, 1982; Redmond and Cubbage, 1988; Webb et al., 1987; Zinkhan et al., 1992.

projecting forest revenues and costs, you should make some effort, however crude, to guess at the probabilities of a few outcomes and calculate expected values of each scenario. For example, a forest manager may project a ⅓ chance of receiving harvest yields in 25 years averaging either \$1,500, \$3,000, or \$5,000. Using equation (10-1) and remembering that probabilities must sum to 1, expected revenues would be:

$$0.33(1{,}500) + 0.33(3{,}000) + 0.33(5{,}000) = \$3{,}135$$

For planting an acre, the cost estimate might be a 100 percent probability of spending \$100 in year 0, *in year 2* a 50 percent chance of having to replant part of the acre at \$35 due to mortality, and a 50 percent chance of no further planting cost beyond the original \$100. Assuming a 3 percent risk-free discount rate, the present value of expected reforestation costs would be:[10]

$$1(100) + [0.5(35) + 0.5(0)]/(1.03)^2 = \$116$$

The previous section notes that appropriate before-tax RADRs for many long-term forestry investments may be well below OMB's recommended 10 percent. The OMB rate is based largely on studies that divide firms' annual before-tax income by the market value of assets.[11] This procedure averages all long- and short-term projects together and thus includes a risk premium that could often be too high for many long-term forestry ventures.

As noted, if we knew certainty-equivalents of revenues, RADRs wouldn't be needed: we could simply discount CEs with a risk-free rate. In the absence of such information, public and private analysts continue to apply RADRs as a means of handling risk in investment analysis. Since correct RADRs depend on probability distributions of cash flows, payoff periods, and investors' degrees of risk aversion, we cannot prescribe a universal forestry RADR. Thus, the RADR approach to handling risk is at best problematic. But since the method is so widely used, investors should be aware of the pitfalls in applying typical short-term discount rates to forestry ventures.

Efficient Capital Allocation

Resource misallocation would result if all projects were required to earn a 10 percent average industrial rate of return. For example, suppose, on the average, the correct forestry risk premium were 2 percentage points; and all projects, including forestry, had to earn 10 percent. That means the *implicit risk-free earning rate (RADR-k)* in forestry would be 8 percent. And RADR-k is the relevant rate for comparing performance of investments with different risks and payoff periods. If the economywide average risk-free earning rate on short-term projects is 3 percent

[10]Note that no risk premium is added to the discount rate for costs that are independent of revenues (see the section titled "Discounting Risky Costs").

[11]See Stockfisch, 1969. A later example is Nordhaus, 1974.

(7 percent risk premium), we'd have serious underinvestment in forestry. Shifting capital from 3 to 8 percent (risk-free) would increase total returns to aggregate capital. A financially optimal equilibrium would be reached when forest output prices declined—due to higher production—to the point where implicit risk-free earning rates (RADR-k) in forestry were equal to the economywide 3 percent at the margin. Then no investment reallocation could increase total returns. Such an optimum leaves room for substantial differences in risk-adjusted earning rates, as seen in Table 10-2.

Thus, one shouldn't necessarily cry "Inefficiency!" if certain forestry investments are evaluated with, say, a 5 or 6 percent RADR, despite higher returns observed elsewhere. Differences in average rates of return can be acceptable if they represent varying risk levels and payoff periods. *The efficiency guide for capital allocation is that rates of return on added investment should be equal for all enterprises on an implicit* ***risk-free*** *basis, not risk-adjusted.* We're following that guide in this chapter when we start with a fixed risk-free real discount rate (here, 3 percent) and add varying risk premiums for discounting revenues with different risks and payoff periods.

PRESENT VALUE PROBABILITY DISTRIBUTIONS

You can avoid the question of risk-adjusted discount rates by calculating *present value probability distributions* of future risky cash flows. Using the risky revenue in Figure 10-1, each of the bars can be considered to occur a given percentage of the time with certainty. For example, the graph presumes that with many repetitions of the investment, you could be confident of receiving $2,000 twenty-five percent of the time. In that sense, each bar is certain and can be discounted with a risk-free discount rate. Discounting for five years at 3 percent gives the present value probability distribution in Figure 10-3. The expected present value for Figure 10-3 is the midpoint value of each bar times its probability of occurrence:

$$E(\text{PV}) = 0.25(1{,}725) + 0.50(5{,}176) + 0.25(8{,}626) = \$5{,}176$$

As noted, the $5,176 exceeds the correct present value for a risk-averse investor because E(PV) conceals the variance.

If you can estimate probability distributions for all positive and negative cash flows of different investments, computer programs can simulate **net present value (NPV)** probability distributions for each venture, using a risk-free discount rate (see the next section, "Simulation"). Looking at the NPV distributions for several projects, a decision maker could subjectively decide which one was the most desirable. Figure 10-4 shows such an NPV distribution in 1967 dollars for a proposed reforestation investment. Each NPV interval is shown below the graph. Remember from Chapter 6 that you'd usually want to compare projects of similar costs and time horizons. Theoretically, NPV distributions are an excellent way to compare present values of risky alternatives, but they usually require complex computer programs. Also, in applying this approach, it's often hard to get the individual cash flow probability distributions, although the next section shows a

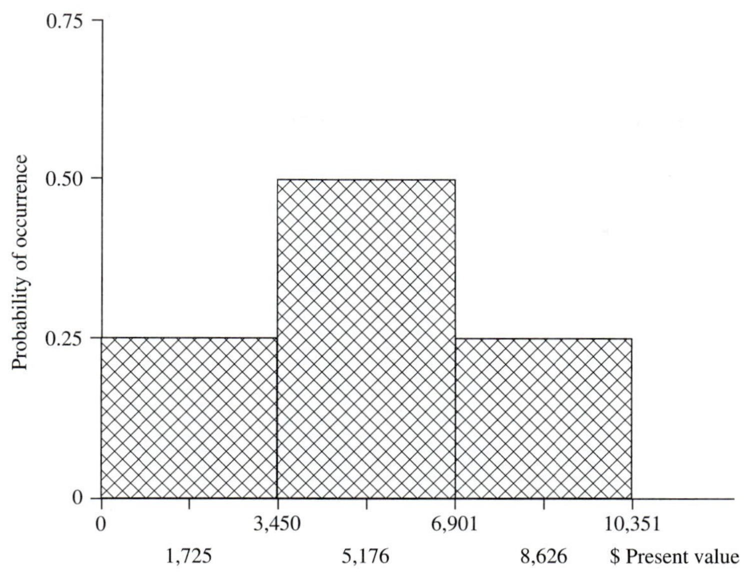

FIGURE 10-3
Present value probability distribution of risky revenue from Figure 10-1 (r_f = .03). (Using the midpoint value of each outcome, the expected present value is E(PV) = 0.25(1,725) + 0.50(5,176) + 0.25(8,626) = $5,176.)

simple estimation trick. On balance, decision makers still prefer the simplicity of a single present value obtained with an RADR. So, for better or for worse, most present values are computed with RADRs, and we're stuck with the problem of determining them.

SIMULATION

Computer simulation is a good way to inform decision makers about risky projects. I'll broadly discuss examples without fine details. One type of simulation generates net present value probability distributions such as that shown in Figure 10-4. But to do so you need probability distributions for all the project's cash flows; say, for a reforestation project these could be planting cost, annual costs and revenues, thinning costs, thinning revenues, fertilization costs, and final harvest revenue. The computer program can simulate the investment hundreds of times, each time calculating and saving the net present value computed with a risk-free discount rate (see the discussion of Figure 10-3). With each simulation, the program randomly selects a value for each of the cash flows according to its distribution. For example, if Figure 10-1 was a harvest value distribution, 25 percent of the time the program would select a value from the $2,000 range, 50 percent of the time from the $6,000 range, and so forth. After computing hundreds of NPVs in this way, the program can construct an NPV probability distribution. This procedure is called **Monte Carlo simulation.**

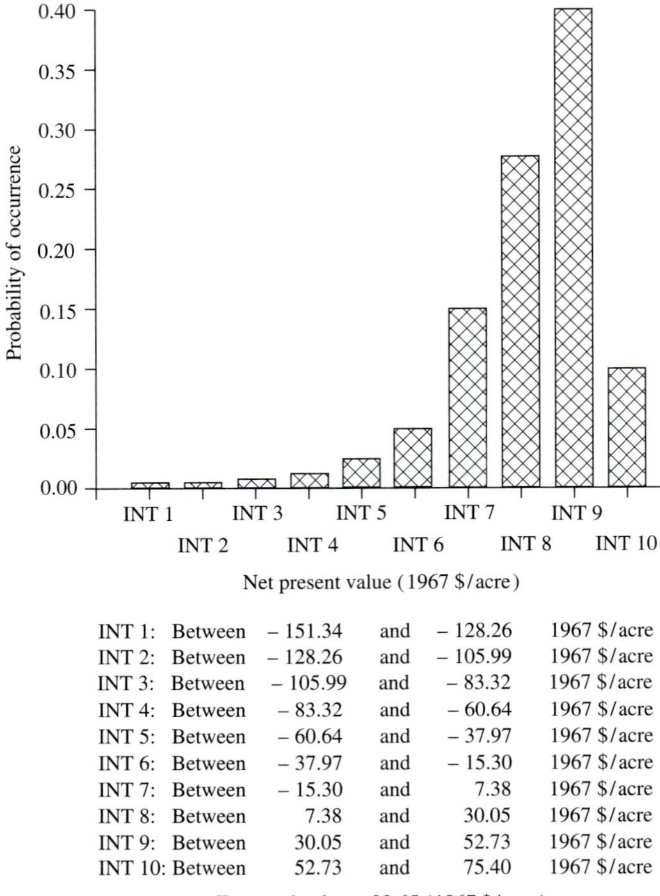

INT 1:	Between	– 151.34	and	– 128.26	1967 $/acre
INT 2:	Between	– 128.26	and	– 105.99	1967 $/acre
INT 3:	Between	– 105.99	and	– 83.32	1967 $/acre
INT 4:	Between	– 83.32	and	– 60.64	1967 $/acre
INT 5:	Between	– 60.64	and	– 37.97	1967 $/acre
INT 6:	Between	– 37.97	and	– 15.30	1967 $/acre
INT 7:	Between	– 15.30	and	7.38	1967 $/acre
INT 8:	Between	7.38	and	30.05	1967 $/acre
INT 9:	Between	30.05	and	52.73	1967 $/acre
INT 10:	Between	52.73	and	75.40	1967 $/acre

Expected value = 22.65 (1967 $/acre)

FIGURE 10-4
Net present value probability distribution for a reforestation investment (r_f = .03). (Adapted from Cathcart, 1989, p. 232.)

An easy way to get probability distributions of cash flows is to ask managers, "What's the lowest value you expect, the highest, and the most likely (or mode)?"[12] Figure 10-5 shows a hypothetical probability distribution resulting from such questions about the real expected value of a 40-year-old Douglas-fir harvest in 40 years. The outcome is a continuous **triangular probability distribution** with relative frequency on the *y* axis. Although shown as symmetrical, it

[12]The extremes can be given in terms of 95 percent probabilities; in other words, you expect values beyond the extremes only 5 percent of the time. But this complicates the distribution construction.

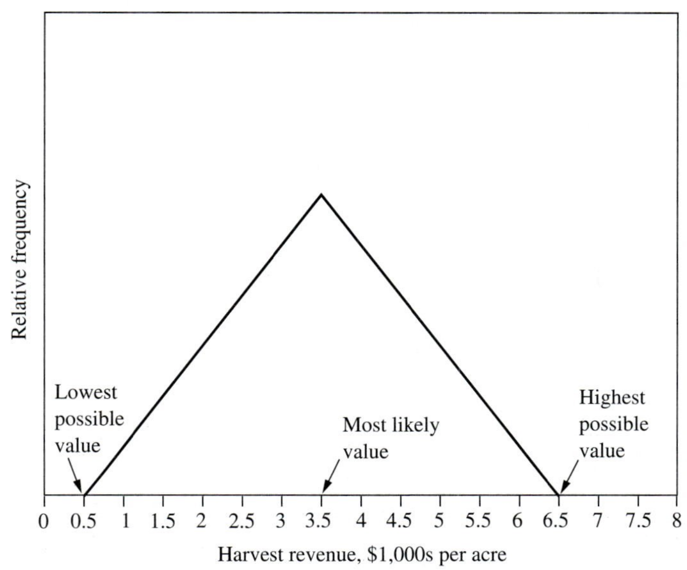

FIGURE 10-5
Hypothetical triangular probability distribution for 40-year-old Douglas-fir harvest revenue in forty years.
(Area under the curve must equal 1, since values fall between $500 and $6,500 with 100 percent certainty. Since the distribution is symmetrical, the most likely value is also the expected value or average.)

wouldn't necessarily have to be. Since there's 100 percent certainty that values would fall between the endpoints, the area (or, in calculus, the integral from one endpoint to the other) under such a continuous distribution is 1. Knowing this and the equation for the area of a triangle, one can derive an equation for the area of a piece of the triangle, say, from 0.5 to 1.5 on the x axis of Figure 10-5. That area is the probability that the harvest value would be between $500 and $1,500. Thus, with continuous triangular distributions, you can divide the x axis into bands (analogous to the bars in a histogram), and the area under the curve in any band is the probability that an x value will be within that band.[13] If all cash flows are expressed as triangular distributions, computer programs can be designed, as above, to generate discrete NPV probability distributions (histograms as in Figure 10-4) that managers could subjectively evaluate.

Instead of computing NPVs, some simulation programs calculate hundreds of **internal rates of return (IRR)** for a risky project and then construct its IRR

[13]In general, for any continuous probability distribution, the probability that a value on the x axis will lie between some x_1 and x_2 is the integral of the function from x_1 to x_2, or the area under the curve in that range.

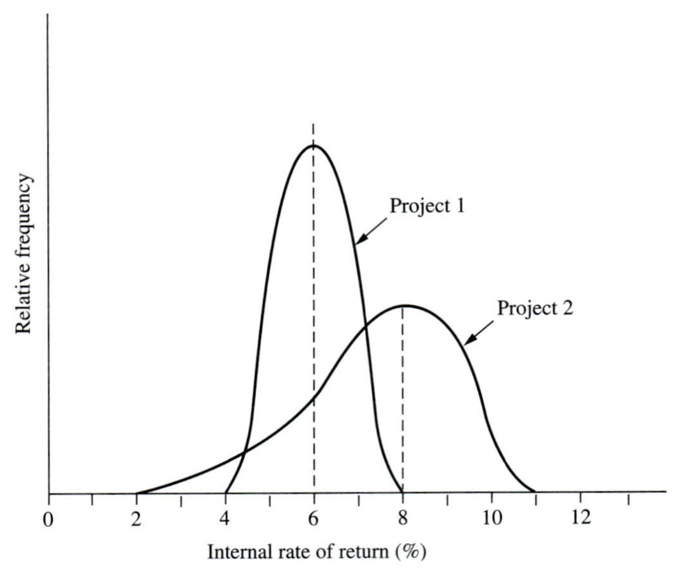

FIGURE 10-6
Internal rate of return probability distributions for two projects of differing risk.

probability distribution. Figure 10-6 shows hypothetical continuous IRR probability distributions for two projects of similar capital requirements and durations. Although project 2 has a higher mode of 8 percent, its risk or expected variation is greater. Managers could subjectively choose between the two. While continuous IRR distributions as in Figure 10-6 are commonly shown in the literature, managers could more easily relate to IRR probability histograms that show the probability of attaining IRRs within certain ranges.

A good way to deal with investment uncertainty, where outcome probabilities are hard to estimate, is to first calculate your profitability criterion, say, NPV or IRR, for an investment's most likely scenario. Then test how profitability varies as you change selected variables from their most likely values to low and high levels. This is called **sensitivity analysis.** Usually you'll find that profitability is much more sensitive to some variables than to others. You can then refine estimates of variables to which returns are the most sensitive.

Appendix 10B shows, in the form of a "decision tree analysis" over time, how you can display and analyze a series of decisions like planting, thinning, and insect control with intervening chance events like drought, fire, insect attack, and changing prices.

DISCOUNTING RISKY COSTS

While this chapter has dealt mostly with revenues, you should be aware that the risk premium for discounting risky *costs* can be *negative* for risk-averse investors (Brown, 1983). This means that the RADR for discounting risky costs can be *less than* the risk-free rate. That appropriately boosts the negative present value of risky costs, reflecting the fact that risk-averse investors dislike cash flow variance. If you don't like variance in a cost (revenue), you make its present value larger (smaller).[14] This applies for cases where probability distributions of costs and revenues are independent. For example, Weyerhaeuser can suffer a fire control cost when timber prices (revenues) are low or high—or the probability of Exxon having a costly oil spill is independent of oil prices. But what if probability distributions of simultaneous revenues and costs are identical and perfectly related, as with revenues and income taxes thereon (when revenue is up, the tax is up, and vice versa)? In that case, revenues exceed costs, and they are inexorably linked: you can subtract costs from revenues in any year and discount net revenue with one risk-adjusted interest rate greater than the risk-free rate.[15] But this is not the case in general.

When calculating bid prices for bare land (WPL in Chapter 7), in present value terms, major costs are usually in year 0, in which case discounting isn't needed. But later risky costs can also be important. And when finding NPVs for entire forest properties, large, risky costs can be crucial in present value terms. So the risk-adjusted discount rate for costs can sometimes be an important issue in forestry—one that hasn't often been addressed. As a first step, I'd recommend using the investor's *risk-free* MAR for discounting future costs with probability distributions independent of revenue distributions. That would lessen understatement of present values for risky costs, which occurs under usual discounting practices. To some extent, past discounting errors might have been offsetting in forestry: using too high an interest rate for revenues underestimates NPV, while discounting costs with too high a rate overestimates NPV.

In an introductory text, I hesitated to mention discounting risky costs with a lower interest rate, especially since the practice hasn't been discussed in the forestry literature. But I feel this is an important issue, especially in the area of major, risky, environmental costs, and we need to begin addressing it. Since future costs in most of this book's discounting examples are relatively small, for simplicity I've used the same discount rate for both costs and revenues, except when finding the borrower's present value of risk-free loan payments (see Chapter 11, "Impact of Loans on Property Valuation").

---

[14]Brigham and Gapenski (1991, p. 416) take a more moderate approach of suggesting that a cost (independent of revenues) with higher-than-average risk should be discounted with a lower-than-average discount rate. But the concept is the same: risk-averse investors should penalize future risky costs by increasing their present values (i.e., by using a lower discount rate).

[15]This is Prince's (1985) argument that when revenues exceed costs in any year, they can both be discounted with the same risk-adjusted rate. Other examples are insect- or disease-caused reductions in timber harvest income, or environmental costs of logging practices. In these cases, the cost isn't felt until the revenue occurs, which makes costs perfectly correlated with revenues. But Prince's argument does *not* apply in cases where probability distributions of costs and revenues are independent. Prime examples of independent, risky costs occur between planting and harvest on a given acre when no income occurs. Examples are costs of possible replanting, precommercial thinning, and insect spraying entered in WPL calculations.

WHERE TO FROM HERE?

At this stage, you're probably wondering what to advise timber investors. There's no simple answer. Unless in some unusually risky situation, such as areas with repeated fire or insect problems, there's defense for using risk-adjusted forestry discount rates lower than typical industrial rates of return, provided you're discounting true expected values of revenues and not optimistic projections. I've heard industry analysts say that they used real after-tax discount rates ranging from 4 to 8 percent for long-term forestry investments, but they're understandably reluctant to be quoted. Remember from the previous chapter that discount rates before income taxes would be higher, but not by much when payoff periods are long. Thus, on a before-tax basis, the lower rates in this 4 to 8 percent range are substantially below the industrial mean 10 percent before-tax rate of return assumed by OMB.

One of the best ways to deal with risk is to estimate probability histograms for all cash flows and then ask an investor to state certainty-equivalents of these. Certainty-equivalents can be discounted with a risk-free discount rate to arrive at a correct net present value for that investor. But this approach takes more data, expertise, and time than may be available. Using risk-adjusted discount rates isn't the best way to account for risk, but since it's one of the easiest and most widely used, it's important to refine the technique. Until we have better approaches that are just as simple, we needn't feel that we're misallocating resources if we use lower RADRs for many long-term forestry revenues than we use for short-term risky gains.

KEY POINTS

- The degree of risk in a revenue is the amount of variation in its possible outcomes. A risk-free return has no variation.
- The *expected value* of any risky variable is the sum of the possible values multiplied by their probabilities of occurrence.
- Most people are *risk-averse:* they prefer less variation in revenues.
- The *certainty-equivalent* of a risky revenue is a sure dollar amount giving the investor the same satisfaction as the risky revenue. For a risk-averse person, the certainty-equivalent of a risky revenue will be less than its expected value.
- Risk-free revenues and certainty-equivalents should be discounted with a risk-free interest rate to arrive at the correct present value.
- The correct present value of a risky revenue is its expected value discounted with a risk-adjusted discount rate (RADR). For a risk-averse investor, this RADR exceeds the risk-free discount rate. Make sure you're discounting expected values, not optimistic values.
- The further in the future a risky revenue is, the lower the correct RADR is, given the same degree of risk and risk aversion. Thus, forestry's long production periods may often require lower RADRs than average short-term industrial RADRs.

- There's no such thing as "the" correct RADR for forestry's expected values. In reality, a different RADR should be used for each cash flow, depending on its probability distribution, on its time from the present, and on the decision maker. We can hope to give only rough guidelines for different situations.
- If you can determine an investor's certainty-equivalent as a percentage of a given expected revenue, in Table 10-2 you can find the correct RADR for discounting that value.
- You should be enormously skeptical of single NPV calculations when comparing alternatives, especially when payoff periods are long. Test how sensitive NPV is to different RADRs and to different estimates of cash flows. When discounting important individual cash flows like major harvest revenues, refer to Table 10-2.
- If data are available, try, using computer simulations, to construct probability distributions of net present value or internal rate of return for projects, and let decision makers compare them (see Figures 10-3, 10-4, and 10-6).
- While it's correct for a risk-averse investor to increase the discount rate for risky future revenues, the discount rate should be lowered for risky future costs that are independent of revenues.
- This is one of the most important chapters on investment analysis—and perhaps one of the most frustrating, because the guidelines are so fuzzy. The essence of risk and uncertainty is that we don't know some things for sure. So while we yearn for precise decision guides, equations, and computer programs to tell us to the nearest penny the value of forestry alternatives, that's not the nature of the world, especially with forestry's long payoff periods.

QUESTIONS AND PROBLEMS

10-1 Define cash flow risk.

10-2 What is a "certainty-equivalent" of a risky, expected revenue?

10-3 Suppose you project a real timber harvest revenue from your woodland in 20 years with the following probabilities: a 5 percent chance that harvest revenue will be between 0 and $500; a 20 percent chance that revenue will be between $500 and $1,000; a 40 percent chance of $1,000 to $2,000; a 35 percent chance of $2,000 to $3,000. What is your "expected value" of this harvest? The next four questions refer to this one.

10-4 If you are risk-averse, will your "certainty-equivalent" of the above expected revenue be greater than, less than, or equal to the expected value? Why? How would you answer this question if you were risk-neutral? If you were risk-seeking?

10-5 What would your certainty-equivalent be for the expected value in question 10-3? (There's no "right" answer; it depends on your view of the risk and your degree of risk aversion. Just make a subjective judgment.) Now, supposing your real risk-free minimum acceptable rate of return is 3 percent, what is your correct present value of the above expected revenue (your certainty-equivalent present value)?

10-6 Given what you answered for the previous question, what is your correct risk-adjusted discount rate (RADR) for discounting the risky *expected value* of the harvest? What is the "risk premium" in this discount rate? (The answers to this and the previous question may be different for each person.) *Hint:* Solve for the

RADR that discounts the expected value to equal the correct present value found in question 10-5 [see equations (10-7) through (10-9)].

10-7 Suppose the risky revenue in question 10-3 occurred 10 years from now. What would be your correct risk-adjusted discount rate? How does this compare with your answer to the previous question?

10-8 In general, for a risk-averse investor, assuming the same variance in a given expected future revenue, what happens to the risk-adjusted discount rate as the payoff period lengthens?

10-9 What is a net present value probability histogram? Sketch a hypothetical one that could help an investor evaluate a risky project.

10-10 Suppose Kurt estimates that the expected value of a timber harvest to be received in 30 years is $4,000. If his certainty-equivalent of this expected value is a sure $2,000, what should his risk-adjusted discount rate be for calculating the present value of the risky $4,000 harvest? Assume that Kurt's risk-free real discount rate is 3 percent. (*Hint:* Use the same procedure as in question 10-6. Or find the certainty-equivalent ratio, and use Table 10-2 to find the risk premium, which you'd add to 3 percent.)

APPENDIX 10A: Risk Aversion

Table 10A-1 shows notation used here in addition to that in Table 10-1.

Figure 10A-1 illustrates risk aversion and shows an investor's total utility from additional wealth. A curve concave to the x axis means declining extra utility with added wealth—a typical situation for most investors.[16] Assuming a sure initial wealth of $W_0 = \$20,000$, adding the $6,000 risky revenue from Figure 10-1 would give three possible wealth positions: $22,000, $26,000, or $30,000, with probabilities of 0.25, 0.50, and 0.25, the expected wealth being $0.25(22) + 0.5(26) + 0.25(30) = \$26,000$. Let the utility function and the expected revenue apply to the same year. The concave utility function portrays risk aversion because receiving, say, $4,000 less than the expected value would bring a utility loss exceeding the utility gain from receiving $4,000 above the expected value. Under such conditions, the $6,000 expected revenue in Figure 10-1 yields less utility than a sure $6,000. Therefore, the certainty-equivalent of the Figure 10-1 revenue must be less than its expected value of $6,000.[17] This is demonstrated below.

---

[16]Utility function concavity is typical of empirical estimates, which are often logarithmic, exponential, or quadratic (for example, see Halter and Dean, 1971; Litzenberger and Ronn, 1986; Wilson and Eidman, 1983).

[17]With a linear utility function (risk neutrality), the certainty-equivalent of a risky revenue equals its expected value, or CE $= E(R)$. A convex utility function (risk-seeking) yields CE $> E(R)$.

TABLE 10A-1
ADDITIONAL NOTATION FOR APPENDIX 10A

$E[U(\text{wealth})]$	= investor's expected utility from risky wealth
$U(\text{wealth})$	= investor's utility (a measure of satisfaction) from wealth
W_0	= the investor's wealth position before receiving a risky revenue

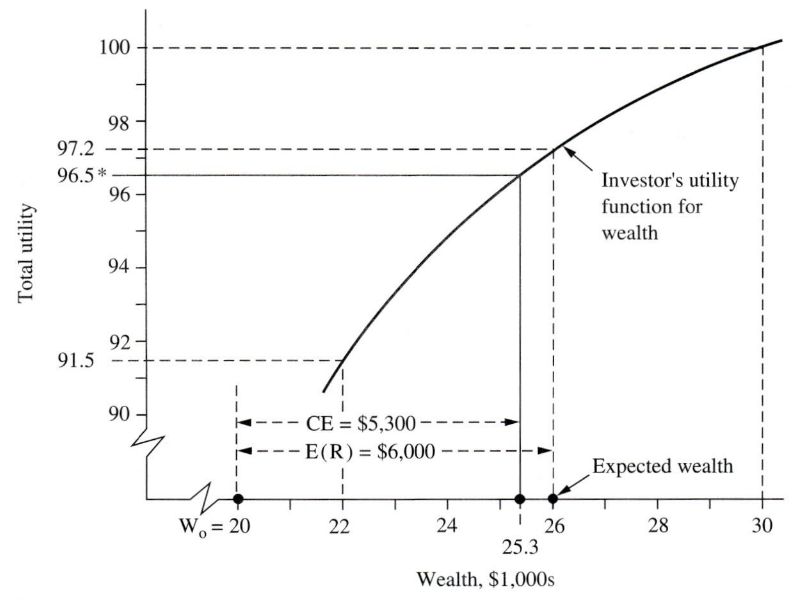

FIGURE 10A-1
Illustration of risk-averse behavior.
(Initial wealth position is W_O = \$20,000. Probability of a \$2,000 gain is 0.25, a \$6,000 gain is 0.50, and a \$10,000 gain is 0.25. Expected utility = .25(91.5) + .50(97.2) + .25(100) = 96.5 and is represented by * on the utility axis. CE = \$5,300 is the certainty equivalent of the expected revenue of \$6,000 from Figure 10–1.) Adapted from Klemperer et al., 1994, by permission.

Reading off the arbitrarily scaled utility for each possible wealth outcome in Figure 10A-1, and applying the above probabilities, the investor's **expected utility** is:

$$E[U(\text{wealth})] = 0.25(91.5) + 0.50(97.2) + 0.25(100) = 96.5 \qquad (10\text{A-}1)$$

Now, find this expected utility of 96.5 on the y axis of Figure 10A-1, go across to the utility curve, and go down to the x axis. There you'll see that a sure wealth of \$25,300 would guarantee a utility of 96.5. Since the initial wealth is \$20,000, the certainty-equivalent (CE) of the risky \$6,000 in Figure 10-1 is \$25,300 − \$20,000 = \$5,300, which is less than \$6,000, as postulated. The \$5,300 CE and the \$6,000 risky revenue both yield the same utility.

The above paragraph says that the utility from W_0 plus the certainty-equivalent of the risky \$6,000 revenue equals the expected utility of risky wealth (W_0 plus the risky revenue in Figure 10-1):

$$U(W_0 + \text{CE}) = E[U(\text{wealth})] \qquad (10\text{A-}2)$$

Substituting the numbers from Figure 10A-1,

$$U(20{,}000 + 5{,}300) = 96.5$$

If the risky \$6,000 revenue is to be received in 5 years, its correct present value, using the left side of equation (10-7), is the certainty-equivalent present value, PV_{CE} =

$5{,}300/(1.03)^5 = \$4{,}572$ (the certainty-equivalent discounted with the risk-free rate). To find the RADR that discounts the expected value of \$6,000 to equal this correct present value, use equation (10-8):

$$\text{RADR} = \sqrt[n]{\frac{E(R)}{\text{PV}_{\text{CE}}}} - 1$$

$$\text{RADR} = \sqrt[5]{\frac{6{,}000}{4{,}572}} - 1 = 0.056 \qquad (10\text{A-}3)$$

Thus, for the 5-year payoff period, the RADR is 5.6 percent, implying a risk premium of $5.6 - 3 = 2.6$ percentage points. The foregoing example serves only to show that the usual case of a concave utility function for wealth yields a certainty-equivalent less than the expected value of a risky revenue. The result is that the RADR must exceed r_f, but this example in no way implies what a risk premium "should" be.

This analysis of risk aversion ties in with Chapter 3 and Figure 3-8, which shows that willingness to pay for some gain is less than willingness to sell it, once obtained. This is what Figure 10A-1 shows: in absolute value terms, the utility from a gain is less than the disutility from a loss of the same dollar value. And that can occur only with a wealth utility function that is concave to the x axis. Chapter 14 covers other reasons, beyond concave utility functions, why this phenomenon occurs: "loss aversion" and the "endowment effect."

APPENDIX 10B: Decision Tree Analysis

In forestry we often make several decisions over many years, and between these decisions are chance events that affect outcomes. An even-aged forestry example is deciding on the type of site preparation at planting date, later deciding whether or not to thin, and then deciding to harvest. Between each decision point can occur economic chance events like stumpage price variations and physical chance events like insect and disease attacks and good or poor weather for timber growth. For simplicity, let's lump the physical events into "good," "medium," or "bad" conditions for timber value growth.

Table 10B-1 outlines a set of hypothetical cases for the above stand management example, assuming a 30-year rotation. Column (1) gives three combinations of site preparation at age 0 and thinning at age 20. Site preparation will be either "heavy" (intensive removal of competing vegetation), costing \$200/acre, or "light" at \$50/acre, both including planting cost. At year 20, the decision will be either to thin or not, yielding the incomes shown under 20 years in column (3). Column (2) shows chance event sequences; the first event in each sequence is timber value growth conditions between ages 0 and 20, and the second is conditions between ages 20 and 30. These events occur with probabilities given in Figure 10B-1. Column (3) gives the cash flows: reforestation costs in year 0, thinning revenue in year 20, and final cut in year 30, if events and decisions actually occur. Other costs and revenues such as lease income, management costs, and taxes are assumed to be the same for all options and thus can be omitted for ranking purposes. In the last column are net present values (NPV), as explained in the table footnote, assuming events would occur with certainty. Since the events don't always occur, the present values aren't true expected values.

For the Table 10B-1 data, Figure 10B-1 displays a decision tree that can help us compute expected NPVs. At each "node," squares are decision points, and circles are

TABLE 10B-1
ALTERNATIVE EVEN-AGED TIMBER MANAGEMENT SCENARIOS, 30-YEAR ROTATION

Decisions (year 0; year 20)	Chance events* Event 1—event 2	Cash flows, $/acre Year 0	Year 20	Year 30	Sure NPV @ 5% $/acre†
Heavy site preparation; thin	good—good	−$200	+$400	+$2,500	$529
	good—bad	−200	+400	+1,500	298
	bad—good	−200	+150	+2,200	366
	bad—bad	−200	+150	+1,200	134
Heavy site preparation; don't thin	good—medium	−200	0	+2,500	378
	bad—medium	−200	0	+2,200	309
Light site preparation; don't thin	good—medium	−50	0	+2,000	413
	bad—medium	−50	0	+1,700	343

*Chance events, with probabilities given in Figure 10B-1, are (1) scenario for the first 20 years and (2) scenario for the last 10 years. Good, medium, and bad refer to timber value growth.

†Year-zero net present value at 5 percent real interest if events occur with certainty. For example, in the first row, NPV = $2{,}500/(1.05)^{30} + 400/(1.05)^{20} - 200$ = $529. Here the investor's interest rate should be nearly risk-free. If the event probabilities applied to these NPVs in Figure 10B-1 are reliable, each result could be considered as risk-free (see discussion of Figure 10-3). But each value point represents a band of values, so some variance or risk exists for midpoint values.

chance events. Starting at square 1, the upper branch is heavy site preparation, and the lower is light site preparation. The first chance events show value growth conditions expected to be good 80 percent of the time ($P = 0.8$) and bad 20 percent of the time ($P = 0.2$). At any chance event node, probabilities P must always sum to 1, so that 100 percent of all possibilities are covered. The second decision points are 2g and 2b: whether or not to thin following good (g) and bad (b) conditions after planting. For simplicity, the tree's lowest branch has only a "don't thin" option.

At the second decision points in Figure 10B-1, we either thin or don't thin in year 20. Thereafter, we have the indicated events—good, medium, or bad—with the probabilities shown. To simplify, for the "don't thin" branches, the probability is 100 percent that value growth conditions are medium. Above the right end of each branch is the harvest income for that scenario. In the far right column is the net present value that would result *if everything occurred along the indicated branch* (from the last column in Table 10B-1).

To find expected net present values for year zero, E(NPV), start with the sure NPV at the right end of each branch and work leftward toward the first decision node (square 1). Remember that expected value is the sum of probable outcomes times their probabilities of occurrence. At node 2g, working from the right side of the upper branch, the expected NPV if you were to thin following heavy site preparation and a good first 20 years is:

$$E(\text{NPV}) = 0.7(529) + 0.3(298) = \$460$$

shown in italics above square 2g. If you don't thin at that point, the expected NPV is 1(378)=$378, shown below square 2g in italics. Above and below square 2b are the expected NPVs for thin and don't thin, computed in the same way. At node 2g, thinning is the best option, since the expected value is highest, indicated with a star. But at node 2b, not thinning is best.

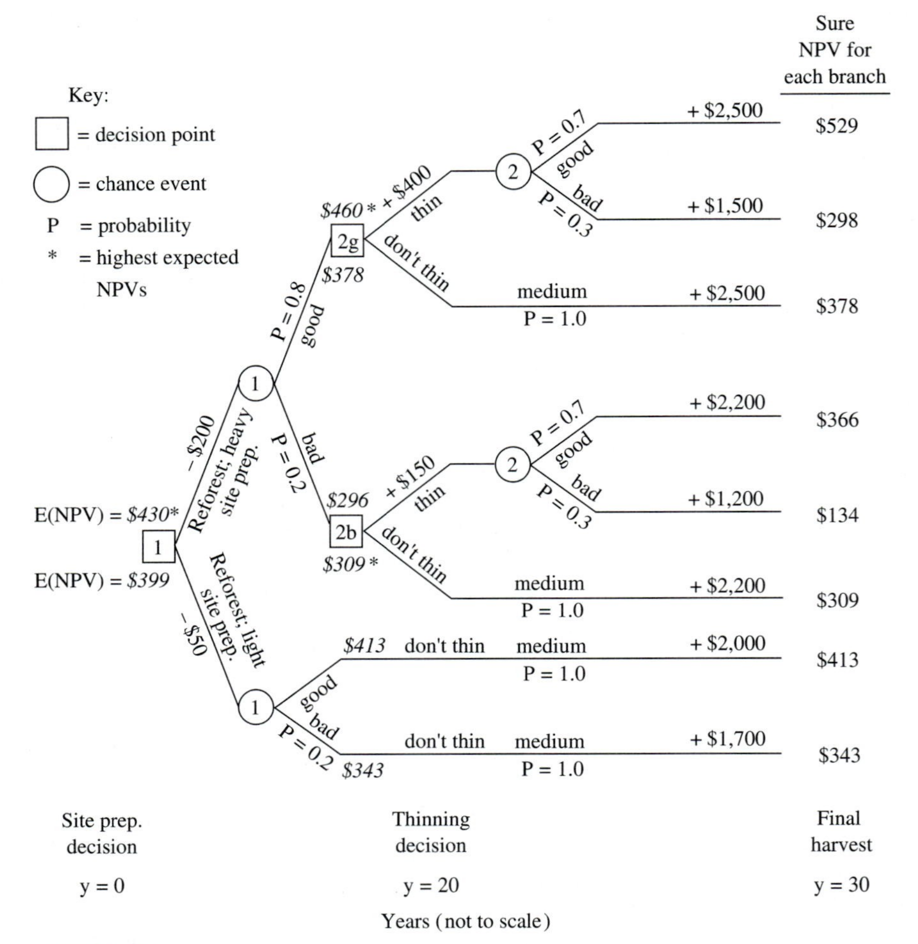

FIGURE 10B-1
Decision tree for stand management alternatives. (See Table 10A-1.)

To get the true expected NPV for heavy site preparation (the tree's top half) we need the expected NPVs of the best thinning choices, following good conditions for the first 20 years ($P = 0.8$) and bad ($P = 0.2$):

$$E(\text{NPV}) = 0.8(460) + 0.2(309) = \$430/\text{acre}$$

shown in italics above square 1. Working leftward in the same way for the light site preparation half of the tree, you get a lower expected NPV of $399/acre. This part is simplified to show only no thinning and the medium conditions after the second decision point.

Based only on expected NPV, the best decision sequence is to choose the $200 regeneration option, to later thin if conditions had been good after planting, or not to thin if conditions had been bad. But, ideally, for each expected NPV, a risk-averse investor wants to know the variance, which could also be computed. If the higher E(NPV) has too great a variance, the lower NPV might be preferred. An analysis estimating NPV probability histograms of options would give information in a format more useful to decision makers than the simple decision tree approach shown here (for example, see Figure 10-4).

Figure 10B-1 is a highly simplified example to show the nature of sequential decision problems under risk. You can see how decision trees could become enormous if they included many decision points and more chance events, each with many outcomes of different probabilities. Thus, large decision trees are best analyzed with computer models.[18]

Joint Probabilities

In Figure 10B-1, a decision point occurs between chance events. But often a sequence of several chance events occurs without intervening decisions, as shown in Figure 10B-2. In those cases you need to compute *joint probabilities* for sequences of events. Figure 10B-2 shows such a case for a hypothetical Christmas tree farm where the manager initially must decide whether to spend $500 planting species 1, or spend $700 on a more costly species 2 that also requires more site preparation. Only the events for species 1 are given. After planting, if drought causes plantation failure ($P = 0.1$), the venture is abandoned. With no drought, $1,000 worth of table-size trees are cut in year 4, and the indicated final harvest incomes occur in year 8 following good growing conditions ($P = 0.7$) or bad conditions ($P = 0.3$). The latter two are *conditional probabilities,* conditioned on the previous events [in this case, no drought ($P = 0.9$)]. As in the previous section, to simplify, other costs and revenues are ignored.

In Figure 10B-2, sure net present values for each branch are in the first column to the tree's right, assuming the events would occur with certainty (see figure footnote for sample calculation). In column (2) are the *joint probabilities,* which are the previous probabilities multiplied together; for example, for the top branch, $(0.9)(0.7) = 0.63$. In the rightmost column are the expected net present values [E(NPV)] for the event probabilities in each branch: the branch's sure NPV multiplied by the joint probability. For the species 1 segment of the tree, at year 0, the expected NPV of $1,684 is the sum of the branch E(NPV)s in the last column (i.e., the sum of probable outcomes times their probabilities of occurrence).

Although data aren't given for the species 2 segment of the tree, E(NPV) calculations would parallel those for species 1. But as noted for Figure 10B-1, the decision maker would need to know the variance, or best of all the NPV probability histograms, before deciding between the two E(NPV)s. Both Figures 10B-1 and 10B-2 show the dangers in making calculations based on optimistic outcomes. For example, in Figure 10B-2, the best possible outcome for species 1 is an NPV of $2,353, if no drought occurs and good conditions follow the first cut. But the average NPV that could be expected with many trials of planting species 1 is $1,684 (the expected value).

REFERENCES

Arrow, K. J., and R. C. Lind. 1970. Uncertainty and the evaluation of public investment decisions. *American Economic Review.* 60(3):364–378.

Berck, P. 1979. The economics of timber: A renewable resource in the long run. *Bell Journal of Economics.* 10(2):447–462.

Brigham, E. F., and L. C. Gapenski. 1991. *Financial Management—Theory and Practice* (6th edition). The Dryden Press, Chicago. 995 pp. plus app.

[18]For reviews of decision tree literature, see Brigham and Gapenski (1991) and Leuschner (1984).

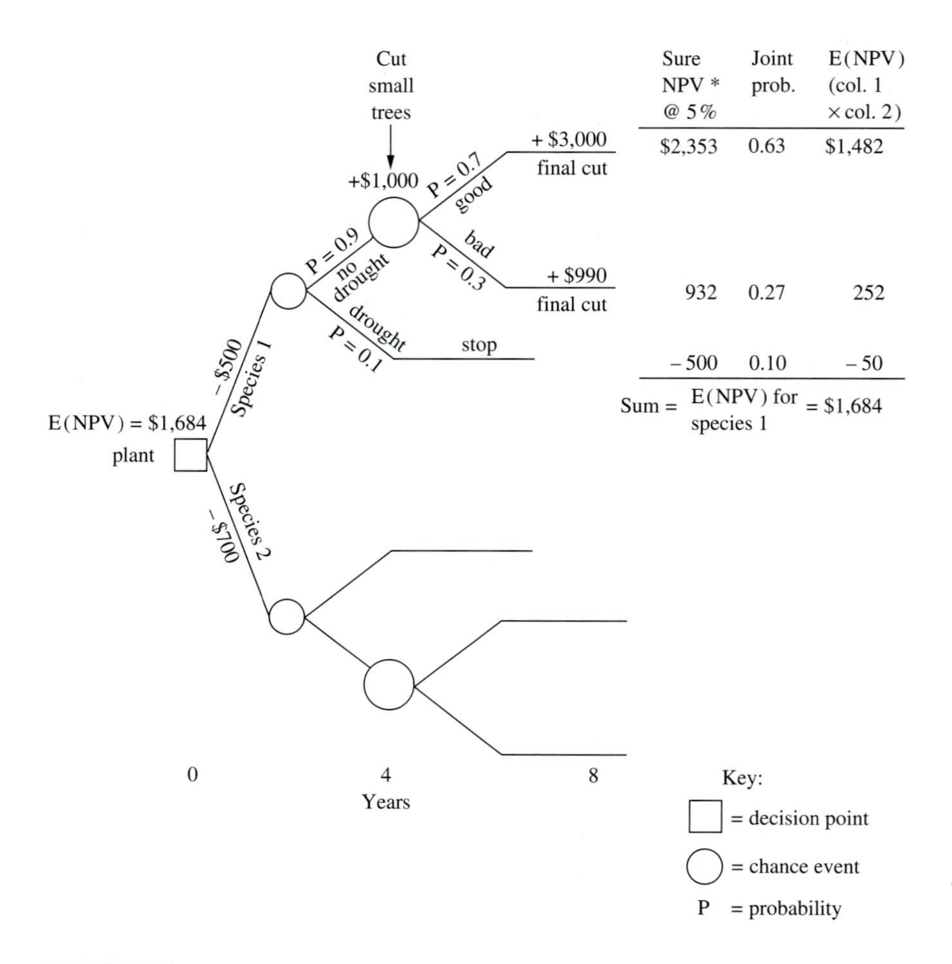

*Net present value at 5% real interest if event occurs with certainty. For example, for the top branch:

$$NPV = 3{,}000/(1.05)^8 + 1{,}000/(1.05)^4 - 500 = \$2{,}353.$$

Also see NPV footnote in Table 10B-1

FIGURE 10B-2
Christmas tree plantation decision tree with joint probabilities.

Brown, S. 1983. A note on environmental risk and the rate of discount. *Journal of Environmental Economics and Management.* 10:282–286.

Carlson, J. A. 1977. Short-term interest rates as predictors of inflation: Comment. *American Economic Review.* 67(3):469–475.

Cathcart, J. F. 1989. *Evaluating Risk-Adjusted Discount Rates in Forest Investment Decision Making.* Ph.D. dissertation. Virginia Polytechnic Institute and State University, Blacksburg. 280 pp.

Cathcart, J. F., and W. D. Klemperer. 1988. *Estimating Uncertain Timber Investment Returns for Landowners.* Proceedings of Southern Forest Economics Workshop. Dept. of Forestry, Univ. of Florida, Gainesville. pp. 44–69.

Chen, H-Y. 1967. Valuation under uncertainty. *Journal of Financial and Quantitative Analysis.* 2(3):313–326.

Conroy, R., and M. Miles. 1989. Commercial forestland in the pension portfolio: The biological beta. *Financial Analysts Journal.* Sept./Oct.

Fortson, J. C. 1986. Factors affecting the discount rate for forestry investments. *Forest Products Journal.* 36(6):67–72.

Haight, R. G. 1991. Stochastic log price, land value, and adaptive stand management: Numerical results for California white fir. *Forest Science.* 37(5):1224–1238.

Halter, A. N., and G. W. Dean. 1971. *Decisions under Uncertainty, with Research Applications.* South-Western Pub. Co., Cincinnati. 266 pp.

Hanke, S. H., and J. B. Anwyll. 1980. On the discount rate controversy. *Public Policy.* 28(2):171–183.

Hirshleifer, J. 1964. Efficient allocation of capital in an uncertain world. *American Economic Review.* 54(3):77–85.

Hirshleifer, J. 1966. Investment decisions under uncertainty: Application of the state-preference approach. *The Quarterly Journal of Economics.* 80(2):252–277.

Hyldahl, C. A., and D. C. Baumgartner. 1991. *Risk Analysis and Timber Investments: A Bibliography of Theory and Applications.* U.S. Forest Service, North Central Forest Experiment Station. Gen. Tech. Rept. NC-143. St. Paul, MN. 29 pp.

Ibbotson, R. G., and R. A. Sinquefield. 1982. *Stocks, Bonds, Bills, and Inflation: The Past and the Future.* The Financial Analysts Research Foundation, Charlottesville. 137 pp.

Klemperer, W. D., J. F. Cathcart, T. Haering, and R. J. Alig. 1994. Risk and the discount rate in forestry. *Canadian Journal of Forest Research.* 24(2):390–397.

Leuschner, W. A. 1984. *Introduction to Forest Resource Management.* John Wiley & Sons, New York. 298 pp.

Litzenberger, R. H., and E. I. Ronn. 1986. A utility based model of common price movements. *The Journal of Finance.* 4(1):67–92.

Mills, W. L., and W. L. Hoover. 1982. Investment in forest land: aspects of risk and diversification. *Land Economics.* 58(1):33–51.

Nordhaus, W. D. 1974. The falling share of profits. In Okun, A. M., and G. L. Perry (Eds.), The Brookings Papers on Economic Activity. The Brookings Institution, Washington, D. C. pp. 169–217.

Pappas, J. L., and M. Hirschey. 1990. *Managerial Economics.* The Dryden Press, Chicago. 826 pp.

Prince, R. 1985. A note on environmental risk and the rate of discount: Comment. *Journal of Environmental Economics and Management.* 12:179–180.

Redmond, C. H., and F. W. Cubbage. 1988. Portfolio risk and returns from timber asset investments. *Land Economics.* 64(4):325–337.

Robichek, A. A., and S. C. Myers. 1966. Conceptual problems in the use of risk-adjusted discount rates. *Journal of Finance.* 21:727–730.

Samuelson, P. A. 1964. Principles of efficiency: Discussion. *American Economic Review.* 54(3):93–96.

Seagraves, J. A. 1970. More on the social rate of discount. *Quarterly Journal of Economics.* 84(3):430–450.

Simon, J. L. 1990. Great and almost-great magnitudes in economics. *Journal of Economic Perspectives.* 4(1):149–156.

Stockfisch, J. A. 1969. The interest rate applicable to government investment projects. In Hinrich, H. H., and G. M. Taylor (eds.). *Program Budgeting and Cost Benefit Analysis.* Goodyear Rubber Co. 420 pp.

Tversky, A., and D. Kahneman. 1992. Advances in prospect theory: cumulative representation of uncertainty. *Journal of Risk and Uncertainty.* 5(4):297–323.

U.S. Office of Management and Budget. 1972. *Discount Rates to be Used in Evaluating Time-Distributed Costs and Benefits.* Circular A-94 (revised). U.S. Govt. Printing Office, Washington, D.C.

Webb, J. F., T. J. Ebner, and M. T. Shearer. 1987. *Risks Associated with Timberland Investments and the Timing of the Investment to Minimize the Risk.* Paper presented at the Forest Products Research Society Conference, "A Clear Look at Timberland Investment," April 27–29. Milwaukee, Wisconsin.

Wilson, P. N., and V. R. Eidman. 1983. An empirical test of the interval approach for estimating risk preferences. *Western Journal of Agricultural Economics.* 8:170–182.

Zinkhan, F. C. 1988. Forestry projects, modern portfolio theory, and discount rate selection. *Southern Journal of Applied Forestry.* 12(2):132–135.

Zinkhan, F. C., W. R. Sizemore, and G. H. Mason. 1992. *Timberland Investments—A Portfolio Perspective.* Timber Press. Portland, OR. 208 pp.

11

FOREST VALUATION AND APPRAISAL

Finding values of forests is vitally important. Buyers and sellers of stumpage for harvesting need to agree on sale prices. Forested properties are often valued not for current harvest but as investments, or to determine compensation in government takings or insurance claims, to compute taxes on property values, or to determine collateral on loans. It's sometimes important to place values on portions of forests such as land of different qualities or separate age classes of timber.

Earlier chapters discuss several types of value: stumpage value, forest holding value, forest liquidation value, bid prices, market value, net present value, and willingness to pay for land or other assets. This chapter delves more deeply into these concepts of value and will help you distinguish between them. Chapter 14 will treat valuation of resources without market prices.

Initially let's consider only monetary income and later add other influences on value. Here we'll first cover values that individual buyers or sellers place on forest assets; then will come appraisal techniques to determine market value, the most likely sale price in the market. *For simplicity, examples will compute only one value; in reality, analysts should do* ***sensitivity analyses,*** *calculating several values based on a range of assumptions.* This approach tests how "sensitive" a value is to changes in certain inputs. Decision makers can then choose the most reasonable scenario and value. Table 11-1 defines notation for the chapter.

AN INVESTOR'S VALUATION OF ASSETS

One objective of this chapter is to distinguish between an asset's market value and its value to one investor. *Here, the procedure for finding an investor's value*

TABLE 11-1
NOTATION FOR CHAPTER 11

a =	equal annual revenue, $/acre	MBF =	thousand board feet
c =	equal annual cost, $/acre	NPV =	net present value or present value of revenues minus present value of costs
$(a - c)$ =	net annual cash flow (may be positive, zero, or negative), $/acre	r =	real discount rate, percent/100
H_t =	clear-cutting revenue at rotation age t, $/acre	t =	rotation age in years
L_t =	market value of land in year t, $/acre	WPL =	willingness to pay for bare land (net present value of future cash flows), $/acre
MAR =	minimum acceptable rate of return (must specify whether before or after taxes)	y =	an index for years

of an item will be called ***valuation.*** Examples are the net present value calculations in earlier chapters. An item's monetary value to one investor—for example, a buyer's willingness to pay or an owner's willingness to sell—may be more than, less than, or equal to market value. **Market value** of an item is its most likely selling price—an average expected selling price for similar items, similarly situated. *The procedure for finding market value is called* ***appraisal.***

Below are *valuation* examples. Most of these will exclude taxes, which were treated in Chapter 9. Here, unless otherwise noted, assume that cash flows are after taxes and discount rates are after-tax **minimum acceptable rates of return (MARs)**. For sources of computer software to calculate some of the valuations discussed in this section, see the Chapter 6 section "A Summary of Capital Budgeting."

Stumpage Valuation

Stumpage value *is what buyers pay for standing timber ready for harvest.* The emphasis is on immediate cutting. As you'll see shortly, investors' willingness to pay for timber to be cut in the future won't necessarily equal stumpage value. Stumpage value may be a price per unit volume, for example, a **stumpage price** in dollars per thousand board feet (MBF), per cord, or per cubic foot. Stumpage value can also be expressed as value per unit area, for example, dollars per acre (the symbol H_t in earlier equations). Since, by definition, stumpage income occurs in the very near future, stumpage valuation involves no discounting.

Starting with a simple example of timber grown for lumber production only, let's consider the maximum amount a logger would be willing to pay for stumpage per MBF on one property. Assume competitive conditions, one region, one species, and a given quality of lumber that mills sell for $300 per MBF. Consider this a price that each mill must accept as a **price taker;** the next chapter discusses price determination. Suppose it costs a mill $100/MBF, *including profit,* to process delivered logs into lumber and distribute it to buyers. Then such a mill would be willing to pay up to $300 − 100 = $200/MBF for logs delivered to the mill.

Suppose a logger can receive $200/MBF for delivered logs, and it costs $70/MBF for logging and hauling of a given type and location of stumpage (this includes building logging roads, administration, felling, skidding, bucking, loading, hauling, and profit). That logger would be willing to pay up to $200 − 70 = $130/MBF for such stumpage. In summary, this maximum **willingness to pay** for stumpage is in the last row below:

Lumber sale price:		$300/MBF*
Cost of milling and distribution:		−100
Delivered log price at mill:	$200	
Cost of logging and hauling:		−70
Residual for stumpage: ("Conversion return")		$130/MBF

*Volumes are based on log scale, and prices are adjusted to account for "overrun" or "underrun"—when mills cut more than (or less than) 1,000 board feet of lumber from 1 MBF of logs.

Or, in general, given any end product like plywood or lumber:

Buyer's maximum willingness to pay for stumpage =
(end product price) − (cost[1] of manufacturing and distributing the end product)
− (cost[1] of logging and hauling logs)
= "conversion return"

(11-1)

Or if the logger sells only delivered logs, the stumpage residual is log price minus logging and hauling cost. Assuming a given type and location of timber, this equation models stumpage valuation as a *residual value,* sometimes called **conversion return.** The seller may value stumpage similarly or could consider prices received by others on comparable sales; or for nonmonetary reasons, the seller may set a very high minimum selling price. The price below which an owner won't sell stumpage is the **reservation price.** Actual stumpage prices are set by bargaining between buyers and sellers, often with competitive bidding among stumpage buyers, as with U.S. Forest Service (USFS) timber sales. Many landowners use variations of equation (11-1) to calculate minimum acceptable bids for their stumpage. Stumpage buyers also adapt equation (11-1) to compute maximum bid-prices for stumpage. If a landowner is willing to sell for this price, it becomes the stumpage price per unit value for that sale. That one sale price could be an input in finding the average price or *market price* of similar stumpage in a given period. Market values are covered later.

If you sell stumpage to a logger, your expected returns will be what the logger is willing to pay. If you log your own timber, the stumpage valuation process will

[1]Including acceptable profits

be through equation (11-1) if you manufacture end products, or if you log your own timber and sell logs, it's log price minus logging and hauling costs.

Equation (11-1) shows that, other things being equal, higher (lower) lumber prices lead to higher (lower) stumpage prices. You can also see the effect of different processing costs at any level; for example, higher log-hauling costs between the forest and mill cause less willingness to pay for stumpage. And this explains why, under competitive conditions, other things being equal, stumpage near mills tends to be worth more than stumpage far away. As another example, on steep lands, higher costs of logging and road building will drive down stumpage prices. Thus, other things being equal, stumpage prices in remote mountainous forests tend to be lower than on easily accessible tracts near mills. When logging and long-distance hauling costs are high enough, stumpage prices are zero or negative, as is the case in remote portions of Alaska and the Rocky Mountains. This is an extremely important concept: you can't assume that a given type of tree—say, a 100-year-old Douglas-fir, 150 feet high and 25 inches DBH, of sawtimber quality—has some fixed stumpage value, even if you assume the same prices of end products. Stumpage value can be radically affected by location.

Equation (11-1) outlines only long-run general tendencies under competition. In reality, for each sale, individual buyers and sellers negotiate prices that may be above or below the conversion return. Sometimes mills use their own logging crews and may have little competition in bidding for stumpage. In that case, a mill may pay the minimum price landowners would accept for their stumpage—the reservation price. If reservation prices average, say, \$70/MBF in such a mill's timbershed, you could find stumpage prices around \$70/MBF on lands near as well as far from the mill. Similar results occur when mills negotiate higher delivered log prices for suppliers bringing logs from farther away. On the other hand, mills sometimes pay above the conversion return for stumpage during log shortages. In those cases, mills may not cover all their fixed costs of processing, but they'd lose even more by shutting down the mill. But owners don't tolerate such scenarios in the long run; they'll invest their capital where it will earn better returns.

The foregoing is a simple example with only one end product and one species. Stumpage valuation often averages values for several species and log types such as veneer logs, sawlogs, poles, and pulpwood. Or standing timber may be valued differently for each log type and species.

Later this chapter shows ways to appraise the market value of different forest components: the most likely price to be paid when an owner offers, say, stumpage, immature timber, or land for sale. But first let's continue with the view of individual investors—owners or buyers of forests.

Valuation of Separate Timber Stands or Bare Land

This section serves to remind you that several of the **net present value (NPV)** equations given in Chapters 7, 8, and 9 show how to calculate an investor's valuation of, or willingness to pay for, separate forest stands from age zero to rotation age. Examples are **willingness to pay for bare land (WPL),** assuming even-aged management [equation (9-7)], WPL with uneven-aged management [equation (8-5)], holding value of immature even-aged timber [equation (7-14)], and net present value of an uneven-aged forest [equation (8-6)]. Thus you've already learned the basics of stand valuation. The objective here is to note the difference between valuing small stands and large forests, and to distinguish between valuation and appraisal.

In an NPV equation for a property (for example, WPL), anything that increases one of the future revenues (R's) will increase the NPV; anything increasing one of the future costs (C's) will decrease the NPV. For example, other things being equal, higher site qualities increase forest NPVs; higher taxes decrease NPVs. And higher real interest rates will reduce NPVs. Other examples are in Chapter 7's section titled "Factors Affecting Forest Land Values." Remember, if investors pay their computed NPVs for assets, they will earn the discount rate used to calculate the NPVs, if the projected cash flows actually occur. *Thus, NPV is an investor's maximum willingness to pay for an income-producing asset, but it's not necessarily the market value.*

When calculating NPVs for individual investors, make sure to remember the following guidelines from earlier chapters:

- Enter cash flows in **real** terms, and discount them with the decision maker's real minimum acceptable rate of return (MAR). (You can also project **nominal** cash flows and use a nominal interest rate, all embodying the same inflation rate, but that's usually more complex.)
- When taxes are not projected, use a before-tax MAR.
- For taxable investments, enter taxes as negative cash flows and use the after-tax MAR.

To review stand valuation, Figure 11-1 (adapted from Figure 7-5) graphs from Table 7-3 the 1-acre loblolly pine stand to be clear-cut at an economically optimal rotation age of $t = 20$ years, assuming only pulpwood markets. The solid curve traces the investor's projected **liquidation value,** or potential sale value of stumpage (H_t) for immediate cutting plus land (L_t) at any age. All values are in **constant dollars.** At any age y, an investor's valuation of the stand for *future* harvest is the dashed curve that is the net present value of the liquidation value in year $t = 20$, including the net present value of any negative or positive cash flows between year t and y (for now, ignore the dotted curve). In Chapter 7, this NPV was called **holding value** [see equation (7-14)]. The NPV curve generally lies above liquidation value, since the latter grows faster than the discount rate at ages less than t.[2] To get timber present value alone, just subtract land value. Recall from Chapter 7 that the inputs on which Figure 11-1 is based are simplified, so the values shown aren't necessarily typical of the loblolly pine region. The object here is to show relationships and procedures.

[2]Remember from Chapter 7 that, before financial maturity, the per acre stumpage value of standing timber must grow faster than the after-tax discount rate (and sometimes substantially faster) in order to offset annual costs, taxes, and the opportunity cost of land.

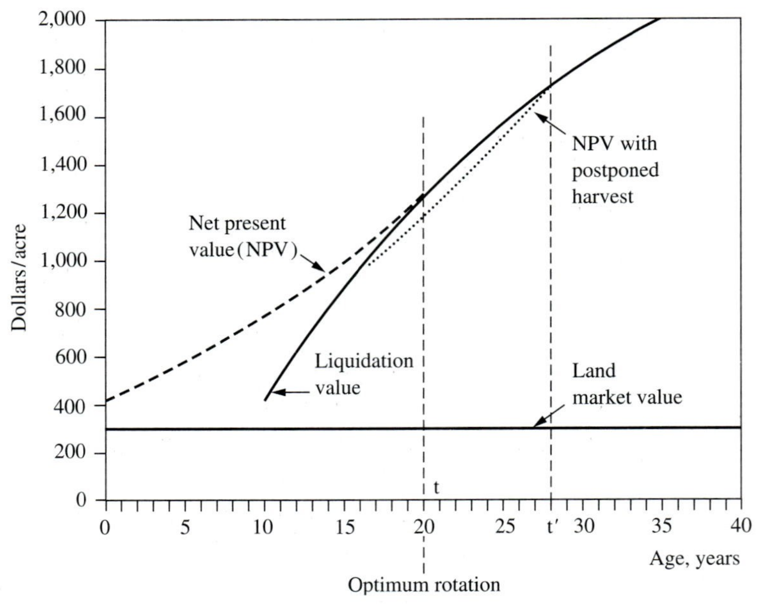

FIGURE 11-1
Loblolly-pine pulpwood forest values.

Because we'll later compare NPV with accumulated costs, let's review the procedure for valuing an even-aged stand at age y (the dashed curve in Figure 11-1), from equation (7-14):

$$\text{Forest net present value at age } y = \frac{H_t + L_t}{(1+r)^{t-y}} + (a-c)\left[\frac{1-(1+r)^{-(t-y)}}{r}\right] \qquad (11\text{-}2)$$

For this simple case, we have a harvest of H, a land value of L, with annual revenue minus cost at $(a - c)$, all per acre. Assuming a clear-cut at age $t = 20$, and using the Table 7-3 values and 5 percent real interest, at age $y = 12$, NPV is:

$$\frac{953.36 + 300}{(1.05)^8} - 4\left[\frac{1-(1.05)^{-8}}{0.05}\right] = \$822.47/\text{acre} \qquad (11\text{-}3)$$

as shown in column (k) of Table 7-3 at age 12 and on the dashed curve in Figure 11-1. Although these values are only rough approximations, they are carried to the nearest penny to later show an equivalent form of valuation.

An investor's valuation of bare land alone would be computed with the relevant WPL equation; for example, see equation (7-9). Here we're valuing land and timber together. Since we're projecting future land sale, land value is entered as an estimated sale value in eight years.

As discussed later, if harvest must be postponed beyond the optimal rotation, say, to age t' in Figure 11-1, the net present value can fall below the apparent liquidation value, as traced by the dotted curve. This models a case with only pulpwood and no sawtimber markets.

Effects of Logging Restrictions Chapter 3 mentions ways to lessen logging damage to water quality, soils, trails, roads, and scenic values. Such practices, compared to those that maximize profits, include seeding more old logging roads and landings, using more culverts and bridges, limiting steepness of roads, scattering slash and reducing its height, reducing clear-cut size, using cable logging on steep slopes, and, in general, using equipment and logging techniques that minimize soil disturbance. Whenever such practices are required and increase logging costs, equation (11-1) shows that stumpage value will decrease. In those cases, the value for H_t in any of the land or timber NPV equations will decrease, assuming no significant change in site quality or stumpage price. That, in turn, decreases a buyer's willingness to pay for the timber or land, *considering only timber income.*

For some forest and land buyers (primarily nonindustrial or public) improved aesthetics from logging constraints can *increase* willingness to pay for forests. But that wouldn't generally apply where the only goal is timber production.

Long-run interactions are complex: Sufficiently severe logging restrictions will reduce current and future harvests enough to increase stumpage prices. (Can you sketch this price-quantity change as a leftward supply curve shift, given a constant demand curve?) The price-increase can partially or completely offset the added costs for some landowners. This is especially true when logging restrictions are applied unevenly. For example, harvest restrictions on public lands in the Pacific Northwest in the early 1990s drove up stumpage prices and increased timberland values for private timber owners whose harvests were unregulated.

Let's now consider valuation of forests composed of many timber stands. We'll continue with investors' values before discussing market values.

Valuing Collections of Stands

Using simple models of large forests to be gradually harvested over time, this section covers basic principles an owner or buyer of timber can use to calculate the net present value of future harvests and how this value can change with assumptions made. The emphasis is still on one investor's valuation, rather than on market value.

Old-Growth Timber Valuation Here we'll consider a buyer's maximum willingness to pay for timber so old that net annual growth is negligible—defined here, from an economic view, as **old-growth.** Unless otherwise stated, further assumptions are that the tract is large enough so that harvesting must occur over a period of several years, the analysis considers only timber and not land, harvest will be in equal annual volumes, stumpage price is the regional trend-line average for timber of similar quality and location, and the only value to the buyer is the NPV of future harvest income.

The five short bars of Figure 11-2 show gross annual harvest incomes from a tract of old-growth timber to be cut over a **depletion period** of 5 years, ignoring price changes, taxes, and other costs. The longer bar at year 0 gives the current stumpage value of $5 million *if* the tract could be clear-cut now at current local stumpage prices. *Assuming no taxes or other costs, no inflation, and no real price increases,* the present value of the five $1 million harvests for an owner with a 6 percent minimum acceptable rate of return (MAR) is [using equation (4-10)]:

$$\text{NPV} = \$1{,}000{,}000\left[\frac{1-(1.06)^{-5}}{0.06}\right] = \$4{,}212{,}364 \qquad (11\text{-}4)$$

To generalize, let's express the NPV as a **valuation factor,** which is the NPV as a proportion of the current stumpage value or, in this case, 4,212,364/5,000,000 = 0.84, or 84 percent. In general:

$$\text{Timber valuation factor} = \frac{\text{net present value of future timber harvests}}{\text{current stumpage value of entire volume}} \qquad (11\text{-}5)$$

Thus, under the above assumptions, any sized old-growth tract to be cut over a 5-year period would have an NPV about 84 percent of the current local stumpage

FIGURE 11-2
Gross present value of old-growth harvesting income.
No inflation and no real stumpage price growth
No volume growth
No taxes or other costs
Five-year depletion period

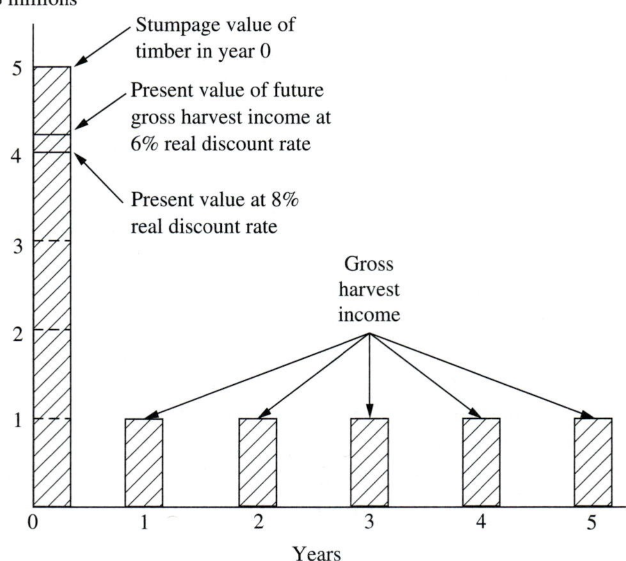

Rising real-harvest value would boost present value
Inflation and taxes would decrease present value
Longer depletion period would decrease present value

value (the latter being stumpage price times volume). As shown in Figure 11-2, at 8 percent interest, the **present value** of harvests drops to about $4 million, implying a valuation factor of 0.80. Other things being equal, the higher the valuation factor, the higher the NPV.

Remember, the tracts in question are so large that immediate clear-cutting would be physically impossible or would depress local stumpage prices. What would that imply about the stumpage demand curve faced by the owner? (This means the demand curve slopes downward, so that significant increases in cutting rates would depress the stumpage price: the seller has some **market power.** Under perfect competition, stumpage sellers face horizontal demand curves so that price is constant no matter how fast they cut.) You must think of local stumpage price as the price received for relatively small percentage additions to local timber harvest per unit of time. "Large" additions drive the price down.

Now suppose real stumpage prices are expected to increase at 6 percent per year. In that case, if the owner has a 6 percent MAR, future harvest incomes in Figure 11-2 increase at the same rate at which they're discounted, and the timber NPV equals the current stumpage value, giving a valuation factor of 1.00. Using equation (11A-6) from Appendix 11A, Table 11-2 shows old-growth valuation factors for different interest rates and depletion periods, given a 6 percent real price growth and *still assuming no taxes or other costs.* At 6 percent interest, you can see the above-mentioned valuation factors of 1.00 for all depletion periods in Table 11-2. If the harvest value increases faster than the interest rate, timber NPV will exceed current stumpage value, as shown by the valuation factors exceeding 1.00 with 3 percent interest. But at interest rates exceeding the 6 percent price growth, the valuation factor is less than 1.

Since private timber owners pay income taxes and other costs, NPVs will be below those in Table 11-2. Table 11-3 revises Table 11-2 to include management costs at 2 percent of harvest income, a 38 percent combined state and federal income tax, and 5 percent inflation (higher inflation increases income tax burdens, as shown in Chapter 5). For the given range of assumptions, valuation factors are

TABLE 11-2
VALUATION FACTORS* FOR OLD-GROWTH TIMBER
(Real Harvest Value Growth = 6 Percent/Yr, No Taxes or Other Costs)

Depletion period, yr	Real interest rate				
	3	**6**	**9**	**12**	**15**
2	1.044	1.000	0.959	0.921	0.886
5	1.091	1.000	0.920	0.850	0.788
8	1.140	1.000	0.884	0.787	0.705
11	1.193	1.000	0.849	0.730	0.634
14	1.249	1.000	0.816	0.678	0.572
17	1.308	1.000	0.785	0.632	0.519

*Investor's NPV as a proportion of current stumpage value. [See equation (11A-6), Appendix 11A.]

TABLE 11-3
VALUATION FACTORS* FOR OLD-GROWTH TIMBER
(Real Harvest Value Growth = 6 Percent/Yr, Income Tax = 38 Percent, Inflation = 5 Percent/Yr, Mgt. Cost = 2 Percent of Harvest Value)

Depletion period, yr	Real interest rate				
	3	6	9	12	15
2	0.947	0.888	0.835	0.788	0.746
5	0.938	0.832	0.744	0.671	0.609
8	0.939	0.791	0.677	0.587	0.515
11	0.948	0.761	0.625	0.523	0.445
14	0.963	0.738	0.583	0.472	0.391
17	0.983	0.719	0.547	0.430	0.348

*Investor's NPV as a proportion of current stumpage value. [See equation (11A-6), Appendix 11A.]

all below 1.00, even for the case in column (1) where the 6 percent rate of harvest value-increase is twice the discount rate of 3 percent. In Table 11-3, longer depletion periods will decrease valuation factors, as long as discount rates are greater than the rate of harvest value-increase. Also, higher discount rates decrease the valuation factor as do lower rates of harvest value-increase. Assuming the Table 11-3 scenario, a buyer with a 9 percent real discount rate and planning to cut the tract over an 8-year period would bid a maximum of 67.7 percent of the current stumpage value for such timber.[3]

The above examples assume no timber volume growth. But the valuation factor formula used here can include a modest volume growth in the harvest value growth rate, as long as you assume that harvest volume increases annually at the volume growth rate. For example, you could let a real 5 percent annual harvest value growth rate be composed of 2 percent volume growth and 3 percent price growth.

I've used the valuation factor approach only to provide generality and to show how the willingness to pay for old-growth timber will vary as you change assumptions and will vary compared to the stumpage value. In reality, buyers will calculate net present values directly from projected after-tax cash flows, rather than as a percentage of current stumpage value. And annual harvests don't have to be equal, as in my simple examples. These net present value approaches have long been used to value mineral and petroleum investments, which are conceptually the same as old-growth timber (Stermole, 1984). Both are resources without any appreciable physical growth: that's why old-growth timber harvesting is sometimes

[3]An extreme example of valuing large, mature timber reserves is Clawson's (1976) attempt to value the entire U.S. Forest Service inventory of sawtimber (over 11 inches in diameter on the Pacific Coast and over 9 inches elsewhere). Since most of this massive volume—1,021 billion board feet in 1970—was growing slowly and could not be liquidated immediately, he estimated market value as 50 percent of the current stumpage value, thus reflecting the above valuation principles.

called "timber mining." The key point here is that the willingness to pay for any resource to be depleted over many years will not usually be the same as the current liquidation value—say, stumpage price for timber, or wellhead price for oil. But the above examples aren't designed for rapidly growing, young timber, which is covered next.

Young Timber Valuation As compared to the case of old-growth timber, it's much more likely that net present values of rapidly growing timber will exceed stumpage values, as shown by the dashed curve in Figure 11-1. So if you have a collection of various aged thrifty young stands to be cut at different times, the timber's net present value is likely to exceed stumpage value. The market confirms this: prices paid for large tracts of young southern pine have often been 5 to 25 percent above current stumpage value when buyers don't harvest immediately. Similar results have been noted in young Douglas-fir sales.[4]

But there are exceptions: when real interest rates are high and buyers are pessimistic about future stumpage prices, even young timber can sometimes sell for less than current stumpage prices (Chambers, 1986; Sizemore, 1984). For the buyer, this can reflect the same effects seen in the old-growth cases of Tables 11-2 and 11-3: *the higher the discount rate and the lower the expected future stumpage price, the lower the NPV.* You might ask, "When NPV is lower than the current local stumpage price, why doesn't the seller clear-cut and get more money?" Sometimes sellers lack good price information. In other cases, as with the old-growth example, you couldn't clear-cut if the tract was so large that immediate harvest was impossible. Then some stands' harvests would be postponed beyond the optimal age t, for example, to age t' in Figure 11-1, where you saw that the dotted curve of NPV was below an apparent (but unrealistic) liquidation value. You'd also avoid rapid harvests large enough to depress stumpage prices significantly. Thus, as with old-growth timber, the local stumpage price may not always indicate current clear-cutting value when properties are large.

As a simple example of an entire young growth forest, picture a 9-acre, 9-age-class regulated forest model with the age classes shown in Figure 11-3 (Figure 9-8 shows a similar case with age classes one year apart). This is a simple hypothetical example to demonstrate valuation principles; a managed forest is unlikely to be so small or so exactly organized. Given a 36-year rotation, suppose net after-tax income from harvesting 36-year-old timber is $2,500 every 4 years, in constant dollars. Reforesting the harvested acre costs $80. Let annual revenue minus cost be $-\$1.50$ per acre for a total of $9(-\$1.50) = -\$13.50$ annually on the whole forest. Assume for the moment that the 36-year rotation is optimal for an owner with a 6 percent real discount rate. A buyer with a real 6 percent MAR would compute the following net present value for the whole forest, land and timber combined, assuming constant real income in perpetuity:

[4]See Fraser (1984) and Bennett and Peters (1984). Douglas-fir conclusions were reported to me by the Oregon Department of Revenue.

age 4	age 8	age 12
age 16	age 20	age 24
age 28	age 32	age 36

Cut 36-year-old timber today and every 4 years

FIGURE 11-3
Regulated forest model—9-Acre, 9-Age Class, 36-Year Rotation.

$$\text{NPV} = 2{,}500 - 80 + \frac{2{,}500 - 80}{(1.06)^4 - 1} - \frac{13.50}{0.06} = \$11{,}415 \qquad (11\text{-}6)$$

Assuming 1 acre in each age class, this is the present value of net harvest income minus reforestation cost today and every 4 years, minus the present value of annual costs, using equations (4-9) and (4-8). With, say, 100 acres in each class, NPV would be 100 times as high. With a higher (lower) interest rate, NPV would be lower (higher), although the optimal rotation would change. And different revenue and cost projections would affect NPV.[5]

Suppose the market value of land was $250 for each acre in the 9-acre model. The investor could assign 9(250) = $2,250 to the land. Subtracting this from the forest NPV gives 11,415 − 2,250 = $9,165 for the timber NPV. Alternatively, an investor could compute an NPV for land, using one of the WPL equations from Chapter 7.

A key point to remember in computing NPVs for forests with delayed harvests, young or old: It's highly unlikely that an investor's timber NPV, for example, the above $9,165, will be the current local stumpage value. If the rotation for young timber is economically optimal, the stumpage value will generally be below the net present value of timber alone, as shown in Figure 11-1. But exceptions exist, as noted above.

Depending on timber value growth rates, projected harvest rates, future costs, and interest rates, *timber NPV may be greater than, less than, or equal to current stumpage value.* Based on projected management plans, a forest buyer, using the MAR, should compute the present value of future revenues minus the present value of future costs for the entire property to get an NPV. If this NPV is the purchase price, and projections are correct, the buyer will earn the MAR on the investment. Actual cases will usually be far more complex than the examples given here, and harvest patterns over time may be irregular, but the same general principles apply.[6] Also note that the harvest plan will affect the NPV, so that

[5]It's an oversimplification to assume the $2,500 net harvest revenue to be after income taxes. In reality, to figure income taxes, the tax-deductible "depletion unit" (see "Depletion" in Chapter 9) should be the timber purchase cost per unit volume (or NPV in this case), which in turn depends on the future tax. To be really precise you need an iterative procedure that keeps using the timber NPV or purchase cost as the total basis until the basis equals the purchase cost.

[6]Although the stand and forest valuations have assumed constant real harvest values, you could also incorporate different rates of change using equations from Appendix 4B.

there's some optimal plan that maximizes NPV. Such issues are usually covered in forest management courses.

Remember from Chapter 4 that not all investors make discounted cash flow calculations exactly following the present value equations referred to here. Many use a seat-of-the-pants discounting where the current valuation is intuitively less than the sum of future cash flows. The equations model such valuations. But sophisticated investors do use present value equations.

Impact of Loans on Property Valuation

Suppose an investor plans to borrow part of the purchase price of a property being valued. How might that affect the NPV? For most loans in the United States, payments stay the same regardless of the inflation rate. Thus the payments are fixed in nominal terms, and you should use a nominal interest rate to compute the present value of loan payments or to solve for the loan payment (see Chapter 5). For example, suppose a bank lends \$100,000 to the Woods Timber Company toward a \$300,000 timberland purchase. The purchase is said to be **leveraged.** Let loan payments be in equal annual amounts for 10 years. Suppose the bank wishes to earn a 5 percent real return (3 percent risk-free plus a 2 percent risk premium due to the risk of default). If the bank projects an average inflation rate of 6 percent over the loan period, it should charge a nominal rate of $(1.05)(1.06) - 1 = 0.113$, or 11.3 percent. Woods' annual loan payment is computed with the capital recovery multiplier [equation (4-12)] and the nominal interest rate, as shown in Chapter 5:

$$\text{Annual loan payment} = 100{,}000\left\{0.113/[1 - (1.113)^{-10}]\right\} = \$17{,}194.31/\text{year}$$

In figuring a net present value of the timberland purchase, Woods should enter the loan amount as a revenue and enter the payments as costs, but a very important warning holds: *The borrower's present value of a fixed-payment loan should be calculated with a **risk-free** interest rate, separately from present value of the risky cash flows.* From the borrower's view, loan cash flows are perfectly certain: Woods knows that the \$100,000 income will occur, and the firm has signed a contract to make payments. The assumed risk-free real rate of 3 percent translates to a nominal risk-free rate of $(1.03)(1.06) - 1 = 0.0918$, or 9.18 percent, given the 6 percent projected inflation rate. From Woods' view, the loan adds the following present value to the timberland purchase, computing with the nominal risk-free discount rate, since the payments are in nominal terms (current dollars):

$$\begin{aligned}\text{Wood's NPV of loan} &= \text{loan amount minus present value of payments} \\ &\quad\ \text{calculated with equation (4-10)} \\ &= 100{,}000 - 17{,}194.31\left[\frac{1-(1.0918)^{-10}}{0.0918}\right] \qquad (11\text{-}7) \\ &= 100{,}000 - 109{,}478.28 = -\$9{,}478.28\end{aligned}$$

Although the NPV to Woods is negative before taxes, the firm can save on its income taxes by deducting loan interest. Thus the net impact of the loan may not be negative. Another twist is that borrowing money makes the timberland venture more risky to Woods, since there's now the obligation to pay off the loan, no matter what the income from the tract will be, thus increasing the risk of bankruptcy. That could boost the appropriate discount rate for the tract's other cash flows and reduce NPV to the buyer.

The literature is rife with disagreement about the effect of debt on a firm's value.[7] The key point to remember here is: *When computing the present value of a loan, the borrower should not use the risk-adjusted discount rate (higher than the risk-free rate).* As you saw in Chapter 10, for risky revenues, we increase the discount rate above the risk-free rate to decrease present value, thus reflecting the disutility from revenue **variance.** But there's no reason for the **borrower** to reduce the present value of loan payments by increasing the discount rate, although it's commonly done.[8] This overstates NPV by understating the present value of loan payments. For example, suppose Woods' nominal MAR (including risk) for the above timberland purchase is 14 percent, and that rate is used in calculating the present value of the loan, according to equation (11-7):

$$\text{Overstated loan present value} = 100{,}000 - (17{,}194.31)[1 - (1.14)^{-10}]/0.14 = \$10{,}312.49$$

Since the borrower's correct loan present value is minus $9,478.28, the above value overstates the loan NPV by the difference between the two values, or $19,790.77, ignoring tax impacts. *Such errors can cause overbidding for properties and overvaluation of leveraged assets in general.* Investors who overpay for assets will not earn their desired rate of return, adjusted for risk.

Keep an Eye on the Market Value

Suppose you own a forest or any other asset and, *including all projected values to you,* you compute an NPV valuation that is below what you could get for it on the open market. In that case, selling could be attractive. This often happens on timberlands, and that's why many forest products firms have real estate divisions in charge of selling property for which someone is willing to pay much more than it's worth in timber production. But this can be a two-edged sword if property sold for residential or recreation use is too close to company logging, which residents may wish to shut down—or if such sales create problems with fire danger

---

[7] See Modigliani and Miller (1958 and 1963).

[8] It's incorrect to assume that, every year, income will offset the borrower's loan payments, and to just discount this net revenue with the risk-adjusted rate (RADR). That would be the same as discounting costs and revenues separately with the RADR. But that view ignores the fact that loan payments are completely independent of revenues. Whether income is high, low, zero, or negative, loan payments are always due (unless a borrower declares bankruptcy). Discounting loan payments with a risk-free rate reflects this certainty.

or liability on company lands. So a firm's lands sold for nontimber uses must be strategically located.

APPRAISING MARKET VALUE

So far we've considered the procedure of finding values to owners or potential buyers—the process of *valuation.* Now let's look at the idea of market value and how to find it, a procedure called *appraisal.* Market value of any asset is its most likely sale price, given a willing buyer and a willing seller, both well informed about values of similar assets.[9] Emphasis is on *most likely;* market value isn't the highest or lowest possible price. Let's examine three appraisal methods, or procedures for finding market value of assets: comparable sales, capitalized income, and replacement cost.

Appraisal by Comparable Sales

Consider the prices per acre paid for tracts of bare forestland of a given size, site quality, location, and slope, all sold in the same year, and **comparable** in all respects, as plotted in Figure 11-4. For that year, the market value of such land is the average of these values, as shown by the dashed line in Figure 11-4. Some sales prices were above the average, and some below; the market value is the

[9] You may sometimes see timber asset values listed in a corporation's balance sheet. By accounting convention, these are usually listed at original purchase cost, also known as **book value,** and may thus be far below market value.

FIGURE 11-4
Prices per acre for comparable bare forest land sales, similar sizes, locations, qualities, and time period (hypothetical data).

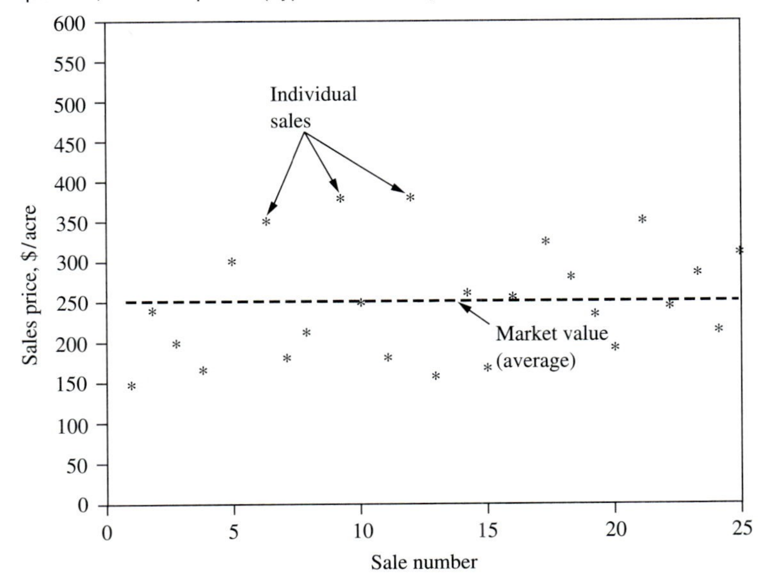

most likely price. Each point in Figure 11-4 results from buyers and sellers making their own valuations and bargaining to reach the selling price. If everyone used net present value to determine bid and asking prices, had perfect information and the same view of the future, and used the same discount rate, all points in Figure 11-4 would be at the same value on the dashed line. In other words, individuals' valuations would be market value. But in reality, sales of similar properties generally yield different values, because the market isn't perfect and properties aren't usually identical.

In principle, the comparable sales approach of Figure 11-4 is the best appraisal method when data are available. It is widely used for appraising buildings of a given type. But with forest properties, we often don't have a large enough sample of truly comparable sales to get a valid average price. For example, rarely is there enough data to construct a bare land price average as shown in Figure 11-4. Most forestland is sold with some timber stocking, or bare land is sold together with other stocked acres. The idea of Figure 11-4 also applies to comparable sales of stumpage, immature timber, and forested properties in general. However, for forests to be truly comparable, they need to be in the same time period and have the same site quality, acreage, timber species and age-class composition, timber quality, distance to market, slope, roading, and other features. That doesn't often happen. So, the problem with averaging sale prices of stumpage or forests is that you often have no sales, or only a few sales of comparable property types, and the spread in prices can be large. Thus, an average price may not be a good estimate of market value for all forest properties of one type.

Appraising Stumpage Price by Comparable Sales Suppose you wish to find the market value per unit volume of your stumpage for immediate harvesting. You'd search for recent sales of similar timber harvested soon after purchase. Ideally, these sales should be comparable to your timber in all respects—same species, quality, average diameter, product mix (e.g., pulpwood, sawlogs, poles, peelers, etc.), terrain, date of sale, distance from mills, road building and logging costs, log rule used, type of harvest (clear-cut or partial harvest), size of sale, terms of sale [e.g., cash up front, pay as cut over several years, who pays taxes (if any) on harvest]—anything that could affect a buyer's willingness to pay for stumpage. If sales were all truly comparable, you'd average them to find market value of your stumpage. That's not saying you'd receive the computed market price when selling your timber for cutting, but it would be the most likely price.

If no truly comparable sales were available, before averaging stumpage prices on timber sales, you could adjust the price on any sale not comparable. For example, if your property was steep, stumpage sale prices on flat land would be adjusted downward by the amount that steepness would increase logging costs [see equation (11-1)]. Or stumpage sales with a smaller percentage of high value poles, compared to your property, would need their prices increased appropriately.

More sophisticated stumpage appraisals use **regression analysis** of a *sample* (or group) of tracts already sold in a common time period and derive an equation that can predict the market price of stumpage not yet sold. The following example of such an equation expresses stumpage price (the dependent variable) as a function of several *predictor variables* (independent variables likely to affect stumpage price), only for sawtimber of one species:

$$\begin{aligned}\text{Market price of stumpage per cubic foot}\\ &= c_0 + c_1(\text{total ft}^3 \text{ cut}) + c_2(\text{acres in sale})\\ &\quad + c_3(\text{average tree DBH}) + c_4(\text{wholesale lumber price})\\ &\quad + c_5(\text{miles of road built}) + c_6(\text{miles to nearest log buyer})\\ &\quad + c_7(\text{average \% slope on logged area})\end{aligned} \tag{11-8}$$

The c_j's are coefficients that are estimated with regression analysis.[10] You need a large enough sample of tracts recently sold as stumpage, with measures of the stumpage price and predictor variables for each sale. All sales in this sample should be of a given class, for example, from the same time period and region, and with certain species and cutting type (e.g., clear-cut or thinning). The less tightly defined your class, the more predictor variables you need and the larger your sample size needs to be. Suppose the estimated equation and coefficients were statistically significant. You could then estimate the market value of unsold stumpage on a forest of the same class and with variable values falling within the range of values in the sample. You'd enter the unsold tract's variable values into the estimated equation and solve for market price of stumpage per cubic foot. Equation (11-8) is just an example; appraisers need to test which variables from the available data can best explain stumpage price.

Appraisers and statisticians often work together on regressions. The statistician can advise on the best functional form for the equation—it may not be strictly linear as above—and help with questions about sample size, tests of significance, goodness of equation fit, etc. The U.S. Forest Service uses regression analysis to appraise many of its stumpage sales. Assessors often use the approach to find market values of real estate for property tax purposes.

A good starting point in stumpage appraisal is to consult local published stumpage price reports about which state extension and forestry departments usually have information. For a list of several stumpage and delivered log price reports in the United States, see Bullard and Straka (1993). These reports are only averages and usually need to be adjusted for individual cases.

Appraising Other Forest Components by Comparable Sales When trying to find market prices of other forest components such as bare land or values of different age classes of immature timber, comparable sales are often hard to find. For example, forestland is usually sold together with trees, so that comparable sales of bare land alone are rare. The same is true for entire properties with

[10] I assume you've learned about regression analysis elsewhere. If not, think of starting with a matrix of data containing the stumpage prices and all the independent variables for all the stumpage sales in the sample. Regression computer programs can, from this data set, estimate the coefficients in equation (11-8) such that the following occurs: If, for any of the sales in the sample, you enter the values for the independent variables and solve for the stumpage price, your estimated stumpage prices will, on average, come as close as possible to the actual values.

unique combinations of features such as timber age classes and species, volumes, location, and land qualities. In spite of problems with finding truly comparable forest sales, you can often find sales values per acre for separate components, such as similarly located land of one site quality with a given volume of 25-year-old Douglas-fir. Comparable sales values for different component values can then be added to appraise the value of an entire property, if component sizes are similar.

Theoretically, the regression approach could be used to find average values per acre for entire forested properties, but for any time period and class of sale, there usually aren't enough sales or enough detailed data on each sale. You'd need information such as number of acres of each site quality; acres and volumes by species, ages, and products; miles of road; and miles to nearest log buyer.

When many truly comparable sales of properties aren't available, two other appraisal approaches are used to supplement sales evidence: the **capitalized** income approach and the **replacement cost** approach.

Appraisal by Capitalized Income

You can appraise an income-producing property by calculating the net present value (NPV) of its most likely future cash flows. This approach is useful if the main reason for purchase is the expected income, as is the case for most business properties. This appraisal procedure is called **income appraisal,** the capitalized income method, or **income approach.** This is arithmetically the same NPV procedure one investor uses for finding bid-prices or asking prices for properties, as discussed above under valuation. *The difference is that an income appraisal seeking market value should compute the* ***most likely*** *sale price, while an individual's NPV valuation uses inputs* ***unique to one investor.*** *The two NPVs can often be quite different.* Because the NPV equations are similar, here I'll discuss only the procedure for finding the most likely inputs and not repeat the NPV computations. The danger with income appraisal is, as shown in Table 11-3, that NPVs can vary greatly, depending on input assumptions such as the discount rate and estimated cash flows.

With income appraisal, the appraiser's job is to find the average investor's most likely inputs into an NPV calculation for a given property and to compute the NPV. With properties like apartment houses, such an NPV can often approximate market value because one can fairly closely estimate rental income, costs, and salvage value over an estimated building life and discount these with typical **risk-adjusted discount rates** found in the rental building industry. Not so for forests, most of which differ substantially. And on each forest, it's hard to estimate the most likely variables such as harvest pattern over time, projected stumpage prices, and buyers' discount rates. It's probably best not to try and simulate NPV impacts of borrowing, since they're ambiguous and highly variable (see the above section on the impact of loans). The section on "Appraisal by Comparable Sales" noted the conditions under which investors' NPVs would be market value. (Can you name them?) Since those conditions don't usually occur, we find endless arguments and confusion about the market value of properties when enough truly comparable sales aren't available.

Despite the problems with forest income appraisal, it's sometimes necessary in the absence of comparable sales. Examples are cases when governments take large old-growth tracts from private owners for parks, and one needs to determine market value to compensate the owner. When no comparable sales are available, appraisers can make their best estimates of future harvest income and reasonable discount rates and compute net present value as an estimate of market value.

To project harvest ages for different sites and species in an income appraisal, you can observe average forestry practices in the region. Local yield tables or simulators can project average harvest volumes, with reasonable reductions made to simulate expected losses. Future stumpage prices can be the current **trend-line** prices increased by annual real percentage rates, as published in projections made by organizations like the U.S. Forest Service. For example, in the late 1980s, the Forest Service was projecting that southern pine sawtimber stumpage prices would increase by about 1.5 percent per year through the year 2020. In that case, a trend-line sawtimber stumpage price of, say, \$130/MBF could be projected for a harvest in 20 years as $130(1.015)^{20} = \$175$, in constant dollars.

More sophisticated stumpage price projections are based on estimated regression equations expressing a particular regional stumpage price as a function of predictor variables like population, housing starts, gross domestic product, number of households, regional timber inventories, and certain forest product output levels. The equation form can be analogous to equation (11-8), but in this case the data points would be for separate years, say, 1965 to 1995. To the extent that these variables can be predicted for any future date, the equation will yield an estimated stumpage price for that date (see Pinkowski et al., 1994).

One of the most controversial income appraisal inputs is the **discount rate.** One can interview forest buyers to find typical discount rates used, but investors are often tight-lipped about such things. And such information by itself can be misleading: one investor may conservatively project no stumpage price-increase and use a low-risk discount rate. Another may project optimistically high prices but use a larger discount rate to reflect the resulting risk. Other approaches are to add a 2 or 3 percentage-point risk premium to average real, risk-free, long-term government bond interest rates or to use moderate-risk Baa average real corporate bond rates of return.

A simple form of income appraisal is the conversion return approach of equation (11-1) to find stumpage value for immediate cutting purposes. The "future" income occurs now, so there's no discounting. The only difference for *appraisal* is that you'd use average expected inputs, rather than those used in an individual's *valuation.* Suppose you owned stumpage and wanted to know its market value, in the absence of comparable sales. You could estimate your stumpage's probable sale price by calculating the conversion return with equation (11-1), using the most likely inputs for your locale. The outcome would be sensitive to assumptions made about inputs. If delivered log prices for your forest's log types were available, you could shorten equation (11-1), so that estimated stumpage price would be a delivered log price minus estimated logging and hauling costs, including profits. However, depending on the degree of competition for stumpage in your area, actual bid prices could be higher or lower than the conversion return.

Income valuation is useless when non-income benefits are a property's major output. For example, when most of a forest property's value stems from the view and recreational benefits to an owner, the present value of potential harvest income isn't likely to indicate the market value. In such cases, appraisers must refer to sale values of other recreational and view properties. Also, nontimber incomes need to be considered, where relevant.

Derived Capitalization Rates One way to make income appraisals more reliable is to derive the discount rate from past property sales. Business buildings provide the simplest example. From a given type of building, say, warehouses or apartment buildings, comparable in all respects, you collect data on recent sales prices, flows of income and costs, and expected lives and salvage values. For each property sold, enter the projected cash flows in an NPV equation (present value of revenues minus present value of costs) with the discount rate unknown. For each property, starting with a low discount rate (or **capitalization rate**), keep calculating NPVs with higher discount rates until the calculated NPV equals the property bid price. Computer programs can do this quickly. The discount rate that brings about the equality is the derived discount rate for that sale. Figure 11-5 graphs the concept. This is the same procedure used for finding an **internal rate of return,** as explained in Chapter 6. An average of many such discount rates for a property class is called the **derived capitalization rate** for that class. Some appraisers call this the "cap rate." *A derived capitalization rate is the discount rate you estimate the average buyer would have used in computing the price*

FIGURE 11-5
Derived capitalization rate for one apartment building.

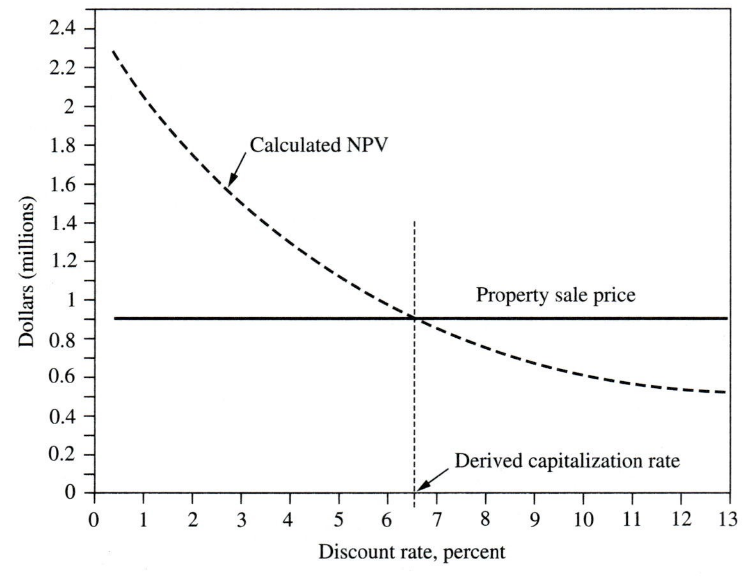

paid for a property. Thus, such derived capitalization rates could be used in an income appraisal of another similar property.

Consider the apartment house graphed in Figure 11-5. Annual real net rental income is $65,000 for the building, which sold for $900,690 last year. Based on an average building life of 50 years, this 15-year-old building is projected to have a salvage value of $100,000 in 35 years. All values include land and are in **constant dollars.** Setting the NPV of income equal to the sale price, and letting the unknown real capitalization rate be r:

$$65{,}000\left[\frac{1-(1+r)^{-35}}{r}\right]+\frac{\$100{,}000}{(1+r)^{35}}=\$900{,}690 \qquad (11\text{-}9)$$

The first term on the left is the present value of annual net income [equation (4-10)]; the second term is the present value of the salvage value. *Iteratively solving equation (11-9) for r, the derived capitalization rate is 0.065, or 6.5 percent.* You would average similarly derived capitalization rates for many properties of the same class to arrive at a derived capitalization rate suitable for income appraisals of other properties in that class.

With the derived capitalization rate approach, it's important to use exactly the same income projection techniques for all sample cases and valuations. For example, if before-tax income is used in all capitalization rate derivations, the discount rate is a before-tax rate and should be used only to value before-tax income. You must be similarly consistent if you choose after-tax income, based on some estimated tax position of the average buyer. Or if *real* income, say, rental per square foot, is projected to increase or decrease at, say, 2 percent annually, you must make that assumption on all properties, both on historical sales and on properties being appraised with the income approach.

A problem with deriving forest capitalization rates is that it's often harder to predict cash flows from forests than from many other business properties.[11] However, in regions where typical harvest ages are predictable, the approach has promise.

The Relevance of Payoff Period One of the major problems with the income approach to appraisal is that long payoff periods make the appraisal highly sensitive to the discount rate chosen. Consider properties where we expect a single net income V_n in year n. With a discount rate of r, the income appraisal is NPV $= V_n/(1 + r)^n$. You can see that as the exponent n gets large, the level of r can affect NPV radically. For example, suppose a property is to be sold for

[11] In deriving a capitalization rate, it's best to ignore cash flows related to borrowing. For example, if you ignore debt in your sample sales, and the average debt in the relevant market increases (decreases) bid-prices of properties, the derived discount rate will be lower (higher), thus automatically reflecting the impact of debt when you use this derived rate on a new income appraisal that ignores debt. But you must be consistent, omitting loan amounts and payments in all sample sales and all properties being appraised.

\$5,000 in 1 year. At 5 percent interest its NPV is $5{,}000/(1.05) = \$4{,}762$. At 10 percent interest the value is $5.000/(1.1) = \$4{,}545$, or only 4.6 percent lower. But suppose you want to value bare forestland yielding a harvest and land sale of \$5,000 in 40 years. Now the two interest rates yield NPVs of $5{,}000/(1.05)^{40} = \$710$ and $5{,}000/(1.10)^{40} = \$111$. Choosing 10 percent interest drops the NPV by 84 percent! Add to this the problem of great uncertainty about harvest values far in the future. Thus, income appraisals of bare forestland and young timber aren't likely to be very close to market values unless you can carefully apply the above-mentioned derived capitalization rate procedure.

Appraisal by Replacement Cost

Applying the replacement cost approach to appraising buildings, you start with what it would cost to build a new structure and then subtract for depreciation or age and wear of the property. The cost approach has limited use in forest appraisal except for recently afforested or purchased timber. For example, if typical site preparation and planting costs for one-year-old plantations of a given species and location are \$93 per acre, the sale prices of such plantations would tend to be \$93 per acre, excluding land. The emphasis is on *typical* expected costs. Thus, you should include the probability of seedling mortality. For example, the above case might involve typical initial establishment costs of \$80/acre with the expectation that 30 percent of the time you'd have to replace dead seedlings one year later for \$22. The accumulated **expected cost** for such one-year-old trees would be $80(1.08) + 0.30(22) = \$93$/acre, assuming the initial cost is accumulated at 8 percent interest. In theory, if the initial investment in land and planting exactly equaled the NPV of future net revenue, you could accumulate all initial and later expenses forward in time at a *nominal* market interest rate, to arrive at a cost appraisal of the forest at some later date. But the further through time you accumulate past costs, the less likely they are to equal market value or NPV at a later date, since the future rarely unfolds the way we originally think it will.

To show how a forest cost appraisal could theoretically reproduce an income appraisal or NPV, consider the \$822.47/acre NPV for a 12-year-old loblolly pine pulpwood stand from equation (11-3), assuming that average market inputs were used. Using the same inputs (from Table 7-3), you can calculate an original owner's age zero NPV immediately after planting as \$422.53/acre, using equation (11-2). Assume for the moment that this \$422.53 was the average price paid for such newly reforested land. For simplicity, let inflation be zero. In reality you should use a nominal interest rate that incorporates the inflation rate over the period in question. Accumulating forward this original cost plus the \$4 net annual costs at the same 5 percent real discount rate for 10 years, you can exactly match the age 12 NPV appraisal:

$$\text{Accumulated costs at age 12} = 422.53(1.05)^{12} + 4\,\frac{(1.05)^{12} - 1}{0.05} = \$822.47 \qquad (11\text{-}10)$$

The first term is the accumulated value of land and planting cost, and the second term is the accumulated value of the annual costs (carrying to the nearest penny only to show the equivalence). You'll note below that this is a highly unrealistic example, but to see why, let's continue. You can view equation (11-10) as a seller's asking price, with all costs accumulated as *positive* amounts, since the seller wishes to recover costs with interest. Conversely, if revenues exceed costs in any year, such revenues are accumulated as *negative* amounts with interest, since the owner, having already received them, should no longer charge for the full accumulated costs. When appraising forests with the income and cost approaches, the guidelines are as follows: *When discounting (using income appraisal), enter revenues as plus and costs as minus. When compounding historical cost (the cost approach), enter costs as plus and revenues as minus.* With cost appraisal, you can think of compounded costs offset by revenues as "net costs." But it's important that these costs be typical for the market in question and not unique to one owner. For example, if one owner atypically had three regeneration failures and reforested four times, those total reforestation costs should not be compounded forward in a cost appraisal: only typical expected costs and revenues should be used.

In the above example, the age-12 cost appraisal exactly equals the income appraisal because the seller's age-0 NPV was based on the same future assumed for the buyer at age 12, and both used the same interest rate. And you could show this equality for any timber age, if you followed the above procedure. Thus, the equivalence of cost and income appraisals depends on buyers and sellers using the same interest rate and having perfect knowledge of the same unchanging future, which, of course, doesn't really happen. For instance, if projected stumpage prices in the foregoing example later became much higher (lower) than perceived at age 0, the forest market value at age 12 could be much higher (lower) than the above cost appraisal. Furthermore, the cost of land and afforestation in year 0 may not equal NPV computed with equation (11-2). Since market value is based on the present value of *future* net income, the further you get from afforestation date, the less original cost has to do with market value. This may seem unjust to you if you own timber whose accumulated cost with interest over many years is, say, $30,000, and no one is willing to pay over $20,000 for it. On the other hand, someone may offer $50,000. *Buyers and the market look forward, not back.* This is why economists speak of the *irrelevance of sunk costs.* Past or **sunk costs** are water over the dam and will affect current asset value only to the extent that they influence future returns.

But cost appraisal can be useful if trees are only a few years from afforestation date. There's likely to be much less argument about compounding typical planting costs, other net costs, and land value forward for, say, three to five years than about discounting distant-future harvest and land values. Also, you don't necessarily have to start with newly reforested land. If the appraised forest was purchased as standing timber for a reasonable market price a few years ago, that value could be accumulated forward in the above manner to arrive at cost appraisal. The problem is that we often don't know what a "reasonable" market value is for immature timber, which is why we're trying to simulate market value in the first place.

You could assume the cash flows and interest rates as being before taxes in the above examples. But you can also compute net present values after income taxes if you can estimate the most likely income tax rate in the market. If correctly done, you ought to get a similar net present value, whether you discount before-tax cash flows with a before-tax interest rate or discount lower after-tax cash flows with a lower after-tax interest rate.[12]

APPRAISING DAMAGE COSTS

Sometimes forests are damaged by things like storms, accidental herbicide drift, or fire. In cases where compensation may be available, appraisers try to find the value of damages. The theory is simple: *The value of damage is the market value before the damage, minus the market value after the damage.* In practice, you may not have market values, and almost certainly not for the damaged forest. Thus, you could use NPVs, which are subject to the problems discussed above: NPVs are highly sensitive to input assumptions. As an example, suppose the above-mentioned 12-year-old loblolly pine stand was killed by an accidental herbicide drift. What should the sprayer's insurance company pay? If available, average sale values for similar 12-year-old pine with land would be the pre-damage value. If not, we could use an NPV such as the $822/acre calculated in equation (11-3). After the damage, suppose it would cost $60/acre to remove the dead trees. Assuming identical bare land sold for $300/acre (ready to plant), the value of the damaged forest would be 300 − 60 = $240—the amount buyers would be willing to pay to use the land for planting trees. Thus, assuming an $822 market value before damage, and following the above guideline, the damage value is 822 − 240 = $582. (Also see the end of Chapter 6.)

Alternatively, you could start with accumulated net costs over the last 12 years, including land and reforestation. For the above example, that was also $822/acre, as calculated under "Appraisal by Replacement Cost." From that, subtract the $240 value per damaged acre to arrive at the same damage value. But remember that the cost approach to pre-damage value becomes increasingly questionable the further you get from afforestation date or timber purchase date. Although this is a very specific damage example, you can apply the same general principles to any case of damage, old or young timber, partially damaged or fully destroyed.

NONTIMBER INFLUENCES ON FOREST MARKET VALUES

When nontimber benefits are predictable revenues such as hunting leases, just build them into the net present value equations when computing investors'

[12]With after-tax calculations, it may be impossible to have a cost appraisal equal an income appraisal, because income taxes affect the seller and buyer differently (Klemperer, 1989).

valuations or income appraisals. Unpriced values such as aesthetic benefits are not so easy to model. For example, when people buy forestlands for personal recreation benefits, they'll often pay more than the present value of potential timber harvesting income. So such values are reflected in a property's market price when they accrue to owners willing to pay for them. Other forests, just as beautiful but bought by timber companies, may not sell for more than timber NPVs if the nontimber values don't accrue to the owners or cannot be sold. Why are some forests being bought at high recreation prices, while other similar tracts are not? Lack of access in part—but in some areas, there aren't nearly enough interested people to buy all the potential recreation sites. Thus, when a few forest tracts sell for high recreation values, appraisers must be careful not to ascribe such values to an entire surrounding area.

When nontimber values can't currently bring income to industrial forest buyers, today's forest bid-price may still reflect such values if buyers project the benefit in a future sale. For example, a forest buyer may project timber income for 20 years and then a land sale at high nonforest values, if development is predicted. Equation (11-2) and the WPL_1 equation in Chapter 7 are NPVs that include future land sales (L_t).

But what about nontimber values that do not accrue to private forest owners? Examples are oxygen production, scenic and open space benefits to travelers, and soil, water, and flood protection benefits to downstream populations. Because such benefits aren't reflected in a forest's market value, private buyers of forestland may not always use it for the socially best use. This is the **market failure** of **unpriced positive side effects** explained in Chapter 3.

KEY POINTS

- A *valuation* is usually a net present value (NPV) calculation, using one investor's unique inputs. Such a value won't necessarily equal market value.
- Valuations are based on NPV equations covered in earlier chapters and can apply to bare land, separate age classes of timber, or collections of stands.
- Valuation of timber to be cut immediately yields the *stumpage value.* Buyers can calculate their willingness to pay for stumpage as the end product price minus all costs of harvesting the standing timber and processing it.
- To value a forest for future cutting, an investor can compute the NPV of all expected cash flows. The faster a forest grows in volume and value, the more likely its NPV is to exceed the stumpage value. When harvest must be postponed significantly, and value growth is slow enough, investors may value timber at less than current local stumpage prices.
- Loans can increase or decrease the amount a buyer is willing to pay for an asset, depending on the loan interest rate. The borrower should discount loan payments with a risk-free interest rate.
- *Appraisal* yields an asset's *market value:* its most likely selling price, given a willing buyer and willing seller.

- The best appraisal method is based on an average of actual sales prices for a given type of asset—the *comparable sales approach.*
- Regression analysis of past stumpage sales can be used to appraise stumpage value.
- When sales evidence is insufficient, appraisers often use the *income method,* which is an NPV based on inputs most likely to be used by the average buyer (rather than one particular investor, as is the case in valuation).
- Appraisals based on *replacement cost* are not generally suitable for forests unless they were planted only a short time ago.
- The value of forest damage is the forest market value before the damage, minus the market value after the damage.
- Nontimber values such as scenic beauty, if they accrue to the buyer, often increase bid-prices for forests. Thus, seemingly unmarketed outputs can actually have market prices per acre. Values accruing to the rest of society, and not to the buyer, will not affect forest market prices.

QUESTIONS AND PROBLEMS

11-1 Define market value.

11-2 What is the difference between valuation and appraisal of a forested property, as defined in this chapter?

11-3 Give reasons why an individual's valuation of a property will not necessarily equal the market value.

11-4 For a forested acre, what is the difference between liquidation value and holding value, as defined here?

11-5 Define stumpage value. Explain the notion of stumpage value as a residual, and give a numerical example. For your example, explain conditions under which the given stumpage price would be a valuation and an appraisal of market value.

11-6 Outline situations where an individual's valuation of standing timber could equal, exceed, or be less than its estimated stumpage value. Discuss likely scenarios for rapidly growing young timber and slowly growing ancient timber.

11-7 What is the comparable sales approach to forest appraisal?

11-8 Explain the income approach to appraisal.

11-9 What is the problem with the cost approach to appraising timber stands? When might the cost approach be valid?

11-10 What is the general procedure for finding the value of damage to a forest?

11-11 Suppose, based on the net present value of expected cash flows, a firm calculates a bid-price for a forest at $700,000, assuming no debt. Now suppose the firm considers borrowing $300,000 toward the purchase. This would be a fixed-payment loan at 9 percent interest with annual payments for a period of 10 years. Suppose the firm's real, risk-free discount rate is 3 percent, and the projected inflation rate is 4 percent. Calculate by how much the loan will increase or decrease the firm's bid-price for the forest.

11-12 Give the conditions under which nontimber values would and would not be likely to affect the market value of a forest.

APPENDIX 11A: Deriving an Old-Growth Valuation Factor Equation[13]

Notation

n = depletion period (number of years over which old-growth timber tract is to be harvested in equal annual volumes or in amounts increasing at the volume growth rate) (start of the depletion period is year 0)

H = year 0 stumpage value of one year's harvest volume using current trend-line stumpage prices

Hn = stumpage value of entire timber tract

V = valuation of timber tract at year 0 (the net present value of future cash flows over the n-year period)

F = valuation factor; the factor by which to multiply stumpage value to arrive at net present value:

$$F = V/Hn \tag{11A-1}$$

t = combined effective federal and state income tax rate, where the state component is multiplied by (1 − federal rate) to reflect deductibility for federal tax purposes

p = percentage of gross stumpage harvest value spent on administration, protection, yield taxes, land taxes, and other costs; expressed to reflect savings resulting from deductibility for income tax calculation

i = nominal discount rate after income taxes (percent/100); includes components for inflation, acceptable real risk-free return, risk allowance, and effective timber property taxes where relevant

m = discounted annual payment multiplier: the present value of a nominal annual payment of $1 for n years, given an i percent discount rate where

$$m = \frac{1 - (1 + i)^{-n}}{i} \tag{11A-2}$$

g_n = nominal annual growth rate (percent/100) of trend line H; includes inflation, real price growth, and volume growth:

$$1 + g_n = (1 + v)(1 + g_r)(1 + f)$$

where v = annual volume growth rate, percent/100
g_r = annual real stumpage price growth rate, percent/100
f = annual inflation rate, percent/100

For consistency, all inflation factors in cash flows and in i must be equal.

d = discount multiplier for n annual payments increasing at rate g_n yearly: the present value of n annual payments starting at $1(1 + g_n) in one year and increasing at g percent annually:[14]

$$d = \frac{1 - \left(\dfrac{1 + i}{1 + g_n}\right)^{-n}}{\left(\dfrac{1 + i}{1 + g_n}\right)} \quad \text{if } i \neq g_n \qquad d = n \quad \text{if } i = g_n \tag{11A-3}$$

If today's stumpage value of one year's harvest is H, and H increases at rate g annually, the present value of n annual harvests (the first occurring next year) is dH.

[13] Adapted from Klemperer (1985), by permission from the Appraisal Institute, Chicago, IL.

[14] The quantity $[(1 + i)/(1 + g_n)] - 1$ is the downward-adjusted interest rate explained in Appendix 4B, here in nominal terms.

Computing Taxable Capital Gain

Capital gains tax rates are applied to harvest income minus the original purchase cost, or **basis,** of harvested timber. The Internal Revenue Code allows several methods for computing this basis or depletion rate per unit volume cut. The depletion rate may be a single weighted average for all timber tracts owned or separate rates for each tract. Either of these rates may be used for separate species groups.

Since it is impossible to predict the buyer's choice of depletion calculation, the following general average approach is taken: the original timber purchase cost FHn is deducted in equal annual amounts (FH per year), assuming roughly equal annual harvests. This allows for full deduction of purchase cost in a manner that closely approximates the range of depletion methods currently allowed. In nominal terms, the annual deductions are equal, regardless of the inflation rate.

Deriving the Valuation Factor

The start of the depletion period is year 0. The year 0 value of future taxable capital gains from harvesting equals the present value of annual gross harvest income (dH) minus the present value of deductions for average purchase cost of harvested timber (mFH, assuming purchase in year 0). Multiplying this by t yields the year 0 value of capital gains taxes:

$$t(dH - mFH) \tag{11A-4}$$

The year 0 value of future net harvest incomes from the tract is the present value of annual gross harvest income minus management cost minus present value of capital gains taxes.

$$V = d(H - pH) - t(dH - mFH) \tag{11A-5}$$

Expanding equation (11A-5), substituting in equation (11A-1), and solving for F:

$$F = \frac{dH - dpH - tdH + tmFH}{Hn}$$

$$FHn - FtmH = H(d - dp - dt)$$

$$F = \frac{H(d - dp - dt)}{H(n - tm)}$$

$$F = \frac{d(1 - p - t)}{n - tm} \tag{11A-6}$$

Since the trend-line stumpage value of the tract in year 0 is Hn, the net present value of the timber is FHn, excluding land, given the assumptions implied in the formula.

Equation (11A-6) is suitable only for cases where annual harvests are equal or increasing at rate v. You'd need different formulas for other harvest patterns.

REFERENCES

Bennett and Peters, Inc. 1984. *Analysis of Large Timberland Transactions in the West Gulf-South Region during the Decade of the 1970s.* Bennett and Peters, Inc. Baton Rouge, LA.

Brigham, E. F., and L. C. Gapenski. 1991. *Financial Management—Theory and Practice* (6th edition). The Dryden Press, Chicago. 995 pp. plus app.

Bullard, S. H., and T. J. Straka. 1993. *Forest Valuation and Investment Analysis.* Starkville, MS: GTR Printing. 70 pp.

Center for Forestry Investment. 1986. *Timberland Marketplace for Buyers, Sellers and Investors and Their Advisors.* Duke University School of Forestry and Environmental Studies, Durham, NC. 303 pp.

Chambers, R. G. 1986. Other financial alternatives concerning timberland investment. In: *Timberland Marketplace for Buyers, Sellers and Investors—Proceedings, 1984 Meetings.* Duke University School of Forestry, Durham, NC. pp. 43–52.

Clawson, M. 1976. *The Economics of National Forest Management.* Resources for the Future working paper EN-6. Washington, DC. 117 pp.

Clephane, T. C. 1978. Ownership of timber: A critical component in industrial success. *Forest Industries.* August. pp. 30–32.

Frazer, E. C. 1984. *Large Timberland Sales in the East Gulf and South Atlantic Areas of the Southern United States during the Period 1976–1980.* F & W Forestry Services, Albany, GA.

Klemperer, W. D. 1977. Forest income valuation revisited. *The Real Estate Appraiser.* 43(1):18–24.

Klemperer, W. D. 1981. Segregating land values from sales of forested properties under even-aged management. *Forest Science.* 27(2):305–315.

Klemperer, W. D. 1985. Effects of inflation and debt on old growth timber appraisal. *The Appraisal Journal.* 53(1):90–114.

Klemperer, W. D. 1987. Valuing young timber scheduled for future harvest. *The Appraisal Journal.* 55(4):535–547.

Klemperer, W. D. 1989. An income tax wedge between buyers' and sellers' values of forests. *Land Economics.* 65(2):146–157.

Modigliani, F., and M. H. Miller. 1958. The cost of capital, corporate finance and the theory of investment. *American Economic Review.* 48(3):261–297.

Modigliani, F., and M. H. Miller. 1963. Corporate income taxes and the cost of capital: A correction. *American Economic Review.* 53(3):433–443.

Pinkowski, R. H., Jr., S. G. Burak, and W. R. Sizemore. 1994. *Stumpage Price Forecasting—An Essential Part of Timberland Appraisal.* Proceedings of Annual Southern Forest Economics Workshop. Savannah, GA. University of Georgia, School of Forest Resources. Athens. pp. 173–194.

Schuster, E. G., and M. J. Niccolucci. 1990. Comparative accuracy of six timber appraisal methods. *The Appraisal Journal.* 58(1):96–108.

Sizemore, W. 1984. Valuation and appraisal. In *Timberland Marketplace for Buyers, Sellers and Investors—Proceedings, 1984 Meetings.* Duke University School of Forestry, Durham, NC. pp. 193–202.

Stermole, F. J. 1984. *Economic Evaluation and Investment Decision Methods.* Investment Evaluations Corporation, Golden, CO. 469 pp.

12

TIMBER DEMAND AND SUPPLY

This chapter applies some of Chapter 2's demand and supply theory to timber. You may wonder, why the gap? The reason is that you can't fully grasp timber supply analysis until you learn about strengths and weaknesses of market economies, interest rates, investment analysis and capital budgeting, inflation, land use economics, taxation, risk, and market value concepts. Now you'll combine these things with basic supply-demand analysis in a timber context.

Forestry's beginning in the United States had its roots in fears of timber famine. As early as 1626, the Plymouth Colony restricted timber exports. In the Province of Massachusetts Bay, the year 1691 saw the British reserving certain trees for their navy's masts. By the late 1800s, worries about forest destruction spawned legislation that later formed the National Forest and National Park Systems.

Concern about timber harvesting rates is in two broad areas: (1) possible damage to nontimber resources like scenic beauty, soil, water, biological diversity, climatic benefits, some types of recreation, fish, and some wildlife species; and (2) assuring future timber supplies. Chapters 3, 14, and 15 cover the first area; this chapter addresses mainly wood outputs, with values based on harvest income. You'll learn about factors that change long- and short-run timber supply and demand. A few sections note how timber and environmental factors interact. Although nontimber benefits are vital and can often exceed wood values, looking first at timber demand and supply will make later sections on multiple-use forestry more meaningful.

In this chapter, **stumpage** refers to standing timber in the forest, and **stumpage price** is the price paid per unit volume for stumpage to be harvested immediately or in the very near future. Table 12-1 lists notation for this chapter.

TABLE 12-1
NOTATION FOR CHAPTER 12

C_y = revenue in year y	Q_L = commodity quantity due to a leftward shift in supply or demand
D = demand, when subscripted with L, R, Lu, Lo, or s, as defined below	Q_R = commodity quantity due to a rightward shift in supply or demand
H_t = clear-cutting revenue at rotation age t	r = real interest rate, percent/100
L_t = market value of land per acre in year t	R_y = revenue in year y
MAR = minimum acceptable rate of return	S = supply, when subscripted with L, R, Lu, Lo, or s, as defined above
MBF = thousand board feet	S^d_{jy} = the jth demand shifter in year y
P_L = commodity price due to a leftward shift in a supply or demand curve	S^s_{jy} = the jth supply shifter in year y
P_{Lo} = price of delivered logs, \$/MBF	t = rotation age, years
P_{Lu} = wholesale lumber price, \$/MBF	V^d_y = stumpage volume demanded (harvested) in year y
P_R = commodity price due to a rightward shift in a supply or demand curve	V^s_y = stumpage volume supplied (harvested) in year y
P_s = stumpage price, \$/MBF	y = a subscript indexing years
P_y = stumpage price in year y	

SHORT-RUN TIMBER DEMAND AND SUPPLY

Timber demand *for a given period refers to stumpage quantities that some group would purchase for harvesting at different stumpage prices.* Graphically, that's a downward-sloping demand curve of stumpage price over quantity demanded, as reviewed in Chapter 2. *In economics, for both demand and supply, the* **short run** *is the period when some inputs are fixed*: for demand, these are inputs like the number of consumers, their incomes and preferences, or advertising. With short-run wood product demand curves for one year, we're holding *all* **shifters** constant and allowing only the product price and quantity consumed to vary.

Timber supply *in a given region and time period alludes to stumpage volumes that forest owners would sell for harvest at different stumpage prices.* Graphically, that's an upward-sloping supply curve of stumpage price over quantity harvested. With short-run stumpage supply curves for any year, we're also holding *all* shifters constant and varying only stumpage price and quantity harvested. In that case, at a given price, new marketable timber could not be created, for example, through supply shifters like afforestation, fertilization, or genetic improvement.

In the **long run,** all inputs are variable, and at any given stumpage price, one can change the amount of harvestable timber as well as the number of consumers and their preferences. Given existing consumers, commercial forests, and technology, let's now see what determines supply of, demand for, and prices of wood products in the short run.

Derived Short-Run Demand for Stumpage

The demand for **stumpage** stems from consumers' demands for wood products like paper, plywood, lumber, poles, composite boards, and all the products made from these. To illustrate, let's start with only one wood product—lumber of one species and quality—sold in competitive markets in one region. Assume that mills are **price takers** selling lumber at one competitive price to wholesalers, that loggers are price takers selling sawlogs to mills, and that landowners are price takers selling standing sawtimber to loggers. For simplicity, let's assume only three *market levels*: (1) wholesale lumber, (2) logs delivered at mills, and (3) stumpage in the forest, as shown in Figure 12-1 for a hypothetical region like a state, for one year. Assume separate businesses: timber growers, loggers, mills, and lumber wholesalers. This diagram could include other market levels like housing, furniture, retail lumber, and roadside logs. *To simplify, let wholesalers represent consumers.* In Figure 12-1, at each market level, the demand and supply curves will remain stable, so there will be no change in shifters like the number of consumers, personal income, production costs, or standing timber inventory. So this is a short-run analysis. For simplicity, wood volumes at all market levels will be in thousands of board feet (MBF) of harvest, log scale, with prices adjusted accordingly. (Due to **overrun**, 1 MBF of standing timber usually yields more than 1 MBF of lumber.)

The upper left panel of Figure 12-1 shows one year's hypothetical regional demand curve for lumber at the wholesale level, aggregating all buyers (as consumers were aggregated to get the market demand curve in Figure 2-1). If

FIGURE 12-1
Deriving short-run regional stumpage demand and end-product supply.

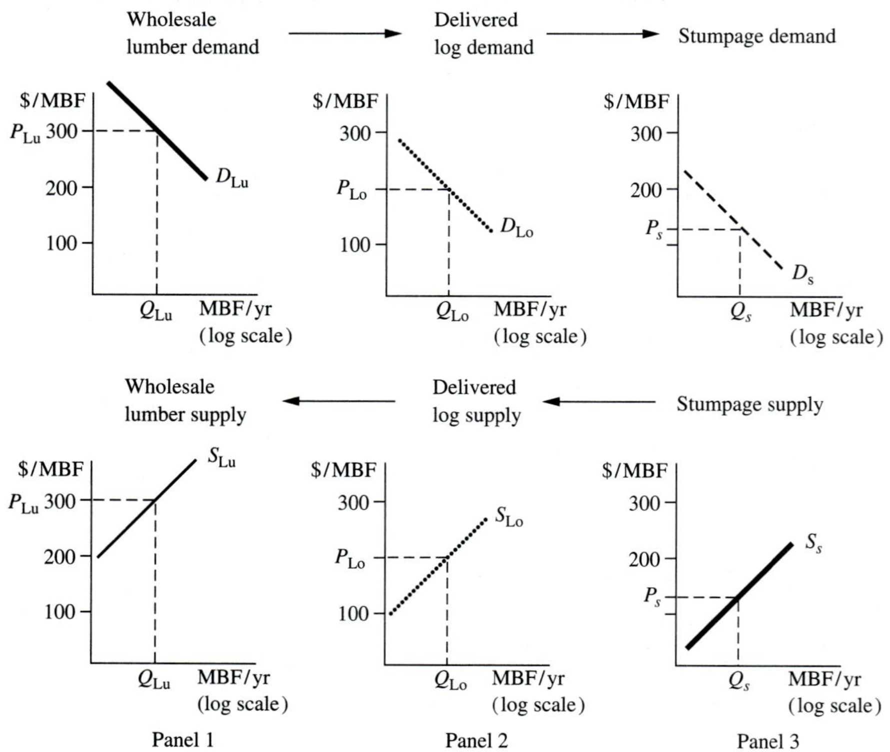

the price is low, consumers (and wholesalers) buy more lumber than if the price is high. Let the current equilibrium lumber price and quantity be P_{Lu} and Q_{Lu}. Using the same numerical example from the last chapter, let $P_{\text{Lu}} = \$300/\text{MBF}$. If sawmills sell lumber to wholesalers for \$300/MBF and incur manufacturing and distribution costs of \$100/MBF including profit, the maximum that mills can afford to pay for delivered logs is $P_{\text{Lo}} = 300 - 100 = \$200/\text{MBF}$. Based on this relationship between lumber and log prices, the upper middle panel of Figure 12-1 traces the mills' demand curve for logs. At any quantity on the horizontal axis, log price is the lumber price minus the manufacturing and distribution costs per MBF (\$100/MBF at current log input Q_{Lo}). *At each production and distribution stage, costs include normal profit, which is an acceptable return to capital.*

If loggers can receive \$200/MBF for logs delivered to the mill, and their logging and hauling costs (including profit) are \$70/MBF, the maximum they can afford to pay landowners for sawtimber stumpage is $P_S = 200 - 70 = \$130/\text{MBF}$. Consider this a regional average stumpage price, since logging and hauling costs will vary depending on the type and location of harvest. At any quantity, stumpage price is the log price minus logging and hauling cost per MBF (\$70/MBF at current harvest Q_s). This repeats the last chapter's example of stumpage price as **conversion return.** Using this relationship between log and stumpage prices, the upper right panel (3) of Figure 12-1 traces the demand curve for stumpage. This curve shows, for the region, how much stumpage will be demanded per year at different average stumpage prices. Since all volumes are in log scale, the assumed equilibrium is $Q_{\text{Lu}} = Q_{\text{Lo}} = Q_s$. The following summarizes from Chapter 11 the maximum willingness to pay for stumpage with our numerical example, as shown in the last row:

Lumber sale price:		\$300/MBF*
Cost of milling and distribution:		−100
Delivered log price at mill:	\$200	
Cost of logging and hauling:		− 70
Residual for stumpage: ("Conversion return")		\$130/MBF

*Volumes are based on log scale, and prices are adjusted to account for "overrun" (or "underrun")—when mills cut more than (or less than) 1,000 board feet of lumber from 1 MBF of logs.

Or, in general, given any end product like plywood or lumber:

$$\begin{aligned}&\text{Buyer's maximum willingness to pay for stumpage} = \\ &\quad (\text{end product price}) - (\text{cost}^1 \text{ of manufacturing and distributing the end product}) \\ &\qquad - (\text{cost}^1 \text{ of logging and hauling logs}) = \text{"conversion return"} \qquad (12\text{-}1)\end{aligned}$$

[1]Including acceptable profits.

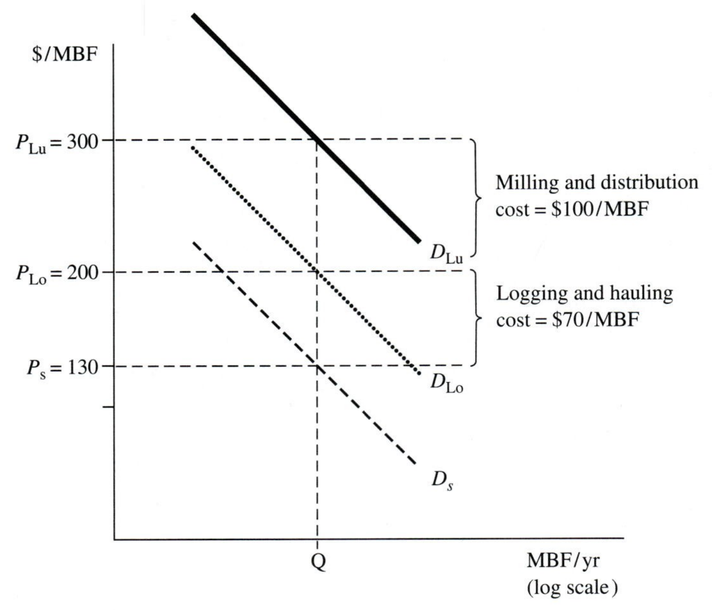

FIGURE 12-2
Regional demand curves for lumber, logs, and stumpage (from Figure 12-1).

Figure 12-2 combines the three demand curves from Figure 12-1. Starting with the lumber demand curve, the vertical distance between the demand curves for lumber and logs is the milling and distribution cost per MBF. The vertical distance between the demand curves for logs and stumpage is the logging and hauling cost per MBF. For simplicity, all processing costs are held constant, so the demand curves are parallel. In reality, such costs per MBF often increase with rising output, in which case the demand curves in Figure 12-2 would diverge. The figure illustrates **derived demand:** *Sawtimber stumpage demand is derived from lumber demand.* But more important, lumber and stumpage demand are derived from consumers' demands for housing and other construction, furniture, shipping containers, and any product or service that uses lumber. And demand for most stumpage is driven by demand for many more wood products in addition to lumber. *It all starts with the consumer*: without final consumer demand, there would be no **willingness to pay** for stumpage. To emphasize that end-product demand is the root of stumpage demand, lumber demand is drawn with a heavy line in Figures 12-1 and 12-2.

Short-Run Supply Curves for Wood Products

In this book, timber supply is the same as stumpage supply and refers to harvest, or *flow supply,* as opposed to timber inventory, or *stock supply.* The lower right corner of Figure 12-1 traces our hypothetical region's stumpage supply curve S_s for one year.

Deriving a forest products supply curve is the mirror image of the demand side. Rather than starting with the end product (lumber, in this case), you start with the resource supply: the regional stumpage supply curve S_s. This is derived from individual forest owners who have different **reservation prices** below which they won't sell stumpage. For example, suppose forest owner Smith won't cut timber unless the price is above $50/MBF, Jones doesn't cut for less than $100/MBF, and others require even more. In most cases, up to a point, the higher the stumpage price rises above the reservation price, the more each forest owner will cut; graphing that relationship gives an owner's timber supply curve. Remember, some owners won't cut for any feasible price because they're mainly interested in aesthetic and recreational values from their forests. Horizontally summing all the individual owners' stumpage supply curves in a region gives the upward-sloping regional stumpage supply curve (following the logic of Figure 2-10). Thus, S_s in the lower right corner of Figure 12-1 shows how much timber per year is offered for sale in the region at different stumpage prices.[2]

Using the same numerical example as above, let the equilibrium stumpage price be $130/MBF at a quantity Q_s in the lower right corner of Figure 12-1. Moving left to panel (2), if loggers incur costs of $70/MBF and have to pay $130/MBF for stumpage, they'll want at least $200/MBF for logs delivered at the mill. By this reasoning, starting at all other points on the stumpage supply curve, you can derive the delivered log supply curve S_{Lo} in panel (2).

Continuing with the same example, if mills have to pay $200/MBF for logs and incur milling and distribution costs of $100/MBF at log input Q_{Lo}, they need at least $300/MBF for lumber delivered to wholesalers. So the lumber supply price at any output tends to be the stumpage price plus the costs of logging, hauling, milling, and distribution (including normal profits). In that way, you can derive the wholesale lumber supply curve in the lower left corner of Figure 12-1. *Thus, end-product supply curves are derived from stumpage supply curves.* Figure 12-3 combines the three supply curves from Figure 12-1. Starting with the stumpage supply curve S_s, the vertical distance between the stumpage and log supply curves is the cost per MBF for logging and hauling. The vertical distance between the log and lumber supply curves (S_{Lo} and S_{Lu}) is the cost of milling and distribution per MBF. Again, with the assumption of constant costs per MBF, the supply curves in Figure 12-3 are parallel. If processing costs rose with increasing output, the supply curves would diverge. To emphasize that stumpage supply is the root of end-product supply, stumpage supply is drawn with a heavy line in Figures 12-1 and 12-3.

The above analysis generates forest products supply curves from an aggregate, regional view. Chapter 2 notes that, from a single buyer's view, under perfect

[2]The above example considers private timber owners. In contrast, U.S. Forest Service stumpage supply curves could sometimes be more elastic in the left portions, due to minimum acceptable bid prices, and more inelastic at the extreme right, due to allowable cut limits. But it's hard to generalize because substantial uncut volumes under contract often exist on public lands (Adams and Haynes, 1989).

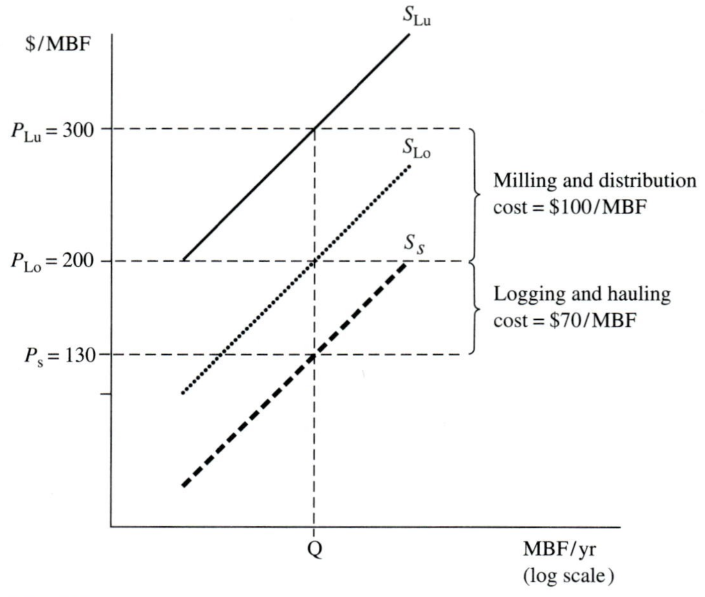

FIGURE 12-3
Regional supply curves for stumpage, logs, and lumber (from Figure 12-1).

competition, a mill would be a price taker and face a horizontal stumpage or log supply curve. But large mills generally face upward-sloping log supply curves, which means they're either **monopsonists** or **oligopsonists** (see Chapter 3). For example, suppose a lumber mill faces the log supply schedule in the first two columns of Table 12-2. If you plotted column (1) (price) over column (2) (quantity), you'd have the positively sloped log supply curve faced by the mill: the more timber the mill wants, the more it must pay per MBF. Suppose the mill currently buys 12,000 MBF of logs per year at $200/MBF, for a total cost of $2.4 million. If the mill wants to increase log purchases to 14,000 MBF, it must offer $250/MBF for a total annual cost of $3.5 million. Note that the added cost per MBF is *not*

TABLE 12-2
HYPOTHETICAL LOG SUPPLY SCHEDULE FACED BY A SAWMILL

Price per MBF paid by mill on all logs of one quality, $	Quantity of delivered logs, MBF/year	Total log cost [column (1) × column (2)], $millions
$100	8,000	0.80
150	10,000	1.50
200	12,000	2.40
250	14,000	3.50
300	16,000	4.80
350	18,000	6.30

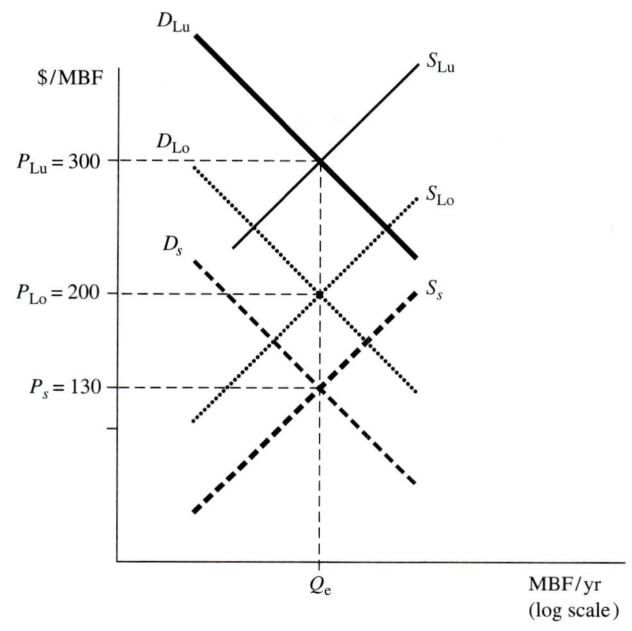

FIGURE 12-4
Short-run regional supply and demand equilibria for lumber, logs, and stumpage (from Figure 12-1).

\$50. To get 2,000 MBF more logs, the firm must pay an added $3.5 - 2.4 =$ \$1.1 million, or a marginal cost of $1{,}100{,}000/2{,}000 =$ \$550/MBF. The added cost is so high because \$250/MBF must be paid not only on the added 2,000 MBF but on the entire input of 14,000 MBF. This example ignores **price discrimination** where a buyer pays different prices to various suppliers.

Combining Market Equilibria

Combining all three regional supply-demand equilibria from Figures 12-2 and 12-3 on one graph, Figure 12-4 shows the above market interactions. Given an equilibrium harvest and product output of Q_e, the stumpage price, delivered log price, and lumber price are P_S, P_{Lo}, and P_{Lu}. Through the mechanisms reviewed in Chapter 2, if any one of the prices is above the equilibrium, a surplus will exist in that market, and sellers will reduce prices until supply equals demand. If prices are below the equilibrium in any of the markets, shortages will exist, and buyers will bid up prices until supply equals demand.

Starting with an equilibrium end-product price (in this case, lumber), Figure 12-4 shows how stumpage price on a given stand is affected by costs of production at all levels. Remember, this is a simple example with only one product. Stumpage prices are often a combined value per unit volume, based on values for several products such as veneer logs, sawlogs, poles, and pulpwood. Or standing timber may be sold at different prices for each product and species. As done above for sawtimber stumpage, you can show how *demand curves for any type of stumpage*—for example, pulpwood, veneer logs, or poles—*are derived from the consumer demand for the end products. If the end-product demand curve shifts, the stumpage demand curve will shift in the same direction.* And likewise, *supply curves of end products* like paper and plywood (or the products made from them) *are derived from stumpage supply curves* (for example, pulpwood or veneer log stumpage). *If the stumpage supply curve shifts, the relevant end-product supply curves will shift in the same direction.*

While you should understand the graphics of supply and demand curves, you should also realize that players in the market are generally not drawing the curves in determining prices and quantities. Prices of, say, lumber, pulpwood, or stumpage are determined by large numbers of buyers and sellers bargaining, wheeling and dealing, cajoling, arguing, and deciding—over the phone, in person, over computer lines, and through price lists and advertisements. The demand and supply curves simply model that procedure. On the other hand, the curves aren't irrelevant; many companies try to estimate the demand curves they face for their outputs and the supply curves they face for their inputs. Such information helps firms know the impacts on total revenues and costs when they change input and output levels. Researchers estimate regional forest products demand and supply functions so that policy analysts can gauge price and quantity effects of proposed actions like new taxes, changes in public harvest levels, private harvest regulations, and new plans for public forest investment or government **subsidies** for private forestry (i.e., government assistance for forest landowners).

It's important to realize that the above log and stumpage price derivations are only general tendencies, based on assumptions of competition. Assuming the short term, the heavy curves of lumber demand and stumpage supply in Figures 12-1 through 12-4 are absolutely fixed, because they depend on the attitudes of a fixed number of consumers and timber growers dealing with unchanged products and resources. But if mills or loggers had higher or lower than normal profits, the lighter drawn supply and demand curves in Figures 12-1 through 12-4 would be positioned differently, because the lumber and log processing costs (shown as \$100 and \$70) would differ.[3] (Recall that these costs include profits.) Thus, higher or lower profits would be shown by redrawing only the lighter curves and changing equilibrium prices and quantities in Figure 12-4. Now let's consider the long run.

LONG-RUN TIMBER SUPPLY

Remember that in the short run, some inputs are fixed, and we can't change the total volume of standing timber. In the examples of Figures 12-3 and 12-4, all

[3]Due to entry and exit of firms, you wouldn't expect unusually high or low profits to persist in the long run under competitive conditions (see Chapter 2). However, in some regions, a mill could earn excess profits if available long-run timber supply wasn't enough to support another efficient-sized mill. In that case, in Figure 12-4, for example, stumpage prices might be less than \$130/MBF. On the other hand, sometimes short-term wood scarcities can cause firms to accept subnormal profits and bid above the competitive equilibrium price for logs or stumpage.

variables are held constant except stumpage price and harvest from commercial forests. In the long run all inputs are variable, including acres of mature timber. From that view, the long run can be one rotation in the future. To increase future acres of financially mature timber, you need to reforest more land today and wait a rotation. For Douglas-fir, that could be 40 or 50 years or more; with southern pine, about 20 to 25 years for pulpwood and 10 to 15 years longer for sawtimber. Amounts of timber on each afforested acre can be increased by genetically improved seedlings and better site preparation. Planning and prediction are very uncertain in forestry because its long run is longer than in almost any other enterprise. Over shorter periods, say, 5 to 15 years, actions like fertilization, improvement cutting, and thinning can increase future timber supply on existing forest acreage.

The above are long-run **biological supply shifters** that change the future volume of timber. Long-run **technological supply shifters** change the amount of salable wood from existing trees. Past examples have been new pulping processes that increase usable wood fiber from pulpwood, the chipping lathe headrig that simultaneously creates chips and cants from small logs, more efficient sawmills that produce less sawdust and take smaller logs, and efficient veneer lathes that can peel logs to very small cores. Researchers have even developed machines that pull trees, taproots and all, like you'd pull carrots, but they're not widely used.

Forest owners' preferences can affect stumpage supply. Over time, if more forest owners prefer partial cutting and longer rotations or no harvesting, less timber is available for cutting at any price. Today, trends in the other direction seem unlikely on nonindustrial private forests (NIPFs) and public lands. But on NIPF lands, these preferences may be temporary: Previously nonharvesting owners often eventually transfer forests to those more willing to cut. However, logging regulations can reduce timber harvests on all forest lands. For example, compared to a no-regulation scenario, Green and Siegel (1994) estimated that by the year 2030, projected forest management regulations would reduce softwood sawtimber harvests by 17 percent in the South, causing stumpage prices to increase by 15 percent.

Figure 12-5 illustrates timber supply shifters. Let D_{2030} and S_{2030} be a region's baseline short-run stumpage demand and supply curves projected for the year 2030, with price and quantity at P and Q. Compared to the baseline, if you project less forest acreage, lower management intensity, or more logging restrictions, the future timber supply curve shifts to the left (S_{2030L}). At any price per cubic foot, less timber will be supplied, thus yielding a higher price and lower output (P_L and Q_L). The opposite occurs with factors shifting supply to the right: Supply in 2030 would be S_{2030R}, with lower price P_R and higher harvest Q_R. Examples of rightward supply shifters are more fertilization and genetic improvement, a projected increase in forest acreage through past reforestation subsidies, or new technology using trees to smaller tops. At any given price, more wood is supplied. Here, demand (D_{2030}) is constant. In reality, leftward and rightward supply shifters occur together with demand-shifts, so end results aren't easy to predict.

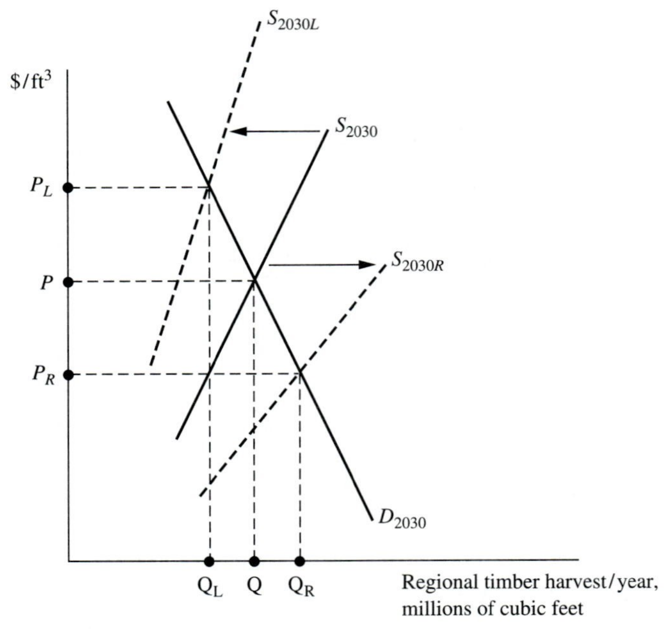

FIGURE 12-5
Regional long-run stumpage supply-demand equilibria with shifting supply.

Some leftward timber supply-shifts can be immediate, for example, cutting restrictions in the Pacific Northwest in the early 1990s to save old-growth forests and spotted owl habitat on public lands. The result is reduced output and higher stumpage prices. Other examples are timber destruction by fires and storms. If timber is partially damaged, supply may be temporarily shifted to the right during salvage harvests, as occurred after Hurricane Hugo in the Southeast in 1989 and after the 1962 Columbus Day storm in the Pacific Northwest.

It's a mistake to assume that wilderness set-asides will always shift the stumpage supply curve to the left. If new wilderness reserves come from areas where timber is economically inaccessible, the stumpage supply curve doesn't shift, because the timber wouldn't be cut anyway. Examples include many wilderness reserves in high elevation areas of the Rockies and Cascades. But if commercially available inventories are reserved, the stumpage supply curve shifts to the left.

Later we'll consider how the market stimulates supply shifters, but first let's look at long-run timber demand.

LONG-RUN TIMBER DEMAND

A short-run regional stumpage demand curve shows the quantity demanded as a function of stumpage price, holding all demand shifters constant, as shown by

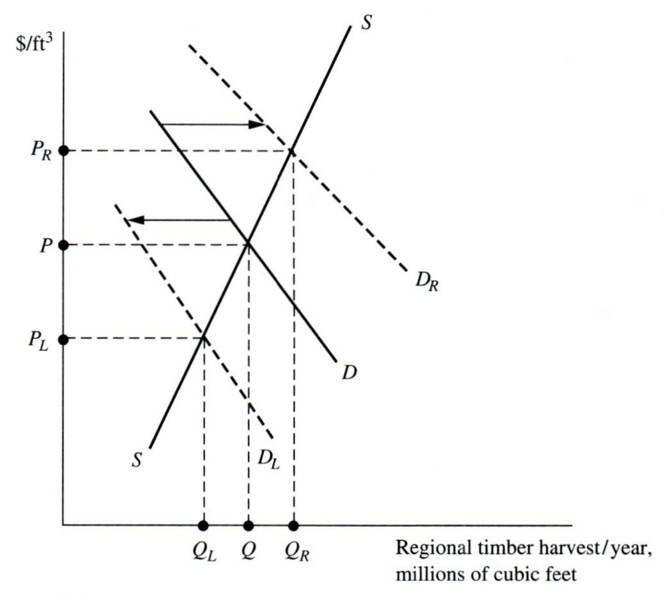

FIGURE 12-6
Regional long-run stumpage supply-demand equilibria with shifting demand.

the demand curve D in Figure 12-6 for one year. Let the stumpage supply curve S remain fixed, with the initial price and quantity at P and Q. In the long run, demand shifters can change, for example, population, real per capita income, consumer preferences, prices of complementary goods or substitutes, and interest rates. Population and income generally increase over time and shift the short-run stumpage demand curve to the right—D_R in Figure 12-6—yielding a higher price and quantity, P_R and Q_R, other things being equal. Increases in timber demand can occur over a much shorter time than most increases in timber supply. Factors like high mortgage interest rates can quickly shift the timber demand curve to the left (D_L), lowering price and quantity to P_L and Q_L in Figure 12-6. That happens because higher interest increases mortgage payments on new houses and depresses housing demand, thus reducing stumpage demand. All stumpage demand-shifts are derived from end-product demand-shifts through the mechanisms shown in Figure 12-2. Rising energy costs could shift timber demand curves to the left by reducing income available for housing and by increasing home heating costs, which prompts purchase of smaller houses. Sometimes a population bulge like the post-World War II baby boom moves through its house-building phase, first exerting a rightward and then a leftward pressure on the timber demand curve.

Both Figures 12-5 and 12-6 show how useful the supply and demand framework is in gauging timber price and quantity effects of policies and events. Now let's see how short-run and long-run supply and demand are interrelated.

LONG-RUN AND SHORT-RUN TIMBER SUPPLY AND DEMAND INTERACTION

Several things complicate short- and long-run timber demand and supply interactions:

- Due to downward-sloping regional stumpage demand curves, faster (slower) harvest today will decrease (increase) current stumpage prices.
- Other things being equal, in a region, the more (less) you cut today, the less (more) is available tomorrow, and the higher (lower) tomorrow's stumpage price is.
- For biological supply shifters to take effect, you need from 5 to 40 years or more.
- Land is continuously shifting into and out of timber production due to changes in prices of wood versus prices of outputs from competing land uses.
- Much timber will grow in the future through natural reproduction, regardless of afforestation investment.
- Log movements from one region to another can vary with things like transportation costs or relative stumpage prices.
- Other factors can shift unpredictably, for example, changes in amounts of timber shipped from Canada and elsewhere, environmental restrictions on harvesting, or sudden changes in demand from other countries.
- The usable timber resource is constantly changing through better utilization practices and changing mixes of species and timber qualities. Over the last century, United States timber quality has declined (e.g., smaller tree diameters, more stem taper, wider growth rings, more knots), but we've increased our ability to process a much wider range of timber types.

Figure 12-7 graphs annual private harvests and stumpage prices over time for the south central United States. From year to year, note how harvest levels rise and fall with stumpage prices. As you'd expect with upward-sloping stumpage supply curves, a higher stumpage price stimulates more cutting. Prices are **nominal** to include any "money illusion" from inflation. (Recall from Chapter 5 that inflation isn't necessarily all illusion: By shifting spending patterns, some buyers can benefit from inflationary income.) In Figure 12-7, in addition to the year-to-year harvest fluctuations, long-run forces of increased planting were also at work, leading to an upward trend in harvests as stumpage prices increased.

Faster cutting with higher stumpage price conflicts with the notion that after a one-time increase in stumpage price, the forest value growth percent and the optimal rotation shouldn't change much (ignoring annual costs). For example, in the forest value growth percent equation (7-10) if land value increases by the same percent as stumpage price, numerator and denominator change by about the same percent, and value growth percent is unchanged. So, in theory, the optimal time to cut is unchanged. But data from all regions show that higher stumpage prices usually stimulate more cutting, perhaps because many owners feel, with good reason, that prices may drop again—an argument often used effectively by

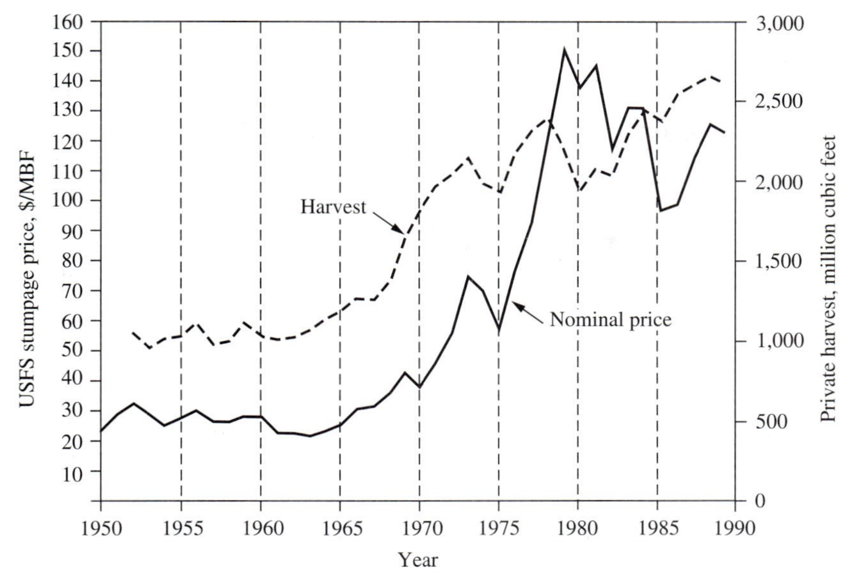

FIGURE 12-7
Nominal stumpage prices and private pine harvest, 1952-1989—south central United States (Data from Richard Haynes, U.S. Forest Service Pacific Northwest Research Station. Forest Service cut softwood prices are a surrogate for private prices since fluctuations are very similar. Prices are nominal to reflect any "money illusion" from inflation.)

loggers convincing landowners to part with their timber. Also, many nonindustrial landowners aren't aware of value growth percent theory.[4]

The triangles in Figure 12-8 plot stumpage prices over private pine sawtimber harvest for the south central United States for selected years from 1955 to 1989. These triangles are also a series of intersection points of short-run supply and demand curves for separate years. The curves are drawn with arbitrary slopes and show how short-run and long-run timber supply and demand are intertwined. It's sometimes hard to tell exactly how the points shift. For example, between 1955 and 1965, did demand actually shift to the right, or were the two points on the same demand curve? This difficulty is called the **identification problem.** On average, the short-run demand and supply curves in Figure 12-8 shifted rightward about the same distance, so real prices didn't change much over the period shown. But longer and more inclusive time series do show an upward trend. If demand

---

[4]High interest rates, which often accompany inflationary spikes in stumpage prices, could stimulate cutting by falsely making the forest value growth percent *appear* less than alternative rates of return (see Chapter 5). Remember from Chapter 7 that, financially, there's pressure to cut if forest value grows more slowly than rates of return available elsewhere. But **real** interest rates don't generally rise with inflation, except for an atypical period during the 1980s (see Figure 5-3). Thus, in theory, inflation doesn't tend to change optimal rotations (see "Make Sure to Correct for Inflation" in Chapter 7).

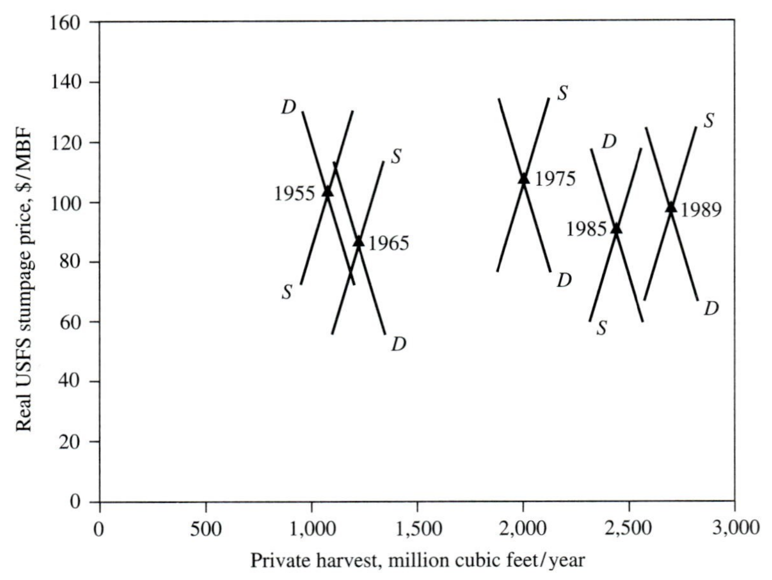

FIGURE 12-8
Real stumpage prices over private pine harvest—south central United States. (Data from Richard Haynes, U.S. Forest Service Pacific Northwest Research Station. Forest Service cut softwood prices are a surrogate for private prices since fluctuations are very similar.)

shifts rightward faster (slower) than supply, prices rise (fall). Can you speculate what the demand and supply shifters were in Figure 12-8?

The simplest shifters in Figure 12-8 are (1) increased population and income moving the demand curve rightward; and (2) timber growth, previous natural and artificial reforestation, and more intensive management shifting supply to the right. But the usable southern pine stumpage resource mix has also changed over time. For example, over the period shown, we've had decreasing minimum acceptable top diameters for sawlogs and peelers. In the ensuing decades, more efficient processing plants stretched the wood resource. These factors shifted the timber supply curve rightward: at any price, more usable wood came from a given stand type. These are the technological supply shifters mentioned earlier. So you can think of the horizontal axis of Figure 12-8 as a changing stumpage resource including a different mix of log sizes and qualities over time.

In the north-central United States, aspen demand curves shifted dramatically to the right during the mid-1980s as new waferboard and oriented strand board (OSB) plants were built. At any given price, more aspen stumpage was demanded. In the Pacific Northwest, red alder, previously considered a weed species, is now sought for furniture and other products. New grading rules and better drying techniques have created a stronger demand for yellow poplar, a once underutilized southeastern species. So new processing technology stretches existing wood resources and creates usable products from previously unmarketed species and qualities of logs.

Demand interactions between market levels are complex. Think back to Figure 12-4. Often demand for lumber and structural panels like plywood and composite board is independent of processing technology. For example, most home buyers don't know or care whether the 2×4s within came from large logs or small logs or whether structural panels were made from veneer or chips, softwoods or hardwoods. Substitution is common; for instance, OSB panels have invaded some plywood markets. So, as structural panel demand increased, demand curves for OSB chip logs shifted rightward, and plywood veneer log demand curves shifted less, or not at all, and in some regions even leftward.

Statistically Estimating Supply and Demand Functions

Econometricians use statistical techniques to model timber supply and demand equilibria over time and to predict timber harvest levels and prices. Based on historical data, they estimate stumpage supply and demand equations for separate regions. For example, a sawlog stumpage supply equation for a given region, species, and ownership could be of the form:

$$V_y^s = c_0 + c_1 P_y + c_2 S_{1y}^s + c_3 S_{2y}^s + c_4 S_{3y}^s + \dots \qquad (12\text{-}2)$$

where subscripts y are years, c_j's are regression coefficients, V_y^s is stumpage volume supplied (harvested), P_y is stumpage price, and S_{jy}^s's are supply curve shifters like acres of forestland by site quality, and timber inventory by site and age and stocking class.

A stumpage demand equation could be:

$$V_y^d = c_0 + c_1 P_y + c_2 S_{1y}^d + c_3 S_{2y}^d + c_4 S_{3y}^d + \dots \qquad (12\text{-}3)$$

where V_y^d is stumpage volume demanded (harvested), and S_{jy}^d's are demand curve shifters like housing starts, real per capita income, population, paper recycling rates, wood-processing costs, and the interest rate. Since supply must equal demand in any time period, another equation should set $V_y^s = V_y^d$. Some studies develop separate supply and demand equations for different market levels (for example, lumber, plywood, and housing). With techniques that are beyond this book's scope, researchers determine appropriate variables, choose functional forms that may not necessarily be linear as shown, and simultaneously estimate coefficients for the equations.

Variations on the above **econometric** models can be used to project price and quantity combinations for forest products under different scenarios (see Figures 12-12, 12-13, and 12-15). For such simulations, you need to project values of independent variables in the equations, for example, areas and volumes of forest types, housing starts, and interest rates. Thus, the above econometric models are often linked with other models projecting forest area and growth as well as certain economic variables.

Such models can also yield short-run price **elasticities** of stumpage supply and demand. Studies have estimated short-run stumpage supply elasticities in the United States ranging from 0.06 to 0.99, with most being below 0.50 (at mean prices in the sampling period). That means, for every 1 percent change in stumpage price, the quantity supplied will change by substantially less than 1 percent, indicating **inelastic** supply, since elasticity is less than 1. Short-run stumpage demand also tends to be inelastic, with estimates ranging from −0.14 to −0.57 (see Cubbage and Haynes, 1988; Hyde and Newman, 1991). For any given scaling of the x and y axes, the more inelastic a curve is, the steeper it is. Thus, when these inelastic stumpage supply and demand curves shift, prices can be volatile, as shown in Figure 12-7. (Try sketching on a price-quantity axis a demand curve shifting the same amount against steep and flat supply curves. Which gives a greater price movement? You can show the same thing shifting a supply curve against demand curves of differing steepness, as in Figures 2-16 and 2-17.) Low timber supply elasticities have led some analysts to worry about the market's ability to provide more timber as prices increase, but as we'll see, that's not necessarily a problem in the long run.

CAN THE MARKET PROVIDE ENOUGH TIMBER?

In the United States, a major timber supply issue is whether the free market can assure adequate timber supplies. Some say that meager private reforestation after rapid timber cutting in the 1800s and early 1900s shows that the market doesn't work well in timber production. To others, this means that the market was stimulating financially efficient resource use. Through the early 1900s, prospective timber prices were so low that expected **rates of return** on reforestation investment were inadequate. One view is that markets in those times were stimulating the most efficient timber management, financially: Harvest and rely on natural regeneration with little or no planting investment, or change the land use. But remember, that view considers only financial returns and ignores possible damage to soil, water, fish and wildlife, scenery, and other resources. After harvest, the rate of return on investment in reforestation, C_0, plus land, L_0, would be the r that brought about the equality in equation (12-4), assuming even-aged forestry and land sold for L_t after a t-year rotation:

$$(L_0 + C_0) = \sum_{y=0}^{t} \frac{R_y - C_y}{(1+r)^y} + \frac{L_t}{(1+r)^t} \qquad (12\text{-}4)$$

The left side of equation (12-4) is the investment in newly reforested land $(L_0 + C_0)$; the right side is the present value of one rotation's future revenues (R_y) minus future costs (C_y), excluding L_0 and C_0, plus the present value of rotation-end land value. Years are indexed with y. Simplifying equation (12-4) to include only the major cash flows of harvest, $R_t = H_t$, and land sale, L_t, we have:

$$L_0 + C_0 = \frac{H_t + L_t}{(1+r)^t} \qquad (12\text{-}5)$$

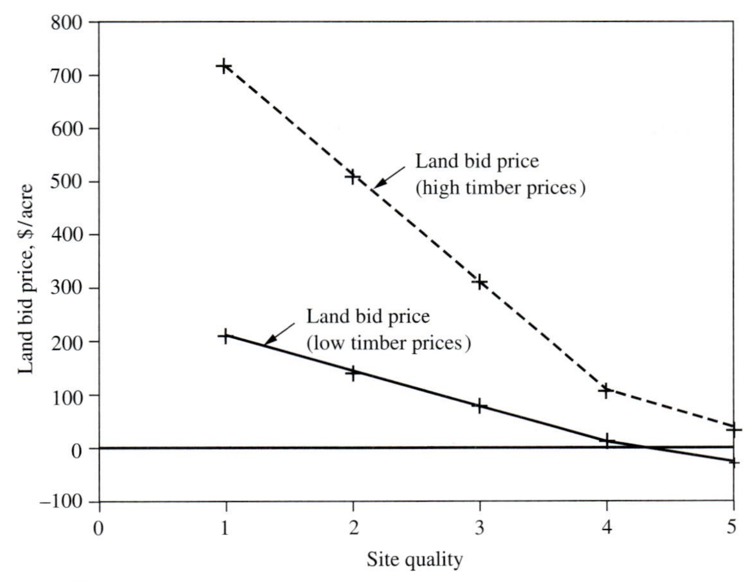

FIGURE 12-9
Hypothetical bid prices for Douglas-fir bare land (WPL).

Since r equates the present value of future revenues with the present value of costs, r is the **internal rate of return** on investment in land and reforestation.

If projected harvest revenue H_t is low enough, the value for r in equation (12-5) will be unacceptable, and heavy private investment in reforestation is unlikely, as was the case in the United States before World War II. Stumpage prices can also affect future wood supply by changing the amount of land devoted to timber crops. For example, in equation (12-5), for a given C_0 and market r, if H_t is low enough, the value for L_0, which is the bid price for bare forestland, falls below that for competing uses (usually different types of agriculture) or may even be negative. Figure 12-9 shows hypothetical bid prices for bare Douglas-fir land in the Pacific Northwest, given two future stumpage price assumptions. Each curve's values are based on after-tax cash flows entered into equation (7-4) for the **willingness to pay for land (WPL),** using 6 percent real interest and actual yield functions for site qualities from 1 (high) to 5 (low). If the WPL values are low enough, timber is no longer the chosen use. The important thing in Figure 12-9 is not the actual values per acre but the fact that they decline with decreasing site quality and lower expected prices.

Changing relative prices of timber and farm crops cause land to shift between timber and agriculture, as shown in Figure 12-10. If wood prices rise enough relative to agricultural prices, some land can change from farm to forest use, and vice versa, either through land being sold to new bidders or through current owners changing to more profitable uses. These shifts influence the supply of and price of timber one rotation later. Governments can also influence the amount of forestland; for example, the "soil bank" program in the late 1950s subsidized planting of farmland to trees. Much of that timber has already been harvested.

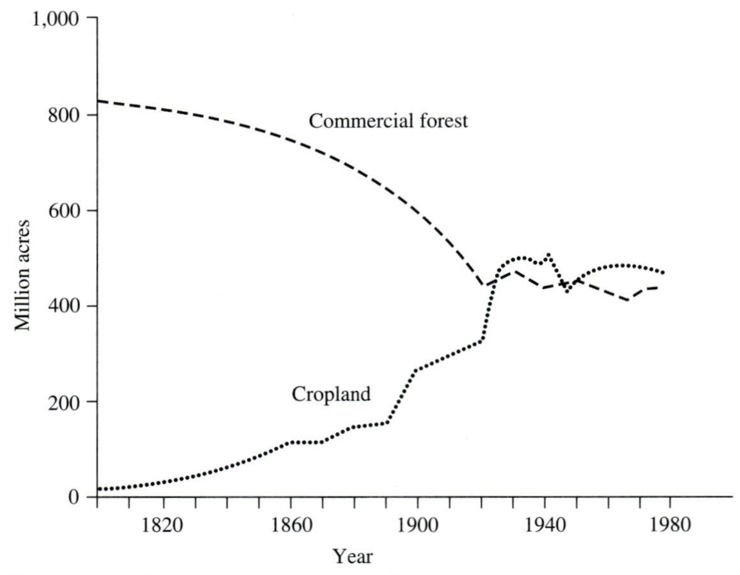

FIGURE 12-10
United States cropland and commercial forest acreage. (Adapted from Clawson [1979], with permission from the American Association for the Advancement of Science.)

Afforestation and Rising Harvest Values

With increasing nominal and real stumpage prices, industrial investment in planting grew, as Figure 12-11 shows for the southern United States. Only industrial planting is shown, since several public programs were subsidizing nonindustrial private planting over the same period. Industrial planting was unsubsidized, and over the period shown, no laws required restocking after harvest in the South. So the general long-term trend suggests a market response, although not every period shows a positive correlation between real stumpage price and planting. Similar graphs can be drawn for other commercial timber-producing regions in the United States.

Figure 12-11 confirms that timber was a good hedge against inflation since World War II. Although the period from 1970 to 1990 shows no real stumpage price increase, the general long-term trend is upward, especially in light of stumpage price increases in the 1990s. The stumpage price lines in Figure 12-11 understate the market stimulus to industrial planting. First, what spurs planting is the *expected* stumpage price at rotation-end, not the current price alone. And since World War II, the U.S. Forest Service and private economic forecasters have been projecting higher real sawtimber stumpage prices in most regions of the United States. Second, Figure 12-11 masks the fact that *harvest values per acre* of identical timber [H_t in equation (12-5)] have been rising faster than stumpage prices. Over time, higher stumpage prices stimulate new technology, which permits selling more wood per acre of similar timber, for example, by using

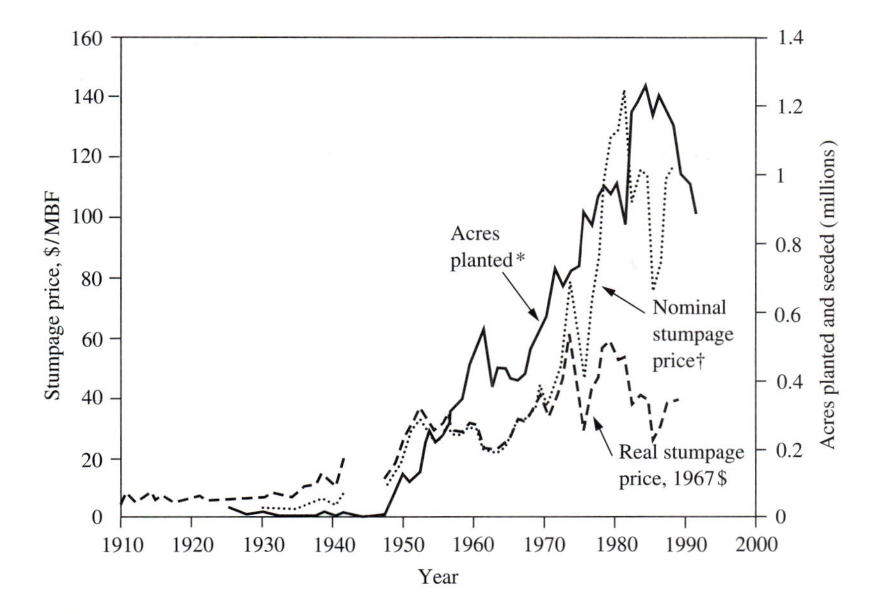

†Dollars/mbf, International 1/4" rule. 1910–34, private second growth; 1935–49, National Forest sales, all species; 1950–88, National Forest sales, pine only (USDA, 1973 & Ulrich, 1990)

*USDA (1986) and USDA (1987–90). Includes a minimal amount of direct seeding.

FIGURE 12-11
Stumpage prices and industry planting—United States southern pine.

smaller logs, more species, more chips, and lower quality wood. And these trends aren't unique to the southern United States.

Figure 12-11 doesn't necessarily prove that higher harvest values *caused* greater industrial reforestation. In fact, other trends also favored plantations in the South: improved fire control, more regulated cattle grazing, and more favorable forest income taxes after 1944. But the dramatic increase in harvest values overshadows these and other positive influences on reforestation effort. Thus, evidence is strong that the market has stimulated increased reforestation activity.[5] You can likewise conclude that higher expected harvest values have increased the rate of return on investments like fertilization, genetic improvement, precommercial thinning, and improvement cuttings, all of which boost long-run timber supplies and have been on the rise in the last 30 years.

Consultants, Investors, and Timber's Future

In recent years, more consulting foresters have been assisting landowners and encouraging reforestation. Consulting is largely driven by higher harvest values,

[5]For empirical evidence in the southeastern United States, see Newman and Wear (1993).

since fees are usually a percentage of a landowner's stumpage receipts. Also, the forest industry leases nonindustrial private forest (NIPF) lands and gives forestry assistance, in the hope of increasing long-run timber supplies. In 12 southern states, by 1984 the forest industry employed over 115 foresters to provide NIPF assistance on 4.2 million acres and retained over 570 procurement foresters who routinely gave reforestation advice to landowners. At that time, the industry also leased about 4.7 million acres of southern NIPF land. In 1985 over 565 independent consulting foresters were available to assist forest owners in the South (Meyer et al., 1986; and Skinner et al., 1990). All of these activities are stimulated by expectations of valuable future timber yields.

It's important to consider how much private reforestation and timber inventory would have existed if stumpage prices had *not* increased and had remained at 1930 levels. There's every indication that, in such a case, private forestry investment over the last 60 years would have been far less. The above trends imply that higher *expected* future harvest values per acre will shift the future timber supply curve to the right. Does that seem confusing? We say that certain things *except current price* will shift the short-run timber supply curve. But if you expect the *future* price—or harvest value per acre—to be higher, you'll plant more today, thus shifting the *future* timber supply curve to the right. This suggests that long-run timber supply in the United States is much more responsive to price than indicated by the above-mentioned short-run elasticities of less than 1. In fact, Newman and Wear (1993) have estimated long-run industry stumpage supply elasticities of 2.5 for sawtimber and 1.8 for pulpwood in the United States South. This means that for every 1 percent change in today's stumpage price, they estimate that industrial stumpage harvest after roughly one rotation will change by 2.5 percent for sawtimber and 1.8 percent for pulpwood.

Long-term timber production also depends on real interest rates. If available real rates of return are low, investors' **minimum acceptable rates of return** (MARs) tend to be low, and forestry's bids for land increase relatively more than those of other uses, due to forestry's long payoff periods (see Chapter 11's "The Relevance of Payoff Period"). In addition, lower real interest rates would lengthen optimal rotations, increase timber inventories and mean annual increments, and boost long-run timber harvests (and the opposite for higher interest rates). Since real interest rates tend to be fairly stable over long time periods, and since timber investments are based on long-term expectations, it's unlikely that changes in real interest rates will cause major long-run timber supply surprises. But high *nominal* interest rates can dampen debt-financed timber investment and shift the future timber supply curve to the left (see Figure 12-5).

Foresters have often complained about low rates of return and high risk in timber crops. But now we're seeing a change. There seems to be an increasing interest in timber as an investment and not just as a mill input. In forestry, investors don't see the risk or high-flying potential of ventures like biotechnology, computer software development, or wildcat oil drilling. They don't see high risks or demand high rates of return in forestry. What's more, markets are fairly efficient, and financial optimism gets capitalized into higher land values, so that new entrants in forestry can expect to earn an average rate of return for modest-risk ventures.

And that's becoming attractive to many investors beyond the wood-processing industry; examples are John Hancock Mutual Life Insurance, Prudential, Travelers, Wachovia Bank, some pension funds, and investor groups from Europe and Asia. Most of these investors aren't just speculating in existing timber inventories; they're actively engaged in timber production and are thus either helping to stabilize the timber supply curve or shifting it rightward. These are added examples of the market at work in long-run timber supply.

As further evidence that timber famine isn't looming in the United States, Figure 12-12 shows past and projected timber harvests increasing over time and accompanied by timber inventories rising until 2020 and leveling off thereafter. Above the harvest line projected in 1993 are pluses marking harvest projections made in 1989 when fewer harvest constraints were predicted on public lands and less paper recycling was projected. The rising inventory shows that timber growth has exceeded cut for many decades, combining all species and ownerships. In 1986 for softwoods alone, growth was 13 percent above cut; for hardwoods, growth was 90 percent above cut. Growth has exceeded cut for hardwoods in all owner groups and for softwoods on all but industry forests. Since 1970, on industrial lands, softwood growth has been from 9 percent to 23 percent below cut. But increasing softwood growth is projected to bring industrial cut and growth into balance by the turn of the century (USDA, 1990).

In spite of these encouraging trends and evidence that the market can stimulate increased timber supplies, worries about timber shortages in the United

FIGURE 12-12
United States commercial timber inventory and harvest. (Inventory from USDA [1990]; actual to 1986, projected thereafter. Solid harvest line from USDA [1994]; actual to 1991, projected thereafter. Pluses show an earlier 1989 projection estimating fewer environmental constraints on public timber harvesting [USDA 1990].)

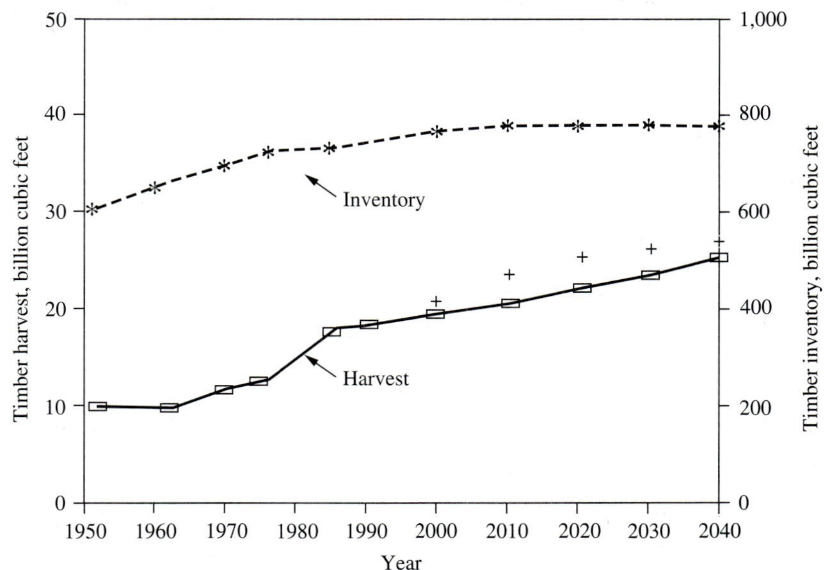

States have persisted. For softwoods, the 1965 *Timber Trends* report warned that by the year 2000, "the projected 'supply' would fall short of the projected cut..." (USDA, 1965, p. 133). Other U.S. Forest Service reports have stated that "the nation is faced with a growing imbalance between supply and the quantity of forest...products that people would like to consume" (USDA, 1981). "The volume of softwood sawtimber supply is a concern, especially over the next 20 years." (USDA, 1994). Yet timber has continued to be abundant. Many recent U.S. Forest Service reports have a stronger economic flavor, recognizing the concept of prices bringing timber supply and demand into balance: "Projections show softwood sawtimber stumpage prices rising substantially in all regions.... Equilibrium prices are those prices at which the amount willingly supplied and the amount willingly demanded are equal" (USDA, 1990, pp. 144 and 131). But the timber famine mentality still prevails in many circles, so let's look at it in greater detail.

Gap Analysis

As an example of a timber shortage mentality, Figure 12-13 shows a graph of projected United States softwood "supply" and "demand" trends published by the U.S. Forest Service in 1982. Quotation marks indicate that these curves are not the same as the economic supply and demand curves discussed so far. The difference between the curves is known as the "gap." For now, ignore the equilibrium line, which is sometimes omitted in such graphs. The supply and demand analysis in

FIGURE 12-13
Projected United States softwood roundwood "demand" and "supply." (Adapted from USDA (1982), p. 204.)

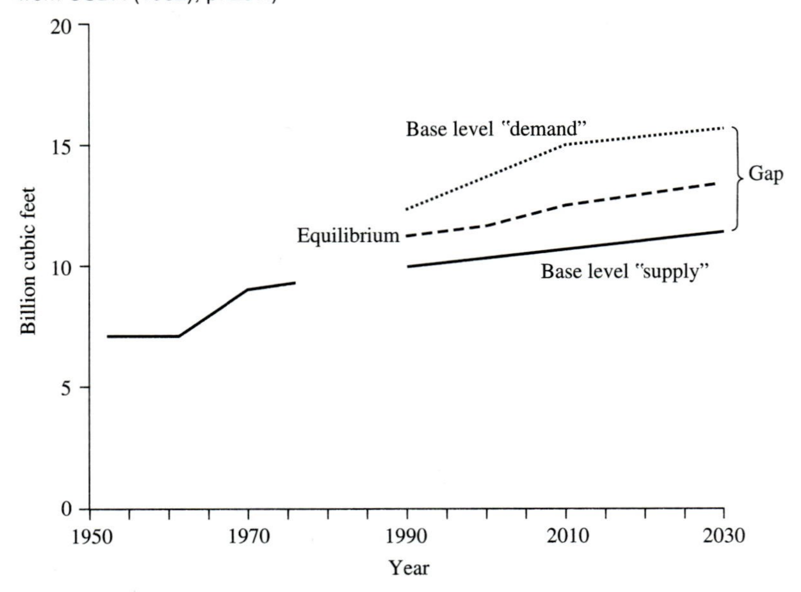

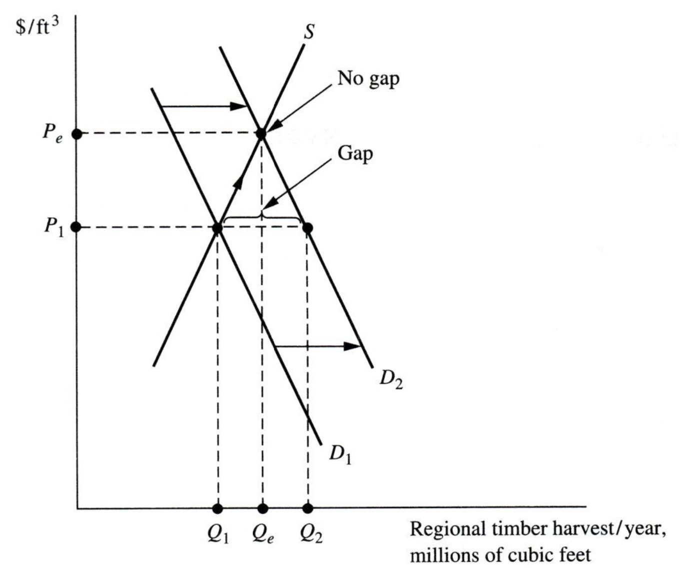

FIGURE 12-14
The disappearing regional timber supply gap.

Figure 12-13 assumes that stumpage prices stay at 1976 levels. In an economic supply and demand framework, the shortage never occurs, as shown in Figure 12-14, where the initial equilibrium harvest is at Q_1 at time 1, given the regional timber demand curve D_1 and a fixed supply curve S. Suppose later demand at time 2 is D_2, due to rising income and population. In that case, if stumpage price were held at the initial P_1, as is the case in the gap analysis, the amount demanded at time 2 would be Q_2, with the difference between supply and demand shown as the shortage or gap. In reality, shortages at one price lead buyers to bid up the price until supply equals demand at an equilibrium price of P_e and quantity of Q_e in Figure 12-14. For simplicity, the supply curve is held constant, although it could shift to the right or left.

For every year in Figure 12-13, you could imagine a stumpage supply and demand equilibrium at a price higher than the initial price, for example at Q_e and P_e in Figure 12-14. Thus, the gap between "supply" and "demand" in Figure 12-13 simply means that stumpage prices will rise enough over time to equate supply and demand, as shown by the equilibrium line and as actually happened historically. So the question isn't whether the market will supply enough timber; supply will equal demand in a well-functioning economy. The important question is, at what price will supply equal demand?

Rising Timber Prices

Figure 12-15 shows U.S. Forest Service projections made in 1989 for softwood sawtimber stumpage prices in 1982 dollars by separate regions. These are the

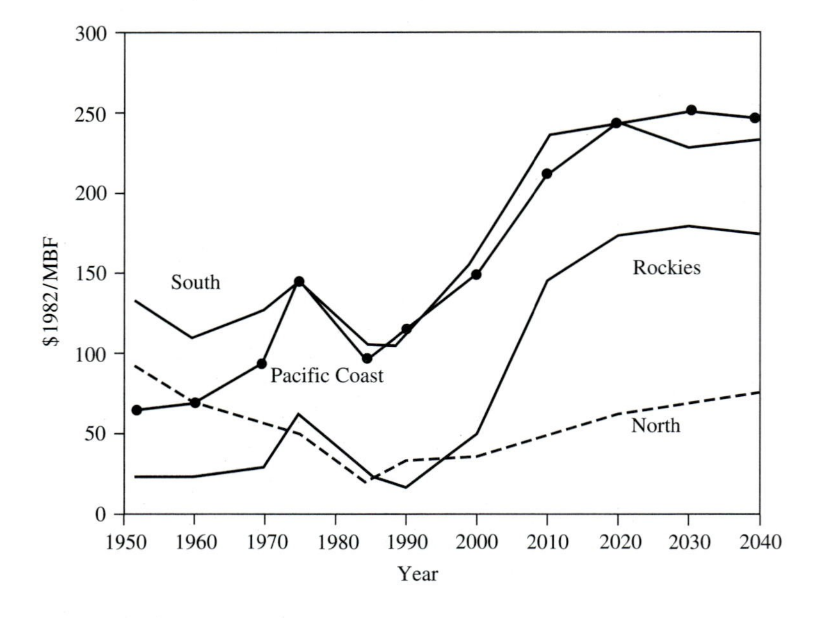

Note: Pacific Coast excludes Alaska and Hawaii.

FIGURE 12-15
Softwood sawtimber stumpage prices, 1952–1986, with projections, 1990–2040. (From USDA [1990], p. 144.)

estimated prices required to close the gap shown in Figure 12-13. Between 1986 and 2040, these prices translate to real annual increases averaging roughly 1.5 percent in the South, 2.5 percent in the North, 3.2 percent in the Rocky Mountains, and 1.7 to 2.0 percent in the Pacific Coast Region. More recent analyses project somewhat slower long-term price-increases in the South, due to more paper recycling (Ince, 1994). But newer projections for the West Coast show higher stumpage prices due to rising pressure to cut less on public forests (USDA, 1994).

Higher real timber prices are damaging to consumers: the more you pay for wood products, the less income is available for other purchases that bring satisfaction. So other things being equal, higher wood prices mean reduced utility for consumers. Also, rising wood prices lead to increased use of plastics and metals, which many analysts believe generate greater negative environmental impacts than forest products manufacturing.

The above reasoning has led the U.S. Forest Service to support government subsidies like financial assistance for NIPF reforestation in order to shift the future wood supply curve to the right and decrease future wood prices (see Figure 12-5). But this raises efficiency questions: If rates of return on added reforestation investment before the subsidy were too low to attract private capital, then why should the government spend public funds to drive down stumpage prices and make the rate of return still lower? This would be taking tax dollars that were earning competitive rates of return in the private sector and investing them where they might earn less—an inefficient allocation of capital. On the other hand, if excellent rates of return on reforestation are being ignored by private investors, then the above case against subsidies is weakened.

While no one likes paying higher wood prices, they aren't all bad; in fact, they're necessary if we want more wood from a market economy. *Some advantages of higher stumpage prices are as follows:*

- Neater logging jobs: Less wood is left on the ground because a higher stumpage price makes it worthwhile to remove logs of a smaller diameter and with a higher percentage of defects like rot and splitting.
- Smaller, more energy-efficient homes: Higher wood prices lead buyers to build smaller homes that require less energy to heat or cool. This conserves fuels and reduces pollution.
- Former "weed trees" become commercially valuable: As wood prices increase, species that formerly had negative stumpage prices now can be harvested profitably, thus stretching the forest resource. Examples are red alder in the Northwest and aspen in many regions.
- More aesthetically pleasing partial cutting becomes economic: Logging costs for thinnings and other types of partial cutting are higher than for clear-cutting. When wood prices are low, partial cuts often generate negative stumpage prices. Other things being equal, as delivered log prices increase, such partial cuts can become economically feasible.
- With the promise of greater harvest income, landowners will invest more in reforestation, better site preparation, genetic improvement, fertilization, and precommercial thinning. Long-term wood supply is increased.
- As wood becomes more valuable, the forest industry makes more effort to help NIPF owners manage their forests.
- Higher wood prices encourage better utilization: for example, chipping, sawing and peeling smaller diameter logs, and using lower quality wood, all of which stretch the existing wood resource.

We should also think about the broader issue of subsidies to lower the price of any good, not just wood. Governments can subsidize or help any sector—for example, with direct payments, free technical assistance, tax relief, low-cost loans—to increase output and reduce prices.While it seems nice to reduce prices—we'd all like cheaper shoes, haircuts, wood, beans, cars, beer—where do you stop? The efficiency argument is to let a well-functioning market reach the equilibria discussed in Chapters 2 and 3 (also see "Tax Policy Issues" in Chapter 9). The emphasis here is on *well-functioning:* **market failures** like pollution and monopoly justify intervention. Remember, too, the problem with intervening to reduce prices because you want to help the poor: reducing market prices helps rich and poor alike. If income redistribution is the goal, many would argue that it's more efficient to help the poor directly, for example, through housing subsidies, rather than distort markets and reduce wood prices for all.

Also, real wood prices are unlikely to rise forever, relative to other commodities. Such increases lead to wood conservation and using more wood substitutes, both of which dampen stumpage price-increases.

How Much Wood Do We Need, and When?

Ignoring market failures, if we wish to maximize total returns from limited capital, we should follow the **equi-marginal principle** and devote more capital to timber production as long as the expected rate of return is at least as great as that on other investments of equal risk. Also, timber growers should (and will) use land for timber production as long as their bids for bare land are higher than those generated by other uses. *Without market failures,* when nontimber uses win the bidding game, the land will yield greater benefits than it could in timber production. These efficiency guidelines mean that the heaviest wood production investment should be on the better sites and in regions where markets are best. Timber investment shouldn't occur where rates of return are unacceptable. And the market does just that: On United States industrial forestlands, the largest timber investments per acre tend to be on the best sites in areas like the South and the Pacific Northwest, and are considerably less, for example, in northern Maine and high mountain regions where site qualities are low.

This efficient pattern of forestland use bothers some foresters who'd like to see more intensive management on all forestlands. They're concerned about idle acres with scrubby hardwoods. But that's not necessarily a problem if it costs more, in **present value** terms, to get softwood regeneration than it's worth to consumers. Think of an agricultural analogy: We wouldn't want all potential farmland in the United States, including people's backyards, to be producing crops. That would severely depress crop prices and rates of return on farming, which are already too low in many areas. Besides, many acres that are "idle" as softwood timber producers have vegetative cover and provide benefits of soil and water protection, scenic beauty, recreation, and perhaps some pulpwood, chips, and low-quality sawlogs. So in the long run, we need as much timber as will be yielded by efficient use of capital.

Nontimber Values and Optimal Harvests over Time In competitive markets, timber growers would manage forests in a way that maximized the present value of harvest income. Ideally, harvests would be distributed over time so that, in present value terms, the added revenue of shifting a unit of harvest from one year to another would exactly equal the resulting cost. Viewing only timber income, such a harvest pattern over time is optimal, because the added cost of changing that pattern would exceed the added revenue. The mechanics of such harvest scheduling are covered in forest management courses: here we'll consider only broad issues.

The financial present value–maximizing pattern of timber harvest on one ownership is often cyclical. For example, computer simulations of financially optimal harvest patterns over time for forests sometimes look like the solid curve in Figure 12-16, especially when starting with large volumes of mature timber. It's optimal to cut slowly growing timber quickly because its value growth percent is lower than required earning rates. In such cases, without an even age-class distribution, rapid timber depletion leads to lean harvests during the long wait for newly planted trees to mature, and the cycle begins anew. What keeps large ma-

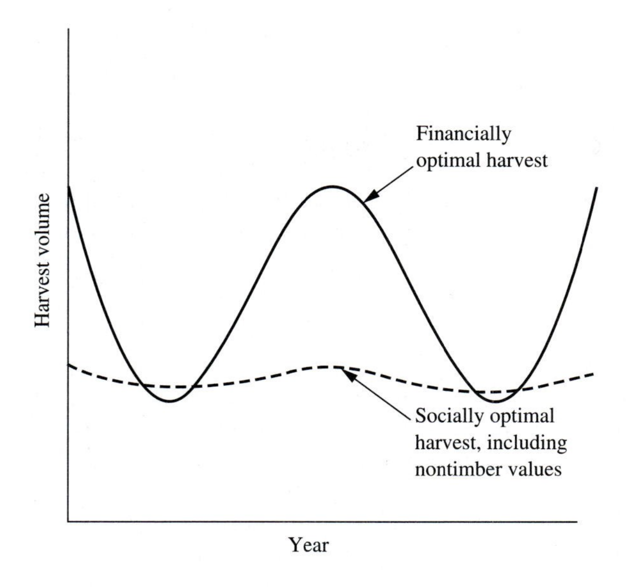

FIGURE 12-16
Hypothetical harvest patterns in region with mature timber reserves—financial and social optima.

ture inventories from immediate liquidation is the downward-sloping stumpage demand curve faced by their owners. Local mills can absorb limited volumes per year, and added timber must be hauled longer distances, thus slashing the owner's stumpage receipts when timber is cut too fast. The financially optimum harvest in Figure 12-16 (the solid curve) is based only on market prices of stumpage and ignores nontimber values.

Suppose we consider the negative side effects of harvest peaks and troughs on large forested areas—environmental damage from too rapid cutting, and economic trauma from gyrating harvests. Including these costs will reduce social benefits when harvests are significantly above and below some average level. In that case, the harvest pattern maximizing present value to society—the dashed curve in Figure 12-16—would fluctuate less than the financial optimum. Also, socially optimal forest management often means partial cutting or no harvesting in certain streamside and scenic protection zones, as well as in wilderness and recreation areas. Thus, as shown in Figure 12-16, socially optimal harvests will generally be at a lower average level than the financial optimum. Changing from the financial to the social optimum would thus shift the future timber supply curve to the left for most periods, reducing harvests and raising wood prices (see Figure 12-5).

Many negative effects of cyclical harvests aren't necessarily relevant on smaller ownerships. Over larger areas, such harvest peaks and troughs often offset one another and cause little regional economic or environmental trauma. Figure 12-16 shouldn't be seen as support for the so-called even-flow harvest policies of some public agencies that seek an equal annual harvest under all conditions. The diagram is more of a plea to consider all impacts of harvest levels when seeking optima. Even flow is arbitrary: Compared to some other pattern, the costs of even flow (in terms of income forgone) can often exceed the benefits.

There's also the question, what's the optimal size of even-flow units? Over a large enough area—a state or a country—you'll often get even flow without trying. Over a small enough area—a single tree, in the extreme—you'll never have even flow. Historically, the United States has had a sustained flow of timber by first clear-cutting the Northeast, then the Lake States, later the South, and finally the West. But some would question the economic and environmental impacts of that pattern. On the other hand, it wouldn't generally make sense to manage a 5-acre woodland to yield equal annual harvests.

NONINDUSTRIAL PRIVATE FORESTS

Many analysts look to the nonindustrial private forest (NIPF) as a major source of timber in the future, since these lands constitute about 59 percent of the United States commercial forest area. But NIPF lands are managed less intensively than industry forests and supply lower harvest volumes per acre. Some see this as a problem, but others point out that, compared with industrial forests, many NIPF lands tend to be on poorer sites and farther from markets. Efficiency dictates that such lands should be managed less intensively. Moreover, many NIPF owners aren't as interested in intensive timber management as they are in owning land for its nontimber outputs. From a utility maximization standpoint, that's not necessarily a problem. Similarly, it's not a problem if people use their backyards for aesthetic and recreational purposes rather than for growing vegetables. The following summarizes why many NIPF lands aren't managed as intensively as industrial forests:

- In many regions the average age of NIPF owners is over 50 years, so forestry's payoff periods may seem too long. (But owners can sell land with immature timber or can sell cutting rights without selling the land.)
- Many woodlot sizes are too small for efficient management.
- The major management objective of many NIPF owners is recreation and aesthetics rather than timber production.
- Many are interested in the land as an investment and plan a change to a more profitable land use before timber could mature.
- Many NIPF lands are located where markets, site qualities, and expected rates of return are poor. Thus heavy timber investment isn't socially desirable. In such locations, industrial forest management is also not very intensive.
- In some cases, NIPF owners lack knowledge about forestry investment opportunities. Here extension forestry programs provide a useful service.

Recall that rising harvest values have stimulated better NIPF management through more consulting foresters, industry-sponsored landowner assistance programs, and industrial leasing of NIPF lands. Ultimately much of the so-called NIPF problem may not actually be socially undesirable, and much of it may disappear with rising timber prices.

KEY POINTS

- Short-run stumpage demand curves are derived from consumers' demand curves for final wood products.
- Short-run final wood products supply curves are derived from stumpage supply curves.
- In the long run, stumpage supply curves can shift to the right or left due to factors increasing or decreasing the salable inventory and acres of timber that can be processed.
- In the long run, stumpage demand curves can shift to the right or left due to things like changes in population, income, consumer preferences, prices of related goods, or interest rates.
- Long-run simultaneous shifts in timber demand and supply functions determine equilibrium stumpage prices.
- Short-run timber demand and supply curves tend to be inelastic, leading to volatile stumpage prices. Harvests tend to rise and fall with stumpage prices.
- Gradually increasing stumpage prices since 1930 have been accompanied by increased tree planting and other forestry investment in the United States. Long-run timber supply is elastic; one study found that a 1 percent increase in real stumpage price leads to roughly a 2 percent or more increase in long-run timber harvest.
- "Gap" analyses in the United States generally project future timber volume demanded to be greater than volume supplied, if you assume fixed timber prices. Stumpage prices will rise until supply equals demand.
- Consumers don't like higher wood prices, but they're needed to stimulate more wood production.
- Efficiency dictates that lands should not be intensively managed for timber if they yield low rates of return or if owners are interested mainly in nontimber outputs. Many nonindustrial forestlands fall into this category.
- Socially optimal harvest levels tend to be lower and fluctuate less than private optima.

QUESTIONS AND PROBLEMS

12-1 Explain how pulpwood stumpage demand curves are derived from consumer demand curves for paper.

12-2 Explain how retail lumber supply curves are derived from sawtimber stumpage supply curves.

12-3 List factors that shift stumpage demand curves in different directions.

12-4 List factors that shift stumpage supply curves in different directions. Explain how market forces can shift timber supply curves to the right.

12-5 Suppose that a government agency predicts a future timber supply "gap" so that the amount of timber demanded will exceed the volume supplied in 20 years. In a supply-demand framework, explain what this means. What will actually happen in 20 years?

12-6 What are the advantages and disadvantages of rising timber prices?

12-7 Suppose that under a subsidy program, the federal government pays 43 percent of Mr. Baum's precommercial thinning (PCT) costs. Consider a forest owner's PCT investment today of $200 (including subsidy), resulting in a real increased harvest value of $530 in 20 years. What is the rate of return on Baum's cash outlay on this PCT investment? [Remember, the government subsidized him at 0.43(200).] What is society's rate of return on the PCT investment? (Both forest owner and government are part of "society.")

12-8 What would the minimum annual value of nonmarket benefits from the investment in the foregoing question have to be, in order to justify the government subsidy? Assume the government takes a broad view that society (a partnership of landowner and the rest of society) should earn a 7 percent rate of return.

12-9 What will increased wilderness set-asides do to the future United States timber supply curve and to timber prices?

12-10 Assume a sawmill has estimated that its supply of logs would be 700 MBF/year if it pays $100/MBF, and 800 MBF/year if it pays $120/MBF. What is the firm's added cost per added MBF if it wants to increase the amount supplied from 700 to 800 MBF per year?

12-11 Why do many analysts predict that stumpage prices for sawlog and peeler production will increase? Illustrate your points graphically. Consider several scenarios.

REFERENCES

Adams, D. A., and R. W. Haynes. 1980. *The 1980 Softwood Timber Assessment Market Model: Structure, Projections, and Policy Simulations.* Forest Science Monograph 22. Bethesda, MD. 64 pp.

Adams, D. A., and R. W. Haynes. 1989. A model of National Forest timber supply and stumpage markets in the western United States. *Forest Science.* 35(2):401–424.

Alig, R. J., B. J. Lewis, and P. A. Morris. 1984. *Aggregate Timber Supply Analysis.* U.S.D.A. Forest Service. Gen. Tech. Rept. RM-106. Fort Collins, CO. 49 pp.

Clawson, M. 1977. *Economic Timber Production Characteristics of Nonindustrial Private Forests in the United States.* Discussion Paper. Resources for the Future. Wash., DC. 77 pp.

Clawson, M. 1979. Forests in the long sweep of American history. *Science.* 204:1168–1174.

Cubbage, F. W., and R. W. Haynes. 1988. *Evaluation of the Effectiveness of Market Responses to Timber Scarcity Problems.* U.S.D.A. Forest Service. Marketing Research Report No. 1149. Washington, DC. 87 pp.

Dana, S. T., and S. K. Fairfax. 1980. *Forest and Range Policy.* McGraw-Hill, New York. 458 pp.

Duerr, W. A. 1974. Timber supply: Goals, prospects, problems. *American Journal of Agricultural Economics.* 56(4):927–935.

Gould, E. M. 1978. Wanted: High-satisfaction forestry. *Journal of Forestry.* 76(11): 715–718.

Greene, J. L., and W. C. Siegel. 1994. *The Status and Impact of State and Local Regulation on Private Timber Supply.* Gen. Tech. Rept. RM-255. U.S. Forest Service. Rocky Mt. Forest and Range Experiment Station. Fort Collins, CO. 22 pp.

Haynes, R. W. 1977. A derived demand approach to estimating the linkage between stumpage and lumber markets. *Forest Science.* 23(2):281–288.

Hyde, W. F. 1980. *Timber Supply, Land Allocation and Economic Efficiency.* The Johns Hopkins University Press, Baltimore. 224 pp.

Hyde, W. F., and D. H. Newman. 1991. *Forest economics and policy analysis: an overview.* Discussion paper 134. The World Bank, Washington, DC. 92 pp.

Ince, P. J. 1994. *Recycling and Long-Range Timber Outlook.* U.S. Forest Service Rocky Mt. Forest and Range Experiment Station. Gen. Tech. Rept. RM-242. Fort Collins, CO. 23 pp.

Johnson, K. N. 1976. *Consequences of Economic Harvest Scheduling Procedures on Five National Forests.* Report to U.S. Forest Service Intermountain Research Station. Utah State Univ., Logan. 55 pp.

Klemperer, W. D. 1981. Is sustained yield an ethical obligation in public forest management planning? *Proceedings of Int'l. Union of For. Res. Organ. World Congress.* September. Kyoto, Japan. pp. 421–432.

McKillop, W. L. M. 1967. Supply and demand for forest products—An econometric study. *Hilgardia.* 38(1):1–132.

Meyer, R. D., W. D. Klemperer, and W. C. Siegel. 1986. Cutting contracts and timberland leasing. *Journal of Forestry.* 84(12):35–38.

Newman, D. H., and D. N. Wear. 1993. Production economics of private forestry: A comparison of industrial and nonindustrial forest owners. *American Journal of Agricultural Economics.* 75:674–684.

Sedjo, R. A., and K. S. Lyon. 1990. *The Long-Term Adequacy of World Timber Supply.* Resources for the Future, Washington, DC.

Skinner, M. D., W. D. Klemperer, and R. J. Moulton. 1990. Impacts of technical assistance on private non-industrial reforestation. *Canadian Journal of Forest Research.* 20(11): 1804–1810.

Sohngen, B. L., and R. W. Haynes. 1994. *The "Great" Price Spike of '93: An Analysis of Lumber and Stumpage Prices in the Pacific Northwest.* Research Paper PNW-RP-476. U.S. Forest Service. Pacific Northwest Research Station. 20 pp.

U.S.D.A. Forest Service. 1965. *Timber Trends in the United States.* Forest Resource Report No. 17. Washington, DC. 235 pp.

U.S.D.A. Forest Service. 1973. *The Outlook for Timber in the United States.* Forest Resource Report No. 20. Washington, DC. 367 pp.

U.S.D.A. Forest Service. 1981. *An Assessment of the Forest and Range Land Situation in the United States.* Forest Resource Report No. 22. Washington, DC. 352 pp.

U.S.D.A. Forest Service. 1982. *An Analysis of the Timber Situation in the United States, 1952–2030.* Forest Resource Report No. 23. Washington, DC. 499 pp.

U.S.D.A. Forest Service. 1986. *A Statistical History of Tree Planting in the South—1925–1985.* Misc. Report SA-MR8. Cooperative Forestry, Atlanta, GA. 23 pp.

U.S.D.A. Forest Service. 1987–1992. *Tree Planting in the United States.* (separate annual issues). Cooperative Forestry, Washington, DC.

U.S.D.A. Forest Service. 1990. *An Analysis of the Timber Situation in the United States: 1989–2040.* Gen. Tech. Report RM-199. Fort Collins, CO. 268 pp.

U.S.D.A. Forest Service. 1994. *RPA Assessment of the Forest and Rangeland Situation in the United States—1993 Update.* Forest Resource Report No. 27. Washington, DC. 75 pp.

Ulrich, A. H. 1990. *U.S. Timber Production, Trade, Consumption, and Price Statistics 1960–88.* U.S.D.A. Forest Service Misc. Publ. 1486. Washington, DC. 80 pp.

Vaux, H. J., and J. A. Zivnuska. 1953. Forest production goals: A critical analysis. *Land Economics.* 29:318–327.

Wear, D. N., and D. H. Newman. 1991. The structure of forestry production: Short-run and long-run results. *Forest Science.* 37(2):540–551.

Zinkhan, F. C. 1991. *A Survey of Practices and Perceptions of Southern Timberland Managers.* Unpublished paper. Campbell University School of Business. Buies Creek, NC. 76 pp.

13

THE UNITED STATES WOOD-PROCESSING INDUSTRY

Let's take a brief look at the business of wood processing. Some foresters think of wood as the forest's primary output. But to many, wood isn't the first thing that comes to mind when viewing forests. On the other hand, it's hard to avoid wood products; for example, paper is practically everywhere (before you now!). Even in the electronic age, you can't escape printouts of some sort. During much of your life, you're surrounded by objects with some wood component—buildings, furniture, art objects, interior decorations, containers, fabrics, musical instruments. The list goes on and on. We have a love affair with wood—with its textures, colors, reflections, sounds and resonances, grain patterns, strength-to-weight ratios, knots, fragrances, paper surfaces. Sometimes wood has a mystical and religious significance: the Japanese use special types of wood in temples and household shrines. Wood color and growth ring width are critical in some interior uses. We also have a love affair with standing trees, forests, and natural areas. The chapter on multiple-use deals with these conflicts. For now, let's accept our craving for wood and consider the industry processing it. How important is the industry? What products does it sell? What types of wood does it need from the forest, and how can tree farmers respond most efficiently? How does the industry adapt to changing wood resources, and how can it secure future wood supplies?

Table 13-1 lists notation for this chapter.

PRIMARY TIMBER PROCESSING[1]

The wood products industry is a major part of the United States economy and contributed 8 percent of all value added by manufacturing in 1986. (**Value added** is

[1]Unless otherwise noted, statistics in this chapter come from USDA (1990) and (1993).

TABLE 13-1
NOTATION FOR CHAPTER 13

c = equal annual cost, $/acre
C_0 = reforestation cost in year 0, $/acre
C_t = reforestation cost in year t, $/acre
f = annual inflation rate, percent/100
H_t = clear-cutting revenue at rotation age t, $/acre
M_j = percentage market share of the jth firm, $j = 1, \ldots, N$
p = fixed, real, annual lease payment, $/acre
r = real risk-adjusted interest rate, percent/100
r_f = real risk-free interest rate (percent/100)
t = rotation age, years
T_i = income tax rate, percent/100

the difference between the sale price of goods sold and the cost of materials and supplies used in production.) Nationwide, in 1986 the industry accounted for 9 percent of all manufacturing employment, but in the South, the figure was 12 percent; in Alaska, 15 percent; and in the Pacific Northwest, 24 percent. Table 13-2 shows employment, payroll, and value of shipments by primary timber processing industries. Paper and lumber have similar numbers of employees. But

TABLE 13-2
U.S. EMPLOYMENT, PAYROLL, AND VALUE OF SHIPMENTS IN THE PRIMARY TIMBER PROCESSING INDUSTRIES, 1986*

Industry	Employees (thousands)	Payroll (million 1982 dollars)	Value of shipments (million 1982 dollars)
Timber harvesting	72.3	$1,247.5	$8,219.3
Lumber manufacturing	191.0	3,238.0	17,911.6
Structural panel manufacturing			
Softwood veneer and plywood	35.9	824.5	4,392.3
OSB/waferboard	2.5	87.1	664.8
Total	38.4	911.6	5,057.1
Nonstructural panel manufacturing			
Hardwood veneer and plywood	17.0	268.2	1,582.0
Hardboard, insulating board, particleboard, and medium-density fiberboard	14.3	274.7	1,587.7
Total	31.3	542.8	3,169.8
Woodpulp manufacturing	15.3	592.5	3,829.6
Paper and paperboard manufacturing			
Newspaper	10.7	321.5	2,524.0
Other paper	118.6	4,100.0	23,202.0
Paperboard	51.0	1,710.5	11,138.0
Total	180.3	6,131.9	36,864.0
Other primary timber manufacturing	79.0	1,117.3	5,502.0
Total, primary timber processing	607.6	13,781.7	†

*USDA (1990). Data may not add to totals because of rounding.
†Total omitted, since shipments from harvesting (logs) and woodpulp are inputs to other listed sectors.

the paper sector's output value is much higher because it's the most capital-intensive: to gain such a high output per worker requires large investments in specialized, costly, automated equipment. Pulp and paper together account for 54 percent of value added by manufacturing in the primary timber processing industry; this proportion is 21 percent for lumber and 10 percent for panels.

Lumber

The lumber industry includes mills sawing hardwood and softwood lumber and specialty products like flooring, pallet lumber, railroad ties, and decorative trim. Trends are toward increased recovery of lumber from decreasing log sizes. For example, "chip-n-saw" mills use chippers to square pulpwood-sized logs into cants, which are cut into lumber. From logs, some plants peel thick sheets (often heavier than typical veneers), glue them into large panels, and cut these into "laminated veneer lumber" (LVL). Large beams are made by gluing smaller pieces of lumber together. Mills are becoming more automated and computerized, producing less sawdust and trimmings. In the most modern mills, automated, laser-guided saws are programmed to seek the most profitable cutting pattern for logs and boards.

In past decades, supplies of premium lumber from fine-grained Douglas-fir, other firs, pines, and spruces were plentiful. With dwindling stocks of the best stumpage, some lumber mills sawed lower quality wood and lesser used species like yellow poplar. With improved drying and sometimes laminating, such species have yielded acceptable lumber. Hardwood lumber production increased nearly 40 percent between 1950 and 1986, and hardwood as a percentage of total production increased from about 20 percent to 23 percent.

New housing consumes over a third of United States lumber output. Average new single-family home size increased 36 percent between 1962 and 1986 (from 1,346 to 1,825 square feet). But the amount of lumber used per house increased only 14 percent over the same period (from 11,385 to 12,975 board feet per house). This reflects more efficient use of wood and substitution of panel products, non-wood flooring, plastics, metals, and masonry for lumber. Below you'll see how panel products have been combined with lumber.

Wood Panels

Structural panels include plywood made of rotary-peeled veneers glued together with grains of adjacent sheets perpendicular. Before the mid-1960s, most softwood plywood came from western old-growth logs, mainly Douglas-fir. After 1964, new techniques allowed smaller southern pine logs to be peeled, and by 1990, the South produced over 55 percent of the nation's softwood plywood.

Other structural panels are **oriented strand board (OSB)** and **waferboard,** which are made of wood chips glued together under heat and pressure. In OSB, thin wood strands about ⅝ by 2 ⅛ inches are oriented in three layers at right angles to one another. Waferboard chips are randomly oriented. Most structural panels are used in wall and roof sheathing and subflooring.

Waferboard and OSB can be made from low-quality wood unsuitable for lumber or plywood, for example, from aspen and low-grade softwoods and hardwoods. But pulp mills also compete for these resources. Excellent I-beams are made with flanges of lumber grooved and glued to webs of waferboard. Length is no problem, strength is uniform, and no one worries about the knots and warping found in some solid beams.

Nonstructural panels are used where high strength isn't important, for example, in furniture, cabinets, floor underlayment, insulating wall sheathing, wall panels, and some types of exterior siding. These panels include many types of hardwood plywood, particleboard, medium-density fiberboard, and insulation boards. The latter three are made from mixes of small wood flakes, sawdust, planer shavings, and other mill residue. Depending on the product density, fibers are bonded into panels with different amounts of resins, heat, and pressure. Such panels utilize the very lowest quality wood, some of which used to be burned as mill waste.

End Uses for Solid Wood Products

By far most solid wood products go into residential and nonresidential buildings. As shown in Table 13-3, the remainder goes into manufacturing and shipping materials. In 1986, including residential upkeep and improvements, building consumed 60 percent of the lumber produced, 74 percent of structural panels, and 51 percent of nonstructural panels. Thus, solid wood products consumption tends to follow building cycles, which are sensitive to interest rates (see Chapter 5). This also applies to furniture output (included in manufacturing), which generally rises and falls with new housing construction.

TABLE 13-3
END USES FOR TIMBER PRODUCTS, UNITED STATES, 1986*

End use	Lumber (million board feet)	Structural panels (million ft^2, ⅜-inch basis)	Nonstructural panels (million ft^2, ⅜-inch basis)
Residential building†	29,250	16,120	7,870
Nonresidential construction†	16,363	8,199	2,374
Manufacturing (incl. furniture)	4,805	1,255	7,750
Shipping materials‡	6,785	375	245
Totals	57,203 (82% softwoods)	25,949	18,239

*USDA (1990).
†Includes upkeep and improvements.
‡Mainly pallets.

Pulp and Paper and Allied Products

Between 1950 and 1986, United States annual paper consumption increased from 16.8 to 46.3 million tons, while per capita consumption grew from 221 to 384 pounds. Table 13-2 shows that in 1986 about 9 percent of the pulp and paper shipment value was from the sector producing woodpulp, known as "market pulp," which is dried and sold to the paper and paperboard industry. Although about 29 percent of all paper consumed in the United States is newsprint, the production figures in Table 13-2 don't reflect this, since most of our newsprint is imported (62 percent in 1986, mostly from Canada). Other printing and writing papers constitute about half of our paper use. Annual paperboard consumption, mainly for packaging and shipping containers, increased from 11 to 32.6 million tons between 1950 and 1986. In such uses, paperboard has replaced much lumber and plywood.

Although the pulp and paper industry uses mainly softwoods, the proportion of hardwood roundwood processed rose from 14 percent in 1950 to 38 percent in 1986. This makes good use of abundant hardwoods where nationwide growth has recently been 90 percent above cut. But in stronger grades of paper, the percentage of hardwood is limited, since its fibers are shorter. The industry's chip use has increased from 6 percent of inputs in 1950 to nearly 39 percent in 1986, mainly from mill and logging residues and whole-tree chips. This was an increase from 1.3 million to 36.3 million cords of chips over that period. Again, these are examples of changing technology stretching the wood resource.

A major uncertainty is the future amount of paper recycling. In 1990, 33.6 percent of all wastepaper in the United States was recycled. The industry goal for 1995 is 40 percent (Weyerhaeuser, 1993). In the late 1980s, based on existing recycling trends, the U.S. Forest Service projected a 0.9 percent per year increase in softwood pulpwood consumption in the South until 2040, with real delivered prices increasing at 0.5 percent annually. But by 1993, new projections showed much higher rates of paper recycling, which would shift the future pulpwood stumpage demand curve to the left: at any pulpwood price, less will be purchased, if more recycled fiber is available at a lower cost. Thus, the latest projections are for real softwood pulpwood prices to *decrease* until 2020 in the South (Ince, 1994). Given the same future pulpwood supply curve, can you sketch demand and supply interactions showing how more paper recycling could lead to projections of lower pulpwood consumption and lower pulpwood prices? (The answer is two sentences back.)

In the early 1990s, less than half of the wood fiber input into the United States pulp and paper industry came from pulpwood. About 26 percent of the fiber came from sawmill and other wood waste, and 29 percent came from recycled paper (Weyerhaeuser, 1993a). Whether or not the projected increase in recycling will save trees in the long run is unclear. Using more wastepaper depresses pulpwood prices and reduces private incentives to plant pulpwood trees. This could lead to larger areas of natural forest or nonforest land uses. Also, decreased pulpwood demand makes more wood available for sawlogs, thus shifting the sawlog supply curve to the right. This would make sawlog prices lower than they would have been and increase sawlog consumption.

Paper recycling is now so widespread that prices for many grades of recovered paper are severely depressed. The Northeast has seen negative prices for some grades like old newspapers: a municipality has to pay for someone to take them. This isn't illogical, since otherwise dumping fees have to be paid. The result is highly inelastic wastepaper supply functions: the supply doesn't change much as prices change (Howard et al., 1992). This makes sense because municipal wastepaper collection programs are driven in part by rising landfill costs and by people's environmental consciences. Wastepaper prices play a minimal role.

Industry Trends

Figure 13-1 shows trends in United States roundwood consumption by products. Much of the increase between 1975 and 1981 occurred in fuelwood, when oil prices were rising dramatically. This is an example of the price system at work: as the price of one commodity, heating oil, rose, many buyers switched to a substitute, fuelwood. When oil prices peaked, fuelwood consumption leveled. The Figure 13-1 data include positive net imports (imports minus exports), which ranged from about 7 to 15 percent of consumption during the 1980s. Figure 13-1 combines all species; hardwoods rose from 24 percent of the total in the mid-1970s to 32 percent by 1986. Roundwood use in pulp products shows roughly a doubling between 1950 and 1986. Over that period, if you include chips from all sources, the wood volume input in this sector increased more than fourfold and continues to climb, despite predictions of the paperless office. Computers, word

FIGURE 13-1
United States roundwood consumption by product, 1950–1986. (USFS [1990], p. 42.)

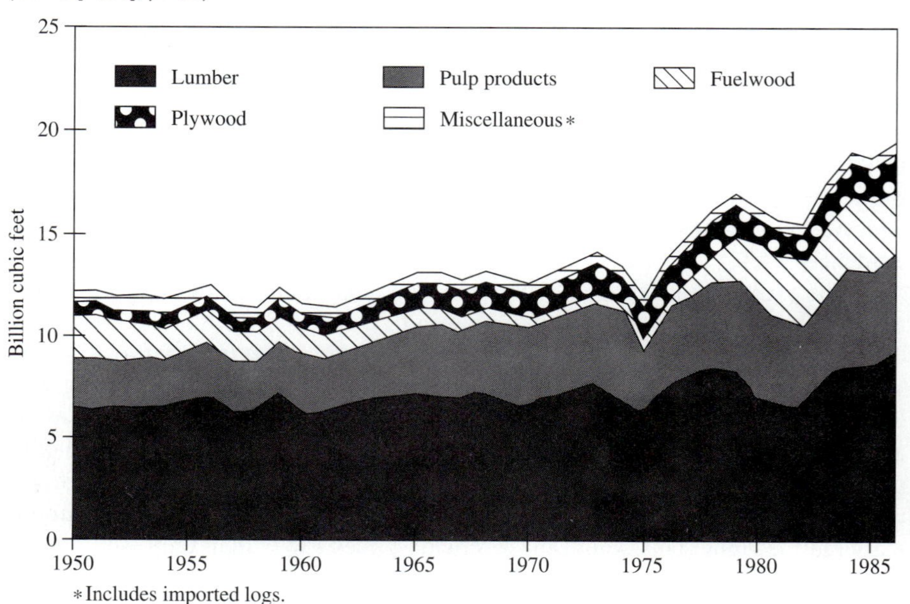

processors, fax machines, and copiers have increased paper use. Also paperboard has displaced wood for shipping containers.

Going back to Figures 5-3 and 5-6, you'll see that around 1974 and 1980, inflation and interest rates spiked, and shortly after each of these peaks, housing starts plummeted. Roundwood consumption troughs occur at the same points in Figure 13-1, driven mainly by changes in roundwood going to building products. Pulpwood output is more stable. Despite a general trend of increasing forest products output shown in Figure 13-1, employment in the industry has fallen, due to increased automation. From 1963 to 1977, employment in the primary timber processing industries hovered around 700,000; by 1986, it was around 607,000, while output was at a record high.

From a timber-growing view, a key trend is that log size and quality are becoming less important. Oriented strand board and waferboard rose from 1.3 percent of structural panels consumed in 1976 to 16.5 percent in 1986. Production of nonstructural panels made of small wood particles doubled between 1965 and 1986. Lumber and plywood, which use the largest and highest quality logs, are the slowest-growing sectors. And even these are using smaller logs. The industry is adapting well to declining log sizes.

I remember in years past bemoaning the scarcity of wide boards, large single-tree beams, clear lumber, and sheets of plywood clear on both sides. Now the best beams are laminated—stronger, more stable, and not unattractive—and they don't require large trees. Plywood is replaced by particleboard for floor underlayment and by various chipboards for structural paneling. Such materials won't make wainscoting in prestigious boardrooms, but they're excellent in unexposed areas, and they don't use big logs. In a furniture showroom, I recently saw my wife lovingly run her hand over an especially fine table and comment on the beauty of solid wood. I peeked underneath and noted a particleboard core covered with veneer and solid edge pieces. But the result, when skillfully done, is extremely attractive, at least as durable as solid wood, and doesn't warp. Moreover, it stretches wood resources.

What do these trends imply for forestry? Does this mean that no one will care about large trees and long rotations? Not at all. You already saw in Chapters 7 and 8 that nontimber values can often justify indefinitely long rotations. And when consumers are willing to pay premiums for larger tree diameters, economically optimal timber rotations lengthen (compare the optimal pulpwood and sawtimber rotations in Chapter 7, Tables 7-3 and 7-4). But strictly for wood production, long rotations won't be common if consumers aren't willing to pay enough for products needing large logs. In that case, cutting slowly-growing trees and reinvesting the proceeds will yield more income than holding the trees to produce large logs. Remember that such things are rarely all one way or the other. A market will probably always exist for some high-priced large logs, especially when they become scarcer. So you could find long rotations and large trees justified strictly for timber purposes on limited areas. But in many forests, nontimber values may become the main reason for longer rotations.

Three trends are helping the industry to squeeze more products from a given sized tree. (1) We're getting better lumber and plywood recovery. For example, in the South, board feet of lumber per cubic foot of logs increased from 5.05 to 6.02 between 1952 and 1985. That's projected to increase to over 7 by the year 2040. In the Inland Empire, square feet of ⅜-inch plywood per cubic foot of logs increased from 13.3 to 17.2 between 1952 and 1985, with a projection to 19 square feet by 2040. (2) Products with the highest recovery are becoming dominant. Plywood and composite lumber often replace lumber, and composite panels can substitute for plywood. (3) Increasingly more mill residue goes to consumer products.

REGIONAL IMPORTANCE OF THE FOREST INDUSTRY

The forest industry tends to own the more productive timberland; it has 14 percent of the United States commercial forest area but produces one-third of the United States harvest. However, for industry's softwoods (roughly 80 percent of its cut) timber harvest was about 30 percent above growth in 1986 and 1991. The forest regions mentioned below are shown in Figure 13-2. Figure 13-3 shows that 55 percent of the industry's 70 million acres of timberland is in the South, where productivity is high. You'll also see that the country's largest concentration of nonindustrial private forest (NIPF) lands is in productive southern lands. There we find the greatest industry interest in NIPF leasing and assistance programs, due to the land's productivity and industry's presence. Nationwide, NIPFs are 59 percent of the commercial forest area, but the percentage is higher in the South.

Although public timber acreage and volume are greatest in the West, future harvesting there is likely to be dampened by economic inaccessibility on some areas (logging and hauling cost exceeding delivered log price) and rising public pressures to limit harvesting or at least limit clear-cutting in other areas. Com-

FIGURE 13-2
Regions for U.S. Forest Service renewable resources planning act assessments. (USFS [1993], p. 3.)

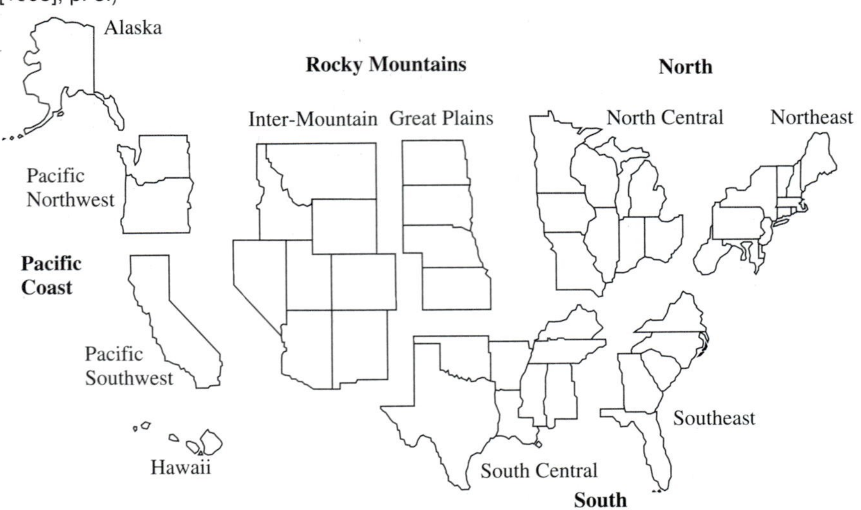

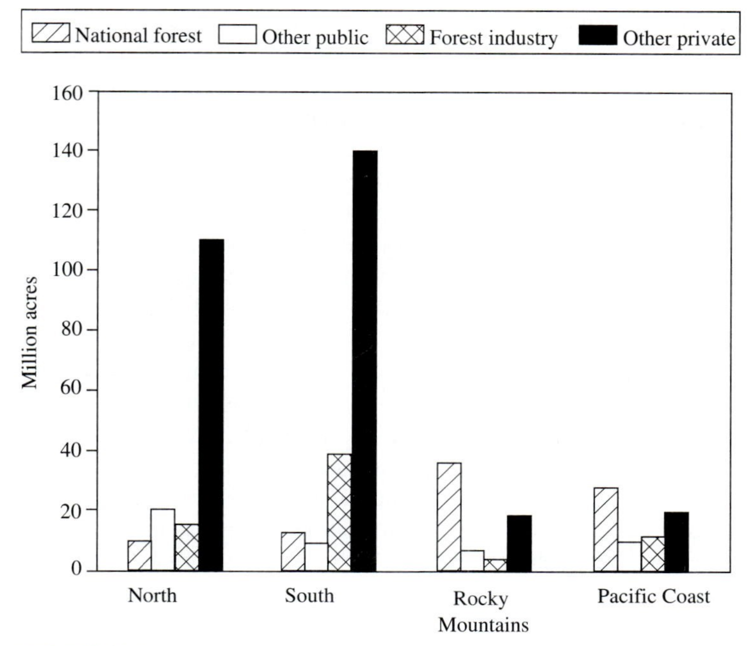

FIGURE 13-3
Timberland ownership by region, 1992. (USDA [1993], p. 8.)

pared to the South, the Northeast has slower timber growth rates and is feeling greater public pressures for nontimber outputs. Thus, the South should continue to be the major industrial wood basket of the United States. The South yielded 55 percent of the nation's cut in 1991—up from 45 percent in 1970.

The first stage of wood processing is to produce logs from trees. Let's now consider how logging managers decide where to market different logs.

LOG MARKETING DECISIONS

Trees are cut for uses like firewood, pulpwood or chipping logs, chip-n-saw logs, sawlogs, veneer logs, and poles. If large enough, poles and veneer logs often bring the highest price per unit volume, delivered to the mill or other final market. This price is called the *delivered price* or *mill price.* In one market, the price usually drops as you move to sawlogs, pulpwood, and firewood. Timber growers try to market logs in a way that maximizes net stumpage revenue per acre. Because higher logging and hauling costs generally reduce stumpage values, the maximum delivered log value doesn't necessarily yield the highest stumpage revenue [see equation (11-1)]. Maximizing stumpage revenue is tricky because each log type must meet certain specifications for diameter, length, species, and quality. Only the straightest trees of certain sizes and species can be sold as high value poles. For the right species and quality, large diameter logs generally bring the highest delivered price as veneer logs for making plywood. Logs not meeting veneer log or pole specifications will usually bring the best price as sawlogs or chip-n-saw logs. Logs too small or of too poor quality to make sawlogs will go for pulpwood, chipping for composite board, or firewood.

Thus, to maximize returns, the butt log of a large tree may yield veneer; the midsection, sawlogs; and the top, pulpwood. The decision also depends on mill locations and quantities of logs. For instance, the veneer mill may be so much farther away than the sawmill that the extra hauling cost would exhaust the veneer price advantage and make sawlogs the best market even for potential veneer logs.

In some cases, tree-length logs are hauled to log concentration yards where they're "bucked" into profit-maximizing combinations of log types. In other cases, felled trees are bucked in the woods. Bucking and sale decisions can be complex and have a huge effect on harvest revenues. Some decisions are easy: where pulpwood values are low and sawlog values high, you wouldn't send potential sawlogs to the pulp mill. Good-quality veneer logs are usually worth more per cubic foot than sawlogs. But sometimes small diameter logs are worth less as veneer logs than as sawlogs. To maximize timber returns, you have to decide exactly where to cut tree stems into logs, how long to make each log while paying attention to quality and size requirements for each log type, and where to market each log. The optimum bucking decision depends on the following factors for each log type: requirements about species, size, and quality; prices for each log type delivered to mills; hauling costs to each mill; bucking and skidding costs. Handheld computers have been programmed to help find the profit-maximizing bucking pattern in the woods. The decision can also be made with the help of tree-length log scanners at modern mills.

Ideally, timber sale managers should know delivered prices of log types and the logging and hauling costs for different log sizes. Below some log diameter, it may cost more to remove logs than they're worth at the mill. Thus, it's sometimes most efficient to leave low-quality logs at the logging site, although that may look "wasteful." Such residue can supply needed organic matter for the soil. As delivered log prices rise, less material is left in the woods. For example, suppose a local mill pays $75/MBF for logs with 5- to 7-inch scaling diameters. If the extra cost (marginal cost) of skidding, loading, and hauling for such logs exceeds $75/MBF, it's most efficient to leave them at the logging site. If delivered prices rise above $75, other things being equal, it may pay to sell these logs to the mill. Note the marginal analysis: You compare the added revenue and added cost of removing certain log types.

Careful log marketing can have a big impact on harvest revenues. But projecting distant-future harvest income presents real problems.

Projecting Future Harvest Revenues

Wood-processing dynamics make it hard to predict future harvest income. Nationwide, the average softwood stumpage price for a mix of sawlogs and veneer logs is currently over three times higher than that for softwood pulpwood. In many regions, the spread is much greater. You'll often find landowners projecting volumes and prices of log types several decades into the future to estimate harvest incomes. For example, for a new plantation on a given site, a stand simulator might project, per acre, 12 cords of pulpwood, 16 MBF of sawlogs, and 10 MBF of veneer logs in 35 years. For these products, if you project stumpage prices of $35/cord, $175/MBF, and $220/MBF in constant dollars, you'll estimate a harvest revenue of:

$$12(\$35) + 16(\$175) + 10(\$220) = \$5{,}420 \text{ per acre}$$

which might be used in computing **net present value** of a forest property. But this may not be too meaningful if you're not even sure what future products will be.

One of the most flexible ways to project harvest values is to simulate scenarios of stumpage prices per unit volume at future dates, using different relationships between average stand diameter and stumpage price, as shown in Figure 13-4. The solid curve projects no premium paid for increasing log diameter, for example, by assuming all logs went to chipboards and pulp. The other two curves model stumpage price increasing with greater average tree diameter, using two different relationships. For each curve shape, higher or lower positions can reflect different average price levels. Information from wood technologists and economists can help to define a range of price-diameter curve shapes and their probabilities of occurrence. The position of the curves, higher or lower, at different dates, would depend on rates of stumpage price-increase, usually projected with regression equations by economists. Using simulators to project average stand diameter and volume, you'd multiply volume by the appropriate price to project harvest value for a given case. This approach lets you easily simulate a wide range of scenarios, economic analyses of which could help managers make decisions.

In Figure 13-4, the more steeply the price-diameter curve rises rightward, the greater the predicted premium paid for increasing log diameter and the longer the financially optimal timber rotation. In the West, where log diameters vary widely, large logs yield much higher product values per unit volume than small logs. One study in the Northwest shows lumber recovery values for 40-inch-diameter logs over twice as high as those for 10-inch logs. For plywood, the value advantage of larger diameter is even greater (Haynes, 1986).

TIMBER SUPPLY STRATEGIES

If timber supply security were paramount, wood processors would get all logs from their own lands. But owning an entire wood supply is a major capital commitment and has an enormous opportunity cost. Suppose a firm owns 500,000 acres of timberland that could be sold for an average of $800 per acre for a total of $400,000,000. Stockholders and managers will seek acceptable **rates of return** on this asset value. In **real** terms, suppose you could earn 4 or 5 percent elsewhere on this capital investment—or $16 million to $20 million forgone

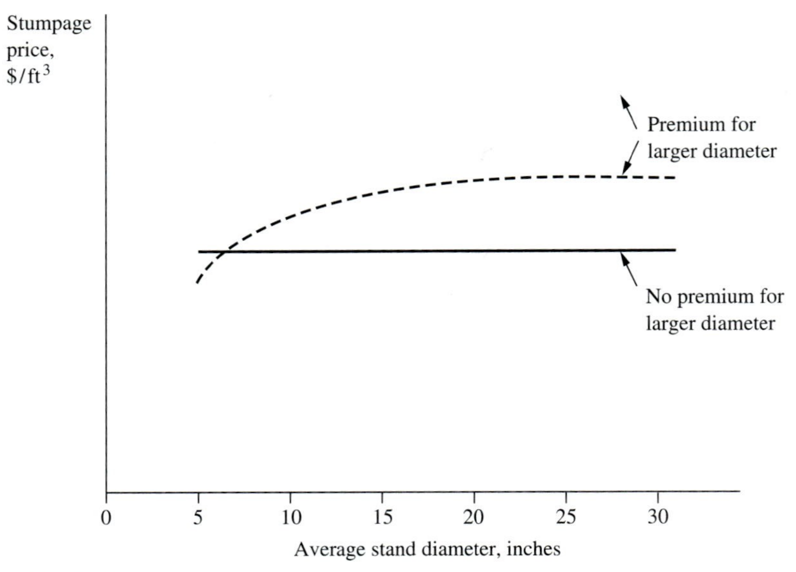

FIGURE 13-4
Hypothetical projected relationships between stumpage price and average-stand diameter.

annually.[2] Often such earnings are more than enough to buy wood on the open market and at the same time provide a major cash flow improvement. Such logic has spurred some wood-processing firms to sell part of their lands—sometimes to timberland investment management organizations (TIMOs). The sellers reasoned that the timber would still be available for purchase in the future. TIMOs are groups of investors who pool their cash to buy large timber tracts that they couldn't afford individually. Such groups manage timber and sell stumpage to the wood-processing industry. Between 1987 and late 1990, TIMOs increased their holdings in the United States by 148 percent to 1.78 million acres, buying mainly from the forest industry (Zinkhan, 1992). By 1994, one TIMO alone, the Hancock Timber Resource Group, owned over 2 million acres of timberland.[3] But as of 1994, TIMOs were probably still less than 2 percent of the private commercial forest area. And overall, there's no evidence that timber processors are shying away from land ownership, although their land acquisition rates have recently slowed: Industry forestland ownership increased gradually by about ½ percent per year from 58.98 to 70.35 million acres between 1952 and 1987. During this time total forest area in all ownerships decreased by about 5 percent. After 1987,

[2]Not understanding the importance of real returns, some might point to **nominal** earning rates of, say, 8 percent, on moderate-risk bonds and incorrectly demand 0.08(400 million) = $32 million. But with bonds, earnings and asset values don't increase with inflation the way they generally do in the forest. Thus, the correct forest opportunity cost is the real interest rate times the forest value (see Chapter 5).

[3]Hancock Timber Resource Group's 1994 Christmas card.

industry acreage was fairly stable, rising slightly to 70.46 million acres by 1992. So timberland ownership is still important in the industry.

Institutional investors such as pension funds and university endowments have recently increased timber holdings. One estimate puts these assets at over \$2.2 billion by the start of 1993 (Binkley and Washburn, 1993). These lands are another part of the nonindustrial private forests, which are a major source of timber for wood processors.

In some parts of the West, so much forestland is publicly owned that many processors buy mainly public timber. But where enough private land is available, mills often maintain a supply mix of stumpage from the open market and from their own land. In 1979–1980, of the top 10 United States forest products firms in sales, 6 obtained roughly 50 percent or more of their timber from their own lands. At that time, the top 40 wood products firms made 40 percent of the industry's sales but owned 80 percent of the industrial forest area. This leaves many firms relying mostly on open market timber (O'Laughlin and Ellefson, 1982).

For a timber processor to buy all wood on the open market can be risky: if many other firms are competing, stumpage prices may be driven up too high. But it can also be expensive for firms to grow their own timber (often called **fee timber** from land owned "in fee simple"). Suppose, to grow pulpwood on a 25-year rotation, a firm would have to buy land for \$300/acre, spend \$120/acre on afforestation, and incur \$5/acre/year costs. At 6 percent real interest, net accumulated costs in 25 years would be:

$$-300(1.06)^{25} - 120(1.06)^{25} - 5\frac{(1.06)^{25} - 1}{0.06} = -\$2{,}077 \qquad (13\text{-}1)$$

This says that if these costs had been invested elsewhere at a 6 percent real rate of return, they would have accumulated to a value of \$2,077 in constant dollars. You can't charge this all to timber, since land value, say at a real \$350, exists at rotation-end when growing your own timber. Thus, the timber opportunity cost is \$2,077 − \$350 = \$1,727. If expected yields are 56 cords, the cost of fee timber is \$1,727/56 = \$31 per cord. The firm could decide whether it could reasonably expect to buy timber at this price in 25 years. More realistic calculations would include risk, taxes, and other costs or revenues.

Landowner Assistance Programs

To assure timber supplies, investors planning to build mills usually analyze available timber inventory and growth/cut ratios within reasonable hauling distances. Firms needing open market timber often increase future wood supplies by helping nonindustrial private forest (NIPF) owners. Often industry foresters give NIPF owners forestry advice and sometimes provide services like reforestation at or below cost. Some firms call these plans "landowner assistance programs" (LAP) or "cooperative forest management" (CFM).[4] Companies sometimes offer planting

[4]Information on LAPs and leasing comes from Meyer (1984).

and other forest management services, either free or at reduced cost, in exchange for an agreement giving the firm certain rights when the landowner decides to harvest. Under one such agreement, the landowner must, at harvest time, notify the assisting firm and allow it to bid competitively on the timber. After bids have been made on a sale, another agreement, called a "right of first refusal," allows the assisting firm to match the highest bid and buy the sale. Because this can discourage other bids, it's a disadvantage to the landowner. In other cases, assisting firms simply ask that the landowner give them an opportunity to bid competitively on any timber sold. Some firms offer assistance with no strings attached, knowing that the LAP will increase the probability of timber being available in the future. In 1984 about 8.7 million acres of NIPF lands were in industry-sponsored LAPs in the southern United States. Another way to get future wood without buying land is to lease timberland.

Timberland Leasing

Wood processors often lease NIPF land for an annual fee giving the firm (the **lessee**) the right to harvest timber at a given date. In the United States, most timberland leasing has been in the South, where acreage under lease declined from about 6 million acres in 1967 to 4.7 million acres by 1984. The decline occurred partly because many contracts didn't allow lease payments to increase with rising timber prices. Thus, as stumpage prices increased over time, many landowners (**lessors**) were dissatisfied. More recent lease contracts solve this problem by having lease payments increase with certain price indexes. Such contracts guarantee future timber for mills and give annual income to landowners who'd otherwise have to wait many years for timber revenue.

The idea of leasing forestland is to adjust annual lease payments so that the net present value of payments to the landowner equals the net present value of future harvest income forgone, given some interest rate, all after taxes. Assuming accurate income predictions, suppose the landowner invests lease payments at the assumed interest rate until the end of the lease period. Then the accumulated value of lease payments with interest would equal the accumulated net value of harvest incomes forgone (and received by the lessee during the lease period). Thus, the lessor is happy. And so is the lessee because the accumulated cost of annual lease payments made is exactly offset by the harvest income received, if both parties use the same interest rate. When timber is very young or land is bare, the attraction of leasing for the landowner is to receive annual income rather than having to wait until harvest, and at the same time to retain land ownership. The lessee finds it attractive to gain future timber without buying land.

The main problem is that rarely does the future unfold as originally expected. Also, lease payment calculations can get complicated if you consider that lessor and lessee will usually have different tax rates, discount rates, ideas for payment escalators, projections of harvest income, and opinions on dealing with future harvest values that differ from projections.

In figuring lease payments, the following equality must hold for the landowner:

$$\text{NPV of owner's timber income forgone} = \text{NPV of lease payments received} \tag{13-2}$$

where NPV is net present value. Thus, landowners get the same present value from lease payments that they would have received from timber harvesting. Since contractual lease payments are practically **risk-free** income, landowners should discount them with an interest rate lower than that for discounting the risky timber income. For the firm, a similar equality must hold:

$$\text{NPV of firm's harvest income received} = \text{NPV of lease payments made} \tag{13-3}$$

In present value terms, the firm is willing to pay no more in lease payments than the value of future income expected. If the payments are contractually set, they are certain, and the firm should discount them with a risk-free interest rate lower than the risky timber discount rate. Using the above equalities, you can solve for the minimum acceptable annual lease payment for the landowner and the maximum annual payment the firm is willing to make.

As an example of a lease contract with all values in **constant dollars** of year 0 and per acre where relevant, suppose a lessee (a firm) leases *bare land,* which, if reforested, could yield an **even-aged** *clear cut revenue* of H_t = \$3,000 in t = 30 years. Assume that lease payments are at the end of each year and are fixed at \$$p$/year in real terms for 30 years, but increasing with some inflation index (often the producer price index). The firm will pay annual costs of c = \$5/year (management and property taxes) and reforestation costs of $C_0 = C_t$ = \$110 at the start and end of the rotation. Newly reforested land is turned back to the lessor (landowner) at rotation-end. For both parties, let the real **risk-adjusted discount rate** r be 7 percent and the risk-free or low-risk discount rate r_f be 4 percent. At state and federal levels, the firm pays income taxes at $T_i = 0.37$, or 37 percent, and the landowner pays at 31 percent. Expressing the equalities of equations (13-2) and (13-3) in constant dollars for both the landowner and the firm:

NPV of timber income =

$$\frac{H_t - T_i[H_t - C_0/(1+f)^t] - C_t}{(1+r)^t} - C_0 - c(1 - T_i)\left[\frac{1-(1+r)^{-t}}{r}\right] \tag{13-4}$$

$$= p\frac{(1-T_i)[1-(1+r_f)^{-t}]}{r_f} = \text{NPV of lease payments}$$

The equation's first term is the after-tax present value of future harvest revenue minus reforestation cost in year t, the second term is the reforestation cost in year 0, and the third term is the present value of tax-deductible annual costs, fixed in real terms but rising at the inflation rate in current dollars. The right side is the present value of the low-risk lease payments discounted with the risk-free rate.

This is a simplified case. Different equations would be needed for each party if their management or tax structure differed.[5] The term $(1 - T_i)$ on the left side reflects deductibility of annual costs for income tax purposes. On the right side, $(1 - T_i)$ shows deductibility of p for the firm and payment of income taxes on p for the landowner. Solving equation (13-4) for the initial lease payment p,

Initial lease payment =

$$p = [\text{NPV of timber income in eq.(13-4)}]\frac{r_f}{(1 - T_i)[1 - (1 + r_f)^{-t}]} \qquad (13\text{-}5)$$

Since *lease payments increase at the inflation rate f*, they are constant in real terms. Substituting the landowner's values into equation (13-5) and projecting 5 percent inflation, the landowner's minimum acceptable lease payment is \$8.86/acre. Substituting the firm's values into equation (13-5), the firm's maximum initial lease payment is \$7.89/acre. If annual inflation were, say, 5 percent, lease payments would increase by 5 percent per year. Here the landowner's minimum required payment exceeds the maximum amount the firm is willing to pay. The calculations are starting points for negotiations between both parties.

The above hypothetical examples shouldn't suggest typical lease payments. And remember that equation (13-5) deals only with a one-rotation lease of initially bare land under even-aged management. Equations and calculated results will vary with each situation. For example, equation (13-4) could model uneven-aged management where clear-cutting never occurs. You could also provide for payments from the lessee (lessor) at harvest dates if real stumpage prices were higher (lower) than originally assumed (but that's not a common provision).[6] Management assumptions and responsibility for taxes and other expenses can vary. Some have suggested adjusting lease payments downward and having the lessee receive, say, 50 percent of the harvest volume, with the landowner selling the remaining 50 percent to the highest bidder.

Variations on the leasing theme are many. Table 13-4 outlines several types of cutting contracts and firms' 1984 ratings of those holding the most promise for the future. Both timberland leasing and the industry's landowner assistance programs are examples of free markets stimulating increased wood production: the higher the expected future stumpage price, the more interest industry shows in such programs and the more wood is grown on private nonindustry lands (see Chapter 12).

[5]For simplicity, the same equation is used for lessor and lessee. For the firm, C_t should be multiplied by $(1 - T_i)$ since it can be immediately deducted for income tax purposes because the firm could never deduct it later. Also some landowners (generally not the larger ownerships) can take advantage of reforestation tax benefits, which could be built into the lessor's equation (see Appendix 9A). The lessee usually buys standing timber for a lump sum. Shaffer et al. (1985) show more detailed payment equations for lessor and lessee.

[6]See Klemperer (1986) for details on different lease payment indexing schemes.

TABLE 13-4
TYPES OF TIMBERLAND LEASE CONTRACTS AS RATED BY FIRMS*

Contract type	Percentage of firms giving top three rating as most acceptable to them and landowners
Lease of land with annual or periodic payments plus initial lump-sum purchase of timber	82%
Lease of both land and timber with annual or periodic payments with no additional payment when timber is cut	45
Sale of timber cutting rights with payment on a volume basis as cut	35
Share crop contract: firm manages land and harvests timber; harvest value at current market price is shared with landowner as contracted	35
Lease of land with annual or periodic payments plus timber cutting rights with timber paid for when cut	33
One initial lump-sum payment that covers both land lease and timber purchase for term of contract	24

* From Meyer (1984).

How Much Timberland Should a Firm Own?

There's no agreement on a wood processor's optimal degree of timber self-sufficiency. Some investors, especially from overseas, have built mills in the United States without owning any timberland. One study of 20 large forest products firms showed a weighted average of 43 percent of timber input coming from company lands in the mid-1970s (Clephane, 1978).

What is a firm's optimum amount of owned timberland? Theoretically, for some needed annual wood volume for a given period, the optimal mix of timber from fee lands, leases, landowner assistance programs, and the open market is that which has the lowest cost (or smallest reduction in **utility** to the firm).[7] As shown in Chapter 10 you could construct present value probability histograms of wood costs over time for different procurement strategies, and the firm's management could choose the one that brought the least reduction in utility. (Rather than the utility from revenues discussed in Chapter 10, here it's the disutility from wood costs.) For example, suppose you estimated that in 10 years there was a 25 percent probability of getting open market wood for about $15/cord, a 50 percent probability of $25, and a 25 percent probability of $45. If the probabilities are correct, each outcome is conceptually risk-free, and the values would be discounted for 10 years with a risk-free rate of, say, 3 percent, giving present values of wood costs at $11.16, $18.60, and $33.48 with the above probabilities.

[7]To analyze the buy versus lease decision for timberlands, see Klemperer and Greber (1986).

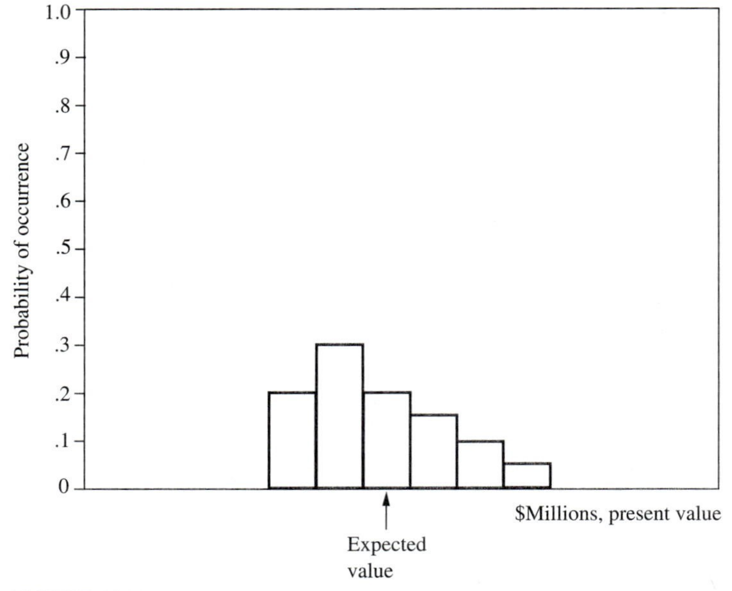

FIGURE 13-5
Hypothetical present-value probability histogram for open-market wood costs—10 million board feet annually for 20 years.

If you could estimate probabilities of getting wood at different prices for each year in the planning period, a simulation program could construct a present value histogram for wood costs such as the hypothetical one in Figure 13-5. Ideally, you'd want to generate present value probability histograms of this type for each mix of annual wood sources for a given period of time and annual wood volume. Examples of mixes are ½ open market and ½ fee—or ⅓ each of open market, LAP, and fee. Since most managers tend to be **risk-averse,** they'd prefer the lowest **variance** in cost for any **expected value** of costs, in present value terms.[8] It would be impossible to find exact probabilities of outcomes for each mix of timber sources. But interviews could yield subjective probabilities for the future costs, and with the simulated histograms, managers could choose a preferred timber supply plan. A simpler approach would use **triangular probability distributions** as discussed in Chapter 10. In practice, decisions about optimal mixes of wood supplies tend to be more intuitive. Also, firms may hold timber for reasons beyond guaranteeing a wood source, for example, as a hedge against inflation, simply as an investment, or to diversify capital holdings and reduce risk.

Let's now consider degrees of competition in the wood-processing industry.

[8]Recall that the expected present value is the sum of outcomes times their probabilities. For the above example, the expected value of costs, in present value terms, is 0.25(11.16) + 0.50(18.60) + 0.25(33.48) = $20.46/cord.

ECONOMIC CONCENTRATION

Chapter 3 mentions the problem of market power where one firm or a small group of firms in one industry can restrict output, keep prices higher than the competitive equilibrium, and earn higher-than-average rates of return. Firms that have some control over product price are said to have **market power.** When sales become more concentrated among a small number of producers, they have more market power. A **monopolist,** being the only producer in an industry, has the most market power.

Chapter 3 shows that too much market power can lead to inefficient resource allocation: If one firm or sector consistently earns a higher-than-average rate of return, total returns can be increased by shifting capital from lower return industries to those with higher returns. Therefore, the U.S. Justice Department breaks up firms or prevents mergers when certain measures indicate too much concentration in one industry like plywood or paper. But as mentioned in Chapter 3, economies of scale mean that some industries may be more efficient when firms are fairly large. Market power is of most concern in industries where substitute products don't compete: generally, for solid wood products, many substitutes exist. With growing world trade in forest products, even the largest firms face increasing competition, for example, in panel products, lumber, pulp, and paper (see Chapter 17).

Measures of Industrial Concentration

One measure of market power is the **concentration ratio,** which gives the percentage of one industry's output value produced by the largest firms, usually the top four or eight (e.g., "four-firm concentration ratio").

Another more sophisticated gauge of concentration is the **Herfindahl index,** or the sum of squared market shares of all firms in an industry:

$$\text{Herfindahl index} = \sum_{j=1}^{N} M_j^2 = M_1^2 + M_2^2 + M_3^2 \ldots \tag{13-6}$$

where M is the jth firm's market share as a percent. When one firm has 100 percent of the market (monopoly), the index is at its highest: $100^2 = 10{,}000$. With perfect competition, the index approaches 0. If three firms each sell 33.33 percent of an industry's output, the index is 3,333. The Antitrust Division of the Department of Justice computes measures of industrial concentration and has guidelines for permissible Herfindahl indexes when ruling on mergers and acquisitions.

Concentration in the Forest Products Industry[9]

To classify the degree of competition based on ranges of four-firm concentration ratios, one arbitrary grouping is as follows (with sample industries from the United States in parentheses, for the mid-1980s):

[9]See Ellefson and Stone, 1984; and U.S.D.C., 1992.

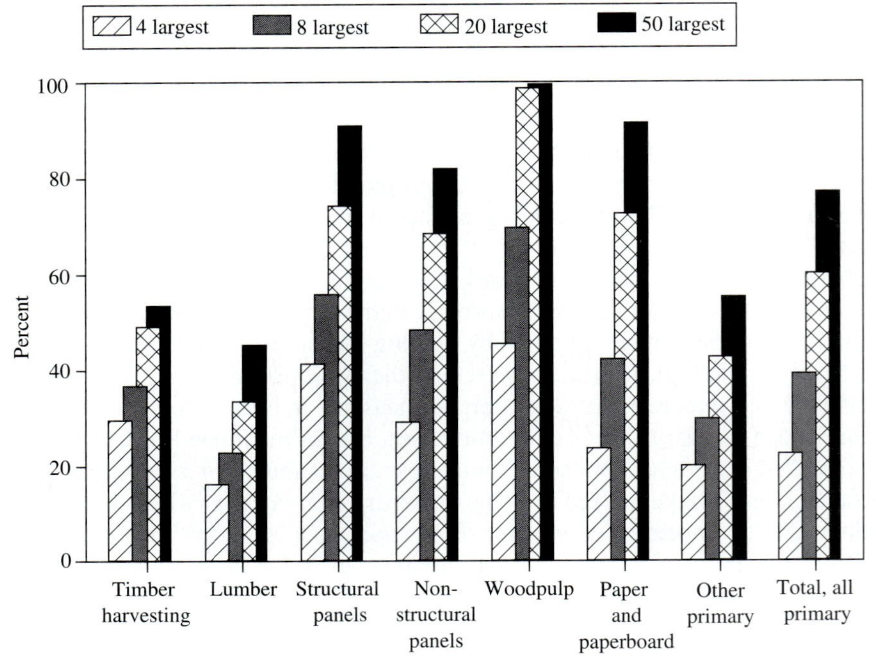

FIGURE 13-6
Market shares of largest United States forest products firms, 1985. (Percent of value of shipments by company size and industry. (USDA [1990], p. 68.)

- Low concentration: 24 percent ratio or lower (sawmills)
- Moderate concentration: 25 to 49 percent ratio (softwood veneer and plywood, pulp mills, and paper mills)
- High concentration: 50 to 74 percent ratio (sanitary paper products)
- Extreme concentration: 75 percent ratio or higher (cigarettes, automobiles, household laundry equipment)

With increasing international competition, United States ratios become less meaningful measures of market power, since foreign firms also compete.

Figure 13-6 gives concentration ratios of the largest firms producing selected forest products in the United States. The figure shows that the highest concentration occurs in pulp and structural panels and the lowest in lumber. By the above grouping, the forest products industry shows moderate to low concentration. Often you'll find the highest concentration where the cost of building plants is greatest and hence where entry is the most difficult. For example, average costs of pulp and paper mills are higher than for sawmills.

High concentration in an industry increases the likelihood that its output will be lower and prices higher than under a competitive equilibrium (see "Imperfect Competition" in Chapter 3). High concentration also makes collusion more likely among major firms to fix output prices and weaken competition. During the 1970s, forest products firms paid over $500 million in fines levied by the Federal

Trade Commission for alleged price-fixing. Most of the claimed antitrust violations were in the paper and allied products industry, where concentration is the highest. However, in the 1980s, several plywood producers paid heavy fines for alleged price-fixing (Ellefson and Stone, 1984).

RESEARCH INVESTMENT[10]

"Technical change," or applying new technology, is influenced by spending on research and adopting foreign technology. Despite major changes in United States wood-processing methods, the industry's research spending is low. In 1986, as a percentage of net sales value, research received 0.5 percent in the solid wood products industry and 1.3 percent in paper and allied products. For all industries in the United States, the figure was 3.2 percent. For public programs, forestry research spending as a percentage of output value was less than ¼ of that in agriculture in 1980. But the United States has borrowed increasingly more technology from overseas, for example, in areas of structural panels, pulp and paper, and containerized seedlings. As a percentage of global forestry research spending, the North American share dropped from 42.5 percent to 32.2 percent between 1970 and 1981. Some of the technology leaders are Scandinavia in mechanical pulping, Germany in structural particleboard, and Japan in papermaking. This technical change has been heavily capital-intensive and labor-saving.

KEY POINTS

- The wood products industry contributes about 8 percent of all value added by manufacturing in the United States.
- Wood products output continues to increase, although employment in the industry has fallen, due to increased automation.
- Log size and quality have been decreasing, and the percentage of hardwood harvest has increased. The industry is adapting to this changing log mix by increasing output of products like oriented strand board, waferboard, particleboard, laminated veneer lumber, and laminated lumber.
- Building consumes about 60 percent of lumber output and 78 percent of panel products. Thus, these industries are sharply affected by housing cycles. Pulp and paper output is somewhat more stable.
- The South should continue to be the nation's wood basket, due to high timberland productivity, a large percentage of private forest ownership, and fewer environmental problems than in most other regions.
- Harvesting and log marketing decisions are driven by relative prices of log types.
- Forest industry land ownership has been fairly stable since 1987. Through leasing and landowner assistance programs, the industry increases timber supplies from nonindustrial private forests.

---

[10]This section draws from Bengston and Gregersen (1992).

- There is no agreement on a mill's optimal degree of timber self-sufficiency. Theoretically, probability theory could be used to help make the decision, but data are hard to find.
- Overall, the forest products industry is fairly competitive, with the lowest concentration in lumber and the highest in woodpulp and structural panels.

QUESTIONS AND PROBLEMS

13-1 Which sector of the wood products industry has the largest output value?

13-2 What are the general trends in United States wood products output and employment? What contributes to these trends?

13-3 What are the general trends in log quality and species composition? How is the forest industry adapting to these trends?

13-4 What sector is the major user of solid wood products? What does this mean for price and output stability in the solid wood products industry?

13-5 Of the United States commercial timber area, the forest industry owns 14 percent, while over half is in nonindustrial private forest (NIPF) ownership. In what ways does the forest industry stimulate more intensive management of NIPF lands?

13-6 Other things being equal, what is the general relationship between tree diameter and stumpage price per unit volume? Why?

13-7 Suppose, for a given species, delivered log prices per MBF are $150 for sawlogs from 6 inches to 10 inches in diameter and $175 for veneer logs from 10 inches to 15 inches. Under what conditions might a logger logically sell veneer-quality logs to a sawmill for $150/MBF?

13-8 A firm wishes to lease bare land from a landowner for timber production over a 25-year rotation. The firm has calculated its net present value of expected harvests at $250/acre. The firm's risk-free, real, after-tax hurdle rate is 3 percent, and its combined income tax rate is 37 percent. What is the firm's maximum fixed, real, annual lease payment? Nominal payments would rise at the inflation rate.

13-9 What is the optimal mix of company timber and open market timber for a wood-processing company? (*Hint:* There's no simple answer: discuss the principles involved.)

13-10 The four-firm concentration ratio for structural wood panels is estimated to be 40 percent. What does this mean? What are the possible problems with excessive concentration in an industry?

REFERENCES

Bengston, D. N., and H. M. Gregersen. 1992. Technical change in the forest-based sector. In Nemetz, P. N. *Emerging Issues in Forest Policy.* Univ. of British Columbia Press, Vancouver. pp. 187–211.

Binkley, C. S., and C. L. Washburn. 1993. Investing in timberland: Does the vehicle matter? *Proceedings of the 1993 Southern Forest Economics Workshop.* U.S. Forest Service Southeast Forest Experiment Station. Asheville, NC. pp. 94–107.

Cleaves, D. A., and J. O'Laughlin. 1985. Forest inventory, plant locations, and firm strategies in the southern plywood industry. *Proceedings of the Southern Forest Economics Workshop.* North Carolina State University, School of Forestry, Raleigh. pp. 35–43.

Clephane, T. P. 1978. Ownership of timber: A critical component in industrial success. *Forest Industries.* August. pp. 30–32.

Ellefson, P. V., and R. N. Stone. 1984. *U.S. Wood-Based Industry—Industrial Organization and Performance.* Praeger, New York. 479 pp.

Haynes, R. W. 1986. *Inventory and Value of Old-Growth in the Douglas-Fir Region.* U.S. Forest Service Pacific Northwest Research Station, Portland, OR. Research Note PNW-437. 19 pp.

Howard, J. L., P. J. Ince, and J. Bickelhaupt. 1992. Economic projections of recovered paper supply in the United States. In Adams, D., R. Haynes, B. Lippke, and J. Perez-Garcia (compilers). *Forest Sector, Trade and Environmental Impact Models: Theory and Applications.* Proceedings of a Symposium. CINTRAFOR. University of Washington, College of Forest Resources, Seattle. pp. 226-229.

Ince, P. J. 1994. *Recycling and Long-Range Timber Outlook.* U.S. Forest Service Rocky Mt. Forest and Range Experiment Station. Gen. Tech. Rept. RM-242. Fort Collins, CO. 23 pp.

Klemperer, W. D. 1986. Adjusting timberland lease payments for stumpage price-changes. *Northern Journal of Applied Forestry.* 3(1):22–25.

Klemperer, W. D. and B. J. Greber. 1986. Economics of buying versus leasing timberlands. *Southern Journal of Applied Forestry.* 10(4):211–214.

Meyer, R. D. 1984. *An Analysis of Long-Term Leasing and Cutting Contracts in the South.* M.S. Thesis. Virginia Polytechnic Institute and State University, Blacksburg. 159 pp.

Meyer, R. D., W. D. Klemperer, and W. C. Siegel. 1986. Cutting contracts and timberland leasing. *Journal of Forestry.* 84(12):35–38.

O'Laughlin, J., and P. V. Ellefson. 1982. *New Diversified Entrants among U.S. Wood-Based Companies: A Study of Economic Structure and Corporate Strategy.* University of Minnesota Agricultural Experiment Station. Bulletin 541. St. Paul. 63 pp.

Pease, D. A., T. Blackman, and J. A. Sowle, (Eds.). 1991. *Forest Industries 1991–92 North American Factbook.* Miller Freeman, Inc., San Francisco.

Shaffer, R. M., W. D. Klemperer, and R. D. Meyer. 1985. Determining the initial annual payment for a long-term timberland lease. *Southern Journal of Applied Forestry.* 9(4):250–253.

U.S. Department of Commerce. 1992. *1987 Census of Manufactures—Concentration Ratios in Manufacturing.* MC87-S-6. Washington, DC. 45 pp. plus app.

U.S.D.A. Forest Service. 1990. *An Analysis of the Timber Situation in the United States: 1989–2040.* Gen. Tech. Report RM-199. Rocky Mt. Forest and Range Experiment Station. Ft. Collins, CO. 268 pp.

U.S.D.A. Forest Service. 1993. *Forest Resources of the United States.* Gen. Tech. Report RM-243. Rocky Mt. Forest and Range Experiment Station. Ft. Collins, CO. 132 pp.

Vardaman, J. F., and Co. 1993. *Vardaman's Green Sheet.* Jackson, MS. July.

Weyerhaeuser. 1993. *Recycling—And Beyond.* Tacoma, WA. 12 pp.

Weyerhaeuser. 1993a. *Answering Common Questions on Forest Resource Management.* Tacoma, WA. 8 pp.

Zinkhan, F. C., W. R. Sizemore, G. H. Mason, and T. J. Ebner. 1992. *Timberland Investments—A Portfolio Perspective.* Timber Press. Portland, OR. 208 pp.

Zinkhan, F. C. 1993. Timberland investment management organizations and other participants in forest asset markets: A survey. *Southern Journal of Applied Forestry.* 17(1):32–38.

14

VALUING NONMARKET FOREST OUTPUTS

Chapter 3 showed how a free market could theoretically maximize welfare. You learned that **market failures** keep unfettered markets from attaining this paradise. One such failure is that there's little private incentive to produce forest outputs that you can't easily sell—for example, scenic beauty—beyond what occurs by chance. We'll call such goods and services **nonmarket** or *unpriced outputs.* Other such forest outputs are clean water, good recreational fishing downstream on others' properties, recreation with high costs of fee collection, nonconsumptive wildlife benefits, endangered species preservation, and improvements in quality of air and climate. Welfare is not maximized if people would be willing to pay more for such added outputs than their production costs (including costs of other income forgone). Other market failures covered in Chapter 3 that are related to unpriced forest outputs are listed below:

- Some types of forest recreation, which could be sold, have been traditionally free and are thus hard to sell.
- Forest operations sometimes cause damage to scenic beauty, fish and wildlife, water, and other resources.
- For unique recreation areas, the following values exist but are not easily sold in the market: option value, existence value, and bequest value.

For these reasons, the private market doesn't provide enough unpriced outputs from forests—more could often be provided at a cost less than the resulting benefits. Here we'll revisit some of these issues and consider why and how analysts try to value nonmarket outputs from forests. This chapter's notation is in Table 14-1.

TABLE 14-1
NOTATION FOR CHAPTER 14

A = attribute of a recreation area (e.g., acres of water)	T_z = travel cost from zone z to a recreation area
c_j = coefficients in a regression equation	V_z = number of visits per year to a recreation area from zone z
CVM = contingent valuation method	WTP = willingness to pay, dollars
I_z = per capita income in zone z	WTS = willingness to sell, dollars
P = price per unit	Z = a zone of origin for recreationists
Q = quantity, units	

WHY VALUE NONMARKET BENEFITS?

Why place monetary values on unpriced forest outputs? If market failures justify publicly providing these outputs, why not "just do it"? We often see graphs like Figure 14-1 predicting shortages of unpriced forest outputs, in this case, nature study opportunities in the United States. As with the timber "gap" analyses in Chapter 12, Figure 14-1 has no economic content because the projected "maximum demand" holds all user costs constant but allows increasing income and population. At the same time, the figure's "expected supply" simply projects past trends of available nature study trips. Without a market, trips would be rationed or discouraged, not by higher prices but by more congestion and higher travel costs to more distant areas.

Within budget constraints, legislators often respond to graphs like Figure 14-1 by spending tax dollars to provide more recreation on public forests. But that approach may not be efficient because the graphs don't show whether users would be willing to pay the costs of supplying more recreation. Think back to the **equimarginal principle:** To maximize returns from limited capital, invest so that the

FIGURE 14-1
Projected "gap" between maximum demand and expected supply for nature study trips. (USFS [1990a], p. 69.)

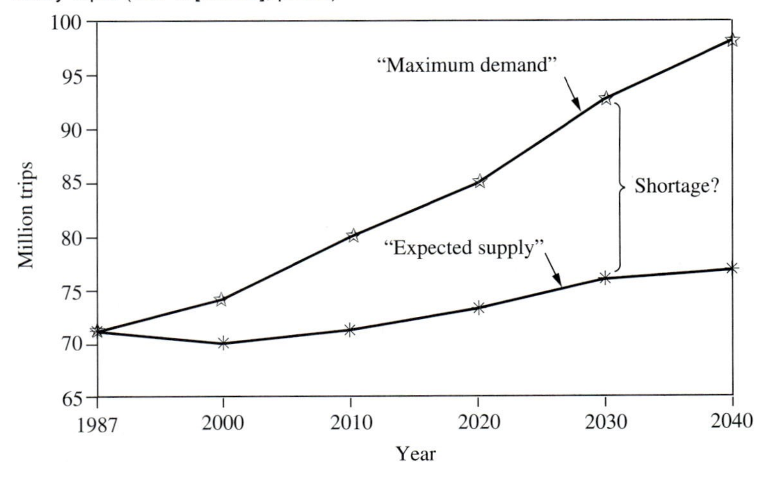

rate of return on the last dollar is equal in all pursuits. Without such equality, you could increase total returns by shifting dollars from low to high rates of return. Ideally, returns should be measured in added **utility,** but since we can't do that, we aim for a dollar return. With nonmarket benefits like the trips in Figure 14-1 valued in dollars, public land managers can calculate rates of return or **net present values** on investments in such outputs. They can then see if investing to increase the **supply** and meet the **demand** in Figure 14-1 is worthwhile. This helps agencies spend taxpayers' dollars efficiently, as lawmakers are demanding. Such analyses are covered in the next chapter.

Private landowners aren't usually interested in valuing a nonmarket good unless they want to sell it and find the public's **willingness to pay.** If governments regulate landowners to prevent environmental damage or gain nonmarket outputs, monetary values are useful to see whether the benefits of regulations exceed the costs. Courts often seek monetary values of nonmarket environmental damages. For example, damaged parties may seek compensation for lost wildlife, fish, or scenic values following fires, logging damage, or oil spills. Public utilities are concerned with placing monetary values on environmental damages, as are international trade negotiators who worry about environmental impacts of trade policies.

Some ask, "How can you put monetary values on something as intangible as a walk in the forest?" Well, how can you place monetary values on concerts, movies, time at a skating rink or ski slope, or visits to art galleries? These are just as intangible, and somehow we arrive at a monetary value of "willingness to pay" an entry fee—a price. With that in mind, some analysts simulate markets and estimate "prices" for nonmarket outputs in order to do benefit/cost analysis of public forestry investments. The idea of simulated prices is easier to imagine for things like visits to campgrounds or hiking areas, somewhat harder for owls or otters, and perhaps most difficult for species preservation or enhancing climate. Before exploring techniques for simulating prices, let's review cases where **economics** can help in deciding how much to spend to gain nonmarket outputs without ever pricing them.

DECISIONS WITHOUT MONETARY VALUES

Here I'll briefly review four ways to organize information to help make investment decisions when outputs are nonmonetary. The first two were covered earlier.

Annualizing Costs of Nonmarket Outputs

We can calculate the costs per unit of some output like recreation activity days and then ask decision makers if they think the benefit is worth the cost. The section on nonmonetary values in Chapter 6 shows how to **annualize** the present value of costs of getting unpriced benefits like campground visits. For a *terminating* case, you just multiply the calculated **present value** of costs by the **equivalent annual annuity** formula to get annualized costs, which have the same present value [equation (6-11)]. For a *perpetual* case, multiplying present value of costs by the interest rate will give the annualized cost [see equation (6-10)]. In either case, divide the annualized cost by estimated visits per year to get the cost per visit. Without actually calculating nonmarket values, this approach lets decision makers ask, "Is the value per visit at least as great as the calculated cost per visit?" If so, make the investment; if not, don't. For example, such an analysis might estimate that the cost of constructing a boat-launching site is $2.40 per launching. Based on intuition, the decision could be that the value per launching is at least that great. But if the cost is $62.00 per launching, the reverse might be true.

Chapter 6 also explains that you can use the above annualizing of costs when a project has an unacceptable criterion like a negative net present value (NPV), too low an **internal rate of return,** or a **benefit/cost ratio** less than 1. Just figure the present value of nonmarket benefits needed at your required rate of return to make the criterion acceptable, and annualize this present value. Then ask, "Are the annual nonmarket benefits worth at least this much?"

The Simple Betterness Method

Another example of organizing costs to make decisions about unvalued nonmarket goods is to use the **simple betterness method** covered in Chapter 8. Figure 8-8 ranks harvesting options in order of decreasing dollar value accompanied by nonmarket benefits, like scenic beauty, ranked by preference. The approach lets you easily eliminate "dominated" options that have the same or less dollar value accompanied by less nonmarket value than some superior option. Figure 8-9 shows how nondominated alternatives could be arranged to pose the question of whether some gain in nonmarket value is worth the resulting cost.

Minimum First Year's Benefits Required

To decide whether or not a Hell's Canyon dam should be built, Krutilla and Fisher (1975) calculated how many dollars the first year's preservation value would have to be in order for the present value of preservation to just equal the present value of development. You could then ask whether preservation benefits were worth at least that much. As above, you don't have to place a specific dollar value on the nonmarket good. The method makes sense under the authors' plausible assumption that the value of preservation benefits would increase at a faster rate than development benefits (in this case, mainly electricity).

The Equi-Marginal Principle

As covered in Chapters 1 and 4, the equi-marginal principle helps us efficiently allocate a scarce input to several activities in order to maximize some common output. Although outputs are dollars in the example at this chapter's start, even with unpriced outputs, we should allocate dollar inputs into different activities until the last dollar spent in each activity brings the same added output from all activities. This principle is enormously useful in making decisions

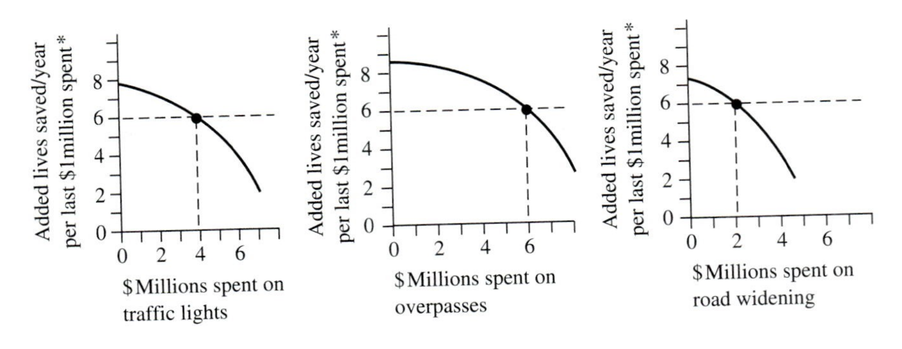

*Marginals, not totals. Total lives saved from each activity is the area under each curve.

FIGURE 14-2
Optimal allocation of capital when outputs are unpriced. (Capital budget = $12 million.)

when the output is unpriced. Although far afield from forestry, it's instructive to apply the theory to saving human lives, to which we hate to assign dollar values. Figure 14-2 shows hypothetical curves of added lives saved per year from the last $1 million spent on three types of highway improvement expenditures for a region. Given a year's budget of $12 million, the maximum number of lives will be saved by spending $4 million on more traffic lights, $6 million on overpasses, and $2 million on road widening. For each of the three activities, the last $1 million spent saves six lives per year: the equi-marginal principle is met, so the budget allocation is **optimal.** (What do I mean by optimal? I've maximized the lives saved; no other spending pattern can save more lives.) As with earlier examples of this principle, try deviating from the optimum, and lives lost will exceed those saved.

As a forestry example, horizontal axes in Figure 14-2 could be dollars spent on different deer habitat improvement activities, and vertical axes could be added numbers of deer harvested annually for the last $1,000 spent. A year's habitat improvement budget would be optimally spent when the last $1,000 allocated to each activity brought the same added increase in deer harvested. The key point is that you can make many welfare-improving investment decisions without ever placing dollar values on outputs. But there's still the question about whether the added deer gained was worth the added cost, and that's where **valuation** becomes important.

RECREATION DEMAND CURVES AND VALUES

Much research on valuing nonmarket goods has tried to derive demand curves for unmarketed outdoor recreation. The quantity units are usually activity days, and the price is a simulated entry fee per activity day or visit. An **activity day** is one person on a recreation site for any part of a calendar day. I will use "activity day" and "visit" interchangeably. (Be aware that other units like "12-hour recreation visitor days" are sometimes used, so values per unit aren't always comparable.) Figure 14-3 is a hypothetical visitation demand curve for Woods Park where

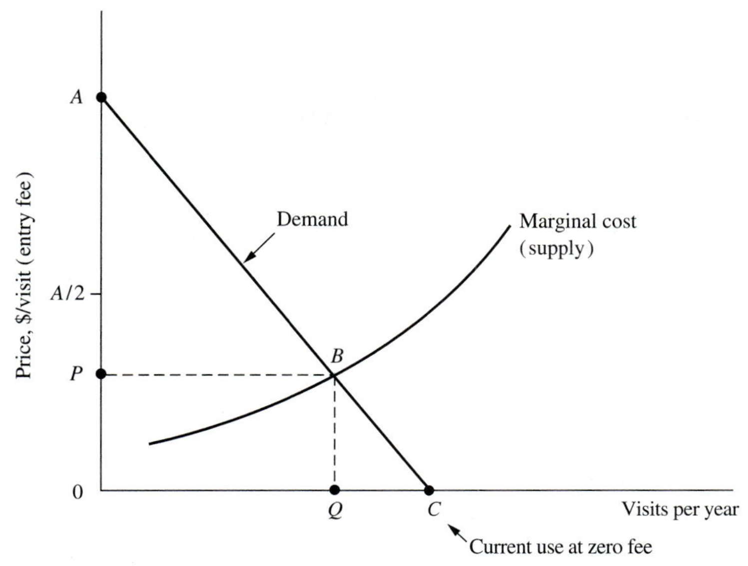

FIGURE 14-3
Site demand curve for visits to Woods Park.

currently no fee is charged. Such curves are called **site demand curves,** since they reflect willingness to pay at the recreation site. For now, ignore the **marginal cost** curve.

Review again the difference between a movement along a demand curve and a shift of the curve (Chapter 2). In Figure 14-3, if fees are actually charged at Woods Park, if you raise (lower) the fee, fewer (more) people will visit. These are movements along the curve. You can think of the Figure 14-3 demand curve as applying to one year. What factors will shift the curve to the right, other things being equal? (Higher per capita income, increased population, increased preference for outdoor recreation, increased advertising for the park, increased prices for substitute recreation, decreased prices of gasoline or travel costs or other costs of using the park, increased leisure time, increased mobility.) Opposite amounts of these **shifters** will shift the demand curve to the left. In most countries, the average effect of shifters is to move recreation demand curves to the right over time. You'll see shortly that some of these shifters are sometimes included in recreation demand curve equations.

Figure 14-3 shows that at the current zero entry fee, visits are C per year. Visitation would be driven to zero at a fee of $\$A$ per visit. The curve traces all other combinations of entry fees and estimated visits per year. Many parks are unique enough that there aren't perfect substitutes nearby, so an operator charging fees would face a downward-sloping demand curve as shown, rather than the horizontal curve faced by perfectly competitive producers. When park operators face downward-sloping demand curves, we have the market failure of monopoly with high fees and restricted output (see Chapter 3). Before deriving recreation demand curves, let's look at how to use them for estimating monetary values of visits.

Interpretations of Gross Value

Arguments abound on how to use a demand curve like the one in Figure 14-3 to value park visits or other nonmarket outputs. From a social view, the optimum output is at point B (or Q visits), where the marginal (added) cost of the last visit equals its price (also see Figure 3-1). Producing an added visit beyond B will cost more (read off the marginal cost curve) than its benefit (an added visitor's willingness to pay, shown on the demand curve). If output is at the optimum Q at price P in Figure 14-3, total gross revenue is $(P)Q$ or (price) $\times$ (quantity): the area of rectangle $0PBQ$, ignoring costs. Many argue that the **consumer surplus,** the triangular area PAB, should also be added to the PQ value to arrive at total social value. This surplus is consumers' extra willingness to pay *above* the price charged (see Chapter 2). Thus, the total social value at use level Q is the area $0ABQ$. In calculus, this is the integral of the demand function from quantity 0 to Q—easily estimated with computer programs and also with some calculators, if you know the function. Dividing this area by the quantity Q gives the value per visit (area $0ABQ)/Q$, which is often called **average willingness to pay.** A problem with this approach is that for most types of unmarketed recreation, marginal cost curves (supply curves) have rarely been estimated, either for one park or regionally, so that the use level Q is hard to approximate.

Including consumer surplus in recreation values is also useful when comparing parks having different demand curve slopes, as shown in the demand curves for park (a) and park (b) in Figure 14-4. Suppose each park operates at level Q and price P. Given the same price and quantity scales, each park has the same $(P)Q$ value, but since park (b) has more consumer surplus, it generates a greater value or satisfaction. Compared to ordinary parks, unique parks with few substitutes have more **inelastic** demand [park (b)], and consumer surplus is a higher percentage of PQ value; examples would be Yellowstone or the Grand Canyon.

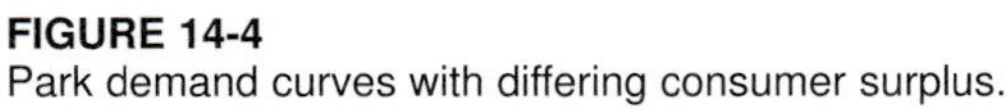

FIGURE 14-4
Park demand curves with differing consumer surplus.

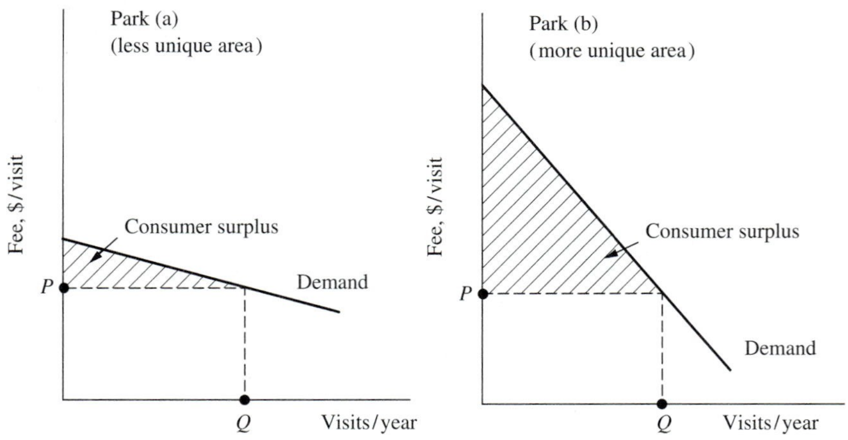

When price for a unique area increases, visitation doesn't drop off quickly because there are no good substitutes. If price increases for an ordinary park (a), visitation falls dramatically, since many substitutes exist at the old price.[1] Thus, for recreation activities with many readily accessible substitutes, consumer surplus may be unimportant.

Note that recreation value derived from a demand curve is strictly a willingness to pay for that good. For nonmarket goods, willingness to pay tends to be less than a consumer's **willingness to sell** the same good, once possessed, an issue addressed shortly. Also, be aware that the above concepts of value in this section are all gross, before subtracting production costs.

Consumer Surplus and Relative Size of Output Change

Some argue that market values of commodities like timber also have consumer surplus, so adding this surplus for recreation and not for timber will tilt benefit/cost analyses in favor of recreation. This argument is valid when comparing total values of recreation and timber for large regions. But for small output changes, timber consumer surplus is negligible, as shown in Figure 14-5, a hypothetical stumpage demand curve for the larger region where Woods Park is located. Let the current

[1]In a private market, if parks (a) and (b) in Figure 14-4 were potential ventures, decisions whether or not to open a park would be based only on $(P)Q$ values, excluding consumer surplus. Assume you regulated for monopoly problems and forced output to some optimal Q; the market still couldn't recognize that park (b) is worth more than (a). Hence the private decision might not be optimal.

FIGURE 14-5
Small changes in regional stumpage demand.

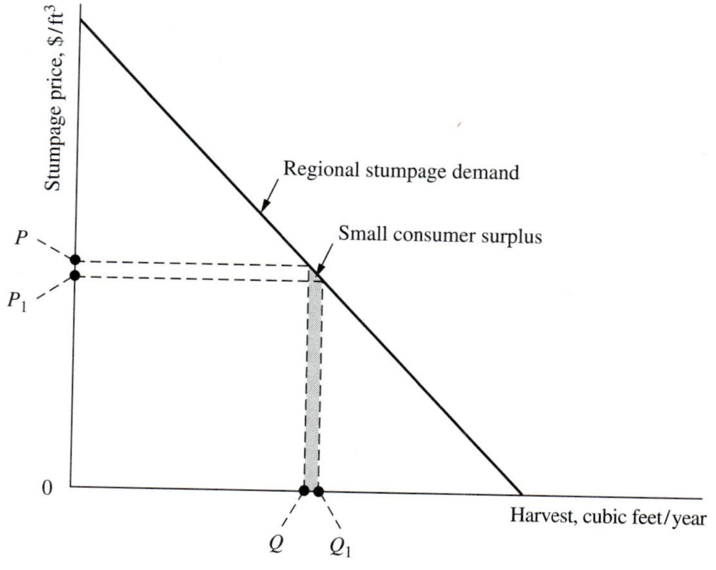

equilibrium be at Q cubic feet and a price of P. Suppose we close Woods Park, where currently no timber harvesting is allowed, and let it supply timber, thus increasing regional stumpage consumed from Q to Q_1. At any quantity consumed, let the social value of timber be the price times quantity plus the consumer surplus (see the previous section). Then, when timber quantity increases from Q to Q_1, the gain in social value of timber is the slender shaded area. But the gain in consumer surplus is only a tiny triangle at the top of the shaded area—an insignificant part of the gain in social value of timber. Thus the timber consumer surplus can be ignored in such cases.[2]

You might be thinking, correctly, that among buyers who bought the quantity of timber $Q_1 - Q$ in Figure 14-5, there could be significant consumer surplus on previous wood bought. But if the price dips just a tad, the only people who buy *more* wood are those who have little or no consumer surplus for *added units*. The units yielding much consumer surplus would already have been bought because buyers could get them for less than they were willing to pay. So for small changes, there's little consumer surplus for marketed goods under competition. But for unique recreation resources where producers face downward-sloping demand curves, consumer surplus can be significant.

The foregoing implies that if you make comparisons between recreation values and timber values from *one relatively small area* like Woods Park, assuming an either/or case, it would be valid to include consumer surplus for the recreation value and *not* for the timber value. But to compare *total* values in a major region, for example, timber and nontimber outputs from the entire National Forest System, you should include consumer surplus for all outputs. As you can see with the regional timber demand curve in Figure 14-5, consumer surplus is significant for entire markets.

Average Consumer Surplus

Another reason for including consumer surplus in recreation values is that without the surplus, (price) × (quantity) values could be unreasonably low at high use levels. For example, in Figure 14-3, at a zero fee and C visits, (price) × (quantity) is zero, while total consumer surplus is the area $0AC$. Dividing this *total* consumer surplus by visits at a *zero* price (C in Figure 14-3) gives another estimated value per visit called the **average consumer surplus** [(area $0AC$)/C]. With a linear demand curve as in Figure 14-3, this average consumer surplus is $A/2$, or half the price that drives use to zero. (That's because twice the triangular area under a linear demand curve would be the price A times the quantity C, so the total consumer surplus area would be half of $(A)C$ or $(A/2)C$. Dividing by C

[2]The small consumer surplus triangle implies that the stumpage demand curve that would be faced by Woods Park if it were a timber supplier would be nearly horizontal—an assumption of perfect competition (the mechanism is explained in Figure 2-3). This raises the point that in cases where stumpage demand curves faced by individual growers are downward-sloping, there can be some timber consumer surplus.

gives the average consumer surplus, $A/2$.) This has led some to suggest that an average recreation value per visit could be very roughly approximated by half the fee estimated to drive visitation to near zero—say, for a campground, half the average per person price of a motel room—admittedly a crude measure, but better than nothing.

Because, for public parks, it's hard to estimate an efficient output level like Q in Figure 14-3, a reasonable estimate of values per activity day is average consumer surplus, as defined above.[3] Examples of dollar values per visit derived from site demand curves are given later.

Theoretical Regional Equilibria for Nonmarket Outputs

Imagine that Figure 14-3 applied to supply and demand of unmarketed recreation like the nature study trips *for the entire United States* shown in Figure 14-1. In Figure 14-3, if trips were allocated in a competitive market, users of nature study areas would pay a price per visit, giving rise to a downward-sloping national demand curve for visits. With visits marketed, an upward-sloping national supply curve would show the number of annual visits to be supplied by landowners at different prices. New supply and demand curves for visits would exist, say, every five years, presumably each to the right of the previous one, given an expanding population. Demand curves would move due to the shifters mentioned earlier, and supply shifts could occur due to landowners making more recreation areas available, if acceptable rates of return could be made. In that case, for the nation or a given region, over time you could trace a series of intersection points between supply and demand curves for nature study trips in the same way as supply-demand equilibrium points are shown for timber in Figure 12-8. The supply-demand intersection points would show the number of visits and price per visit at each five-year interval, and the "shortage" shown in Figure 14-1 wouldn't exist. (Can you sketch a theoretical series of regional market supply-demand equilibrium points over time, every five years, for some nonmarket output like hiking, showing price per visit on the y axis and number of visits per year on the x axis?)

The problem is that, for nonmonetary forest outputs like water, endangered species, nonconsumptive game uses, fish, and many forms of recreation, markets don't exist, so this kind of equilibrium is strictly theoretical. Instead of prices stimulating output and rationing consumption, the output of unpriced goods is often lower than the amount people would be willing to buy. Rationing occurs, in the case of recreation, through congestion, queuing, and having to drive longer distances to substitute areas.

You saw above how demand curves are used to estimate values for recreation and other amenities, but how do we derive such curves for unmarketed recreation?

---

[3]Note that producing beyond output Q is inefficient unless marginal costs are lower than shown. Thus, total area under the demand curve could be an overvaluation.

THE TRAVEL COST MODEL

A common way to derive demand curves for recreation requiring travel from different distances is the **travel cost method.** The approach uses simulated increases in travel costs to estimate a demand curve for visits. First, let's look at the role that travel costs play in recreation site values.

The Role of Travel Costs

Our objective is to estimate a willingness to pay for recreation *at the site,* in the sense of an entry fee. This is like obtaining values of stumpage in the woods, *on-site.* In this way we can make valid value comparisons between different land uses. Thus, logging and hauling cost for stumpage is analogous to travel cost for recreation, the difference being that with stumpage, you transport trees from the site to the users, but with recreation, you transport users to the site.

Figure 14-6 diagrams this comparison. Using the example from Chapter 12, the upper panel shows that if mills pay $200 per thousand board feet (MBF) for logs, and logging and hauling costs are $70/MBF, the maximum willingness to pay for stumpage standing in the forest is $130/MBF. The figure's lower panel shows a recreationist at home willing to pay $40 to visit a park. If she has to spend $30 on travel and other costs to use the park, she is willing to pay only $10 to actually enter the park. The $30 could include recreation "processing costs" allocated per trip for recreation equipment. Just as the stumpage value isn't the logging and hauling cost, the value of recreation at the site is not the travel cost. Depending on the case, these transport costs may be greater than, less than, or equal to the on-site values. Other things being equal, the higher the hauling cost, the lower the stumpage value; and the higher the recreation travel cost, the lower the willingness to pay a park entry fee. So you can think of on-site recreation values as a "stumpage-like" value for recreation. In the above case, if travel costs increase to $35, the willingness to pay a park entry fee drops to $5, other things being equal.

Deriving the Site Demand Curve

The travel cost model derives recreation site demand curves, assuming that recreationists would react to increases in travel costs the same way they'd react to increases in recreation area entry fees. A highly simplified example starts with Figure 14-7, which shows Picnic Park, where *no entry fee is currently paid.* The surrounding region is divided into three zones including all visitors. The number of zones is normally greater, and areas needn't be rings; for example, clusters of counties are often used. For each zone, Table 14-2 gives hypothetical data on population, travel cost per visit from the midpoint of each zone to Picnic Park, number of visits per year, and visits per 1,000 population. Population and travel cost can be gathered from existing sources; a park entrance survey would yield the number of visitors from each zone, and the visits per 1,000 population is column (4) divided by column (2).

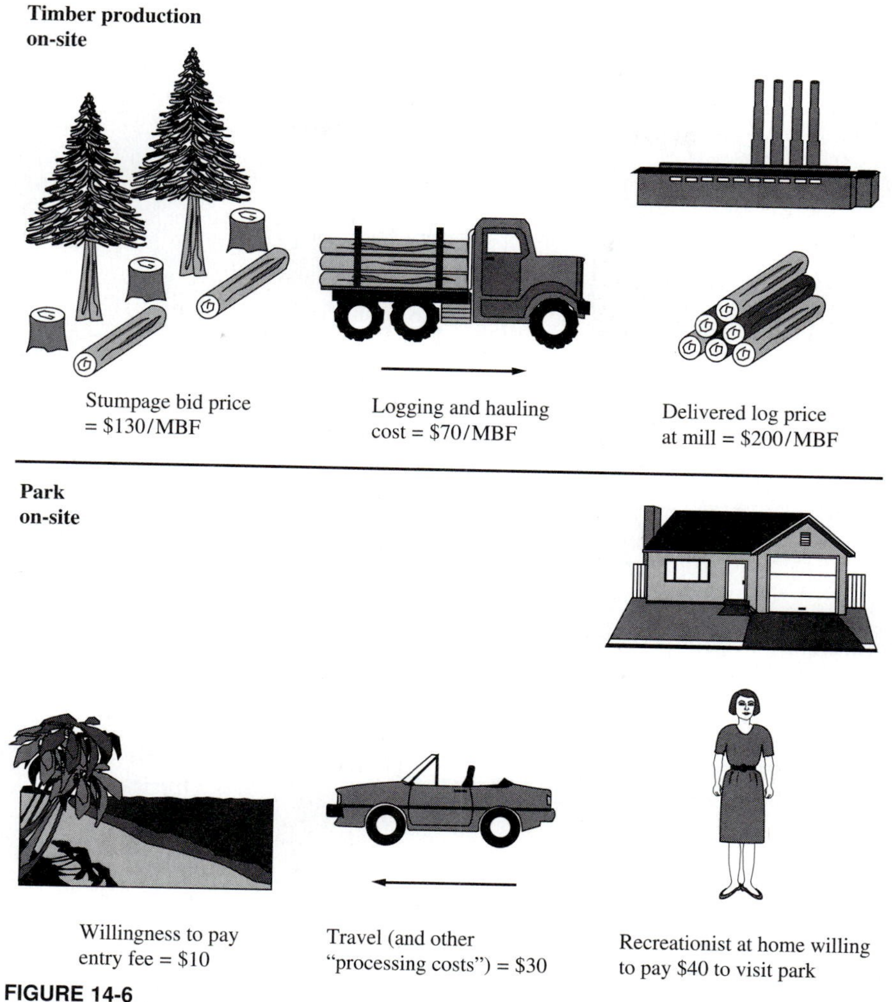

FIGURE 14-6
The role of transport and processing costs in on-site values of timber and recreation.

Deriving the park demand curve requires first constructing a **trip demand curve,** as shown in Figure 14-8. From Table 14-2, for each zone, this curve plots travel cost per visit over annual visits per 1,000 population. Actual data are unlikely to fall so neatly on a straight line, but visits per thousand population should decline with zones farther from the park (with higher travel costs). This is not a true site demand curve because it doesn't plot price per visit over total visits to the site. To get the site demand curve, we'll simulate increases in travel cost and assume that these would affect visitors the same as a park entry fee. You can do this with Figure 14-8 and Table 14-3, which shows simulated increases in travel costs from $0 to $3 across the top: these are the "entry fees." Now let's calculate the numbers in the center of the table: the visits from each zone at each "fee"

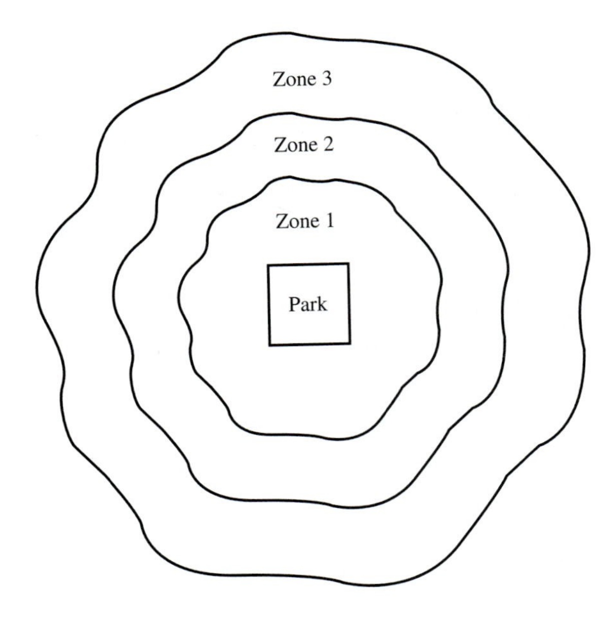

FIGURE 14-7
Zones of travel origin for Picnic Park.

TABLE 14-2
VISITATION DATA FOR PICNIC PARK

Zone	Population (thousands)	Travel cost per visit	No. of visits per year	Visits per 1,000 population
1	2	$1	600	300
2	1	2	200	200
3	4	3	400	100

level. At a zero increase in travel cost, you record the current number of visits (at no fee) from column (4) of Table 14-2, the total being 1,200, as shown in the last row of Table 14-3. At a $1 simulated increase in travel cost, for zone 1, you go to Figure 14-8 and add $1 to travel cost for zone 1 travelers. That gives a $2 travel cost, and assuming that zone 1 people behave like zone 2 people, you read off the trip demand curve that zone 2 people visit at the rate of 200 per 1,000 population. Since there are 2,000 people in zone 1, that's 200(2) = 400 visits, which you enter in column (3) of Table 14-3.

For a $1 "fee" in zone 2, go to Figure 14-8 at zone 2 and add $1 to the travel cost. That pushes you up the curve to a $3 travel cost at the zone 3 visitation rate of 100 per 1,000 population, assuming that zone 2 people behave like zone 3 people. Since Table 14-2 shows 1,000 people in zone 2, that's 100(1) = 100 visits, which you enter in column (3) of Table 14-3. Adding $1 to the zone 3 travel costs in Figure 14-8 puts you at a $4 travel cost, where the figure shows that visits drop to zero. So you enter a zero for zone 3 in column (3) of Table 14-3. Column (3) shows 500 total visits from all zones at a $1 entry fee.

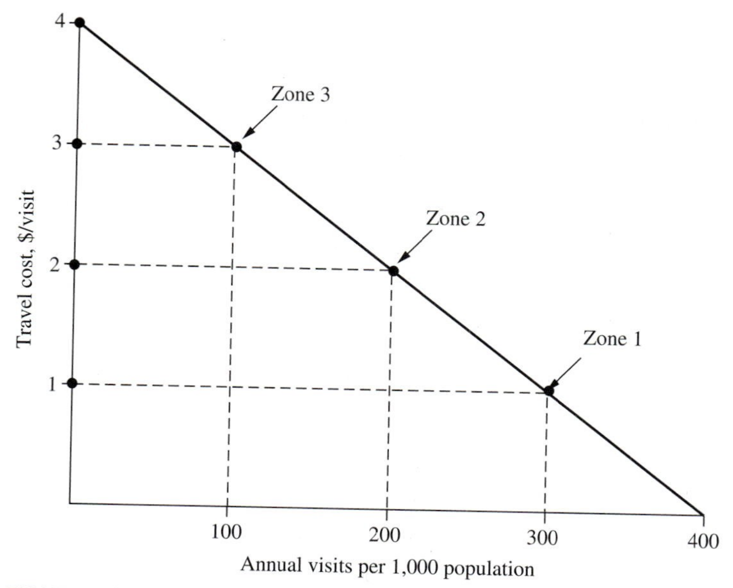

FIGURE 14-8
Trip demand curve for Picnic Park.

TABLE 14-3
TOTAL VISITS PER YEAR WITH SIMULATED INCREASES IN TRAVEL COSTS

	Simulated increases in travel cost (like an entry fee)			
Zone	**$0 [from Table 14-2, col. (4)]**	**$1**	**$2**	**$3**
1	600 visits	400 visits	200 visits	0
2	200	100	0	0
3	400	0	0	0
Total visits	1,200	500	200	0

Simulating a $2 fee in column (4) of Table 14-3, add $2 to the zone 1 visitors in Figure 14-8, and you'll be up to the zone 3 point of $3 travel costs, where visits are 100 per 1,000 population. With 2,000 people in zone 1, that's a total of 100(2) = 200 visits, as shown in column (4) of Table 14-3. In the same column, if you add $2 to travel costs of zones 2 or 3, you're up against the *y* axis in Figure 14-8, at zero visits. So you enter zeros in column (4) for zones 2 and 3. The same thing happens in the last column of Table 14-3: adding $3 to travel costs of all zones in Figure 14-8 puts you against the *y* axis. So you enter 0 visits from all zones at a $3 fee in Table 14-3.

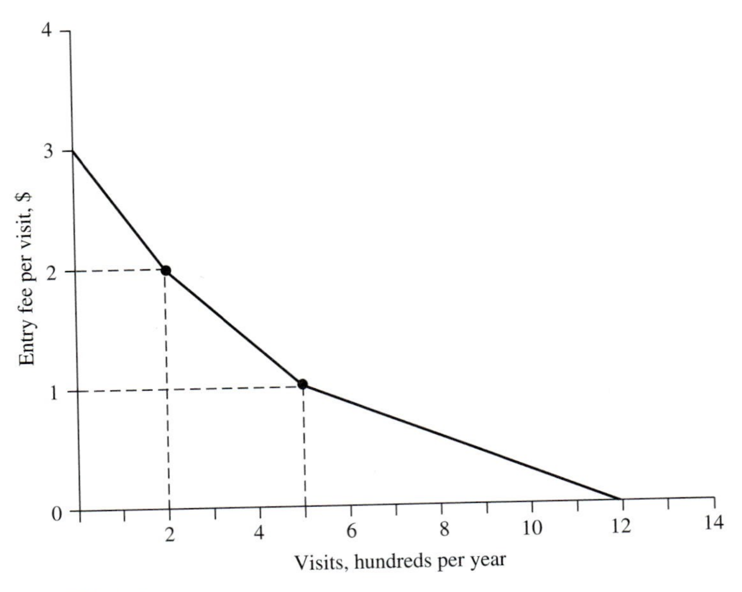

FIGURE 14-9
Site demand curve for Picnic Park.

In Table 14-3, the header row and last row give demand curve points for fees and visits. At a zero price, total visits are 1,200; a \$1 price yields 500 visits, etc. Plotting these points gives Figure 14-9, the estimated *site demand curve* for Picnic Park. This is analogous to an on-site stumpage demand curve; both are net of all transport costs. In Figure 14-9, what would be the on-site recreation value at 500 visits per year? Price times quantity is \$500, and you can eyeball the consumer surplus areas (two triangles and a rectangle), summing them as $100 + 200 + 150 = \$450$, giving a total willingness to pay of \$950 per year at 500 visits.[4] At this visitation level, "average willingness to pay" is $950/500 = \$1.90$ per visit. By similarly adding up triangle and rectangle areas, the total area under the demand curve at a zero price is \$1,300, giving an "average consumer surplus" of $\$1{,}300/(1{,}200 \text{ visits}) = \$1.08$. So, you can see that even if you have a demand curve, interpretations of recreation value can vary greatly. To clarify procedures, these hypothetical data are very simple; thus the values are low. Two actual travel cost valuations of average consumer surplus for picnicking in the Midwest are about \$7 and \$9 per activity day, in 1987 dollars (Walsh et al., 1988). Others have been higher.

You may be thinking of weaknesses in the model: Population characteristics could differ among zones, the travel experience may be positive or negative, travel

[4] With a statistically estimated curve you'd have the mathematical function and could calculate the integral (with calculus or a computer program) from a use level of 0 to some point to get the total area under the curve (or value) up to that point.

time has a cost, users shift to competing sites when the price gets too high at one area, congestion can change park quality, and trips may be very long and have multiple destinations. Many of these points are handled in more complex travel cost models. For example, you could start with the following multiple regression equation for a trip demand function for one recreation area. Each data point is one of many zones:

$$\text{Visits from zone Z} = V_z = c_0 + c_1 T_z + c_2 P_z + c_3 I_z + \ldots \qquad (14\text{-}1)$$

The c_j's are coefficients, T_z is cost of travel from zone Z to the park (this could include a component for time cost of travel), P_z is thousands of people in zone Z, and I_z is per capita income in zone Z. In this linear form, you'd expect a negative sign for the travel cost coefficient and a positive one for population and income (the latter two are shifters mentioned earlier). After estimating the equation's coefficients with regression analysis, you could substitute the first zone's independent variable values into the equation and simulate increases in travel cost (the entry fee), say, dollar by dollar, and store the resulting visitation (V_z) for each fee. You'd do this for each zone. Then for each increase in travel cost, you'd sum the resulting visits across all zones to get total visits at each simulated entry fee. These would provide the recreation site demand curve points. Computer programs exist for such analyses.

More detailed travel cost models include things like multiple recreation sites with attractiveness indexes, more population characteristics, travel cost to substitute sites, nonlinear equations, multiple destinations, and "cutoff points" so trips drop to zero above some fee. But the end products are the same: recreation site demand curves from which values can be derived.

Other kinds of recreation valued with the travel cost model have included hiking, wilderness use, hunting, fishing, boating, nature study, national park visits, and camping. You can also estimate such demand curves for one type of recreation with and without certain amenities, thereby valuing the amenities. For example, for similar campgrounds with different amenity levels, you could derive demand curves, given varying amounts of insect damage, old-growth timber, water recreation, scenic vistas, or other amenities. But the travel cost model isn't suitable when travel costs are roughly the same for all users or when valuing small changes in quality for one recreation area. For those cases, the "contingent valuation" approach is often used.

CONTINGENT VALUATION METHODS

Using either interviews or mail surveys, another way to value nonmarket goods and services is to simply ask users what is the maximum they'd be willing to pay for them (*willingness to pay*) or what minimum compensation they'd require to willingly give them up (*willingness to sell*). This is the **contingent valuation method** (CVM), so called because values are *contingent* on a hypothetical market constructed in a survey. The *survey method* is an earlier name for CVM.

Willingness to Pay

In asking people their maximum willingness to pay for some nonmarket good or service, gaming can be a problem unless questionnaires are carefully designed. For example, if respondents, say, wilderness hikers, felt they wouldn't actually have to pay, they might cite very high values if they thought it would increase the chance of saving a wilderness area in jeopardy. But if users felt it was highly likely they'd eventually have to pay, they might cite low values to ensure low user fees on an area already earmarked for recreation. To get average CVM values per person, divide the total cited willingness to pay by the number of people surveyed. If questions elicit the *maximum* willingness to pay, this approach includes consumer surplus and should approximate the *average consumer surplus* mentioned earlier.[5] Be careful in interpreting willingness to pay (WTP) values: they're sometimes total values (partial or complete areas under demand curves), average values per unit consumed (including consumer surplus), or marginal WTP values comparable to market prices per unit.

Gaming is minimized when situations are realistic. Randall et al. (1974) had such realism when asking electric power customers about their willingness to pay higher electric bills in exchange for different levels of reduced air pollution from coal-fired power plants. Respondents knew they'd actually have to bear the cost. Perhaps the most realistic willingness-to-pay estimate is a "*simulated market value*" derived from experiments where money actually changes hands (compared with contingent valuations that are simply answers to hypothetical questions). One simulated market study allowed unsuccessful applicants for single-deer hunting permits to bid competitively for permits, with high bidders actually buying a permit. The resulting simulated market value averaged $31 per permit in 1984 (Bishop and Heberlein, 1990). Contingent values collected from surveys of other hunters from the same group were slightly higher at $35. For several WTP tests, the study showed no statistically significant difference between simulated market values and contingent values. So at least in that case, contingent values could roughly mirror simulated market values.

It's important to carefully phrase contingent valuation questions. For example, if you're aiming for willingness to pay on-site user fees, you have to assure that respondents don't give their willingness to pay for all costs of participating in the recreation experience, including travel costs (like the $40 willingness to pay at home in Figure 14-6).

Willingness to Sell

Surveys can also estimate people's minimum compensation required to willingly forgo certain nonmarket goods: "willingness to sell" (WTS). Chapter 3 explained

[5]Bulow and Klemperer (1994) argue that in actual bidding for goods, buyers' stated willingness to pay is likely to be less than their maximum valuation based on their demand curves. This occurs because buyers are unlikely to bid significantly above the market price they can buy the good for elsewhere. If such mechanisms are involved in contingent valuation studies, consumer surplus would be under-estimated.

how willingness to sell can exceed willingness to pay for nonmarket amenities, for people with a declining marginal utility of money (see Figure 3-8). Again, gaming can be a problem if questions aren't carefully phrased. If you're interested in preserving a resource and feel that actual compensation is highly unlikely, you might name an unrealistically high compensation payment for hypothetically losing a nonmarket amenity, in the hope of saving it. Someone's willingness to sell a good should be the *minimum* compensation required in order to maintain the same level of satisfaction existing before the loss. Willingness to sell can be overstated if respondents are thinking of the maximum compensation they believe could be extracted.

Surveys are more likely to yield plausible WTS estimates if compensation is really expected. Two studies realistically simulated market values for willingness to sell by actually purchasing (with real money) single-goose hunting and deer hunting permits from hunters (Bishop and Heberlein, 1990). The same studies also compared these simulated markets with contingent valuation surveys asking what compensation hunters would require to willingly give up their permits. For the deer permits, the simulated market value of willingness to sell (WTS) was $153, which, as expected, exceeded the above-cited simulated market value of willingness to pay (WTP) of $31. But simply asking hunters the minimum compensation they'd require to give up their deer permits yielded a WTS of $420, which was 2.75 times higher than the simulated market value for WTS. This suggests, at least from that example, that hypothetical contingent valuations of WTS may be overestimates. The most realistic WTS values are probably from the few studies that have used real cash.

Overall, the excess of WTS over WTP has averaged about 3:1 or 4:1 for studies of environmental values, some using real money (Gregory and Bishop, 1988). But in many cases the divergence has been far greater. Overall, the excess of WTS over WTP may be much more than the difference we'd expect based only on the declining marginal utility of money reviewed in Figure 3-8, especially for small changes in income. So there must be something else at work: some of the divergence between WTS and WTP can be explained by "loss aversion" (Tversky and Kahneman, 1991). Also, Kahneman et al. (1990) have found an "endowment effect," where people resist exchanging what they consider to be their own. Hanemann (1991) gives a theoretical defense for large observed differences between WTP and WTS for public goods with few substitutes. The key is lack of substitutes: it seems that if perfect substitutes are available, people's required compensation for loss of one good should simply be the increased cost (if any) of consuming a substitute.[6] In an experimental bidding game with actual cash, Boyce et al. (1992) found WTS three times higher than WTP for ornamental trees that would be killed if sold. They speculate that feelings of "moral responsibility" make people loath to see environmental amenities destroyed, thus driving a wedge between WTP and WTS. Since large differences between WTP and WTS can have

[6]Although disagreement exists, Shogren et al. (1994) find little difference between WTP and WTS, and thus no endowment effect, for goods with perfect substitutes. But for a nonmarket good with imperfect substitutes, they show WTS consistently above WTP.

important impacts on how we value amenities and how much we intervene to reduce environmental damage, more research in this area is crucial.

Which Value to Use?

Values of willingness to sell can be much higher than those for willingness to pay, even if contingent valuations are replaced by market simulations. Which value should be used when? One approach is to use willingness to pay in benefit/cost analyses of providing new nonmarket goods and to use willingness to sell when estimating values of amenities lost or damaged. An example of the former would be a public agency computing rates of return on investments in new campgrounds. Willingness to sell would be more relevant in calculating the fine on a firm causing damage to wildlife or in finding the benefits of restoring damaged amenities like recreational salmon fishing. In the latter case, to make damaged anglers as well off as they were before the loss, you'd have to compensate them at their full WTS value. Thus, if it costs less to restore the fishery, restoration is optimal.

General CVM Procedures

With contingent valuation, the resource to be valued is described verbally, or with photographs, drawings, or on-site observation. Photographs are often useful to describe changes in characteristics like insect damage or air quality.

Depending on the market being simulated, the means of payment should be clarified, for example, higher taxes of a given type, an entry fee, a license fee, or higher prices of some product. You need to choose a payment vehicle to which respondents react least negatively.

To elicit hypothetical values of willingness to pay, you can simply ask for the maximum amount a respondent is willing to pay in dollars (open-ended question). A take-it-or-leave-it format asks whether or not respondents would pay a specified amount for some nonmarket good (closed-end). A popular approach is "iterative bidding," where the interviewer first suggests a price, and if the respondent is willing to pay it, the amount is repeatedly raised until it is refused. The highest accepted amount is the maximum willingness to pay. Similarly, a "payment card" approach lists brackets of values from which the respondent chooses. The question format can sharply affect valuations. For example, in measuring willingness to pay to protect Appalachian spruce-fir forest quality, Haefele et al. (1992) found that take-it-or-leave-it questions gave values 3.3 to 4.8 times higher than payment card questions.

To clarify and to avoid biases, contingent valuation questionnaires are often lengthy and at times hard to answer. Examples follow.

From the above-cited spruce-fir survey:

"Suppose the only way to provide for these tree protection programs is to start a special conservation fund financed by increased taxes. Although most of the southern Appalachian spruce-fir forests are like those shown in photo A, without these programs most of the forests will eventually decline to the level seen in photo C. The whole forested area is at risk from the insect and pollution damage.

Question 13. What is the most money you would pay *each year* to provide protection programs for spruce-fir forests along roads and trails in the southern Appalachian Mountains (which is about one-third of the remaining forest areas)? (Circle one amount.)

$0 $2 $4 $6 $8 $10 $15 $20 $25 $30 $40 $50
$75 $100 $125 $150 $175 $200 $250 $300 $350 $400 $450 $500
other $________" (Haefele et al., 1992)

From a 12-question waterfowl hunting survey, a question to elicit the maximum willingness to pay for a season of waterfowl hunting above current costs:

"Question 8. Suppose that your waterfowl hunting costs for the 1968–69 season were greater than you estimated in Question 7. *Assume these increased costs in no way affected general hunting conditions.*

ABOUT HOW MUCH GREATER DO YOU THINK YOUR COSTS WOULD HAVE HAD TO HAVE BEEN BEFORE YOU WOULD HAVE DECIDED NOT TO HAVE GONE HUNTING *AT ALL* DURING THAT SEASON?

We emphasize that the dollar amounts given below are intended to represent imaginary increased costs, that is, costs over and above the actual costs you estimated in Question 7. Please check the answer below that you consider most appropriate.

________ $0.00 to $2.49
________ $2.50 to $4.99
________ $5.00 to $9.99
________ $10.00 to $19.99
________ $20.00 to $29.99
________ $30.00 to $49.99
________ $50.00 to $74.99
________ $75.00 to $99.99
________ $100 to $199
________ $200 to $299
________ $300 to $399
________ $400 to $499
________ $500 to $749
________ $750 to $1000
________ Over $1000
(please specify amount
$________________)"

(Hammack and Brown, 1974, pg. 92, by permission.)

From the foregoing survey, a question to elicit the minimum willingness to sell a season's waterfowl hunting:

"Question 6. Suppose you have the right to hunt waterfowl just as you have had that right during the last season and the seasons before it. But also suppose that you could sell your right to hunt waterfowl for *a season,* and if you did sell the right, you yourself could not hunt waterfowl during that season. You would set your own price and the choice would be entirely up to you whether or not you sold this right. We emphasize that this situation is entirely fictitious—no one is going to restrict waterfowl hunting on the basis of this questionnaire and no one could actually buy or sell this natural right.

But, WHAT IS THE *SMALLEST* AMOUNT YOU THINK YOU WOULD TAKE TO GIVE UP YOUR RIGHT TO HUNT WATERFOWL FOR A SEASON—SAY, 1968–1969?

Please check the most appropriate answer below.

________ $0.00 to $2.49
________ $2.50 to $4.99
________ $5.00 to $9.99
________ $10.00 to $19.99
________ $20.00 to $29.99
________ $30.00 to $49.99
________ $50.00 to $74.99
________ $75.00 to $99.99
________ $100 to $199
________ $200 to $299
________ $300 to $399
________ $400 to $499
________ $500 to $749
________ $750 to $1000
________ Over $1000
(please specify amount
$________________)"

(Hammack and Brown, 1974, pg. 91, by permission.)

Incorporating Price and Employment Effects

When being asked willingness to pay for increasing or maintaining certain amenities, people should also be given information about resulting increases in prices of goods whose supplies would decrease. For example, if you're asked about willingness to pay for more Pacific Northwest salmon, northern spotted owls, or old-growth forests, you should know the resulting increase in forest products prices and temporary decreases in employment and regional income, if any. Such knowledge could reduce willingness to pay. Further reductions could accompany an awareness that greater environmental gains in the United States could occur at the expense of more environmental damage in other countries exporting more to replace, say, reduced United States timber output. But willingness to pay is only relevant under the **victim liability** framework where we don't seek damage reductions unless victims' willingness to pay to reduce environmental damage exceeds the damager's abatement costs (see Chapter 3).

What about **damager liability,** which says that landowners have no right to impose uncompensated damage? In the United States we apply both liability frameworks. We use the damager liability approach when regulating to prevent the most blatant forms of pollution, levying fines on violators, and not even considering people's willingness to pay to reduce damage. But for less obvious forms of environmental damage, we've not committed ourselves consistently to regulate under damager liability which values amenities at the higher willingness to sell. We often think in terms of victim liability and worry whether people would be willing to pay to compensate landowners for lost income. This results in higher levels of environmental damage (see Chapter 3). Which framework we choose depends in part on the economic effects. If choosing damager liability, with its more stringent regulations, brings painful price-increases and unemployment, we might not choose it. And if we do, and ask people their required compensation to accept added damage, we should concurrently let them know the accompanying price-declines and employment-increases, which could decrease the compensation required.

The questions raised above and in the next two sections about the contingent valuation method shouldn't imply that the technique should be abandoned. But we do need to interpret resulting values carefully. (See Portnoy, 1994; Hanemann, 1994; and Diamond and Hausmann, 1994, for a good review of pros and cons of the CVM.)

USE VALUES AND NON-USE VALUES

So far we've considered "use values" like activity days at a picnic ground or campsite. Harder to quantify are "non-use values" like option value, existence value, and bequest value, mentioned in Chapter 3. **Option value** is the willingness to pay for the option to use some resource in the future. **Existence value** is our willingness to pay for the assurance that something will exist, even if we never plan to use or visit it. And **bequest value** is the willingness to pay to preserve some resource for future generations. These apply mainly to unique resources for which there are few or no substitutes, for example, endangered species, certain old-growth forests, and unusual scenic resources. In theory, non-use values are zero for goods having perfect substitutes.

Although use values are readily measured with the travel cost model or contingent valuation method (CVM), non-use values can be measured only with the CVM. Thus, questions raised in the CVM section will also apply here. The more generalized a benefit becomes, the harder it is for people to respond to questions about willingness to pay. For example, respondents can more easily state how much they're willing to pay for a day of hunting deer than for preserving scenic rivers in their state or for saving old-growth forests in the Northwest. Nevertheless, due to the substantial levels of non-use values, research continues in this area. An example is the previously mentioned study where respondents divided their $18 annual willingness to pay for spruce-fir quality protection into 8.2 percent for use value, 29.6 percent for bequest value, and 58.5 percent for existence value (Haefele et al., 1992). Aylward (1992) cites a study of Australian parks where non-use values per visitor exceeded use values by about 35 percent. A Colorado study found that, out of a $55 per household annual willingness to pay for wilderness areas, 24.5 percent was for on-site use value, 21.7 percent for option value, 24 percent for existence value, and 29.8 percent for bequest value (Walsh, 1986).

If you think non-use values are unimportant, imagine selling a unique area like Yellowstone National Park to private industry. Suppose they developed the park to produce and sell geothermal power. There would no doubt be an international protest, and surveys could probably show that option, existence, and bequest values would be more than enough, worldwide, to offset the development income. This means that preservation would be socially optimal. But inability to fully collect such values could often preclude preservation in a private market. Or suppose projected use values aren't quite high enough to justify preventing hydroelectric dam construction in a unique canyon. Even minimal non-use values could be enough to change the decision to favoring preservation.

Be careful not to assume that the above non-use values are important only for unmarketed goods like natural environments. Many marketed outputs, if unique with few substitutes, have non-use values that producers can't easily sell. Examples are certain music groups, private art museums, and unique amusement parks. Also, non-use values exist for general classes of goods like wood, coal, and oil.

In thinking about use and non-use values of forests, note that scenic, open space, and recreation values yield external benefits too (the positive side effects mentioned in Chapter 3). Among these are a society that is more productive, happier, healthier, longer lived, less stressed, and more peaceful. Psychologists argue that natural amenities have superior stress-reducing qualities (Ulrich, 1988). Although these benefits are poorly accounted for by markets, they're by no means unique to forest recreation. They are also provided by many marketed outputs like music, sports, theater, cinema, art, and other entertainment, for both spectators and participants. So in making arguments for forest recreation values, we shouldn't assume that certain benefits apply only to nonmarket amenities and not to marketed outputs.

MARGINAL VERSUS TOTAL VALUES

Countless separate surveys or simulations have gauged people's willingness to pay for currently unmarketed amenities like owls, wilderness, recreation visits, wild river preservation, otters, salmon, old-growth timber, and woodpeckers. These are all separate or marginal valuations. But we never approach the same people and ask them about their willingness to pay for all things simultaneously. Then after several questions, we might hear, "You mean I have to pay for the recreation visits *and* the owls *and* the otters *and* the salmon?" Marginal values commonly cited are likely to be overstatements, since there probably isn't enough personal income to make all payments. Thus, if we found willingness to pay for all possible environmental benefits concurrently, it's likely the total value would be far less than the sum of stated marginal values. Since incomes are finite, the only way to learn for sure how people would spend their money would be to place all goods in the market, which isn't possible. Thus, despite the problems in valuing nonmarket goods, we continue to muddle through, knowing that to ignore them would be worse than making crude estimates. For the above reasons, many analysts feel that public agencies should charge more often for recreation.

Interestingly, if consumers had to buy all environmental amenities, less income would be available to spend on goods whose production damages the environment. The argument is partly academic, since many of these amenities are **public goods,** which, by definition, cannot be marketed to individuals. But it's just another way of seeing that unregulated markets tend to give us more than the optimal amount of environmental damage.

Remember that forest output values discussed here are derived from fairly small additions to, or subtractions from, total use levels. From these values, you can't make inferences about large changes in output mixes. For example, suppose CVM and travel cost studies estimated that per acre wilderness values for a given region generally exceeded stumpage values. To use an extreme example, this wouldn't necessarily mean that you should set aside an entire multistate region as a wilderness area. Such major changes would drive stumpage values up and presumably reduce the willingness to pay for wilderness. This shows the importance of distinguishing between what economists call **partial equilibrium analyses** (the most common analyses in forestry, which look at relatively small marginal changes) and **general equilibrium analyses** (less common, quite difficult, and comprehensive approaches, which account for all interactions over time).

HEDONIC TRAVEL COST MODEL

The **hedonic travel cost model** focuses on valuing site characteristics. The approach estimates regression equations of travel costs, T_z, from each origin zone, z, as a function of several attributes, A_j, at several recreation sites. For example, with fishing areas, attributes might be acres of scenic areas, average size of fish in pounds, average daily catch per person, and acres of water. The equations, one for each origin, could take the form:

$$T_z = c_0 + c_1A_1 + c_2A_2 + c_3A_3 + \ldots$$

Since the A's are usually attractive attributes, the estimated coefficients, c_j, should have positive signs and are each origin's price of one added unit of each attribute. For example, if A_2 is acres of water, c_2 is the added willingness to pay travel costs (T_z) per added acre of water, or the per acre "hedonic price" of water for users from origin z.

With information from the above equations, for each attribute, analysts statistically estimate separate demand functions showing the amount of the attribute consumed across all origins at different hedonic prices. This permits estimating the added willingness to pay per recreation day for an increase in the amount of an attribute like number of scenic overlooks or average daily fish catch. Hedonic methods can estimate willingness to pay for certain site attributes and can simulate the effects on recreation site values when attributes are changed. But theoretical questions have been raised about the approach, it is not yet widely used, it can be fairly complicated, and it requires very detailed data on site attributes and visitation rates (Freeman, 1993; Mendelsohn and Markstrom, 1988).

EXAMPLES OF DERIVED NONMARKET VALUES

For several types of recreation, Table 14-4 gives values per activity day, averaged from many studies in the United States. These are approximate average consumer surplus values, or the total area under recreation site demand curves divided by total use at a zero fee. The Table 14-4 values were used in the U.S. Forest Service 1990 Renewable Resources Planning Act program analyses. Most of the valuations were from travel cost models and contingent valuations, with some hedonic models included. The first two rows of Table 14-4 show higher values per activity day in the Northeast, where population pressures are greatest. But in the second two rows, for hunting and fishing, the Northwest shows slightly higher values.

TABLE 14-4
AVERAGE CONSUMER SURPLUS VALUES PER ACTIVITY DAY FOR FOREST RECREATION*

	Region	
Activity and year	**Oregon and Washington**	**Northeastern and north-central U.S.**
Camping, picnicking, and swimming (1989)	$7.76	$13.40
Hiking, horseback, and water travel (1989)	6.03	8.92
Hunting (1987)	39.08	38.22
Fishing (1987)	49.92	45.84

*Values are averages from many studies in the United States and are shown in USFS (1990) as "market clearing price plus consumer surplus." However, they are effectively the total area under the demand curve divided by the number of visits at a zero price, which yields "average consumer surplus" (phone conversation with Linda Langner, U.S. Forest Service, RPA Staff, Washington, D.C., 2/28/94).

Don't be fooled by the seeming accuracy of recreation values given here to the nearest penny; these result from averaging many studies, all of them approximations. For any recreation site, valuations can vary greatly depending on techniques and assumptions. For example, with travel cost approaches, values can vary with the time cost of travel assumed, the functional form used in the trip demand curve estimation, incorporation of substitute sites, car pool factors, and the method of interpreting values from the derived demand curve, among other things. With contingent valuations, answers can be affected by gaming, method of asking questions, background information given, realism of questions, payment vehicle, and a host of other factors.

Other activity day values averaged from many United States studies from 1983 to 1988, in 1987 dollars, are nonconsumptive fish and wildlife, $20.98; wilderness use, $16.83; sightseeing and off-road driving, $16.57; and nonmotorized boating, $44.64 (Walsh et al., 1988).

Imperfect as nonmarket valuations are, we can ill afford to ignore them, since they often exceed values of marketed forest outputs. As an example of comparative values, including consumer surplus in all cases, 1989 annual benefits from the National Forests in the South (Region 8) were estimated as 34.0 percent fish and wildlife, 30.8 percent timber, 21.6 percent recreation, 13.4 percent minerals, and 0.2 percent other (Pearse and Holmes, 1993). While we can argue at length about valuation techniques, one conclusion is inescapable: nontimber values are significant.

WATER VALUES

Where water is scarce, especially in the southwestern United States, public forest managers worry about the effects of forestry practices on the amount of water flowing from their lands. For example, on some watersheds, a lighter timber stocking will yield more water but less timber. The next chapter will discuss such trade-offs. In order to see if water gains exceed timber losses, we need to know water values.

The broadest categories of water use are municipal and industrial, agricultural, and use of water left in rivers and lakes for boating, fisheries enhancement, hydroelectric power, and other uses. Although water isn't widely sold, market prices are sometimes available when municipal and industrial users buy water from agricultural water districts. Studies can estimate the increase in net agricultural income from irrigation at different levels and arrive at a value per acre-foot of added water (43,560 cubic feet). The travel cost model and contingent valuation method can estimate increases in recreation values resulting from differing amounts of water.

Municipal and industrial water users have purchased water rights in parts of Colorado for $100 to over $200 per acre-foot per year. Studies show that added water in agricultural irrigation brings a net increase in income of about $10 per acre-foot per year in Colorado and from $25 to $35 in California and Arizona. Estimated water values for added hydroelectric power have ranged from about $17 to $27 per acre-foot in the West. All values apply for the early to mid-1980s (Bowes and Krutilla, 1989). Although some studies value added water at its lowest valued use, others try to estimate allocations to different uses.

COMMON MISUNDERSTANDINGS ABOUT NONMARKET VALUES

When we're interested in how outputs affect land value and land use, we value the outputs on-site, as done in this chapter. Even when land buyers don't actually calculate dollar values of nonmarket benefits, they often implicitly assign present values to such amenities by paying more for land from which they personally receive these benefits. For example, nonindustrial forest buyers often pay much more for properties than the present value of future monetary income. But for industrial forest buyers, forest bid prices are less likely to be affected by nondollar values, unless they increase future sale price. Public agencies, in finding the social value of bare or stocked land, can enter estimated nonmarket, on-site values per year as positive "cash flows" (a) in any of the present value equations given in earlier chapters. Or dollar estimates of nonmarket values can be entered into equations for internal rate of return or benefit/cost ratio.

Figure 14-6 explained that travel costs for recreation, or logging and hauling costs for timber, cannot be included in on-site values. By the same reasoning, other expenditures by recreationists, for example, on sporting equipment, motels, recreational vehicles, and guides, cannot be included in on-site recreation values. Such costs are "recreation processing costs" analogous to processing costs for, say, logs, lumber, and furniture. For the reasons that we can't ascribe retail maple furniture prices to an equivalent volume of maple stumpage in the forest, we can't say that total recreation expenditures are the willingness to pay for using the recreation site. These are important points because you'll sometimes see multi-million-dollar "values" of recreation estimated for a park, based on total expenditures by users. Although these are invalid as on-site values, they are used in some types of regional economic analysis (see Chapter 16).

KEY POINTS

- Private firms have no monetary incentives to produce unpriced forest outputs that people desire. Examples are scenic beauty, species and ecosystem diversity, certain types of recreation, improvements in air and water quality, and wilderness.
- Governments, when supplying unpriced outputs, seek their monetary values to see if benefits exceed costs of provision.
- The following techniques aid decision making about goods without monetary values: annualizing production costs, applying the simple betterness method, finding minimum first year's benefit required to meet some financial criterion, and using the equi-marginal principle.
- Using an estimated demand curve for unpriced recreation, an average value per visit is the area under the entire curve (consumer surplus at a zero price) divided by total visits. This is *average consumer surplus.*
- Demand and supply curves don't exist for unmarketed forest outputs like water, endangered species, game, fish, and many forms of recreation. In these cases, supply-demand equilibria cannot occur as they do with timber (see Chapter 12). The result is that people would be willing to pay for many more nonmonetary goods than the market provides.
- By interpreting simulated increases in travel costs as recreation entry fees, the *travel cost method* can derive recreation site demand curves when no market exists. Values estimated from such curves can yield benefit estimates for making public land use decisions.
- The *contingent valuation method* uses surveys to find people's maximum *willingness to pay* (WTP) for environmental amenities. An average of these values is another estimate of average consumer surplus. Surveys can also estimate *willingness to sell* (WTS), or the minimum compensation someone requires to willingly forgo a nonmarket amenity.
- Average WTS is often at least three to four times average WTP. WTS might be more appropriate to value nonmarket losses, while WTP may be best for evaluating investments to provide new unpriced amenities.
- In surveys of WTP and WTS, respondents should be told about major estimated changes in prices and employment resulting from hypothetical scenarios.
- Averages from a number of studies in the 1980s report values per visitor day from about $17 for wilderness use to $40 or $50 for big game hunting and freshwater fishing.
- Non-use values such as option value, existence value, and bequest value can be important for unique environmental amenities.
- We need to consider what it implies to keep adding simulated values of people's willingness to pay when they may not have the income to make these payments.
- In arid areas of the United States, estimated values of increased water yields have ranged from about $10 to well over $100 per acre-foot.
- Recreation values in this chapter have been on-site, net of all travel and equipment costs. Total recreationists' expenditures are inappropriate to use in evaluating land use investments.

QUESTIONS AND PROBLEMS

14-1 Chapter 3 explains how a competitive free market could theoretically maximize social welfare under certain assumptions. With reference to forest recreation, what types of "market failures" might prevent the above welfare maximum, and how would they prevent it?

14-2 Why are public agencies interested in valuing unpriced forest outputs?

14-3 Describe two ways of making efficient decisions about providing nonmarket goods without actually measuring their dollar values.

14-4 What factors will influence the future quantities of forest recreation (visitor days) demanded and will shift recreation demand curves to the right? To the left? Distinguish between movements along a recreation demand curve and shifts in this curve.

14-5 Given an estimated demand curve for unpriced recreation visits, discuss different ways of arriving at total value and value per visit.

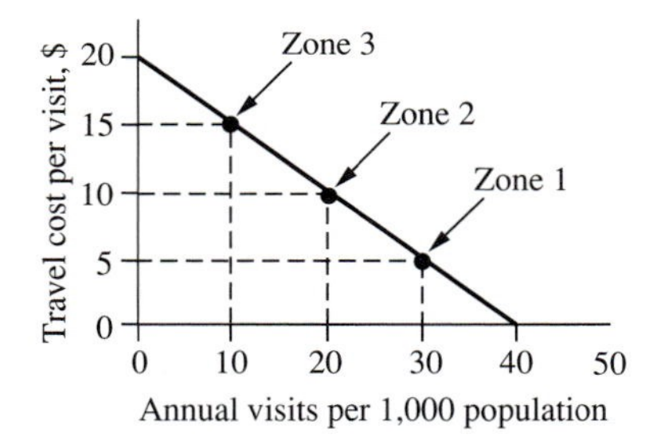

14-6 You are given the above "trip demand curve" for a park at which no fee is charged. The following populations are in each zone.

Zone	Population
1	200,000
2	30,000
3	120,000

If you simulate an increase of $5 in the travel cost per visitor (like a fee), what is the total number of visitors per year you estimate would come to the park from all zones (based on the travel cost model)? Make the same estimate for increases of $10 and $15. Then draw a site demand curve for the park, assuming there's currently no entry fee and that the increased travel costs are like entry fees.

14-7 For your site demand curve, what is the total consumer surplus at a zero price? What is the average consumer surplus per visit?

14-8 What are some weaknesses of the above simple form of travel cost model? How are these weaknesses overcome in more complex models?

14-9 In using contingent valuation methods for valuing nonmarket goods, what differences are you likely to find between willingness to pay and willingness to sell? Why?

14-10 What are different types of "non-use values" for unique environmental amenities? Why is uniqueness important in this regard?

14-11 Discuss the difference between a recreationist's willingness to pay to enter a recreation area and the willingness to pay for the entire experience, including travel. Which value is most appropriate for on-site recreation benefit/cost analysis? Why?

14-12 Following the approach for timber in Figure 12-8, sketch a theoretical series of regional market supply-demand equilibrium points over time, every five years, for some nonmarket output like hiking. Clearly label x and y axes, and all curves. Assume rising prices per visit over time.

REFERENCES

Arnold, J. R., G. L. Peterson, K. E. Watkins, G. E. Brink, and N. E. Merritt. 1991. *User's Guide to RMM Software: A Short-Run Partial Equilibrium Model for Economic Valuation of Wildland Resource Benefits.* GTR RM-202. U.S. Forest Service. Rocky Mountain Forest and Range Experiment Station. Fort Collins, CO. 57 pp.

Aylward, B. A. 1992. Appropriating the value of wildlife and wildlands. In Swanson, T. M. and E. B. Barbier. *Economics for the Wilds.* Island Press. Washington, DC. pp. 34–64.

Bishop, R. C., and T. A. Heberlein. 1990. The contingent valuation method. In Johnson and Johnson (1990), pp 81–104.

Bishop, R. C., and T. A. Heberlein. 1979. Measuring values of extramarket goods: Are indirect measures biased? *American Journal of Agricultural Economics.* 61:926–930.

Bockstael, N. E., W. M. Hanemann, and C. L. Kling. 1987. Estimating the value of water quality improvements in a recreational demand framework. *Water Resources Research.* 23(5):951–960.

Bowes, M. D., and J. V. Krutilla. 1989. *Multiple-Use Management: The Economics of Public Forestlands.* Resources for the Future. Washington, DC. 357 pp.

Boyce, R. R., T. C. Brown, G. H. McClelland, G. L. Peterson, and W. D. Schulze. 1992. An experimental examination of intrinsic values as a source of the WTA-WTP disparity. *American Economic Review.* 82(5):1366–1373.

Bulow, J., and P. Klemperer. 1994. Rational frenzies and crashes. *Journal of Political Economy.* 102(1):1–23.

Clawson, M. 1959. *Methods of Measuring the Demand for and Value of Outdoor Recreation.* Reprint No. 10. Resources for the Future. Washington, DC.

Diamond, P. A., and J. A. Hausmann. 1994. Contingent valuation: Is some number better than no number? *Journal of Economic Perspectives.* 8(4):45–64.

Freeman III, A. M. 1993. *The Measurement of Environmental and Resource Values.* Resources for the Future. Washington, DC. 516 pp.

Gregory, R., and R. C. Bishop. 1988. Willingness to pay or compensation demanded: Issues in applied resource valuation. In Peterson et al. (1988), pp. 135–141.

Haefele, M., R. A. Kramer, and T. Holmes. 1992. Estimating the total value of forest quality in high-elevation spruce-fir forests. In U.S.F.S. (1992), pp. 91–96.

Hammack, J., and G. M. Brown. 1974. *Waterfowl and Wetlands: Toward Bioeconomic Analysis.* Resources for the Future. Washington, DC. 95 pp.

Hanemann, W. M. 1991. Willingness to pay and willingness to accept: How much can they differ? *American Economic Review.* 81(3):635–647.

Hanemann, W. M. 1994. Valuing the environment through contingent valuation. *Journal of Economic Perspectives.* 8(4):19–43.

Johnson, R. L., and G. V. Johnson (Eds.). 1990. *Economic Valuation of Natural Resources: Issues, Theory, and Applications.* Westview Press. Boulder, CO. 220 pp.

Kahneman, D., J. L. Knetsch, and R. H. Thaler. 1990. Experimental tests of the endowment effect and the Coase theorem. *Journal of Political Economy.* 98(6):1325–1348.

Krutilla, J. V., and A. C. Fisher. 1975. *The Economics of Natural Environments.* Johns Hopkins University Press. Baltimore. 292 pp.

Mendelsohn, R., and D. Markstrom. 1988. The use of travel cost and hedonic methods in assessing environmental benefits. In Peterson et al. (1988), pp. 159–166.

Pearse, P. H. 1990. *Introduction to Forestry Economics.* University of British Columbia Press. Vancouver. 226 pp.

Pearse, P. H., and T. P. Holmes. 1993. Accounting for nonmarket benefits in southern forest management. *Southern Journal of Applied Forestry.* 17(2):84–89.

Peterson, G. L., B. L. Driver, and R. Gregory (Eds.). 1988. *Amenity Resource Valuation: Integrating Economics with Other Disciplines.* Venture Publishing, Inc. State College, PA. 260 pp.

Portnoy, P. R. 1994. The contingent valuation debate: Why economists should care. *Journal of Economic Perspectives.* 8(4):3–17.

Randall, A., B. Ives, and C. Eastman. 1974. Bidding games for valuation of aesthetic environmental improvements. *Journal of Environmental Economics and Management.* 1:132–149.

Rickard, W. M., J. M. Hughes, and C. A. Newport. 1967. *Economic Evaluation and Choice in Old-Growth Douglas-Fir Landscape Management.* U.S.D.A. Forest Service Pacific Northwest Forest and Range Exp. Sta. Research paper PNW 49. Portland, OR. 33 pp.

Rosenthal, D. H., and J. B. Loomis. 1984. *The Travel Cost Model: Concepts and Applications.* General Technical Report RM-109. Ft. Collins, CO. 10 pp.

Rosenthal, D. H., D. M. Donnelly, and M. B. Schiffhauer. 1986. *User's Guide to RMTCM: Software for Travel Cost Analysis.* USDA Forest Service General Technical Report RM-132. Ft. Collins, CO. 32 pp.

Rubin, J., G. Helfand, and J. Loomis. 1991. A benefit-cost analysis of the northern spotted owl. *Journal of Forestry.* 89(12):25–30.

Sedjo, R. A. 1993. Global consequences of U.S. environmental policies. *Journal of Forestry.* 91(4):19–21.

Shogren, J. F., S. Y. Shin, D. J. Haynes, and H. B. Kliebenstein. 1994. Resolving differences in willingness to pay and willingness to accept. *American Economic Review.* 84(1):255–270.

Tversky, A., and D. Kahneman. 1991. Loss aversion in riskless choice: A reference-dependent model. *Quarterly Journal of Economics.* 106:1039–1061.

Ulrich, R. S. 1988. Toward integrated valuations of amenity resources using nonverbal measures. In Peterson et al. (1988). pp. 87–100.

U.S. Forest Service. 1990. *The Forest Service Program for Forest and Rangeland Resources: A Long-Term Strategic Plan.* Washington, DC.

U.S. Forest Service. 1990a. *An Analysis of the Outdoor Recreation and Wilderness Situation in the United States: 1989–2040.* GTR RM-189. Ft. Collins, CO. 112 pp.

U.S. Forest Service. 1992. *The Economic Value of Wilderness.* Proceedings of the conference. Southeastern Forest Experiment Station. GTR SE-78. Asheville, NC. 330 pp.

Walsh, R. G. 1986. *Recreation Economic Decisions: Comparing Benefits and Costs.* Venture Publishing, Inc. State College, PA. 637 pp.

Walsh, R. G., D. M. Johnson, and J. R. McKean. 1988. *Review of Outdoor Recreation Economic Demand Studies with Nonmarket Benefit Estimates—1978–1988.* Colorado State University. Colorado Water Resources Research Institute, Fort Collins. Technical Report No. 54. 131 pp., plus app.

15

MULTIPLE-USE FORESTRY

That multiple-use appears toward the end of our journey doesn't mean the topic is an afterthought. In fact it's the essence of modern forestry from society's view. But before pulling everything together, you needed to understand the separate parts and analytic tools. Here you'll learn about different views and theories of multiple-use, as well as the basics (and frustrations!) of applying these concepts. More detailed, computer-based, forestwide applications are generally covered in forest management courses.

When I was a forestry student from 1958 to 1962, we gave lip service to multiple-use, based on the Multiple-Use Sustained Yield Act passed by the U.S. Congress in 1960, but underneath, most of us were still timber beasts. It seemed like other uses, if considered, would usually have to fit in with timber production. But over time, the public and most foresters have become much more attuned to nonwood outputs from forests. Through voting, legislation, court actions, journalism, television, and interest groups, society is getting more involved in public and private forestry decisions. One of the biggest problems with multiple-use is that it has so many different meanings.

THE MULTIPLE MEANINGS OF MULTIPLE-USE

Multiple-use forestry. The number of times we use the term, you'd think we knew what it meant! The truth is, multiple-use has many meanings, some of which follow:

1. *Many outputs from each forest acre*, for example, soil and watershed protection, wildlife, water, fish, many forms of recreation, scenic beauty, carbon

sequestering, climatic benefits, oxygen production, as well as minerals and timber. To produce all outputs on each acre is impossible. For example, many uses preclude wilderness.

2 *A mosaic of single uses on separate areas*, say, wilderness, developed recreation, a protected watershed, wildlife refuge, and intensive timber production. In a sense, true single-use forestry is impossible: no matter what your single output objective, a forested area can't help but produce other outputs.
3 *Various forms of multiple-use, with smaller but highly intensive timber management areas.* This approach has smaller timber production areas of highly intensive plantation management using elaborate site preparation, genetic improvement, thinning, pruning, fertilization, and herbicides—subject to environmental constraints—thus leaving a larger percentage of our forests devoted to nontimber uses. This would eliminate public timber sales with low or negative stumpage receipts, which have often occurred in mountainous and remote regions where much of the high value recreation areas are located. It could allow us to have our cake and eat it too—to get more timber as well as more forest recreation and ecosystem protection.
4 *Management for a "dominant use" and all other compatible uses;* for example, the dominant use could be timber, and you'd accept other coincidental outputs like hiking, hunting, and watershed benefits, so long as they didn't significantly reduce the dominant output.
5 *Many uses over time.* Although not commonly thought of as multiple-use, a newly planted clear-cut can provide deer hunting, and over the ensuing decades can become a general recreation area, and after 40 years another timber harvest area.

Figure 15-1 diagrams three of these concepts. So which kind of multiple-use is best? There's no easy answer. Some think that single-use is "bad" and multiple-use is "good." But often a mosaic of single uses brings society more satisfaction than trying to produce many outputs on each forest acre. In fact, for uses like wilderness, which yields great social satisfaction, it's impossible to produce certain other outputs like timber, hunting, minerals, or motorized vehicle trails. And on many single-use areas such as wildlife preserves, it's impossible *not* to produce other outputs like soil and water protection, scenic beauty, air and climate improvement, and fish. So there's no simple definition of multiple-use, and on any forest it's possible to mix all the above forms. In addition to a continuum of such mixes, within each form of multiple-use there's a continuum of variations and thus an infinite number of ways to manage forests.

What about all the other types of comprehensive management we hear about in forestry—sustained yield or sustainable forestry (sustaining not just timber but all outputs), "new perspectives," the "new forestry," ecosystem management, biodiversity? All of these are multiple-use if they consider many outputs of forests, both on the forest and elsewhere, and if they recognize relationships between physical, biological, and social systems on and off the forest.

Two views of forestry are **anthropocentric**, or human centered, and **biocentric**, or centered on biology and ecology. That may be a needless division. Most

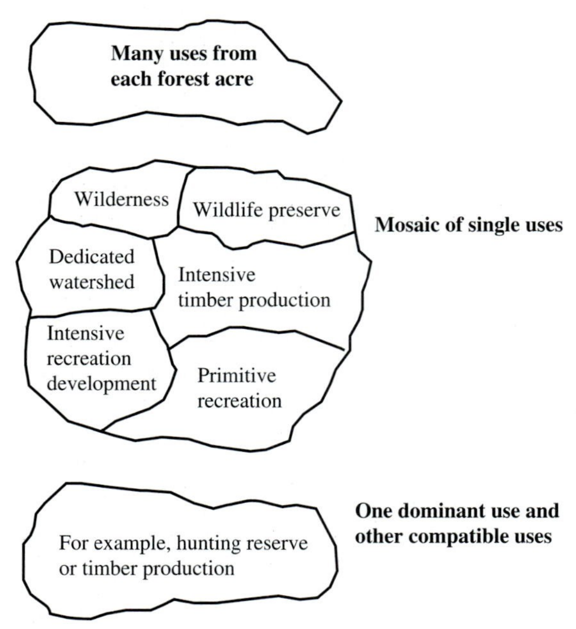

FIGURE 15-1
Aerial maps of three multiple-use forestry concepts.

people are concerned about endangered species and ecosystems, and those with a strong ecological bent get enormous satisfaction from that pursuit. So ultimately we all seek human satisfaction from forests. And there's a gradual continuum of views, from those who get the most enjoyment extracting and consuming material goods from nature, to those whose greatest pleasure comes from restraining consumption and saving ecosystems or treating them gently. The common threads are that we *all* consume, and virtually *all* of us enjoy nature. In the broadest sense, perhaps we are all anthropocentric—the view in this book.

The most detailed multiple-use planning usually occurs for public forests. Private owners tend to have less concern for outputs they can't market. But private forest owners also face multiple output questions. For example, what is the optimal mix of grazing, timber, and wildlife production for fee hunting on a forest? Overall, I'll be taking a public multiple-use view because, even on private lands, the public is concerned about management outcomes, especially negative side effects. This leads us to multiple-use goals.

MULTIPLE-USE GOALS

Forest managers are plagued with conflicting goals. You saw in Chapter 1 the problem with trying to provide "the greatest good for the greatest number in the long run." You simply can't do all three at once. Thus, it's useless to aim for things like simultaneously maximizing outputs of recreation, timber, wildlife, water, and soil protection. Furthermore, maximizing any physical output is generally unsuitable, since you invariably get to the point where producing one more unit of output will cost more than it's worth (**marginal cost** will exceed **marginal revenue**). So we optimize, which is what this book is all about.

Ultimately, people want the pattern of land uses that maximizes satisfaction or **utility**. Since we can't measure utility, analysts often assign dollar values to outputs, as in the previous chapter, and then seek the pattern of inputs and outputs over time that maximizes **net present value (NPV)** at some interest rate. That's the essence of forestwide optimization models like the U.S. Forest Service's FORPLAN (USDA, 1987). This is a computerized "linear programming" model that seeks to maximize some forestry objective, usually net present value, subject to constraints like minimum and maximum levels of certain inputs and outputs. Such models are studied in forest management courses. One of the problems with NPV maximization is that often widely differing mixes of outputs can yield similar **present values**. For example, Table 15-1 shows very similar net present values of four widely differing U.S. Forest Service policy alternatives prepared for the 1980 "program" under the Forest and Rangeland Renewable Resources Planning Act. Given the uncertainty in such calculations, there's no meaningful percentage difference between the four present values. Yet the plans have very different levels of costs as well as inputs and outputs in areas like timber, wilderness, fish and wildlife, range, water, and minerals. Each alternative yields different levels of benefits in various parts of the country, as well as different distributions of benefits over time and among user groups. So NPV maximization doesn't necessarily tell us anything about who should get what, where, and when: the *distributional questions*. To answer these, we need political processes, regional economic analyses (next chapter), negotiated consensus-building between groups, and sometimes litigation.

To show the net present value quandary, Figure 15-2 gives simple examples of six hypothetical multiple-use output patterns all with *exactly* the same $2,000 NPV, calculated at 5 percent real interest. For consistency, the 10-year time horizon is the same for each example, as is the initial $10,000 cost (which looks different in some panels because scales differ). In each case, let net benefits be multiple and valued correctly in dollars (benefits in any year are net of costs). Panel

TABLE 15-1
NET PRESENT VALUES OF FOUR U.S. FOREST SERVICE PROGRAM ALTERNATIVES DISCOUNTED AT 7-⅛ PERCENT INTEREST—1980*

	Higher market and nonmarket outputs	Lower market and nonmarket outputs	Medium market and nonmarket outputs	Lower market and higher nonmarket outputs
Net present values, $millions†	45,754	44,894	44,918	46,200

*USFS (1980).
†Present value of revenues minus present value of costs. Present value of costs varies from about $18 billion to $38 billion, another reason why NPV alone isn't a suitable criterion when budgets are limited. Fixing the costs for all options still doesn't solve the problem of identical NPVs for widely differing alternatives (see Figure 15-2).

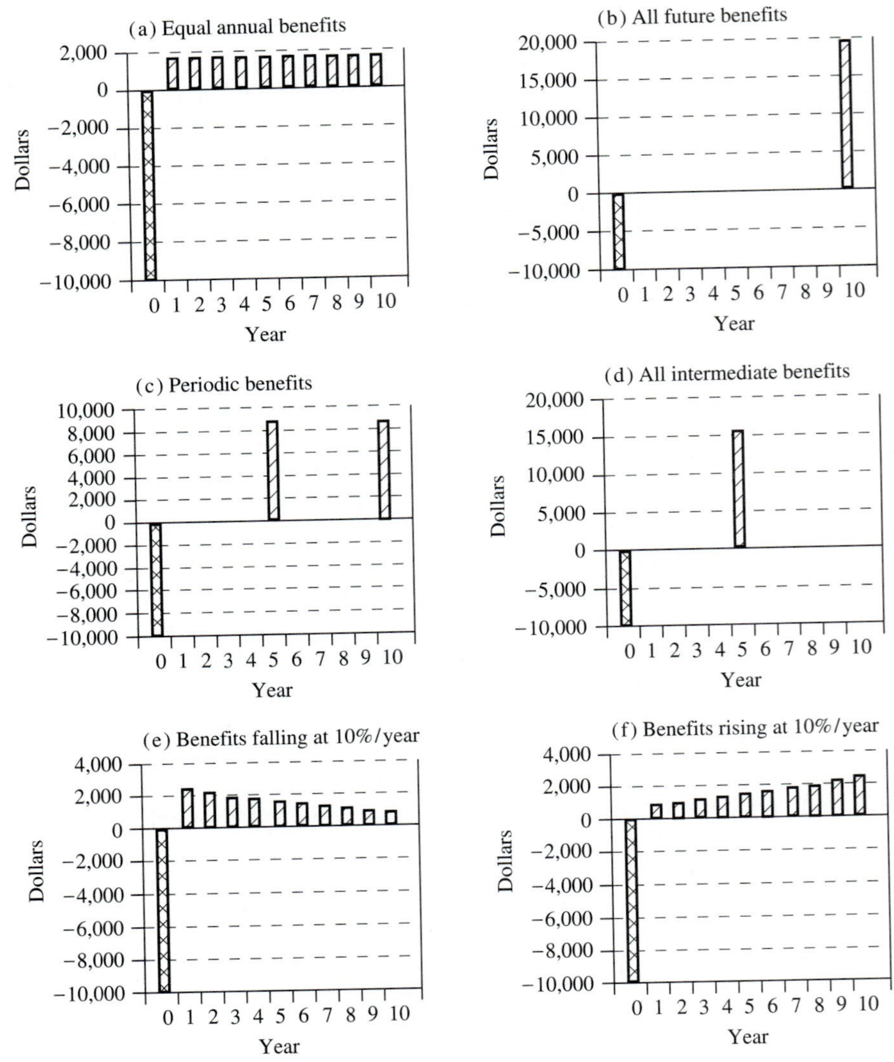

FIGURE 15-2
Hypothetical multiple-use net benefit patterns, all with the same $2,000 net present value, $10,000 cost, 5 percent discount rate, and 10-year life.

(a) shows equal annual benefits of several types accruing to different groups: these could be diverse user groups in the same region or dissimilar groups in separate regions. Panel (b) shows all benefits accruing at one future date. The remaining panels show benefits that are periodic, intermediate, falling, and rising. *For all six cases, outputs differ radically, but net present values are identical.* This result holds only for the 5 percent interest rate. In present value terms, lower discount rates favor projects with more distant benefits; higher rates do the opposite. Figure 15-2 shows separate patterns of benefit distribution over time; in reality, patterns are likely to be mixed. And they all can face the problem raised for panel (a): benefit distribution can differ—across regions, groups, and types of benefits. The combinations are mind-boggling, especially when you consider that each of the panels refers to only a fixed area; what about all combinations of other areas?

Figure 15-2 glosses over the environmental cost and benefit estimation problems discussed in Chapters 3 and 14. It's hard to assign dollar values to scenic and ecological benefits and to costs of certain environmental damage from harvesting. But these are needed for calculating NPVs.

All this doesn't mean that net present values or models for their maximization are useless. Other things being equal, we should still prefer high to low NPVs. And we want to avoid a negative NPV, which means that capital isn't earning the required **rate of return** used to calculate the NPV. Also, NPV-maximizing mathematical programming or simulation models are extremely practical for organizing resource data and playing "what would happen if" games to test results of different plans. So NPVs are often very useful for weeding out bad projects and ranking others, especially separate segments of larger multiple-use plans. Thus, this chapter looks at ways that NPV can help with multiple-use decisions. But NPV maximization alone cannot solve multiple-use problems because (1) several very different alternatives can have similar net present values, (2) for all outputs simultaneously, we don't usually have good data on physical **trade-offs** (step by step, as you change one or more forest outputs, how do all others change?), (3) there's much disagreement on the right discount rate for public agencies, (4) many benefits and costs might be ignored, and (5) even if benefits and costs are somehow "correctly" measured, NPV can't address distributional questions. So we're back to a goal of somehow maximizing satisfactions, while trying to maintain positive net present values of projects and their separable parts. And if NPVs are negative, we need to ask if the results are somehow "worth it" [equations (6-10) and (6-11)].

THE EASY DECISIONS

Along the lines of the **simple betterness method** mentioned in Chapters 8 and 14, you can often rule out certain management plans as obviously unacceptable or dominated by other options. For example, some plans may be infeasible for the following reasons: requirements exceed available resources of money, labor, etc.; biological infeasibility (e.g., planting a species that wouldn't survive); strong public opposition (an important archaeological site, an endangered species, or unique scenic resource might be destroyed); equity problems (e.g., a plan might provide abundant recreation for one group and none for another of equal size and importance). After eliminating dominated options, the tougher choices remain.

TRADE-OFF ANALYSIS

Consider a simple case of producing two outputs simultaneously from one forest area to maximize net present value. For selected ranges of outputs, examples could be:

1 Deer and timber: Opening a stand through partial cuts or clear-cuts yields more low vegetation that deer can use as browse, thus increasing the herd. If the initial stand is overstocked, then reducing stocking can increase timber growth and long-term harvest as well as deer. But if the stand is at or below optimal stocking, reducing it further will lower timber growth but still increase browse and deer.
2 Timber and water: Reducing timber stocking will increase water flow by lowering the forest's water transpiration, increasing surface runoff, and (where relevant) increasing the snowpack and Spring runoff. As above, depending on initial stocking, long-run timber growth and harvest may increase or decrease.
3 Timber and grazing: Reducing timber stocking will raise forage and beef production by increasing sunlight on the ground. As above, depending on initial stocking, long-run timber growth and harvest can increase or decrease.
4 Timber and recreation: For most, but not all, types of forest recreation, more intensive timber production will reduce the number of visitor days and reduce the value per day.

In each case, *within some range*, to get more of one output, you sacrifice some of the other: you "trade off" one against the other. In that range, outputs are *competitive*. When you get more of both outputs, the trade-off is positive, and outputs are *complementary*.

A Timber and Beef Example

Figure 15-3 is a simple example of three hypothetical combinations of **uneven-aged** timber and grazing. While mixing timber and grazing isn't feasible in many forest types, it can be done, for example, in the ponderosa pine region with beef cattle, and in Australia and New Zealand with radiata pine and sheep. At first, the only outputs for our example will be net incomes from timber harvests and beef production. Moving from scenario A to C, timber stocking increases as does timber growth and harvest income. At the same time, the area supports fewer cattle and yields lower beef income because less sunlight is available for forage. Below each diagram, graphs show the perpetual net cash flow patterns (gross income minus costs), assuming constant physical outputs per year and fixed real prices. To simplify, benefits are constant and perpetual, but that needn't be. Incomes are **annualized**, but they could also be, say, every 5 or 10 years and still have the same present values shown [see equations (6-10) and (6-11)]. The figure's last row gives the net present values (NPV) for timber plus beef at 5 percent interest: scenario B brings the greatest net present value at $900 per acre.[1]

---

[1]Assume in all cases that neither timber nor cattle stocking is so high that clear-cutting, selling land and cattle, and investing the proceeds at 5 percent interest could bring more present value than shown. Also assume that inputs such as fertilizer and labor are used efficiently, to the point where the last unit's cost equals its contribution to gross revenue.

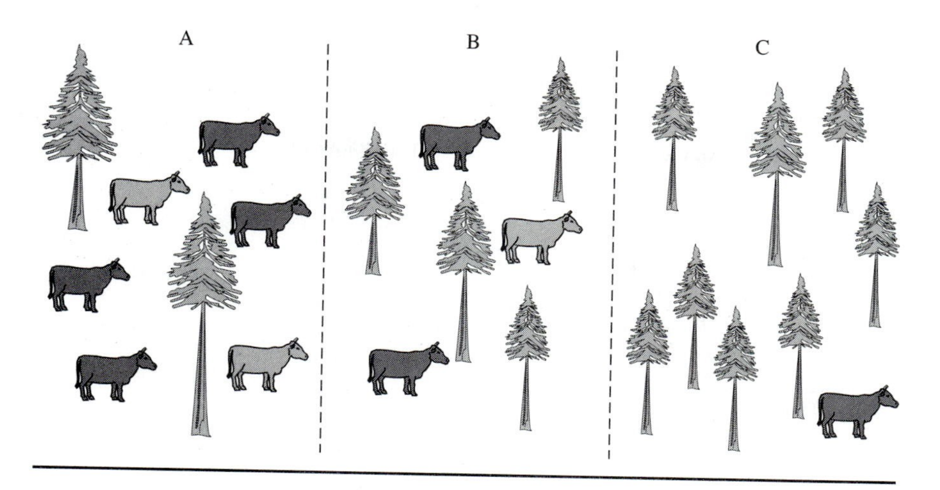

Perpetual annual income from each scenario:

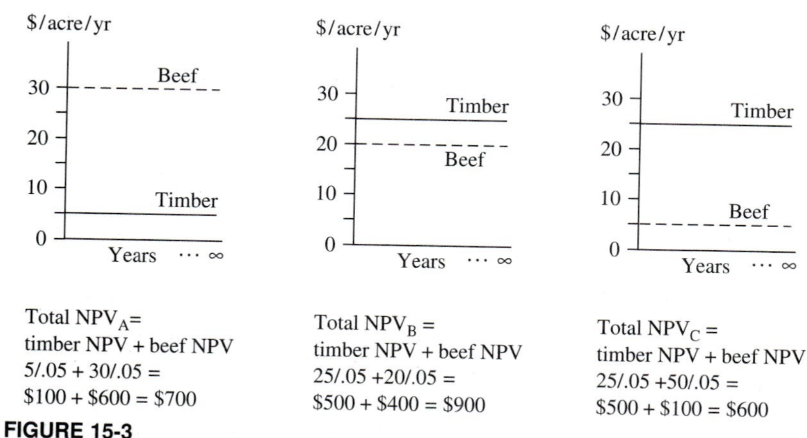

FIGURE 15-3
Three hypothetical timber and beef-cattle scenarios.

Since only three options are simulated, we don't know for sure whether B is the best of all possible combinations. Figure 15-4 smooths the NPVs for timber, beef, and totals and shows that the true maximum total net present value is at a timber stocking level slightly below B. The horizontal axis is units of timber stocking, say, cubic feet per acre, with the accompanying number of beef-cattle per acre at each timber stocking from Figure 15-3. Mathematical programming approaches might bring us nearer to the optimum. But often simple simulations like the above can get us close because net present value functions can be quite flat over a wide range of options.

In Figure 15-4, at timber stocking levels above point X, if you reduce timber stocking, timber and beef values are *complementary*. Reducing stocking from, say, C to X will increase incomes from both timber and beef, since the stand is overstocked in that range. Below timber stocking X, outputs are *competitive*. As

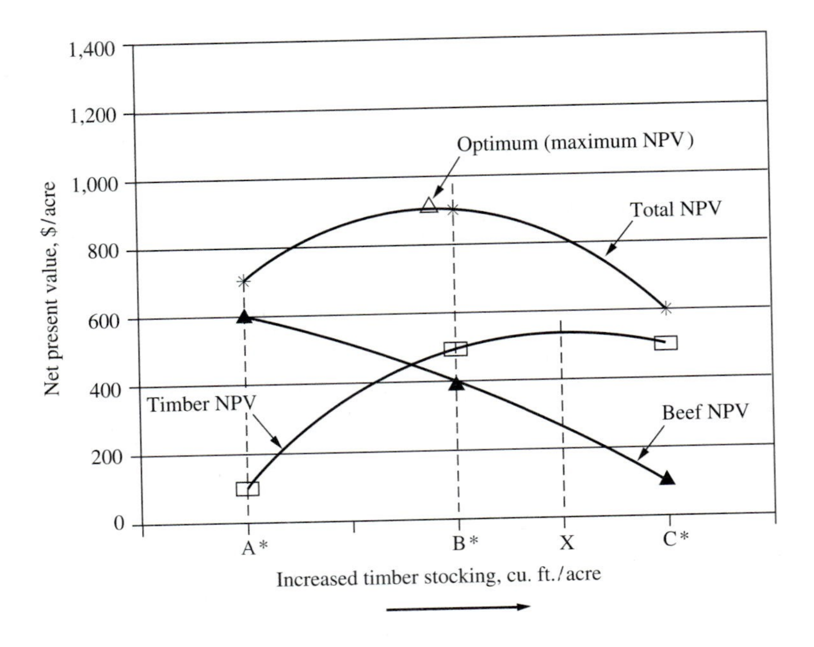

*Scenarios from Figure 15-3. As timber-stocking volume increases linearly (e.g., 2, 4, 6, 8...), the number of cattle decreases, but not necessarily linearly. As you move from left to right, up to point X, think of increasing timber harvests and decreasing beef production (and vice versa from X leftward).

FIGURE 15-4
Hypothetical beef and timber net present values.

you decrease timber stocking below *X*, annual growth and timber revenue will decline, while beef production will increase. Moving from case *A* to *B*, in present value terms, the trade-off is to gain $400 in timber value and lose $200 in beef value, obviously a good move.[2] Starting at *A* in Figure 15-4, imagine splitting added stocking increments more finely. You keep increasing timber stocking as long as the added timber NPV exceeds the loss in beef NPV. Total NPV is maximized when the added timber NPV just equals the loss in beef NPV. This is the same basic maximization principle found throughout much of economics. You go to the point where added benefits equal added costs, in terms of inputs, outputs, present value, dollars, satisfaction, utility, whatever. One economist has mentioned that it's almost embarrassing to admit that such a large part of economics is so transparently simple. But, as you'll see, things do get complex when you think of simultaneously optimizing many outputs.

What Will Change the Optimum? Concentrating on the range where output values are competitive (between *A* and *X* in Figure 15-4), what happens to

[2]Multiple-use optimization is often explained with joint production theory (see Gregory, 1955). But the approach here is simpler and more clearly shows present value maximization.

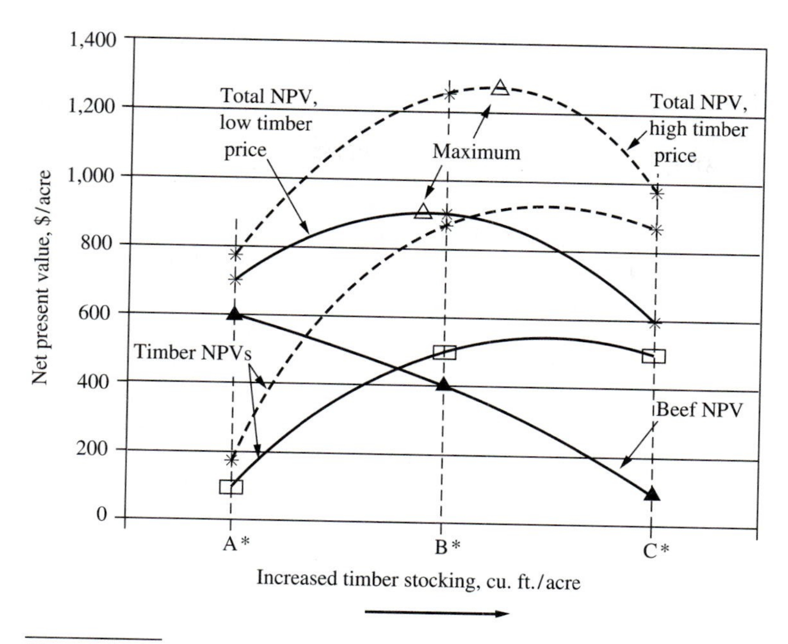

*Low timber price scenarios from Figure 15-3. High timber price is low price NPV times 1.75 (dashed lines).

FIGURE 15-5
Beef and timber NPVs with low and high timber prices.

the optimum if timber prices are higher at all points in time? Intuitively, if timber is worth more, other things being equal, you'll want to produce more timber. To illustrate, Figure 15-5 shows a dashed timber NPV curve with the timber price 75 percent higher (timber income and NPV are multiplied by 1.75). Given unchanged costs, the timber NPV curve is higher, rising and falling more steeply. That shifts the total NPV maximum to the right, as shown on the dashed total NPV curve, meaning more timber stocking and harvest and less beef income.

What if beef prices are higher? You'll want to produce more beef and less timber. In Figure 15-4, you'd multiply beef values by a factor of increase (or pounds of beef by a higher price). Other things being equal, that would make the beef NPV curve higher and more steeply sloped, thus shifting the total NPV curve's maximum point to the left, meaning more beef and less timber output. If beef prices increase enough relative to timber, the optimum is all beef and no timber. However, due to beef and timber complementarity between *C* and *X* in Figure 15-4, no matter how high timber prices are relative to beef, the optimum will include some beef, as long as NPVs are positive. Other scenarios could have timber prices rising over time and beef prices falling, or vice versa (as opposed to the zero-slope curves in Figure 15-3), which could change the optimum now and over time.

Even when prices stay fixed, if timber production becomes more efficient through, say, genetic improvement, the optimum can shift to higher timber stocking and fewer cattle. Or if cattle production becomes more efficient, the optimum moves in the opposite direction. And costs of inputs like tree fertilizer or supplemental beef feed can affect the optimum. So, what determines the optimum multiple-use pattern is the output and input prices and the physical trade-offs between outputs. For the above reasons, *multiple-use optima usually change over time, and you have to keep rerunning analyses.*

For the hypothetical example, multiple-use gave the highest value, but often values could be higher if certain areas were set aside for timber and others for cattle. Also, for true multiple-use, you must consider all other effects of alternatives, both on and off the forest: for instance, impacts on scenic and recreational qualities, fertilizer runoff, erosion, wildlife and other organisms, endangered plants and animals, stream sedimentation and turbidity, increased stream temperature due to less shade, and impacts of the foregoing on fish and other aquatic life. In Figure 15-4, you could include other net present value curves, for example, for water, recreation, and hunting, in addition to grazing and timber. These would all be vertically summed to get a total net present value curve.

For most multiple-use questions, physical trade-off data aren't very accurate over a wide range of options; for example, with stocking reductions, if you decrease annual timber harvest by 20, 40, or 60 cubic feet per acre, by how much can beef production increase? And at the same time, you have other outputs: by how much will water-yields or deer populations and a host of other factors increase or decrease? But rough estimates can be made, and it's still instructive to first look at the anatomy of the problem with the above example.

Published Studies of Trade-Offs

To further stress that trade-offs are important, let's look at a few published studies of how one forest output changes as you vary another. Accurate measures of timber management trade-offs are few, but research continues and data are improving. The following examples are all two-output comparisons—either hypothetical or simulations based on actual data. As you review these, keep in mind that decisions should not be based on only two outputs: this is just a prelude to a broader view.

Trade-Offs in Southeast Alaska Figures 15-6 and 15-7 show effects of clear-cutting 20 percent of an unmanaged **old-growth** spruce/hemlock forest on a southeast Alaska watershed of several thousand acres in the twentieth year of a computer-based simulation. Figure 15-6 estimates initial reductions in pink salmon spawners due mainly to more stream sedimentation following several logging and road-building options. Notice how numbers of fish are later increased due to protection and pool formation provided by large logging debris in streams.

Although we usually think of deer populations increasing after logging, Figure 15-7 simulates a decrease because deeper snow and slash in harvest areas make less browse available. After valuing estimated changes in hunting and fishing success, you could compare the gain in net timber income with fish and

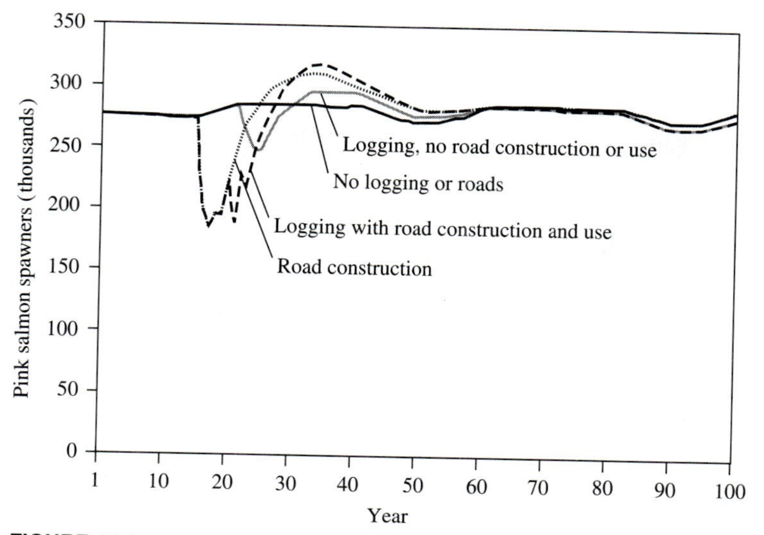

FIGURE 15-6
Impact of logging and road construction on pink salmon spawners. (Southeast Alaska simulations from Fight et al., 1990, pp. 87–88.)

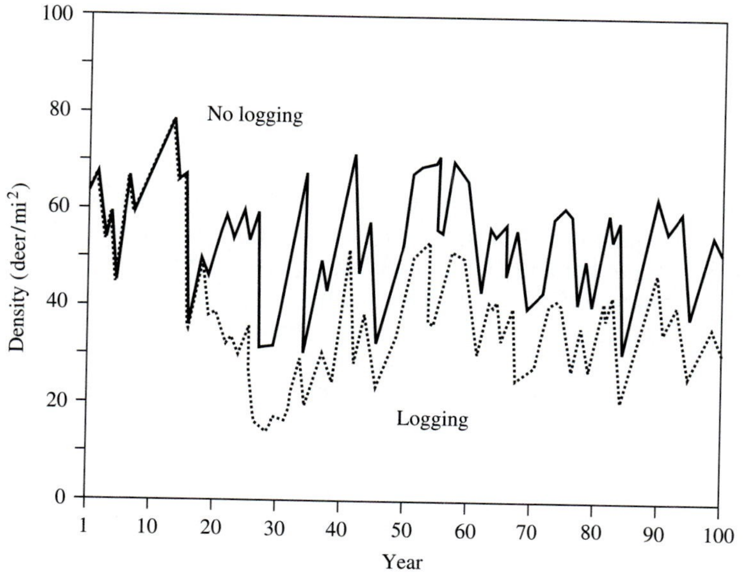

FIGURE 15-7
Effect of logging on deer densities in a watershed with snowfall. (Southeast Alaska simulations from Fight et al., 1990. pp. 87–88.)

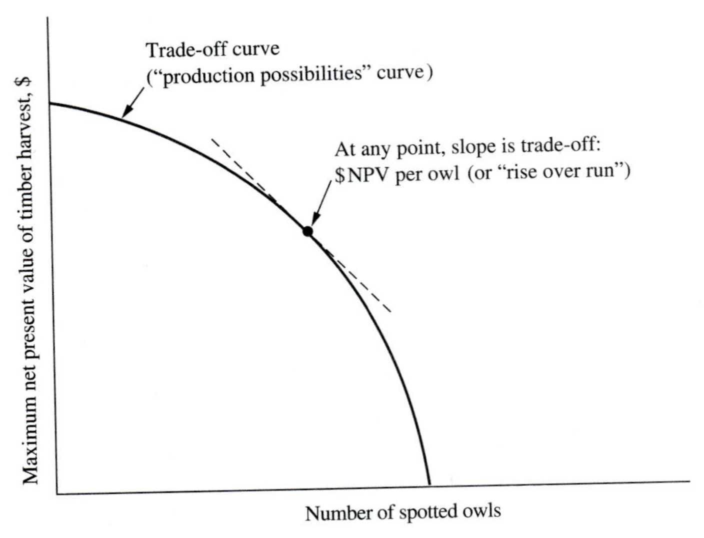

FIGURE 15-8
Trade-off between spotted owls and timber harvest values. (Adapted from Connaughton and Fight, 1984.)

wildlife losses, in present value terms. But these are only two of the impacts, and you'd have to consider many others, as mentioned later.

Owls and Timber in the Northwest Figure 15-8 outlines the likely shape of a *production possibilities curve* showing a hypothetical trade-off between maximum possible net present value of timber harvests and numbers of northern spotted owls on a Pacific Northwest old-growth forest area (given a fixed production technology). Starting with no harvest, you don't lose many owls with initial increases in harvests, especially without clear-cutting. But, beyond a point, the number of owls drops quickly as you approach the unconstrained maximum NPV of timber harvest. At any combination of owls and timber harvest, the slope of the trade-off curve tells you the marginal rate at which owls are traded for NPV: "rise over run," or dollars per owl. You can apply this line of reasoning to any two outputs from a forest. The toughest part is gathering data to define the physical trade-offs: In a given ecosystem, exactly how many owls do you lose as you increase harvests in a certain way? And the trade-off curve's shape changes when you alter the level of other outputs maintained, like the number of elk or recreation visitor days. To date no such information is available in any detail. And there's the added issue of jobs and income in the wood-processing industry with reduced harvests (see the next chapter). This shows the problem in looking only at two outputs.

The above analysis ignores a major concern in the spotted owl controversy. Jack Ward Thomas, who became chief of the U.S. Forest Service in 1993 and had earlier directed the Service's study of the northern spotted owl, has said,

> The issue is far bigger than the spotted owl and jobs. That trivializes it. The important topic is really the larger issue of whether we are going to preserve a particular ecosystem. About 10 percent of the old-growth forest is left in the Pacific Northwest. How much of that do we want to preserve? (TECO, 1993)

A *Time* article reflects the same sentiment in its title, "Whose Woods Are These? The fight is not just about spotted owls anymore. Conservationists step up an all-fronts campaign to save America's ancient forests" (Dawson, 1991). This issue is especially poignant when we realize that in most forested regions of the southern and eastern United States, little or no old-growth forest remains. Photographs reveal that some former old-growth southern pine forests appeared much like ancient ponderosa pine in the Inland Empire, somewhat smaller, but nonetheless magnificent. Virtually none remain. Thus, in the Northwest, another crucial trade-off is between added harvests of high-value timber and losing increments of a unique and dwindling ecosystem.

Wood Flow and Wildlife Reserve Area From a linear programming analysis of a 1,600-acre forested portion of the Naval Surface Warfare Center in eastern Virginia, Table 15-2 outlines the maximum net present value of timber harvests when different numbers of acres are devoted to meeting wildlife habitat needs. No timber harvests are allowed on wildlife reserves. In the first two rows, the trade-off is to gain 200 acres of reserve and lose an NPV of $594,000 – $563,000 = $31,000. Over this range, the trade-off averages $31,000/200 = $155 per acre of wildlife reserve. In the last two rows, the trade-off averages ($395,000 – $323,000)/200 = $360 per acre reserved between 800 and 1,000 acres. The cost rises due to increasing unevenness of age classes and loss of better site qualities.

More Wood and Water in Colorado In the above examples, outputs are competitive: you get more of one at the expense of another. Table 15-3 shows a positive trade-off where heavier cutting brings increased water from a Colorado national forest. The outputs, expressed in present value, are complementary.

TABLE 15-2
TRADE-OFFS BETWEEN TIMBER PRODUCTION AND WILDLIFE RESERVES ON AN EASTERN VIRGINIA FOREST*

Wildlife reserve area, acres	Net present value of timber harvests, $thousands
0	594
200	563
400	510
600	456
800	395
1,000	323

*From Cox and Sullivan (1995).

TABLE 15-3
PER ACRE PRESENT VALUES OF INCREASED TIMBER AND WATER YIELDS UNDER THREE SIMULATED HARVEST OPTIONS—GUNNISON NATIONAL FOREST, COLORADO*

Harvest schedule (rotation/years between cuts), years	Timber value $/acre	Increased water value $/acre†	Road costs $/acre	Increased net present value $/acre†
120/30	30	295	237	88
90/30	36	298	237	97
60/20	53	332	263	122

*Discounted at 4 percent real interest. From Bowes and Krutilla (1989, p. 171), by permission.
†Increases, compared to no harvesting.

Under the three harvest schedules shown, every 20 or 30 years one-third of a watershed is clear-cut in small patches, using the rotation ages shown, with thinnings also done. From top to bottom, rows in the table show increasing harvest levels. With the resulting stand openings, trees consume less groundwater, and less snow and rain is caught in tree branches where much of it would evaporate. Thus, simulated downstream water yields are increased—based on hydrologic research results in Colorado. Water values per acre-foot result from estimated allocations of the added water to hydroelectricity, irrigation, and municipal and industrial uses, valued as given in Chapter 14.

In Table 15-3, the joint costs of roads haven't been subtracted from timber values, since the roads give rise to both timber and added water. Subtracting road costs from timber income alone would give large, negative timber NPVs. But deducting road costs from the sum of timber and water values gives a positive net present value for each option. Water present values are *increases* from harvesting, compared to no harvesting. Thus, NPVs are *increases* in forest value resulting from harvest options. Based only on the data shown, the third option is best, since it yields the highest increase in NPV. You can't really decide which option, if any, is socially best, since only two outputs are given, but the analysis nicely shows how to handle joint costs of roads (see "Joint Costs in Multiple-Use"). Final decisions would have to include negative visual impacts, stream siltation, and other ecological damage, as well as possible increases in big game. Since water in the West is scarce, it's fairly certain that most of the increased flows would be used, at least for irrigation. But questions surround estimates of how much would go to high-value municipal and industrial uses, due to uncertain industrial growth rates and potentials for urban water conservation programs.

Net Present Value and Scenic Beauty in Arizona For uneven-aged ponderosa pine forests in northern Arizona, Figure 15-9 shows trade-off curves for net present value per acre and scenic beauty. Numbers along the curves are timber stocking levels in square feet of basal area per acre. The x axis is based on an average annual "scenic beauty estimate" (SBE) score given by respondents

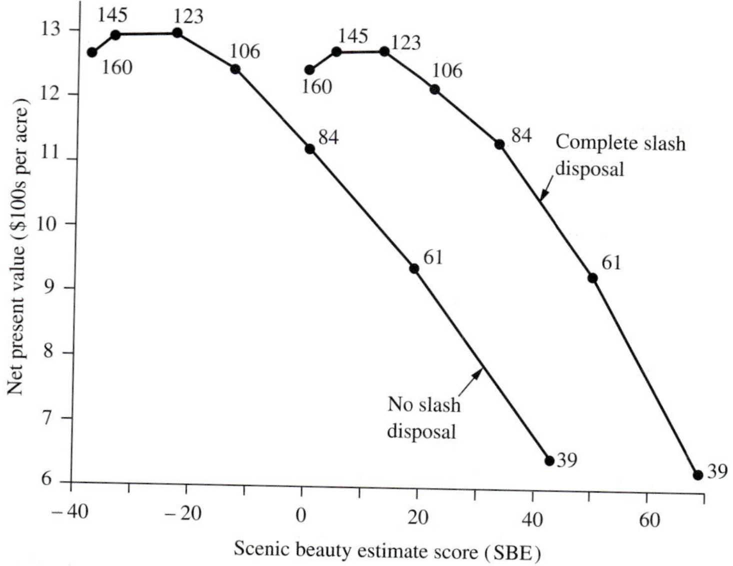

FIGURE 15-9
Trade-off between scenic beauty scale and net present value from timber, forage, and water—northern Arizona ponderosa pine. (From Brown, 1987, by permission. Numbers along curves are average annual stocking levels in square feet of basal area per acre. Real discount rate = 4 percent.)

who evaluated the scenic beauty of photographs showing different ponderosa pine management options. The NPVs are based on simulated yields of sawtimber, pulpwood, grazing income, and water, with local values applied, given different cutting regimes and timber stocking levels.

In Figure 15-9, the no slash disposal curve shows that managing for an NPV of \$1,250 per acre gives an SBE of about −15, while a \$920 NPV option yields an SBE of +20. In this range, the trade-off is to get 35 scenic beauty units at an NPV cost of \$330/acre. Going back to the photographs, and using the figure's 4 percent real interest rate, decision makers could ask if such a gain in scenic beauty in perpetuity is annually worth the annualized cost of 0.04(330) = \$13.20 per acre per year [see equation (6-10)]. The Arizona study notes that slopes of both slash disposal curves are similar at each basal area level. (Recall from Figure 15-8 that the slope of a trade-off curve is the number of y axis units given up to gain one x axis unit.) Between stocking levels of 123 and 106 square feet of basal area, the slopes of both curves average \$4.50 in NPV loss for each unit of SBE gained. Annualizing this, as above, gives a cost of 0.04(4.50) = \$0.18 per SBE unit gained. Between stocking levels of 84 and 39 square feet, this annualized cost rises to \$0.48 per SBE unit. In each of these cases, you could ask, "Is the gain in scenic beauty worth the added annual cost?" Or starting with high levels of scenic beauty, as you allow more intensive management for grazing, timber, and water, you could ask, "Are the additions to NPV enough to compensate those who lose scenic beauty?" The first approach stresses **victim liability**, and the second, **damager liability** (see Chapter 3).

THE EQUI-MARGINAL PRINCIPLE (AGAIN!)

If you can separate output values for several forest investments, the **equi-marginal principle** can be a rough guide for allocating multiple-use budgets. Chapter 4 notes that you maximize investment returns by allocating a limited budget so that the rate of return on the last dollar spent in each activity brings the same rate of return. Figure 15-10 applies the principle to one forest's budget of $140,000 that is efficiently allocated by spending $20,000 on planting trees, $70,000 on developed recreation, and $50,000 on wildlife management. At that point, the last dollar spent in each activity brings the same rate of return of 5 percent. As with earlier examples, try to reallocate the budget—say, put $10,000 more in one activity and $10,000 less in another—and you'll see that the loss in the marginal rate of return will exceed the gain. While only three activities are shown, you could have many more. (If you need to review the reasoning behind the equi-marginal principle, see the discussion of Figure 4-4.)

The above approach is useful only if (1) you know the physical production responses, say, how many more hunter days or harvested deer you'll get from a given wildlife management expenditure; (2) outputs can be measured in dollars; (3) spending is great enough that rates of return on more investment in an activity will fall; and (4) distributional concerns are met (who gets what, where, and when). If these things hold, this technique of budgeting can be simple and can handle a wide range of alternatives at once. And even if the last two conditions aren't met, it's still useful for decision makers to know the rate of return on more spending in separate areas.

The equi-marginal principle has drawbacks with large input changes. Major investment-changes in one activity can often affect the shape and position of marginal rate of return curves for other activities. For example, by investing heavily in timber production, you can change responses to wildlife investments (e.g., large timber investments can create so much forest cover that little space is left for wildlife food patches). But that's a critique common to many multiple-use

FIGURE 15-10
Optimal allocation of a $140,000 multiple-use budget.

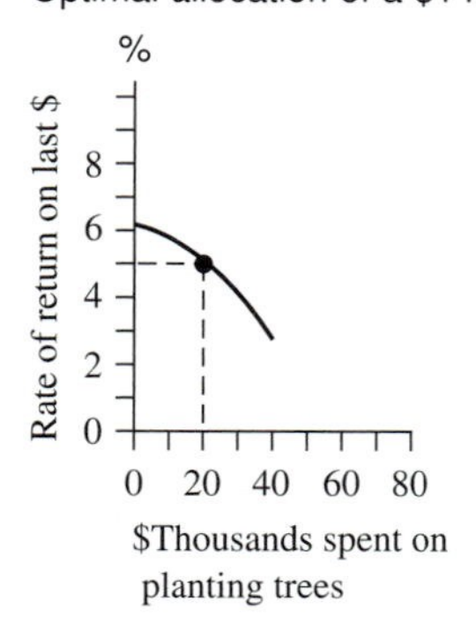

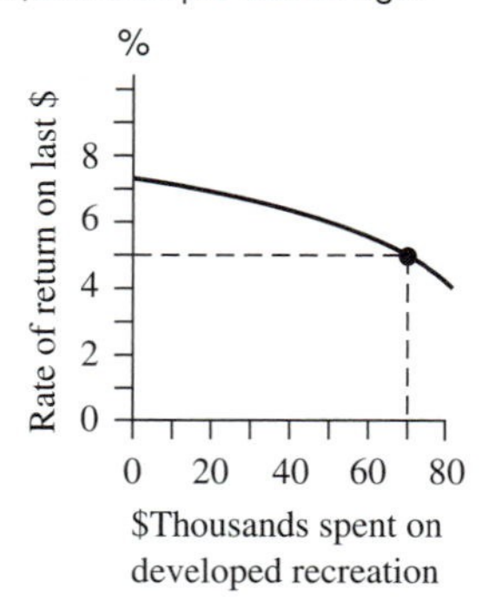

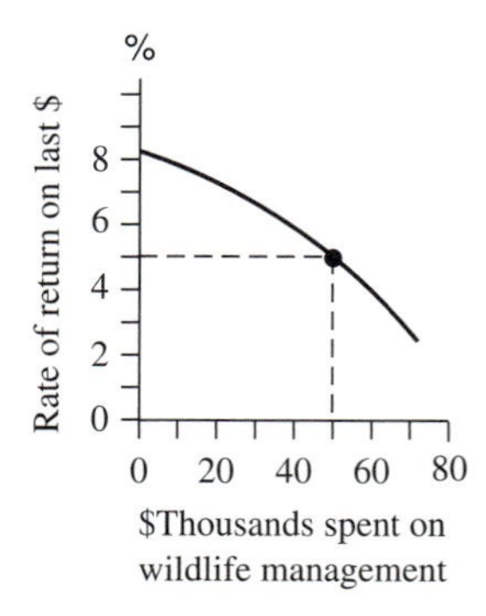

guidelines. So the equi-marginal principle isn't a cure for all multiple-use problems but simply another tool to help in the process. *Nothing can replace the need to carefully look at all consequences of each option.*

As you saw in Chapter 1, the equi-marginal principle is also useful in allocating budgets even when outputs from each activity are all in different units. You can still maximize total satisfaction by spending in a way that the last, say, $10,000 in each activity brings the same added satisfaction. At the same time, decision makers should note whether spending some of the forestry budget on a nonforestry project might bring even greater benefits, in which case money should be diverted from forestry. In Figure 15-10, this means that, if real rates of return exceed 5 percent elsewhere at the same risk (say, they're 7 percent), you should invest less in the three projects, even though you've met the equi-marginal principle on the forest. Now you should meet it at 7 percent.

The general format for applying the equi-marginal principle is always the same: The horizontal axes have the same input like time or money spent in different pursuits (as many as you want), and vertical axes are some common, *added* benefit (like utility, rate of return, income, lives saved, fish caught, recreationists served) per added unit of the input.

PROPERLY LINKING BENEFITS AND COSTS

When comparing benefits and costs of forestry actions, sometimes benefits are improperly linked to practices. For example, softwood planting investments often bring benefits of recreation, scenic beauty, and protection of soil and water. But in some regions, little or no investment on these sites will yield hardwoods or other vegetation that will also bring the same nontimber benefits. In such cases, you shouldn't include the nontimber outputs in the planting benefit/cost analysis to boost the computed rate of return or net present value. The softwood planting would have to stand on its own virtues, since the nontimber benefits would have occurred without the softwoods. But it *is* valid to attribute to forestry investments any nonmarket benefits that wouldn't otherwise be available.

Another example is the *allowable cut effect* whereby the U.S. Forest Service has attributed immediate increases in allowable cuts to current reforestation. This has occurred when current harvest rates for large mature inventories were set to equal the long-run annual timber growth of a National Forest. Suppose more planting investment is made, and today's annual harvest is therefore increased by the amount that the planting boosts long-run annual growth. If this immediate, continuing, higher annual harvest income is ascribed to one planting investment, its rate of return can become phenomenally high, even for a poor reforestation investment with high costs on a low site. But this is deceptive accounting because the mature timber harvest could have been increased anyway without the reforestation.[3] The only harvest income physically tied to a planting is the harvest later

[3]I realize the argument is that on U.S. Forest Service lands, you couldn't *legally* increase the allowable cut without added planting commitments, according to congressional constraints. My point is that these constraints can lead to nonsensical investments that could reduce society's well-being (see Klemperer, 1975).

yielded by the planted trees. Thus, the planting investment should be analyzed as a separate project. The rate of harvesting today's inventory really has nothing to do physically with planting on other acres. In fact, we might accelerate certain inventory harvests for insect or disease control, regardless of how much planting occurs on other acres.

The above two cases stress the **with and without principle** in benefit/cost analysis: *The true net benefit from any added investment in a multiple-use system is the net present value of the system with the investment minus the net present value without the investment.* Remember from Chapter 6 that an NPV addition of zero means that an investment is just barely acceptable: the added capital earns exactly the interest rate used in the NPV calculations. Applying this principle is vital, and sometimes not easy, when evaluating separate segments of complex multiple-use plans.

A BROADER MULTIPLE-USE VIEW

So far we've looked at incremental decisions and trade-offs for a few outputs, but large resource systems are obviously more complex. Figure 15-11 shows six responses to changes in tree basal area per acre on the hypothetical "Ponderosa National Forest" in the southwestern United States, where values of water, grazing, hunting, and other recreation are significant. Responses fall within reasonable ranges for ponderosa pine sites in northern Arizona on Brolliar soils with average annual precipitation of 20 to 22 inches (Brown, 1981). Basal area is the cross-section area of tree stems 4.5 feet from the ground in square feet per acre, a measure of stocking density. The graph in each panel shows a separate response to increasing basal area of trees, all responses occurring simultaneously. The top two panels are the kinds of timber and forage responses displayed in Figure 15-4; the next two show first rising and then falling quality of scenery[4] and deer habitat. Squirrel habitat rises continuously with heavier tree stocking, while stream flow and soil erosion decrease. Further outputs that might be affected by basal area and harvest activity could be numbers of other plants, fish, songbirds, and other wildlife; probabilities of fires; and visitor days of general recreation, fishing, and hunting. Then, *within* each response type are other effects; for example, outputs of timber, forage, and recreation will affect regional employment, income, and taxes. Also, stocking could be changed in many ways, for example, by clear-cutting in various patterns, by shelterwood cutting, by changing areas of wilderness, and through developing two-aged or uneven-aged stands.

Based on the physical response information and assigned values, simulation or optimization models could generate alternatives like the four given in Figure 15-12 for the Ponderosa Forest. For simplicity, the vertical axes show relative scales of maximum and minimum average annual output levels attainable over

[4]For good examples of scenic concerns in forestry, see the February, 1995, issue of the *Journal of Forestry*, which is devoted to forest aesthetics.

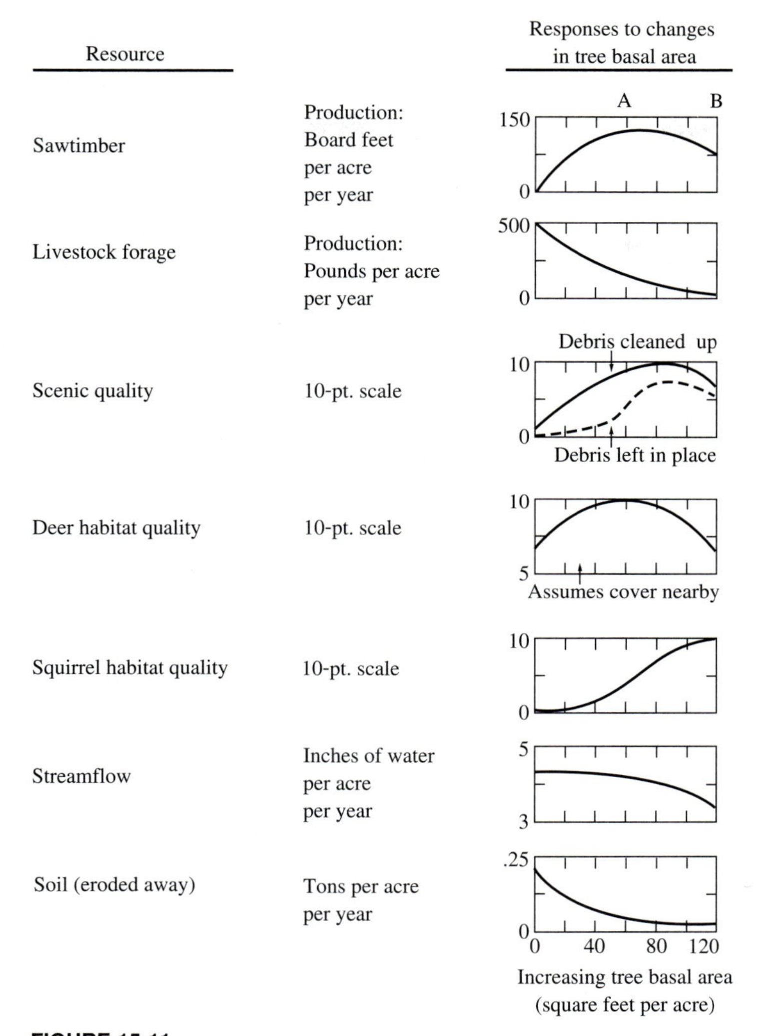

FIGURE 15-11
Typical resource responses on ponderosa pine sites in northern Arizona. (From Brown, 1981.)

the next 40 years. In reality, outputs would be displayed in the appropriate units. These scenarios could have been developed by maximizing net present value subject to certain maximum and minimum output requirements by interest groups. For each alternative, the net present value, costs, and the other information mentioned above should also be listed as flows over time, with all outputs quantified as well as possible, and valued in dollars where appropriate. We should also know who and where the recipients are. Interest groups can then use these arrays of information to help them decide on preferred alternatives. Or the arrays could spawn requests for other analyses.

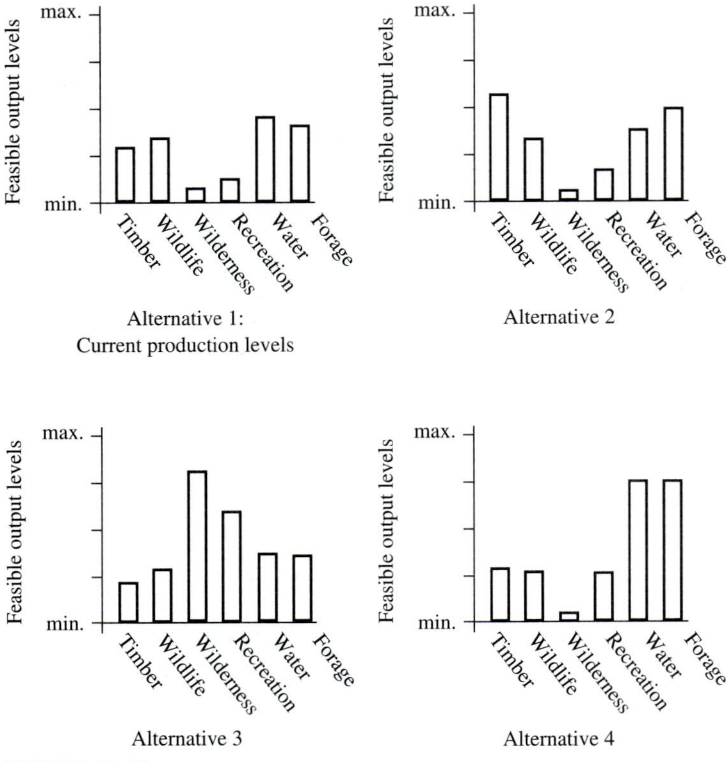

FIGURE 15-12
Relative average annual outputs for four alternatives on the Ponderosa Forest over the next 40 years. (From Brown, 1981. Bars do not show units of output, but only relationships. Max. is the maximum physical production possible for one output without considering other outputs.)

Joint Costs in Multiple-Use

In multiple-use, one cost, such as road building, often produces several benefits like timber harvests, recreation access, and fire protection. Such costs are called **joint costs** if, by dropping one or more of the outputs, the cost doesn't change. Any allocation of joint costs to separate outputs, although often done, will be arbitrary. You can't tell for sure whether, say, 15 or 30 percent of road costs should be allocated to recreation (unless you've invested in special road features needed *only* for recreation—such spending is a *separable cost*, not joint). Since there's no precise way to divide joint costs, economic analyses should consider the entire multiple-use forest and assure that net present value (NPV) is positive (that the present value of benefits exceeds the present value of costs). If you're concerned about one separable segment like timber sales, figure the NPV of the entire forest *with* sales and *without*. If a timber sale increases NPV above that for any other option, the sale is economic; if not, it isn't. This is another example of the "with and without principle." Applying it could settle many of the current arguments about "below cost" timber sales that some allege lose money on National Forests.

NEGOTIATED CONSENSUS BUILDING

The U.S. Forest Service and other agencies often help groups to negotiate a consensus on public forest plans, thus reducing appeals and expensive legal proceedings. Groups are informal or organized and can include loggers, mill owners and other businesses, wood-processing employees, wilderness advocates, consumptive and nonconsumptive wildlife users, anglers, hikers, off-road vehicle users, mountain bikers, snowmobilers, conservation groups, educators, researchers, and other interested people. At a series of local and regional meetings, groups consider different forest management options and first try to find common points of agreement. They then negotiate what they're willing to give up in order to gain certain benefits for their group. Meetings often yield "win-win" solutions where all groups gain, for example, separate trails for hikers, horses, and mechanized equipment, or no harvesting in certain recreation areas while having more intensive timber management on less attractive areas.

In presenting alternatives and gauging impacts of proposals, computer-based simulations and optimizations are useful. Also helpful are geographic information systems ("GIS") and computer visualizations that can help show what different land management actions will look like on the ground and aerially. I suspect that "virtual reality" systems will some day let us simulate a stroll through timber stands managed in different ways. A key point is that computer-based models don't give the answer, but they help parties reach a consensus.[5]

A possible problem is that groups with the best organization, the most time and money, and the most eloquent spokespersons may have the greatest influence. But if carefully done, with all parties represented, negotiation seems to be a far better approach than the adversarial conflict and litigation of the past. Those negotiating will probably recognize that multiple-use conflicts often cause strife even within the same person. A good example is the conservationist and actor Robert Redford and his inner turmoil about his Sundance ski area in Utah. He admits to feeling torn between hating clear-cutting and loving to ski on the clear-cut slopes of Sundance. He finally said to himself, "Okay, why don't you just live with the fact that you're a hypocrite?" (Oliver, 1992). Open dialogue about such feelings within people and groups and between groups should pave the way to more peacefully solving multiple-use forestry conflicts.

The answer to multiple-use problems on public forests lies not in mathematical models giving some "optimal" solution like maximum net present value, which excludes too many variables. The best we can hope for is to let models simulate what will happen if we manage under different scenarios and efficiency criteria. Groups can then use this information in their consensus building.

---

[5]For reviews of visual imaging and data visualization in forest and environmental management, see Orland (1992), Buhyoff and Fuller (1993), and Buhyoff et al. (1994). Bergen et al. (1995) review a computer-based method of predicting visual impacts of harvesting.

THE PROBLEM OF UNEQUAL CONCENTRATION OF LOSSES AND GAINS

Many multiple-use plans emphasizing nontimber values will have gains exceeding losses. But, as noted in Chapter 9, a common problem is that losses are concentrated painfully in certain sectors like the timber industry, which can easily muster the funds to launch public relations and lobbying efforts against the policy. But larger environmental gains are often scattered over so many millions of people at low benefits per person that it's hard to organize and raise funds to support the policy. The problem is compounded when gains, for example, from genetic diversity or climatic improvement, extend across international boundaries. This is the problem of *concentrated losses and diffuse gains*. When it occurs, as it often does in forestry, environmental policies with net gains are hard to implement.

A well-quantified example of the above case is a U.S. Department of Agriculture analysis of a marketing order restricting quantities of California-Arizona oranges shipped. The study showed that repealing the marketing order would cost producers $13 million annually, while saving consumers $30 million. But among the millions of consumers, each would gain about $0.12, while each producer would lose around $3,150 (C.E.A., 1992). Lobbying against is often stronger than lobbying for such policies, thus making them hard to implement even though gains exceed losses.

Somewhat less precise are analyses of policies to change timber harvesting plans to protect plant and animal species, improve climate and scenery, or protect soil and water resources: the results are likely to mean very concentrated losses to small groups and diffuse gains scattered regionally and often worldwide. An example is the spotted owl/old-growth preservation controversy in the Northwest, where losses are sharply concentrated among the forest products industry, its employees, and dependent communities, while gains are diffuse and not easy to measure.

The opposite of the above cases are policies that yield *concentrated gains and diffuse losses*. This is a problem if, in total, losses exceed gains. A possible example of this case is the former U.S. National Forest policy allowing large clear-cuts. The benefits in terms of low logging costs were concentrated in small groups, so gainers could easily mobilize to support their position. These gains could well have been offset by diffuse losses to millions of recreationists and others who were not organized well enough to resist the policy. The problem is that very little work had been done to quantify the diffuse losses, precisely because they were diffuse. Responding to environmental groups, the U.S. Forest Service started a "new perspectives" program in the early 1990s stressing partial cutting and other practices more sensitive to the environment (Bartlett and Jones, 1991).

Unfortunately, "new perspectives" may be an example of too little too late in many mountainous forests. Forest harvesting debates in the United States have become unnecessarily polarized. For years many foresters remained wedded to the most profitable harvesting methods, which usually meant clear-cutting. This may not be a problem on many flat areas out of the public view. But failure to stop clear-cutting in easily visible, mountainous forests has led to major conflicts. For example, had the U.S. Forest Service historically done more partial cutting rather than clear-cutting in visually sensitive areas, we might never have had the controversies over scenic damage on the Monongahela and Bitterroot National Forests in the late 1960s. These resulted in temporary bans on harvesting in many National Forest areas. Greater trauma from sudden cutting restrictions occurred recently in the Pacific Northwest. For years, the U.S. Forest Service's aggressive clear-cutting had been insensitive to scenic and other values in the region. Thus the public and the courts mistakenly believed the choice was between clear-cutting and no harvesting at all. They chose the latter on many public forest areas in the Northwest, with the result of sudden job losses. Had more partial cutting been done in recent decades, we might never have had the economic and ecological disturbances that occurred in the Pacific Northwest.

Twenty years ago I was a forest economist for the forest industry lobby in one of the nation's largest timber states and saw firsthand the tremendous influence that businesses with concentrated gains or losses have in the legislative process. But in recent decades, people involved with diffuse gains and losses have increased their political clout. Citizen lobby groups such as Common Cause have become much more effective. And environmental groups—where gains from lobbying are often diffuse—have also learned the lobbying trade. Examples are the recent successes of groups supporting endangered species and opposing clear-cutting in unique ecosystems in the United States. We are now seeing this power develop on an international scale. Although lobbying sometimes gets a bad name, it's an important part of the democratic process if emphasis is on giving information, not gifts. But lobbying needs to be open, honest, and balanced, so that all parties have influence, whether gains and losses are diffuse or concentrated.

The Need for International Cooperation

Some of the biggest chances for welfare improvement lie in a global vision. Millions and sometimes billions of people are often left out of regional benefit/cost analyses. For example, we need greater concern about how United States land management affects rivers flowing into Mexico, how tropical forest management affects world climate and genetic resources, or how United States energy consumption affects the atmosphere and resource prices. In terms of well-being, it's natural to worry more about our own children than strangers, more about our townspeople than those elsewhere, and more about our countrymen than foreigners. But as a world citizen, can I really say that the value of some environmental or economic benefit or cost is worth any more to me than to an Asian, an African, a European, or a South American? We're all in this game together! We need to be concerned about the effects of our actions on those far beyond our shores.

Unfortunately, everyone wants someone else to make sacrifices. In the United States we decry the loss of rain forests in other countries and the resulting loss of biodiversity and increases in atmospheric CO_2 levels. And this after we've removed nearly all of our old-growth forests! I'm not suggesting that we should be unconcerned about rain forest losses. But we in the United States might have more credibility if we made a greater effort to curb our energy appetite, which makes huge contributions to the world's greenhouse gases.

INDIVIDUAL FREEDOM AND MULTIPLE-USE

To some, true multiple-use seems mainly a public forest concern. But governments often take actions to protect multiple-use values on private lands. Examples are reducing property taxes to encourage forested open space near cities, making payments or reducing income taxes to encourage conservation practices like reforestation (see Chapter 9 for pros and cons of tax policies), educating landowners on conservation practices, buying private forestlands for public nontimber uses, levying taxes or fines on units of environmental damage, setting environmental quality standards, and zoning or regulating land uses to protect certain ecosystems.

In areas of fines, regulations, or setting environmental standards, issues of property rights and individual freedom are paramount, as mentioned in Chapter 3. Such intervention can reduce private property income and the owner's freedom. This raises the sensitive question, "Should we maximize individual land use freedom or optimize it?" The optimization view suggests that as population pressures mount, optimal levels of individual freedom seem to decrease, due to the rising costs of too much freedom. For example, urban areas were first to renounce the liberty to drive without a license; urbanites more readily curb freedoms to own firearms or to smoke in public. Land use planning, zoning, and regulations are more widely accepted in populous areas. Due to unpriced negative side effects, too much freedom in certain realms entails costs, just like too much of anything, so some will argue that we benefit from curbing certain land use freedoms. Others wish to maximize land use freedoms.

What is the optimal amount of land use freedom? Governments could apply marginal analysis by incrementally increasing actions to protect multiple-use values on private lands as long as the resulting benefits exceed the added costs, including costs of losing freedom. The optimal level of freedom occurs when the added benefit from some intervention just equals the added cost. This concept can provoke heated arguments: to some, the costs of losing freedom are so high that they want no government intervention; others see benefits to limited actions like damage fines, environmental standards, and regulations. We also need to include all administrative costs of freedom-limiting actions and to consider other approaches like education, incentives, or tax adjustments that might yield the same result at lower cost. Obviously, none of this can be very precise, but it's what we seem to be doing intuitively in the area of government intervention. In most forested areas where populations are low, the added benefits of limiting land use freedoms are less than in urban areas, so optimal levels of rural intervention are theoretically lower, which they are in practice. But as population increases, private forest landowners are subject to more government intervention to protect multiple-use benefits. Applying some of the notions of this chapter and Chapter 3 can help us approach the optimal degree of government intervention in private forestland use.

Remember that in a smoothly running, competitive economy one of the rationales for government intervention is the presence of unpriced positive and negative outputs. With all outputs priced, there's no need, for example, to tell landowners whether to plant Douglas-fir, ponderosa pine, or alfalfa—whether to raise cattle or kangaroos. Market signals should do the trick. And these signals are improving in many areas. For example, we now see more hunting leases giving landowners financial incentives to provide wildlife values. An Alabama survey showed large forest owners receiving hunting lease incomes averaging $5 per acre per year, with a range of $1 to $21 (McKee, 1987). Forest landowners sometimes sell **conservation easements** to environmental groups or government agencies. With these agreements, a landowner, while still maintaining timber harvesting rights, sells other property rights, thus preserving a given land character. Examples are selling rights to develop and build, rights for recreational use, water rights, and rights to clear-cut in certain scenic zones (see HTRG, 1993). These easements are often perpetual, so that when a property is sold, the new owner is obligated to the agreement. While far from perfect, the market is gradually becoming more efficient in yielding previously unmarketed outputs.

With nonmarket aspects of forestry, at one extreme, land use freedom is an uncaring pursuit of a private owner's well-being, and any public action is big, bad government interfering. On the other hand, you could view land use freedom as the liberty to practice responsible management that is sensitive to society's needs. Here the government's role could be educational, fostering the ultimate freedom: self-regulation. Nowadays most of us need no laws to keep from spitting in cities, although laws against public spitting have long existed. Ideally, we'd need no government actions to foster responsible land use. That's something nice to look forward to, but we're not there yet. However, things are changing: several forest products firms in the United States are now voluntarily leaving forested buffer strips around scenically and environmentally sensitive areas. And in states with few regulations, forest owners and loggers often voluntarily abide by "best management practices" to soften environmental impacts of logging.

Remember the stepwise framework for justifying any type of government intervention in land use, say, a damage fine, enforcing an environmental standard, or a regulation: *Institute the action only if the present value of resulting benefits exceeds the present value of costs, including the loss of freedom.*

WHERE TO FROM HERE?

Let me close with a touch of what may sound like heresy. We humans have an uncanny ability to adapt to new situations, forget old ones, develop new tastes, substitute goods and services, and learn new technologies. Across generations this ability is truly awesome. In the long run, consumers can quite happily change consumption patterns—from lumber to plywood to waferboard, from pristine wilderness to less natural "wilderness" or vice versa, from rivers to lakes, from softwoods to hardwoods, from wood to concrete, or from hoola hoops to Frisbees.

Does all this mean that nothing matters and anything goes? Not at all. But it does imply that within a wide range of options, it's impossible to prove that society would be better off in the long run with one multiple-use scenario than another. And with changing preferences, no management plan can stay optimal forever. It's tempting to think that voting on broad multiple-use guidelines would solve the problem. But there's no way for a voter to register intensity of preference. Someone with a flaming passion for or against an issue has the same power in the voting booth as one who is nearly indifferent. So there's no assurance that simple majority voting on multiple-use matters will necessarily maximize our well-being (Tullock, 1959). Although I hate to admit it, economic thinking, even if broadly applied, can't solve all multiple-use problems. To improve decisions, we'll need to apply forestry, sociology, psychology, economics, political science, ecology, and many other branches of the physical, social, and biological sciences.

You've seen that some multiple-use decision guides don't address equity issues very well. For example, a plan maximizing the National Forest System's present value could theoretically have all timber investment in the West and all recreation in the East (or vice versa). That's obviously not acceptable. We address these equity questions by trying to at least roughly distribute benefits throughout the country.

The Problem of Sudden Change

Sudden changes in land use usually cause problems. Damming a river in a lovely valley or canyon causes an outcry, no matter how beautiful the resulting lake. But draining an attractive lake to create a valley would be just as shocking. Switches from forestry to agriculture or other land uses bring resistance, as do sudden landscape changes from clear-cutting. We can't say that forests or lakes or meadows or valleys are inherently good or bad. But when people get accustomed to enjoying certain land use benefits, suddenly losing them is an obvious trauma. Some change is inevitable, and one guide can be to make changes gradual where possible. This shows an advantage of partial cutting over clear-cutting. But, as always, benefits and costs need to be compared.

Perhaps the broadest multiple-use forestry guideline should be to maintain variety, flexibility, and diversity so that we don't preclude future options. This is similar to Ciriacy-Wantrup's (1963) "safe minimum standard of conservation," or avoiding irreversible destruction of resource systems. But even here there's a catch! Unless the survival of humanity is at stake, we might accept some irreversibilities—especially with resources that aren't unique—if it costs "too much" to prevent the loss. This applies even to the sacred notions of species preservation and biological diversity. An example is the debate over whether to destroy the last remaining supplies of the smallpox virus, variola. Many scientists feel that the costs of possible disease outbreaks from preserving the virus would exceed the potential research benefits. In September 1994 a World Health Organization (WHO) committee recommended destroying variola by June 30, 1995. But in January 1995 the WHO executive board gave variola a stay of execution (Toner, 1995). While an extreme example, this shows that even the guideline of avoiding irreversibilities has problems.

KEY POINTS

- Multiple-use can mean (1) many uses from each forest acre, (2) a mosaic of single uses, (3) a combination of (1) and (2) with smaller areas of very intensive forest management, (4) a dominant use and other compatible uses, (5) many uses over time on the same area.
- You can't maximize everything at once, but you can optimize output levels so as to maximize net present value or, better yet, satisfaction.
- Net present value (NPV) maximization is often an unsuitable goal when comparing complex multiple-use plans, since very different plans often have the same NPV, unvalued benefits and costs might be ignored, and NPV can't tell us who should get what, when, and where.
- For two-output cases, trade-off analysis tells us how much of one output we have to give up to gain more of another. Sometimes we can get more of both outputs, for example, timber and water. For simple cases, marginal analysis can show the output combination that maximizes net present value.
- Usually two-output optimizations are limiting because they can ignore many other valuable yields.
- A limited budget is efficiently allocated to multiple-use investments when the last dollar in each activity brings the same rate of return.
- Make sure that benefits attributed to an expenditure like planting are not something that could occur anyway without the expenditure, for example, scenic beauty from natural regeneration, or harvests from existing inventory.
- The true benefit from any added investment in a multiple-use system is the net present value of the system with the investment minus the NPV without the investment. This is how "below cost" timber sales should be evaluated.
- For large multiple-use forest plans, we should simulate many alternatives and evaluate options by considering as many costs and benefits as possible. Net present value should be one variable to consider, but not the only one.
- Through negotiated consensus building, groups try to agree on the best multiple-use plan for public forests, after considering alternatives, often simulated with computer programs.
- Beneficial multiple-use policies are hard to implement if large losses accrue to a few parties and larger gains are diffuse. Harmful policies are sometimes enacted when benefits are concentrated and larger losses are diffuse. These are especially important points on an international scale.
- Rising population pressure is increasing the optimal level of government intervention to gain multiple-use benefits from private forests.
- A good long-term multiple-use guideline is to maintain flexibility and avoid sudden changes or irreversible losses of ecosystems, unless the costs of doing so are inordinately high.

QUESTIONS AND PROBLEMS

15-1 Discuss various concepts of multiple-use forestry.

15-2 Theoretically, what is the socially optimum pattern of land use?

15-3 From society's view, what are the problems with net present value maximization as a multiple-use guideline?

15-4 For a simple two-output land use, considering only dollar values, what factors will influence the present value maximizing output combination?

15-5 What is "trade-off analysis," and how can it help in making land use decisions?

15-6 You are a member of a legislative committee voting on the allocation of one year's $10 million budget between recreation and timber production for a public forest. You have evaluated the added benefits per additional million dollars spent on each activity as shown below. What budget allocation would you vote for to maximize benefits as you see them (i.e., what would be the optimal allocation of the $10 million from your view)? *Hint:* Use the equi-marginal principle.

$Millions spent on recreation management	Added recreation benefit from last $million spent ($millions)	$Millions spent on timber production	Added timber benefits from last $million spent ($millions)
1	7	1	5
2	8	2	6
3	9	3	7
4	8	4	6
5	7	5	5
6	6	6	4
7	5	7	3
8	4	8	2
9	3	9	1
10	2	10	0

15-7 Discuss how you could apply the "with and without principle" in evaluating added investments in a multiple-use system.

15-8 Although there are problems in using net present value (NPV) as the only criterion to make public multiple-use forestry decisions, we shouldn't ignore NPV. Discuss the useful role that NPV can play in multiple-use decisions.

15-9 Discuss the situations under which "concentrated gains and diffuse losses" or "concentrated losses and diffuse gains" can cause problems in the public multiple-use policy arena.

15-10 Discuss the notion of a "safe minimum standard of conservation."

REFERENCES

Bartlett, W. T., and J. R. Jones, editors. 1991. *Rocky Mountain New Perspectives—Proceedings of a Regional Workshop.* U.S. Forest Service. Rocky Mountain Forest and Range Experiment Station. Gen. Tech. Rept. RM-220. Fort Collins, CO. 51 pp.

Behan, R. W. 1994. Multiresource management and planning with EZ-IMPACT. *Journal of Forestry.* 92(2):32–36.

Bergen, S. D., J. L. Fridley, M. A. Ganter, and P. Schiess. 1995. Predicting the visual effect of forest operations. *Journal of Forestry.* 93(2):33–37.

Bowes, M. D., and J. V. Krutilla. 1989. *Multiple-Use Management: The Economics of Public Forestlands.* Resources for the Future. Washington, DC. 357 pp.

Brown, T. C. 1981. *Tradeoff Analysis in Local Land Management Planning.* Gen. Tech. Rep. RM-82. Rocky Mountain Forest and Range Experiment Station. U.S. Forest Service. Ft. Collins, CO. 12 pp.

Brown, T. C. 1987. Production and cost of scenic beauty: Examples for a ponderosa pine forest. *Forest Science.* 33(2):394–410.

Buchanan, J. M., and G. Tullock. 1962. *The Calculus of Consent.* University of Michigan Press, Ann Arbor. 361 pp.

Buhyoff, G. J., P. A. Miller, J. W. Roach, D. Zhou, and L. G. Fuller. 1994. An artificial intelligence methodology for landscape visual assessments. *Artificial Intelligence Applications.* 8(1):1–14.

Buhyoff, G. J., and L. G. Fuller. 1993. Explanation of quantitative models: A need and an example. *Artificial Intelligence Applications.* 7(1):37–43.

Ciriacy-Wantrup, S. V. 1963. *Resource Conservation.* University of California. Agricultural Experiment Station, Berkeley. 395 pp.

Clawson, M. 1975. *Forests for Whom and for What?* Johns Hopkins University Press. Baltimore, MD. 175 pp.

Connaughton, K. P., and R. D. Fight. 1984. Applying trade-off analysis to national forest planning. *Journal of Forestry.* 82(11):680–683.

Council of Economic Advisors. 1992. *Economic Report of the President.* U.S. Govt. Printing Office. Washington, DC. 423 pp.

Cox, E. S., and B. J. Sullivan. 1995. Harvest scheduling with spatial wildlife constraints: An empirical examination of tradeoffs. *Journal of Environmental Management.* 43:333–348.

Dawson, P., A. Dorfman, and E. Shannon. 1991. Whose woods are these? *Time.* Dec. 9, pp. 70–75.

Fight, R. D., L. D. Garrett, and D. L. Weyermann (Eds.). 1990. *SAMM: A Prototype Southeast Alaska Multiresource Model.* U.S.D.A. Forest Service PNW Research Station. Gen. Tech. Rep. PNW-GTR-255. Portland, OR. 109 pp.

Gregory, G. R. 1955. An economic approach to multiple use. *Forest Science.* 1(1):6–13.

Hancock Timber Resources Group. 1993. Conservation easements enhance timberland returns. *ForesTree Investor.* Vol. 7. December, pp. 1–3.

Haney, H. L., Jr. 1980. Economics of integrated cattle-timber land use. In *Southern Forest Range and Pasture Symposium.* Winrock International. Morrilton, AR. pp. 165–183.

Johnson, K. N. 1987. Reflections on the development of FORPLAN. In USDA (1987). pp. 45–52.

Joyce, L. A., B. McKinnon, J. G. Hof, and T. W. Hoekstra. 1986. *Analysis of Multiresource Production for National Assessments and Appraisals.* U.S. Forest Service Rocky Mt. Forest and Range Experiment Station. Gen. Tech. Rept. RM-101. Ft. Collins, CO. 20 pp.

Klemperer, W. D. 1975. The parable of the allowable pump effect. *Journal of Forestry.* 73(10):640–641.

Loomis, J. B. 1993. *Integrated Public Lands Management.* Columbia University Press, New York. 474 pp.

McKee, C. W. 1987. Economics of accommodating wildlife. In *Managing Southern Forests for Wildlife and Fish—A Proceedings.* U.S.D.A. Forest Service Southern Forest Experiment Station. Gen. Tech. Rep. SO-65. New Orleans, LA. pp. 1–5.

Oliver, P. 1992. Dancing to a different drummer. *Skiing.* 44(7):96–99.

Orland, B. (Ed.). 1992. *Data Visualization Techniques in Environmental Management.* Proceedings of a workshop. *Landscape and Urban Planning.* Special Issue. 21(4).

Schuster, E. G., L. A. Leefers, and J. E. Thompson. 1993. *A Guide to Computer-Based Analytical Tools for Implementing National Forest Plans.* USDA Forest Service Intermountain Research Station. Gen. Tech. Rep. INT-296. Ogden, UT. 269 pp.

Schuster, E. R. 1988. Apportioning joint costs in multiple-use forestry. *Western Journal of Applied Forestry.* 3(1):23–25.

Schweitzer, D. L., and R. A. Sassaman. 1973. Harvest volume regulation affects investment value—A look at the "allowable cut effect." *Forest Industries.* March.

Stankey, G. H. 1992. *Social Aspects of New Perspectives in Forestry.* Grey Towers Press, Milford, PA. 33 pp.

The Environmental Careers Organization. 1993. *The Complete Guide to Environmental Careers.* Island press, Washington, DC. 364 pp.

Toner, M. 1995. Fear of smallpox as terrorist tool spurs research proposals. *Atlanta Constitution.* May 19. p. A3.

Tullock, G. 1959. Problems of majority voting. *Journal of Political Economy.* 67: 571–579.

U.S.D.A. Forest Service. 1987. *FORPLAN: An Evaluation of a Forest Planning Tool.* Symposium Proceedings. Gen. Tech. Rep. RM-140. Rocky Mt. Forest and Range Experiment Station. Ft. Collins, CO. 164 pp.

U.S.D.A. Forest Service. 1980. *A Recommended Renewable Resources Program—1980 Update.* Publ. FS-346. Washington, DC. 539 pp. plus Appendixes.

Vincent, J. R., and C. S. Binkley. 1993. Efficient multiple-use forestry may require land use specialization. *Land Economics.* 69(4):370–376.

16

FORESTRY AND REGIONAL ECONOMIC ANALYSIS

Some of the most emotional issues in forestry have an economic development side. For example, many residents within New York State's Adirondack Forest Preserve wish to see more logging and wood processing, recreation development, and highways within the preserve to boost employment, income, and real estate values, and in general spur a flagging economy in northern New York. Others, especially city dwellers outside the region, fiercely defend the "forever wild" status of state lands within the preserve. In the Pacific Northwest, some groups seek to reduce **old-growth** harvesting to save selected stands of ancient forests and to protect northern spotted owl habitat. Others worry about the resulting loss of jobs and income. Tempers flare, rhetoric flows, and bumper stickers multiply. Similar conflicts occur throughout the world's forested regions.

Parties in these debates might be conservationists, business groups, government agencies, individuals—the same people involved in multiple-use controversies. Groups often consult economists to gauge economic impacts of forest management choices. These options include developed recreation areas, wilderness areas, **agroforestry**, more (or less) intensive timber management and harvesting, and more (or fewer) mills. Economic impacts of interest might be numbers of jobs, household income, tax receipts, local property tax rates, business sales, numbers of tourists, and land values. Other development impacts, not covered in detail here, include population growth rates, income distribution, traffic and congestion, changes in ecosystems, air and water pollution, crime, and visual impacts.

Groups gathering economic data often have an agenda: either to foster or discourage economic development, or to promote their group in some other way like generating goodwill, pleading for tax breaks and government assistance, or (in the case of public agencies) trying to increase their budgets. Thus, you often need to look at regional economic analyses critically because the deck might be stacked, not necessarily through malice but simply through zeal and misunderstanding. Rather than deeply probe the mathematics of such analyses, this chapter will help you interpret results of economic impact studies and better understand some of the issues.

STIMULATING REGIONAL ECONOMIES

Every forest management activity will have some economic impact. That obviously applies to harvesting and wood processing, but also to planting trees, doing timber stand improvement, and even setting aside wilderness areas to which recreationists come and spend money locally. While this chapter takes a narrower view than the last, we should always be aware of environmental impacts of alternatives and not look only at employment, income, taxes, property values, and the like. As ecologists warn us, "Never ever can you do just one thing."

I'll refer to any economic activity like forest management or wood processing as a "project." New projects have direct and indirect benefits. **Direct effects** are those resulting directly from constructing or operating the project—jobs created, wages paid, tax revenues generated. **Indirect effects** are those occurring not directly from the project but elsewhere as a result of the direct benefits. Examples are increased employment and wages in industries supplying the project; in services like grocery stores, gas stations, and restaurants; and in an expanded public sector that includes schools, police, roads, and other services. You can see how this mushrooms, since each of the new supply and service activities generates its own direct and indirect benefits. So a new project has a **multiplier effect**.

Regional Income Multipliers

Let's start by looking at the mythical "Forest County" with \$100 of new wages paid to tree planters on a reforestation project. The workers don't spend all of this in a way that immediately becomes new income in the county. Part of the \$100 goes to catalog sales and traveling outside the county, to savings, or to state and federal taxes: these are called **leakages**. Suppose that, of every dollar earned in the county, on average 75 percent is leakage, and 25 percent is spent in the county at local businesses. Suppose again that 75 percent of that business income is leakage, and 25 percent stays in the county, mainly as wages. These *respendings* continue until virtually all the original new income has leaked out of the county. The original \$100 is the *direct effect*, and the respending that stays in the county is the *multiplier effect*.[1] So the original \$100 of new income eventually becomes more than \$100 locally if you include the multiplier effects.

[1]Most input-output studies separate multiplier effects into *indirect effects* of industry interactions and *induced effects* of continued consumer respending. Since all examples here include both effects, I'll not separate the two. Multipliers including only indirect effects are "type I," and multipliers including both indirect and induced effects are "type II." Here all multipliers will be type II.

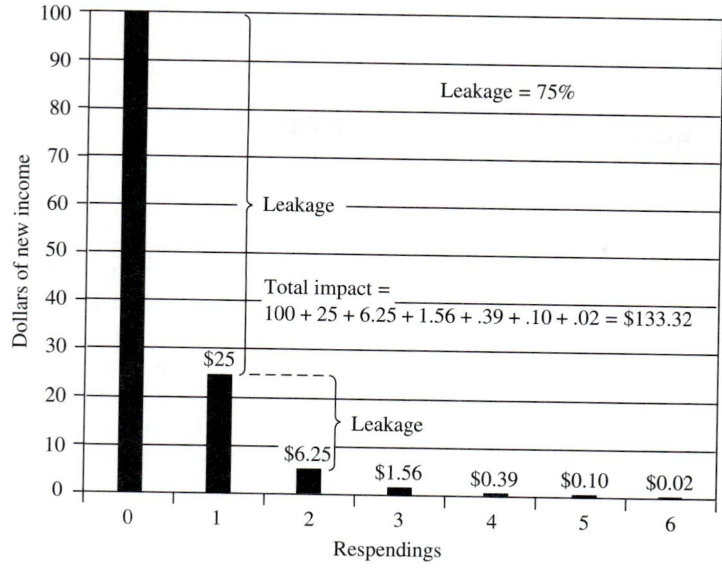

FIGURE 16-1
Direct and multiplier effect of $100 new income in Forest County.

For Forest County, starting with the $100 of new wages, Figure 16-1 shows how each respending yields a smaller amount of new local income, as 75 percent leaks out of the county. The direct or "first round" effect is $100. On the first respending, $75 is leakage, and $25 remains in the county; on the second respending, $18.75 leaks out, and $6.25 remains, and so on, until only $0.02 remains after the sixth respending. So eventually, the multiplier effects reach a limit. As shown in Figure 16-1, summing all the multiplier effects, you get $33.32. Adding this to the original $100 of new income, the direct and multiplier effects come to $133.32 of income in Forest County. That means the *income multiplier* is about 1.33. Thus, to find the direct and multiplied effect of new income in a region, just multiply the new income by the region's income multiplier. One expression for this multiplier is:

$$\text{Multiplier} = \frac{\text{direct effect plus multiplier effect}}{\text{direct effect}} \qquad (16\text{-}1)$$

This serves as a basic format for multipliers applying to income, employment, and sales. For the above example, the income multiplier is $(100 + 33.32)/100 = 1.33$. More specifically, for household income:

$$\text{Regional income multiplier} = \frac{1}{(\text{percent leakage})/100} \qquad (16\text{-}2)$$

Substituting the Forest County example of 75 percent leakage (25 percent remaining in the county) with each respending, the multiplier is:

$$\text{Forest County income multiplier} = \frac{1}{0.75} = 1.333 \qquad (16\text{-}3)$$

With each respending, the leakage can vary. The percent leakage for household income can differ greatly from leakage for, say, the income of a grocery store, gas station, or farmers' market where workers spend their wages. So the leakage proportion is a regional average, making the above income multiplier a rough approximation. More detailed multiplier formulas include differences in leakages for businesses and consumers. Other types of multipliers are covered later.

Equation (16-2) reveals that the smaller the leakage, the larger the multiplier. For example, if the above leakage is 50 percent, the multiplier becomes 2. What can you conclude about forested areas' income multipliers? Forested regions are rural; they import most of their goods. Therefore, they have high leakages and low multipliers. So forested, rural economies are often harder to stimulate with development than some urban areas. Planting trees, while often the best use for many rural areas, won't be a large economic stimulus, because timber-growing doesn't use much labor. Its major inputs are capital and time. But when timber matures, wood processing can be a vital economic boost.

Region size is important. Other things being equal, the larger a region, the more self-sufficient it is, and the lower its leakage. Thus, states will generally have larger income multipliers than counties, and multistate regions' multipliers sometimes exceed those for states. Every multiplier is based on a specific region. So it's really important to know where a multiplier was calculated and not to apply, say, a state's multiplier to a county.

Warnings about Multipliers Some analysts warn that we should be suspicious about rural county-level multipliers much over 2. That's not to say they're never higher, just not usually. Larger regions can have bigger multipliers. Project promoters sometimes quote multipliers of 5 or more. These are probably improperly computed, or they may be something like the estimated number of times income is respent before it all leaks out of a region (see Figure 16-1). But that's not a multiplier.

Sometimes people look at business input like standing timber worth $100 that is locally manufactured into furniture retailing for $700. This illustrates a **value added** of $600 to the stumpage. This is not a multiplier of 7.

Leakages Again

Often promoters will tout spending generated by, say, an existing or proposed park: "Visitors to Fairview Park spend $2.5 million annually in our county, a major stimulus to the economy!" How much of this is a real stimulus depends on the leakage for recreationists' spending. If it's mainly on imported gasoline, sporting equipment, or food, then most of the money simply passes through the county. The only part of imports that matters locally is the *value added*, for example, retailing-costs—returns to labor and capital used locally. A project's stimulus comes from wages and local taxes paid, from local purchases made from project sales income, and from the portion of any spending and respending that stays in the local area.

It's even more misleading to call total park user expenditures the "value" of a park. As shown in Chapter 14, the on-site value of a park should be the **present value** of users' **willingness to pay** to enter the park minus the present value of maintenance costs. But it *is* valid to speak of any project's regional economic impact, which is separate from the project's market value. The true local impact can be either new multiplied income or jobs.

Employment Multipliers

Analogous to income multipliers, we can also compute *employment multipliers*. These are of great interest in areas where unemployment is high. An employment multiplier of, say, 1.4 means that 100 new employees at a project will eventually create a multiplied effect of 40 more jobs, for a total of 140 jobs. The original new employees put greater pressure on stores, private and public services, schools, recreation facilities, and project suppliers, which all have to hire more employees. The section on input-output analysis shows one way to compute short-term employment multipliers.

Employment multipliers can vary greatly between industries, especially if you look at long-term effects. For example, if a new hardwood lumber mill attracts furniture manufacturers and outlet-stores locally, and both generate multiplied employment, you can get a major increase in jobs over several years. In contrast, adding more fast-food outlets may have a smaller long-term employment multiplier, especially if they cannibalize sales of existing restaurants. Also, whether an industry is labor-intensive or capital-intensive makes a big difference in its local employment impacts.

LOCAL BENEFITS VERSUS NET NATIONAL BENEFITS

Employment and income benefits that can be very real in one locality may or may not be significant at the national level, depending on the nationwide and regional unemployment levels. Consider two scenarios:

1 **Full employment.** At any time, part of the work force will be between jobs, the so-called frictional unemployment, so zero unemployment is impossible. Economists speak of the **natural rate of unemployment**, currently around 6 percent, below which inflation accelerates and becomes unacceptable (Samuelson and Nordhaus, 1989). When unemployment gets much below the natural rate, labor is scarce, employers bid up wages above increases in productivity, and prices tend to rise. Not all agree that the natural rate necessarily has to be so high; in the early 1960s it was thought to be 4 percent

(Levy, 1963). A number of developed economies have had even lower unemployment without excessive inflation.[2]

2 **Unemployment.** Let this be unemployment above the "natural rate"—say, above 6 percent.

If workers on a new project were previously unemployed, gains exist locally and nationally. But under full employment, if localities expand by hiring people formerly in other jobs, there's little national gain. Sure, the local economic expansion—if citizens truly want it—can be a gain *locally*. But if new workers leave other jobs, those sectors contract because they can't find new workers, given full employment. In this case, one measure of net national gain is the wage difference, if any, between the old and the new employment, which attracted the employees.

Understandably, local promoters of economic growth have little concern about whether new workers from elsewhere were previously employed, as long as local growth occurs. But federal agencies, when touting benefits of new projects, need to note whether or not net *nationwide* benefits will actually exist.

How Long Do Benefits Last?

During a recession with significant unemployment, suppose a new project hires unemployed workers and generates direct and multiplied income of $2 million per year. If you make the *unrealistic* assumption that this annual net national benefit lasts forever, the present value at, say, 5 percent real interest is 2/0.05 = $40 million [equation (4-8)]. More realistically, benefits should be counted for the recession's duration, after which national income would have returned to former levels without the project. This is another application of the **with and without principle.** If analysts project that regional unemployment would have lasted for, say, 3 years, the present value of benefits will be [from equation (4-10)]:

$$\text{Present value} = 2\,\frac{1-(1.05)^{-3}}{0.05} = \$5.45 \text{ million}$$

But in a developing economy with persistent unemployment, benefits of new income can be counted for long periods.

Multipliers in Reverse

Something often overlooked is a *reverse multiplier effect* when a sector contracts. Suppose 100 new jobs in a region eventually become 150 through the multiplier.

[2]The natural unemployment rate may have increased due to a rising gap between high-tech job requirements and a less specialized work force. Also, when unemployed, the growing number of highly specialized workers may have trouble finding new jobs quickly. But this shouldn't suggest that unemployment as high as 6 percent is unavoidable: government programs like education and job training could well reduce unemployment without fueling inflation (see Weiner, 1994).

If 100 jobs are lost through, say, closing a mill or reducing employment on public forests, other jobs will be lost in the supporting goods and service sectors, for a total of 150 jobs. As above, how severe and long this impact is depends on the unemployment level. In a healthy, full-employment economy, laid-off workers are re-employed fairly soon, and the negative direct and multiplied employment effects aren't serious. With more persistent unemployment, the reverse multiplier effect can be traumatic, as we've seen with many Pacific Northwest mill closures. Empirical examples are given later.

COMMUNITY STABILITY

Communities want economic stability, which usually means diversity. That avoids the economic roller-coaster rides felt by "one-industry towns." Take Seattle during the 1970s and to a lesser extent today: When Boeing did well, incomes were high, unemployment low, housing scarce, and home values high. When aircraft orders slumped, so did the entire region. And that's when the reverse multiplier hits hard. Not only was Boeing's payroll down, but so were the payrolls of all the supporting sectors. And as people left the region to find jobs elsewhere, many homes were for sale without enough buyers, and housing prices slumped. These things happen in timber-dependent communities when mills close or scale back.

Diversified employment can soften this yo-yo effect. When one sector is down, often another is up and can carry some of the slack. That obviously doesn't work when we're in a nationwide recession. But even then, some businesses are more recession-proof than others. For example, recreation-based economies can be unstable because travel is one of the first things people limit during tough times. Food processing is more stable. People always have to eat.

Communities also want to avoid seasonal employment instability, which you get with, say, ski resorts. That's why towns like Vail and Aspen in Colorado's ski country also promote year-round convention trade, golf, tennis, music festivals, and hiking. Unfortunately, many rural forested areas don't have much choice in development options and can't easily diversify. This also applies to diversification within the timber industry. Sparsely populated, heavily timbered states have large primary wood-processing sectors—lumber, wood panels, pulp, and paper. But often secondary processing—e.g., boxes, furniture and cabinets, mobile homes, and prefabricated buildings—tends to be in states closer to markets, to reduce transportation costs. Thus, expanding secondary wood processing may not always be a viable option for rural timber-dependent states (Polzin, 1994).

Community stability has long been a rationale for the U.S. Forest Service even-flow harvest policy. The idea is to provide timber-dependent communities with roughly an equal annual harvest of federal timber in order to stabilize their economies. But that's an elusive goal. Even with this policy, numbers of wood-processing employees decline as mills automate. Also, forest products demand fluctuations based on housing cycles are unavoidable (or at least not controllable at the resource end—see Chapter 5). Some timber-dependent communities with ties to buoyant regional centers have grown, despite stable or declining harvests.

So there's no assurance that even flow will foster community stability, although it has in some cases. Moreover, buyers have some flexibility in how fast they cut purchased timber, so economic cycles still influence U.S. Forest Service harvest levels and wood-processing employment.

Chapter 12 notes that harvest "instability" depends on the spatial scale of your view. Over a small enough area, harvest flows always look unstable. But over a broader area, an uneven timber flow from public lands can often offset private harvest fluctuations, thus providing a more even total harvest. Also, many have argued that even-flow harvests can worsen stumpage price instability: constraining timber sales during peak demand pushes prices higher, while keeping up sale offerings during slumps will depress prices further. This leads some analysts to advise a more market responsive harvest policy on federal lands: sell more timber when the price is high, less when it's low.[3]

Assuming an 18 percent wood products price decline, a Montana simulation showed that stabilizing federal harvests would cause decreased U.S. Treasury receipts far exceeding the benefits to local communities. Under a best-case scenario, these local benefits—greater forest products income and employment than would otherwise have occurred—were 1/7 to 1/10 of the loss to the U.S. Treasury, compared to a market responsive harvest (Daniels et al., 1991).

Some forest industry instability is unavoidable: all timber dependent areas are affected by housing cycles, and many western communities will continue seeing a drop in forest industry employment as old-growth timber supplies decline or are set aside, and as mill automation continues. So we have to live with a certain amount of instability. Counter to popular belief, Berck et al. (1992) find that employment in forest products is no more unstable than in other sectors but that many communities rely heavily on the forest industry. Thus they don't benefit from the stabilizing effect of diversity: not all sectors fluctuate together. Rather than try to foster community stability with steady public harvests, some analysts suggest that governments should help communities broaden their economic bases and retrain workers when wood products employment declines permanently. The latter was part of the federal government's response to unemployment caused by old-growth harvest restraints in the Northwest during the early 1990s. Another view is to hope that market forces will create new jobs after periods of unemployment.

INCOME DISTRIBUTION[4]

Most forest policies influencing economic growth have some effect on income distribution. An example: less timber harvesting and more recreation development on public lands will reduce the number of higher-wage wood products jobs and increase employment in the lower-paying recreation service sector. The models discussed next can quantify such impacts. But we shouldn't look at income levels alone. For example, rural wages tend to be lower than in urban areas, but forested

[3]For other aspects of uneven- versus even-flow policies, see Chapter 12's section "Nontimber Values and Optimal Harvests over Time."

[4]Parts of this section draw from Wear and Hyde (1992).

areas have offsetting advantages of lower living costs and more environmental amenities.

Another distribution question is, what income groups benefit from government assistance? Under the federal Forestry Incentives Program, qualifying private nonindustrial forest landowners receive money to help them do planting and timber stand improvement. Studies show that these payments to stimulate private nonindustrial forestry go to people with above-average incomes. This transfers money from poorer taxpayers to richer ones. Furthermore, these subsidies increase future timber supplies and reduce future wood prices for rich and poor alike. Where no **market failures** exist, subsidies can actually reduce total economic welfare.[5] In those cases, if the objective is to reduce costs of wood products for the poor, some argue that it would be more efficient to assist that group directly, for example, through tax-reductions, rather than distort the market and also assist upper-income groups. A related matter is that users of publicly funded forest recreation areas, especially wilderness areas, have above-average incomes (Roggenbuck and Watson, 1988). Yet for many U.S. National Forests, the largest deficit category is recreation, followed by timber, grazing, and minerals, in decreasing order (Shaw, 1994). Again, this transfers revenues from average-income taxpayers to benefit those with above-average earnings, since Forest Service deficits are funded with taxes. These issues need more attention.

INPUT-OUTPUT ANALYSIS

Input-output analysis is widely used for measuring impacts of forest-related activities on employment, income, taxes, exports, and imports in different sectors. This is useful for explaining how important a segment of the forest industry is in a region or for gauging the effects of, say, new mills or recreation areas, or changes in public timber harvests. Without getting into the underlying mathematics, I'll summarize the general outcomes of input-output analysis and how the model can show interdependence between sectors of an economy.

The foundation of input-output analysis is the **input-output table,** or "interindustry transactions table": Table 16-1 is a simple hypothetical example for a region. The table shows the output (in $millions/year) sold *from* each industry or sector in the left column *to* each industry or sector in the top row. For example, going across row 1, industry A sells $7 million worth of its output to itself in column (1), say, for operations and remanufacturing. It sells $1 million worth of goods to industry B in column (2), $2 million to C, $5 million to households in the region, and $21 million for export. Here, "exports" will be any sales outside the region. In this highly aggregated example, the three industries might be

[5]For explanations of how subsidies can be inefficient, see "Tax Policy Issues" in Chapter 9 and "Rising Timber Prices" in Chapter 12. In a smoothly functioning economy *without market failures*, the **equi-marginal principle** is met in allocating investments to different enterprises, and total returns are maximized (see Figure 4-4). Then, using tax revenues to subsidize one sector can increase its output and drive down its prices and rate of return. In that case, losses from subsidies (the competitive return the tax monies could have earned elsewhere) will exceed gains (the reduced rate of return on the subsidy). Economists refer to this as a "deadweight loss."

TABLE 16-1
HYPOTHETICAL REGIONAL TRANSACTIONS TABLE—OUTPUT FROM SECTOR ON THE LEFT TO SECTOR AT THE TOP ($MILLIONS/YEAR)

		Buying			Final demand		
	Selling	**Ind. A**	**Ind. B**	**Ind. C**	**Households**	**Exports**	**Total output**
Processing	Industry A	7	1	2	5	21	36
	Industry B	0	6	3	22	29	60
	Industry C	6	9	0	23	14	52
Payments	Households	13	25	20	1	57	116
	Imports	10	19	27	65	0	121
	Total outlays	36	60	52	116	121	385

forest products, other producers, and services. Transactions tables usually show many more industries—up to several hundred in the largest analyses—as well as separate sectors for government and changes in inventories and capital value. Often industries are divided; for example, forest products may show separate sectors like logging, pulp, paper, paperboard, hardwood mills, millwork, plywood, or particleboard. Tables may apply to counties, groups of counties, states, regions, or an entire nation.[6]

The transactions table shows interdependence throughout the economy. Note that total outlays across the bottom and total output at the right are equal for each sector. In the payments and final demand blocks, this equality wouldn't necessarily occur, due to trade imbalances and debt. The payments category normally includes other entries than those in Table 16-1; for example, taxes to government and capital investment (or **depreciation**). Returns to capital are, in part, dividends and interest included in household income and also retained business earnings spent on capital investment, which is usually shown as depreciation. Except for imports, the payments block of Table 16-1 shows payments to "factors of production" (labor and capital): this is *value added* by the regional economy.

Table 16-2 shows direct purchases per dollar of output or sales by the sector in the top row from the sector on the left. These **technical coefficients** are calculated from each column of Table 16-1 as the purchases from each sector divided by the total outlay at the bottom (which is the same as total output). For example, going down column (1) in Table 16-1, industry A buys $7 million worth of its own output, and 7/36 = 0.19, the coefficient in the corresponding cell in Table 16-2. So for each $1.00 of sales, A buys $0.17 worth of its own output. Industry A buys nothing from B, so a zero is in row 2 of column (1) in Table 16-2. Continuing down column (1) in Table 16-1, A buys $6 million worth

[6]In the most accurate analyses, data for interindustry transactions tables are gathered through expensive surveys of individual firms' purchases and sales. But often, for smaller regional studies, the proportions of an industry's total output spent on each category are based on national averages with some local adjustments.

TABLE 16-2
HYPOTHETICAL TABLE OF DIRECT PURCHASES PER DOLLAR OF SALES (TECHNICAL COEFFICIENTS MATRIX)—BASED ON TABLE 16-1*

	Selling	Buying			Final demand	
		Ind. A	Ind. B	Ind. C	Households	Exports
Processing	Industry A	0.19	0.02	0.04	0.04	0.17
	Industry B	0	0.10	0.06	0.19	0.24
	Industry C	0.17	0.15	0	0.20	0.12
Payments	Households	0.36	0.42	0.38	0.01	0.47
	Imports	0.28	0.32	0.52	0.56	0.00
	Total outlays†	1.00	1.00	1.00	1.00	1.00

*Direct purchases per dollar of output by the industry at the top from the industry on the left (from Table 16-1, purchases in each cell divided by its column's total outlays).
†Due to rounding, coefficients shown may not total to 1.

of C's output, and 6/36 = 0.17, the coefficient in the corresponding cell in Table 16-2. Next, A spends $13 million on wages to households (13/36 = $0.36 per $1.00 of sales) and $10 million on imports from outside the region (10/36 = 0.28). Since this is all first-round spending, there's no sales multiplier effect, and coefficients in each column of Table 16-2 sum to the $1.00 of outlays in the last row (it's also $1.00 of sales, since outlays equal sales).

The next step from the first round to the predicted outcome is Table 16-3, which shows the multiplied effects of export sales in each sector. Input-output analysis computes these by applying computer-based matrix algebra to the industry and household technical coefficients matrix to simulate the effects of respendings by households and by industries purchasing from one another. For example, in

TABLE 16-3
REGIONAL OUTPUT OR SALES MULTIPLIERS: MULTIPLIED OUTPUT OF SECTOR ON THE LEFT PER DOLLAR OF EXPORT SALES FROM SECTOR AT THE TOP—BASED ON TABLE 16-2*

	Multiplied purchases (dollars per dollar of sales)			
Selling	**Ind. A**	**Ind. B**	**Ind. C**	**Households**
Industry A	1.30	0.08	0.09	0.09
Industry B	0.17	1.28	0.19	0.29
Industry C	0.38	0.34	1.16	0.31
Households	0.69	0.70	0.56	1.29
Output or sales multipliers†	2.54	2.41	2.00	1.98

*In input-output parlance, where Table 16-2 is the **A** matrix, and **I** is an identity matrix, Table 16-3 is the inverse of (**I** minus **A**), or $(\mathbf{I} - \mathbf{A})^{-1}$. With households included, impacts include direct and indirect effects, and effects induced by household respendings.
†Due to rounding, coefficients may not sum to the output multipliers.

Table 16-2, recall that when industry A increases sales by \$1.00, it buys \$0.17 worth of C's output as a first-round effect. Industry C must then buy more outputs of other industries, which in turn buy more from others, and so on. These successive boosts get smaller and smaller and reach a limit. With a \$1.00 increase in export sales of a sector across the top, Table 16-3 shows the multiplied effects on eventual outputs of the sectors on the left.

To show full multiplier effects, Table 16-3 includes households as an "industry"; thus the relevant sales in the table are to exports.[7] With the multiplier effects in Table 16-3, when industry A increases export sales by \$1.00, the output of industry C now rises to \$0.38 rather than the \$0.17 shown in Table 16-2. Summing the multiplied effects in each column gives the *output multipliers* (or *sales multipliers*) in the last row of Table 16-3. For example, if industry B's export sales increase by \$1.00, total regional business sales and household income will eventually increase to \$2.41, including the original dollar of export sales.[8] Export sales are a major boost locally because they bring in new money from outside the region. In a sense you can think of this as "money importing": other examples are tourism and "transfer payments" from nonlocal governments.

Table 16-3 shows how multiplier effects on sales can vary, depending on the sector expanding. To the extent that regions have a choice, if their economies are faltering, they may wish to stimulate development in sectors with large sales multipliers. But even more important are multipliers for income and employment. The sector with the largest sales multiplier may not always have the greatest employment or income multiplier. *Notice that the Table 16-3 multipliers apply to business sales and household income combined: they are not household income multipliers.* Rough income multipliers can be calculated for increases in household income with equation (16-2). Calculating more precise household income multipliers from input-output tables takes extra steps, as shown below.

Household Income Multipliers in Input-Output Analysis

In Table 16-2, the household *row* shows *direct coefficients* for household income generated for each dollar of sales by the sector at the top (0.36, 0.42, 0.38, 0.01). Multiplying these coefficients by those in any *column* of Table 16-3 will give the multiplied effects of household income per dollar of export sales in that column. For example, in the industry A column of Table 16-3, multiplying each coefficient by the above direct coefficients and summing the result gives 0.6907 dollars of direct and multiplied household income per dollar of export sales in industry A:

$$0.36(1.30) + 0.42(0.17) + 0.38(0.38) + 0.01(0.69) = 0.6907$$

[7] A table like Table 16-3 *without* households would have smaller ("type I") multipliers and would show *direct* and *indirect* effects. When household respendings are included, we say that *induced* effects are added, giving larger ("type II") multipliers, as shown (see footnote 1).

[8] It's important to note that this sales multiplier applies only to sales outside the region, not total sales. But in general, a region's sales-increase will be nearly all exports, since local markets in equilibrium will be saturated. Or a major increase in one firm's sales will be at the expense of reduced local sales elsewhere, giving no net gain.

Remembering that industry A's direct household income effect is $0.36 per $1.00 of sales (from Table 16-2), the household income multiplier for industry A, using equation (16-1), is 0.6907/0.36 = 1.92. This means that, say, $10,000 of household income generated by industry A eventually becomes $19,200, including all multiplier effects. Note how this differs from industry A's sales multiplier of 2.54. It's very important to understand the different multipliers and apply them correctly.

Employment Multipliers in Input-Output Analysis

If you know the number of employees needed per $1,000 of sales in each industry, you can compute employment multipliers with Table 16-3, in a way similar to the above income multiplier procedure. Suppose industry A employs 0.05 people per $1,000 of annual sales, industry B 0.03, industry C 0.02, and households employ 0.002 people per $1,000 of sales: these are *direct employment coefficients* showing first-round effects. In each *column* of Table 16-3, multiply the coefficients for industries A, B, and C by the foregoing employment coefficients. That gives the number of people employed in each industry with $1,000 added export sales for the industry in the top row. Summing these numbers in each column gives the multiplied employment effect per $1,000 of added export sales. Performing these steps in the industry B column gives a total multiplied employment effect of 0.0506 jobs per $1,000 of sales in industry B. Recalling that industry B's direct employment is 0.03 per $1,000 of sales, the employment multiplier, according to equation (16-1), is 0.0506/0.03 = 1.7 for industry B. So if industry B hires 100 new people, the effect will be to produce a total of 170 jobs.

Input-output multipliers for sales, income, and employment are fairly short-term effects within about a year or two and don't include long-term impacts of, say, new factories, which change the character of the economy. As with the sales multipliers in Table 16-3, income and employment multipliers can vary greatly between industries. So these analyses allow planners to see which industries have the greatest economic impacts.

Cautions

While input-output analysis is an excellent way to measure impacts on regional economies, you should be aware of its limitations. One is that the "technical coefficients" in Table 16-2 are fixed for all output levels, so the input proportions stay constant for any industry (economists would say that "the production functions are linear"). This also means that input-output (I-O) tables assume perfectly **elastic** (horizontal) supply curves so that prices remain constant for outputs (which are someone else's inputs). That's not a problem when modeling small changes. But with major changes in demand, say, an increase, you'd expect input prices to rise and input mixes for some industries to change (remember the upward-sloping supply curve?). Since I-O, in its simplest form, gives us a snapshot in time, it's useful for short-term analyses in which industry structures don't change. Although not covered here, some researchers have adjusted the model to make it more dynamic.

Gathering detailed information for the technical coefficients in I-O analysis is very costly. While much of these data are available in terms of national averages, the coefficients usually need to be adjusted to fit local situations. Also, as industry structures change, data must be updated. In local I-O analyses, some data can be unreliable, especially when national coefficients are used or adjusted, and thus certain outputs can be questionable.

Staring at input-output tables and multipliers can lull some policy makers into a bigger-is-better growthmania mentality. Sure, it's nice to think of rising income levels and living standards. But you need to remember that typical input-output analyses tell us nothing about environmental costs of economic growth. Some analysts have adapted input-output analyses to include "pollution coefficients" like units of sulfur dioxide in the air per $1,000 of each sector's output. That way you can estimate environmental effects of certain policies or economic projections (Miller and Blair, 1985). In the same way that employment multipliers are derived from input-output tables, you could also compute "pollution multipliers" for types of air and water pollution. Such analyses would let you estimate that 10 more units of hydrocarbon emissions from increased sales in industry X would lead to an eventual total release of, say, 21 units in all industries combined, including direct and multiplier effects: in this case, the pollution multiplier is 2.1.

In an interindustry transactions table (see Table 16-1) for the entire United States, the lower right cell would approximate **gross domestic product (GDP)**. Some feel that the GDP overstates national well-being because it ignores environmental damage (see Daly, Mishan, and Galbraith, cited in Chapter 3).[9] In fact, some minuses are actually recorded as pluses—to wit, the billions of dollars spent cleaning up the Exxon Valdez oil spill and other pollution. Bearing in mind the previous paragraph, we need to look beyond the raw numbers in input-output tables when thinking about overall regional or national well-being.

You also have to keep in mind the earlier warning that large regional benefits may translate into someone else's loss during times of fairly high employment. That also applies to benefits of increased tourist expenditures if spending declines elsewhere.

Remember also the warning that rural multipliers for single counties are usually less than 2. Groups promoting rural development projects may be tempted to tinker with I-O models in the hope of getting larger (and possibly bogus) multipliers that can advance their agenda.

While you can raise questions about input-output analysis, it's still one of the best ways to analyze regional economies. It's also far easier to use than approaches like multi-equation **econometric** models. As with any model, though, you should know its limitations.

[9]The San Francisco–based organization Redefining Progress has developed a measure of social well-being called the Genuine Progress Indicator (GPI). Compared to a historically rising GDP, the GPI has been falling since 1970 because it accounts for things like environmental damage, resource depletion, declining capital stock, increased income inequality, unemployment, and costs of automobile accidents (Cobb and Halstead, 1994).

APPLICATIONS

Much forestry-related input-output analysis has been done with the U.S. Forest Service IMPLAN model, which runs on desktop computers (Taylor et al., 1992). One study measured, for each of five Georgia State Parks, employment multipliers ranging from 1.21 to 1.52 in local areas covering five to seven counties. One park generated 415 jobs directly, for a total of 630, including all multiplier effects. *Value added multipliers* ranged from 1.66 to 2.12. These apply to the sum of wages, taxes, and property income that are true local income, not simply dollars passing through the area. The largest park's direct value added was multiplied from $9.7 million to $16.2 million for 1986 (Bergstrom et al., 1989).

A 1985 input-output analysis of the Lake States (Michigan, Minnesota, and Wisconsin) gives an employment multiplier of 1.97 for the forest products industry, including all multiplier effects. These multipliers varied among sectors, ranging from 1.5 for flooring and special product sawmills to 3.3 for sanitary paper products. Personal income multipliers ranged from 1.7 for household wood furniture to 2.3 for sanitary paper products. For outdoor recreation in forested areas, the employment multiplier was 1.6, and the personal income multiplier was 1.9, all cases including full multiplier effects (Pedersen et al., 1989).

Another study estimates interregional impacts of changes in United States forest products demand in the South, West, Northeast, and Midwest. It shows that some multiplier effects can be far-reaching: for example, of a 2.36 total output multiplier for sawmills and planing mills in the Midwest, 23 percent of this occurs outside the region; of that amount, 45 percent occurs in the South, 45 percent in the West, and 10 percent in the Northeast. These effects result from an industry buying supplies from outside its region. The same study finds that, of multiplied income generated by the entire forest products industry in the Midwest, 15.4 percent occurs outside the region. The corresponding figure for the South is 9.2 percent (Teeter et al., 1989).

A 1989 study looks at proposed harvest reductions on two National Forests in California. The input-output analysis measures income effects throughout the following California regions: Northern Interior, North Coast, Sacramento, and San Joaquin. In one four-region simulation, a 30.8 percent proposed harvest reduction on the Shasta-Trinity forest in the Northern Interior region would reduce personal income by $5.29 million in 1982 dollars, including multiplier effects. Of this total impact, 36 percent would be felt outside the Northern Interior region. The direct first-round income loss is $3.87 million, making the reverse multiplier 5.29/3.87 = 1.4. Although these are large regions, they're rural, so the multiplier is not large. As in the Teeter study above, this analysis, on a smaller scale, shows the importance of impacts outside the local region (Sullivan and Gilless, 1990).

For Washington, Oregon, and northern California, a 1990 study simulates economic effects of a projected 24 percent reduction in timber harvests below 1983–1987 levels in order to protect the spotted owl and old-growth ecosystems. Regional job losses are estimated at 48,000, about 23,000 of these directly in forest products and the rest in indirectly related jobs—the "reverse multiplier" mentioned earlier. Here this multiplier is 48,000/23,000 = 2.09, using equation (16-1). Estimated household income lost is $1.1 billion per year, including direct and reverse multiplier effects (Gilless et al., 1990).

In the last two studies of lost income and jobs, the hardest part is to gauge how long it takes for such effects to dissipate as people get re-employed. In reducing harvests to make environmental gains, we need to be aware that the financial losses may be temporary, but the benefits can be perpetual. Thus, in measuring the **annualized costs** to compare with perpetual benefits, the relevant equation is (6-10): (the interest rate) times (the present value of costs). For the above spotted owl harvest reductions, suppose the first year's $1.1 billion lost income declines by $0.22 billion per year to zero in the sixth year due to re-employment. At 6 percent real interest, the present value of losses is $2.89 billion. The annualized cost in perpetuity is 0.06(2.89) = 0.173, or $173 million. If the gains are of national interest, with a United States population of 263 million (estimated for 1995), the necessary annual gain to offset the cost would have to be worth 173/263 = $0.66 for every adult and child. This is another example of concentrated losses and diffuse gains, and not much consolation to the losers. But it does give support for the idea of federally funded job retraining programs when saving ecosystems of truly national concern. The foregoing just shows a way to look at a problem and isn't meant to be precise. Also be aware that if you use such reasoning thousands of times and accumulate results, pretty soon you run out of willingness to pay (see Chapter 14).

KEY POINTS

- Using forests will have some kind of regional economic impact like increased income or employment. The impact will depend on the use or mix of uses, for example, developed recreation, wilderness, or harvesting and processing timber.
- Direct economic impacts of any project are those that arise directly from project construction and operation, for example, people employed or wages paid.
- Added to direct effects are multiplier effects of consumer respendings and increased demand for services and supplies.
- A general multiplier formula for sales, income, or employment is (direct effect plus multiplier effect)/(direct effect). If an employment multiplier is 1.7, the total employment impact of 100 new forest recreation jobs is 1.7(100) = 170 jobs.
- One form of regional income multiplier is the reciprocal of (leakage/100), where leakage is the percent of income that leaks out of (or leaves) the region with each respending.
- The smaller a region and the more rural it is, generally the larger its leakage and the smaller its income multiplier. Multipliers for individual rural counties are usually less than 2.

- During times of full employment, increased local economic benefits may yield little net national benefit if new employees left existing jobs elsewhere.
- When one sector contracts, say, due to less timber harvesting, supporting sectors also contract, giving rise to a reverse multiplier effect, which applies to income and employment.
- Many analysts conclude that stabilizing public timber harvests cannot stabilize timber-dependent communities.
- Input-output (I-O) analysis can show interactions between sectors of an economy—who buys what from whom. I-O analysis can separate direct effects and multiplier effects of changes in sales from different sectors.
- For a given region, I-O analysis can yield multipliers for sales, personal income, employment, and value added for separate industries. These let you know which sectors have the greatest potential for boosting an economy.
- Because I-O analysis assumes fixed input proportions, it may not be appropriate for modeling large changes in an economy. Also, applying national data to local analyses can sometimes give questionable results.
- I-O analysis has been widely used to model local and regional economic impacts of forest recreation, separate wood-processing sectors, and proposed changes in public and private timber harvests.

QUESTIONS AND PROBLEMS

16-1 List some regional economic impacts of forest uses.

16-2 Distinguish between direct economic impacts of a project and multiplier effects.

16-3 Suppose the direct employment effect of a project is 200 new jobs and the multiplier effect is 180 additional jobs in a region. What is the region's employment multiplier? Interpret this multiplier.

16-4 Suppose that 57 percent of any new income leaks out of a region. What is this region's income multiplier? What does this multiplier mean?

16-5 Compared to urban regions, do rural areas tend to have income and employment multipliers that are larger, smaller, or about the same? Explain why.

16-6 Discuss the role of regional and national employment levels in determining the net national benefits of new projects.

16-7 Explain the notion of the "multiplier in reverse" in regional economic analysis.

16-8 What are some of the problems in stabilizing public timber harvests in order to foster stability in timber-dependent communities?

16-9 In Table 16-1, explain the industry B column of numbers.

16-10 The household row of Table 16-2 shows the following direct coefficients of household income created per dollar of sales by the sectors at the top: 0.36, 0.42, 0.38, 0.01. Use these coefficients and the industry C column in Table 16-3 to calculate the direct and multiplied household income per dollar of export sales in industry C. Then use equation (16-1) to find the household income multiplier for industry C. Interpret this number.

16-11 Forested counties often have income multipliers of 3 or more. True or false?

16-12 How can forest policy analysts use the results of input-output analysis?

16-13 What are some weaknesses of input-output analysis?

REFERENCES

Bendavid-Val, A. 1991. *Regional and Local Economic Analysis for Practitioners* (4th edition). Praeger, New York. 238 pp.

Berck, P., D. Burton, G. Goldman, and J. Geoghegan. 1992. Instability in forestry and forestry communities. In Nemetz (1992), pp. 315–338.

Bergstrom, J. C., H. K. Cordell, B. A. Ashley, D. B. English, and A. E. Watson. 1989. *Rural Economic Development Impacts of Outdoor Recreation in Georgia.* GA Agric. Experiment Sta. Res. Rept. 567. University of Georgia, Athens. 10 pp.

Beuter, J. H., and D. C. Olson. 1980. *Lakeview Sustained Yield Unit, Freemont National Forest.* Report to the U.S. Forest Service. Oregon State University, Corvallis.

Cobb, C. and T. Halstead. 1994. *The Genuine Progress Indicator: Summary of Data and Methodology.* Redefining Progress, San Francisco. 46 pp.

Coppedge, R. O., and R. C. Youmans. 1970. *Income Multipliers in Economic Impact Analysis.* Special Report 294. Oregon State Univ. Cooperative Extension Service, Corvallis. 3 pp.

Daniels, S. E., W. F. Hyde, and D. N. Ware. 1991. Distributive effects of Forest Service attempts to maintain community stability. *Forest Science.* 37(1):245–260.

Gilless, K., R. Lee, B. Lippke, and P. Sommers. 1990. *Three-State Impact of Spotted Owl Conservation and Other Timber Harvest Reductions.* Institute of Forest Resources Contribution #69. University of Washington, Seattle. 43 pp.

Levy, M. E. 1963. *Fiscal Policy, Cycles and Growth.* Studies in Business Economics No. 81. National Industrial Conference Board, New York. 141 pp.

Miernyk, W. H. 1965. *The Elements of Input-Output Analysis.* Random House, New York. 56 pp.

Miller, R. E., and P. D. Blair. 1985. *Input-Output Analysis: Foundations and Extensions.* Prentice-Hall, Inc. Englewood Cliffs, NJ. 464 pp.

Nemetz, P. N. (Ed.). 1992. *Emerging Issues in Forest Policy.* University of British Columbia Press, Vancouver. 573 pp.

Pedersen, L., D. E. Chapelle, and D. C. Lothner. 1989. *The Economic Impacts of Lake States Forestry: An Input-Output Study.* U.S. Forest Service North Central Forest Experiment Station. Gen. Tech. Rept. NC-136. St. Paul, MN. 32 pp.

Polzin, P. E. 1994. Spatial distribution of wood products industries. *Journal of Forestry.* 92(5):38–42.

Probst, D. B. (Compiler). 1985. *Assessing the Economic Impacts of Recreation and Tourism.* Proceedings of a Conference and Workshop. U.S. Forest Service Southeastern Forest Experiment Station. Asheville, NC. 64 pp.

Roggenbuck, J. W., and A. E. Watson. 1988. Wilderness recreation use: the current situation. *Outdoor Recreation Benchmark 1988: Proceedings of the National Outdoor Recreation Forum.* U.S. Forest Service Southeastern Forest Experiment Station. Gen. Tech. Rept. SE-52. Asheville, NC. pp. 346–356.

Samuelson, P. A., and W. D. Nordhaus. 1989. *Economics* (13th edition). McGraw-Hill, New York. 1013 pp.

Schallau, C. H., and R. M. Alston. 1987. The commitment of community stability: A policy or shibboleth? *Environmental Law.* 17(3):429–482.

Schallau, C. H., W. Maki, and J. Beuter. 1969. Economic impact projections for alternative levels of timber production in the Douglas-fir region. *Annals of Regional Science.* 3(1):96–106.

Schuster, E. G., and E. L. Medema. 1989. Accuracy of economic impact analysis. *Journal of Forestry.* 87(8):27–32.

Shaw, J. S. 1994. Subsidies for the backpackers. *PERC Reports.* 12(4):10.

Sullivan, J., and J. K. Gilless. 1990. Hybrid econometric/input-output modeling of the cumulative economic impact of national forest harvest levels. *Forest Science.* 36(4): 863–877.

Taylor, C., S. Winter, G. Alward, E. Siverts. 1992. *Micro IMPLAN User's Guide.* U.S. Forest Service Land Management Planning Systems Group, Ft. Collins, CO.

Teeter, L., G. S. Alward, and W. A. Flick. 1989. Interregional impacts of forest-based economic activity. *Forest Science.* 35(2):515–531.

University of Montana. 1983. *Western Wildlands.* Special Issue: Timber harvesting policies and community stability. 8(4).

Waggener, T. R. 1977. Community stability as a forest management objective. *Journal of Forestry.* 75(11):710–714.

Wear, D. N., and W. F. Hyde. 1992. Distributive issues in forest policy. In Nemetz (1992), pp. 297–314.

Weiner, S. E. 1994. The natural rate and inflationary pressures. *Economic Review—Federal Reserve Bank of Kansas City.* 79(3):5–9.

17

FORESTS IN A WORLD ECONOMY

Human activities, including forestry, are becoming ever more global. **Capital** moves easily throughout the world, and today we find foreigners investing in timberlands and wood processing in the United States and elsewhere. And United States firms invest in wood growing and processing in areas like South America, Asia, and New Zealand. Logs and processed wood products enter world trade. Local and global environmental impacts of this trade are alarming international conservation groups. Industrialized nations give forestry assistance to developing regions. And a worldwide network of lending for forestry projects is available through organizations like the World Bank, the International Monetary Fund, the Inter-American Development Bank, and the Asian Development Bank. This chapter explores some of these global aspects of forestry.

WORLD FOREST RESOURCES

Columns (2) and (3) of Table 17-1 give forest areas and roundwood production by regions. World forest area estimates range from about 3 billion to over 5 billion hectares because definitions of "forest" vary. For some countries, column (2) includes "open forests" with as little as 10 to 20 percent tree cover. Column (4) shows that Africa, South America, and Asia use about 70 to 90 percent of roundwood harvests for fuel and charcoal. In more developed regions, fuelwood is about 20 percent or less of total production. Although not shown, you can deduce from this that North America produces the most industrial wood (nonfuelwood). Industrial wood harvest per unit area of forest is by far the greatest in the United States and Europe. This occurs because these regions have highly desirable and

TABLE 17-1
WORLD FOREST STATISTICS*

Region	Forest area, thousands of hectares†	Roundwood production, millions m^3 1991	Percent fuelwood and charcoal
Africa	541,076	527.2	89%
North & Central America	864,500	737.4	20
(U.S.A.; incl. in above)	(295,989)	(495.8)	17
South America	880,614	345.4	70
Asia	517,995	1,086.1	77
Europe (excl. former U.S.S.R.)	174,351	335.5	14
Former U.S.S.R.	941,530	355.4	22
Oceania‡	192,382	42.5	21
Total world§	4,112,448	3,429.4	53

*Column (2) WRI (1994); others from FAO (1993).
†Areas mainly for 1990, with 1980 areas for North Africa, temperate southern Africa, and nontropical South America.
‡Primarily Australia, New Zealand, and Papua New Guinea.
§Totals may differ due to rounding. For some regions, column (2) includes open forests with as little as 10 to 20 percent tree cover.

accessible species for construction and pulping, and they have efficient harvesting, processing, and transportation systems.

Table 17-2 shows that the United States, the former U.S.S.R., and Canada supply over half the world's industrial wood (sawlogs, veneer logs, poles, pulpwood, and chips). Notice the large forest area in the former U.S.S.R. This includes 52 percent of the world's coniferous forest area and 57 percent of the world's coniferous volume, most of which is in Russia. Earlier studies noted that about

TABLE 17-2
INDUSTRIAL ROUNDWOOD PRODUCTION, 1990—TOP 10 NATIONS AND WORLD*

Area	Volume, million cubic meters†	Percent of world output	Percent change since 1979–1981
United States	417.9	25.2%	+26%
Former Soviet Union	294.3	17.7	+6
Canada	172.2	10.4	+15
China	92.9	5.6	+20
Brazil	76.0	4.6	+29
Germany	54.8	3.3	+48
Sweden	49.3	3.0	+6
Malaysia	41.2	2.5	+37
Finland	38.3	2.3	−7
France	34.5	2.1	+22
Others	389.7	23.5	
World	1,661.1	100	+15

*WRI (1994). Excludes wood for fuel and charcoal.
†Average annual figures for 1989–1991.

half of the former Soviet timber volume wasn't economically accessible (Backman and Waggener, 1990). But this could change as the world shows more interest in these resources. For example, in 1993 and 1994, wood exports from eastern Russia to Japan increased, following a decade of declining shipments. With rising wood prices, greater political stability, and favorable exchange rates, some analysts predict that larger volumes of Russia's vast timber inventories will be exported (Sedjo et al., 1994).

Nearly all countries increased their wood output during the 1980s. You saw in Chapter 12 that the United States rate of harvest is sustainable. But this isn't the case everywhere. For example, China's harvest exceeds growth, and some analysts predict that China will become a major wood importer in the future (Brown et al., 1991; Perez-Garcia, 1995). Let's now look at global timber supply prospects.

Wood Production Trends and Prospects[1]

About 70 percent of the world's industrial wood harvest is softwood, and 90 percent of softwood forests are in the northern temperate regions. Thus, these areas supply most of the industrial roundwood (see Table 17-2). About 60 percent of the earth's forested area is hardwoods, roughly 70 to 85 percent of which is tropical forest covering 1.76 billion hectares in 1990 (WRI, 1994). Since industrial wood is 70 percent softwoods, and more than half the world's forests are hardwoods, the latter receive less pressure from commercial logging than softwood forests. Less than 10 percent of the world's industrial wood harvest is tropical timber. Thus, tropical deforestation, while a serious problem, isn't likely to affect global industrial timber supply much.

The upper curve in Figure 17-1 shows total global roundwood production—industrial wood and fuelwood—increasing from 1969 to 1991 at an average rate of 1.3 percent annually. Table 17-3 shows percentage annual rates of increase in worldwide industrial wood consumption for periods between 1950 and 1985. Over this period, the average rate of increase was about 2 percent per year. Notice how the rate of increase in wood consumption has lessened over time. This is probably due to (1) improved utilization that stretches usable wood fiber per cubic volume of trees, (2) use of more wood substitutes in response to rising wood prices, and (3) a slowing of the postwar housing boom in developed countries. The percentage of pulpwood in the industrial wood mix has changed from about 23 percent in 1950 to 34 percent by 1980.

Data on average worldwide stumpage prices are not available, but we can use an index of world real roundwood export prices as a proxy. Figure 17-2 graphs this price index over world industrial roundwood production at five-year intervals: the points (shown as triangles) are intersections of supply and demand curves for the years shown. Curve slopes are arbitrarily drawn. The slight rise in real prices, on average, means that world wood demand curves have been shifting rightward slightly faster than supply curves (if they shift at the same rate, prices are constant).

---

[1]This section draws from Sedjo and Lyon (1990), except where noted otherwise.

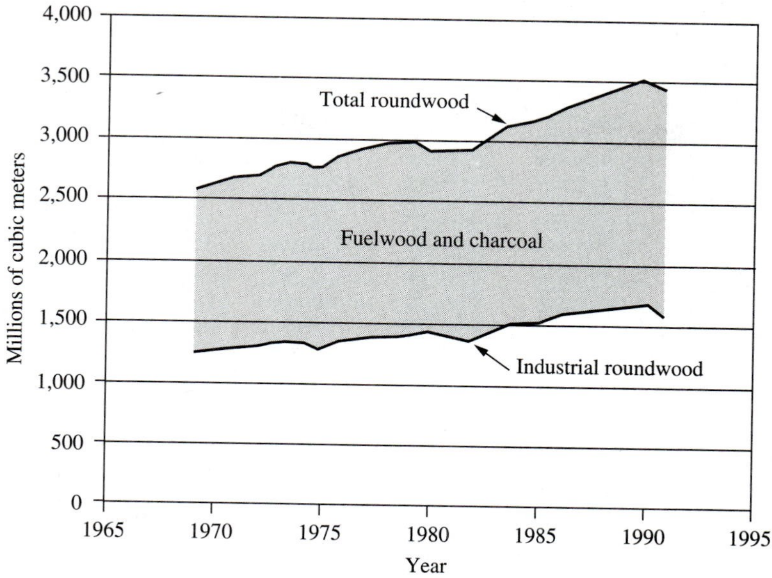

FIGURE 17-1
World roundwood production. (FAO, 1982 and 1993.)

TABLE 17-3
GROWTH RATES OF GLOBAL INDUSTRIAL WOOD CONSUMPTION*

Period	Consumption, annual rate of increase
1950–1960	3.54%
1960–1970	2.20
1970–1985	0.94
1950–1985	2.01

*Sedjo and Lyon (1990, p. 56), by permission.

Annual rates of increase in average log prices in the United States were about 1.37 percent between 1900 and 1950 and 0.34 percent between 1950 and 1985. Trends were similar in other parts of the world. This declining rate of increase in wood prices since World War II suggests that wood has not become increasingly scarce. As the world's **old-growth** stocks declined, wood prices rose, which stimulated more intensive forest management and better utilization of existing volumes—the major **supply shifters.** This in turn dampened price increases. Rising population and incomes were the major wood **demand shifters:** between 1950 and 1990, world population grew at 1.88 percent per year.

The forecasts in Table 17-4 project that the world's forests can meet increasing demands for industrial wood over the next few decades and that harvests will

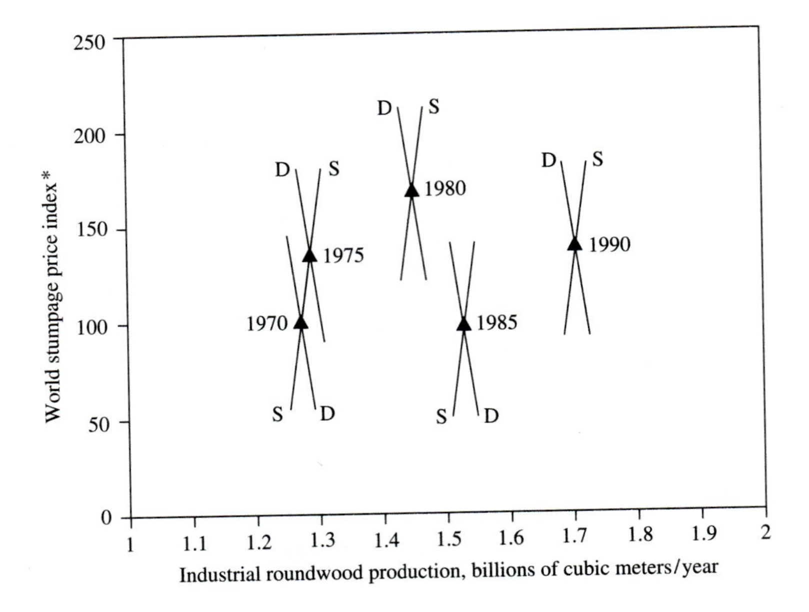

* Surrogate stumpage prices are real United States dollar values of world round wood exports divided by export volumes and translated into an index with 1970 = 100 (from FAO 1982 & 1993). Nominal prices were deflated with the United States producer price index.

FIGURE 17-2
World real stumpage price index over industrial roundwood production. (Wood production from FAO, 1982 and 1993. Curves S and D are world stumpage supply and demand drawn with arbitrary elasticities for the years 1970–1990.)

TABLE 17-4
FORECASTS OF INDUSTRIAL ROUNDWOOD PRODUCTION*

Organization	Year made	Projected to year:	Future harvest, billions of m^3/yr	Average annual growth %/yr†
International Institute for Applied Systems Analysis (IIASA)	1987	2000	1.8	1.2
	1987	2030	2.6	1.2
Resources for the Future (RFF)				
Base case	1988	2000	1.7	0.8
	1988	2035	2.0	0.6
High demand	1988	2000	1.8	1.2
	1988	2035	2.3	0.9

*Sedjo and Lyon (1990, p. 177), by permission.
† From 1.5 billion m^3 in 1985.

increase at 0.6 to 1.2 percent per year for different scenarios. At the 1.2 percent annual growth rate, industrial roundwood harvests will rise from 1.5 to 1.8 billion cubic meters between the years 1985 and 2000, according to forecasts by the two organizations shown in Table 17-4 (IIASA and RFF). The RFF model's base case forecasts average real timber prices rising a modest 0.2 percent annually until the year 2000 and 1.3 percent per year until 2035 under a high demand scenario. A recent projection shows average southern United States softwood *sawtimber* stumpage prices rising at about a real 1.3 percent annual rate between 1992 and 2040 and softwood *pulpwood* prices falling over the same period (Ince, 1994). Price trends are likely to vary among regions, but increasing world trade should lessen these differences. Under reasonable scenarios, a number of projections show that over the next few decades, the world's forests should be able to meet increasing demands for industrial wood without large rates of price growth.

Perhaps the biggest unknowns in projecting timber supply-demand balances and prices over time are environmental constraints on harvesting, which are increasing worldwide. These could cause lower timber production and higher prices than those in the above projections. On the other hand, opportunities exist to reduce logging in primary forests, manage other forests better, and still increase wood output.

Forest Plantations[2]

Information on the world's plantation area is conflicting, but combining several sources yields an estimated 122 million hectares, most of which is in North America, Europe and the former Soviet Union, and South Asia (Laarman and Sedjo, 1992; WRI, 1994). Brazil's plantation area is increasing rapidly and contains some of the most productive sites. While plantations are only about 3 percent of the world's forest area, they yield rapid growth of high-quality industrial wood for pulping or construction. Often plantation volume growth can be over 10 times higher than in natural forests. Plantation areas continue to increase, and one estimate is that they may supply up to half the world's wood by the year 2000. Introduced species, often genetically improved, fertilized, thinned, and sometimes pruned, are common; examples are radiata pine in New Zealand and Australia, and eucalyptus in Brazil.

Unfortunately, plantations, while yielding much wood volume, aren't likely to slow tropical deforestation, which is caused mainly by clearing for agriculture, as discussed shortly. Also, plantations usually yield species that aren't substitutes for the high-quality hardwoods logged in tropical forests. In dry areas where wood is scarce, plantations can supply much needed fuelwood and building poles, but seedling survival is often poorest in these regions.

For 12 regions, Table 17-5 shows rotation-start **net present values** per hectare for typical, well-managed plantations on good sites yielding pulpwood and standard quality sawtimber (from Sedjo, 1983). The values were computed with a

[2]This section draws from Kanowski et al. (1992), except where otherwise noted.

TABLE 17-5
ROTATION-START NET PRESENT VALUE AND MEAN ANNUAL INCREMENT FOR FOREST PLANTATIONS GROWING PULPWOOD AND SAWTIMBER—5 PERCENT REAL DISCOUNT RATE, CONSTANT PRICES IN PERPETUITY*

Region/species	Net present value 1979 U.S. $/ha	Mean annual increment, m^3/ha/yr
North America		
U.S. South		
Pinus taeda, average site	2,474	12.40
Pinus taeda, high site	3,742	17.72
Pacific Northwest		
Pseudotsuga menziesii, average site	902	14.42
Pseudotsuga menziesii, high site	2,248	20.03
South America		
Brazil, Amazonia		
Pinus caribaea	4,080	16.00
Gmelina spp.	3,184	18.00
Brazil, Central		
Eucalyptus spp.	4,027	25.00
Brazil, Southern		
Pinus taeda	4,715	20.00
Chile		
Pinus radiata	4,509	21.98
Oceania		
Australia		
Pinus radiata	2,141	17.40
New Zealand		
Pinus radiata	4,118	25.00
Africa		
South Africa		
Pinus patula	3,727	16.00
Gambia-Senegal		
Gmelina spp.	2,622	15.00
Eucalyptus spp.	3,262	17.00
Europe		
Nordic		
Picea abies	154	5.00
Asia		
Borneo		
Pinus caribaea	2,364	13.50

*From Sedjo (1983, pp. 37 and 45), by permission. Dollar values only for comparisons—numerous costs excluded.

5 percent real **discount rate**, assuming perpetual rotations with constant prices in 1979 dollars, given exchange rates around 1979. The net present values are before taxes and are analogous to the values for **willingness to pay for land** explained in Chapter 7 [see equation (7-3)]. These values are designed to compare economic productivity and not to reflect market values of land. The Table 17-5 values show that some of the most productive plantations in the United States are in the southern pine region. But these are surpassed by radiata pine in Chile and

New Zealand, and by Pinus caribaea, eucalyptus, and loblolly pine in Brazil. In general, the southern hemisphere plantations are the most profitable, with Nordic spruce showing the lowest values of the 12 regions. Typical physical productivity of plantations is shown in cubic meters per hectare per year in the last column of Table 17-5 (**mean annual increment**). Some studies report eucalyptus mean annual increments as high as 40 cubic meters per hectare per year.

The Table 17-5 values show that the forest plantation regions surveyed can economically supply timber for world markets. The same study also notes that forestry is economically the best use for the lands in question, most of which are poor for agriculture. Often, land abandoned after slash-and-burn farming in the tropics can yield excellent forest plantations. Given the very high productivity of southern hemisphere plantations, Sedjo (1983) predicts a southward shift in world wood production and an increasing volume of such wood entering Europe, the United States, and Japan. But more research is needed on the sustainability of intensive, short-rotation plantation forestry: will productivity remain the same in perpetuity, as assumed in Table 17-5? Or will site qualities of intensively managed plantations decline after several rotations, thus requiring more costly fertilization?

GLOBAL ENVIRONMENTAL ISSUES

Many environmental groups operate worldwide: examples are the World Wildlife Fund, Nature Conservancy, the Rainforest Alliance, Greenpeace, the World Resources Institute, and several organizations promoting "green certification" of wood products from sustainably managed forests (Brown and Hammel, 1994).[3] These groups have brought attention to environmental damage resulting from misuse of forestlands throughout the world. Such concerns often lead to timber harvest restrictions and to controversy about their economic impacts.

Environmental constraints on timber harvests in one country can have worldwide effects on forest management and wood trade. Using the University of Washington CINTRAFOR global trade model, a 1992 study simulated the following global impacts if Pacific Northwest harvests were reduced by 4.7 billion board feet to preserve old growth and spotted owl habitat. Fourteen percent of the shortfall would be covered by reduced demand due to rising wood prices, and the rest would come from increased supplies elsewhere: 3 percent from the western United States, 21 percent from other United States regions, 27 percent from Canada, and 36 percent from other countries. Using the same model, another simulation gauged the effects of a hypothetical 33.3 million cubic meter per year total harvest reduction for environmental gains in the western United States and Canada and in Malaysia. Compared to pre-1990 supply levels, by 1995, 37.5 percent of the shortfall would be covered by reduced worldwide demand due to higher wood prices. About 36.3 percent would come from other United States lands, 3.6 percent from eastern Canada, and 22.5 percent from other countries,

[3] See the April 1995 *Journal of Forestry* (vol. 93, no. 4) for several articles on forest certification.

mainly in Europe. Another scenario softens the impact on these regions if 23 percent of the shortfall could come from Siberia (Perez-Garcia, 1992; 1993). So one country's forest policies shouldn't be viewed in isolation.

The above analyses imply that harvest restrictions in one region can shift forest destruction by increasing harvests elsewhere. So the key to worldwide forest damage problems is in instituting sustainable forestry practices that treat ecosystems gently in all regions of the world.

Tropical Deforestation

Many natural tropical forests aren't expected to be major suppliers of industrial wood for export in the near future. Although the resource is vast, much of it is hardwood species, few of which are known in export markets. Some Latin American and Asian tropical forests have volumes of 200 to 300 cubic meters per hectare, of which only 5 to 10 percent may yield lumber or plywood suitable for export (Bethel, 1988). This raises extraction costs. Thus, while tropical forests are roughly half the world's forest area, they yield less than 10 percent of the world's industrial wood.

Nevertheless, tropical forest losses are significant. Between 1980 and 1990, their area declined by 15.4 million hectares (0.8 percent) per year (WRI, 1994). About 26 percent of the loss is from commercial logging, 15 percent from fuelwood harvesting, 12 percent from cattle ranching, and 47 percent from shifting agriculture. In the last two cases, forests are clear-cut and burned. Counting partial clearing, the portion of tropical deforestation from shifting agriculture is 64 percent, with the other percentages then being lower (Meyers, 1992). Thus, while much tropical forest is lost to commercial logging, clearing for agriculture is the leading problem. Unfortunately, much of the tropical forest area cleared for crops isn't well suited for agriculture and bears crops only temporarily. And tropical clearings often don't reforest well naturally, as most temperate forests in the United States have. Figure 17-3 shows annual deforestation rates worldwide. The following regions have rates of decline in tropical forests above the world average of 0.8 percent per year (annual rates of decline in parentheses): Continental Southeast Asia (1.5 percent), Central America and Mexico (1.4 percent), Insular Southeast Asia (1.2 percent).

Most tropical timber is publicly owned. Several studies have shown that such timber is often sold for less than market value. The effect is to accelerate deforestation and aggravate boom-and-bust development. Repetto and Gillis (1988) find that underpricing of public timber occurs not only in the tropics but also in parts of the United States and Canada. In many tropical forests, logging roads give wider access to shifting cultivators, who continue the clearing.

Laarman and Sedjo (1992) point out that government policies often encourage deforestation. Examples are colonization schemes to develop forested regions in Brazil, Indonesia, and elsewhere. Such policies often include government subsidies for logging and agriculture in forested areas. Another problem is that private harvest contracts on public forests are sometimes of such short duration that concessionaires have no incentive to practice long-term management.

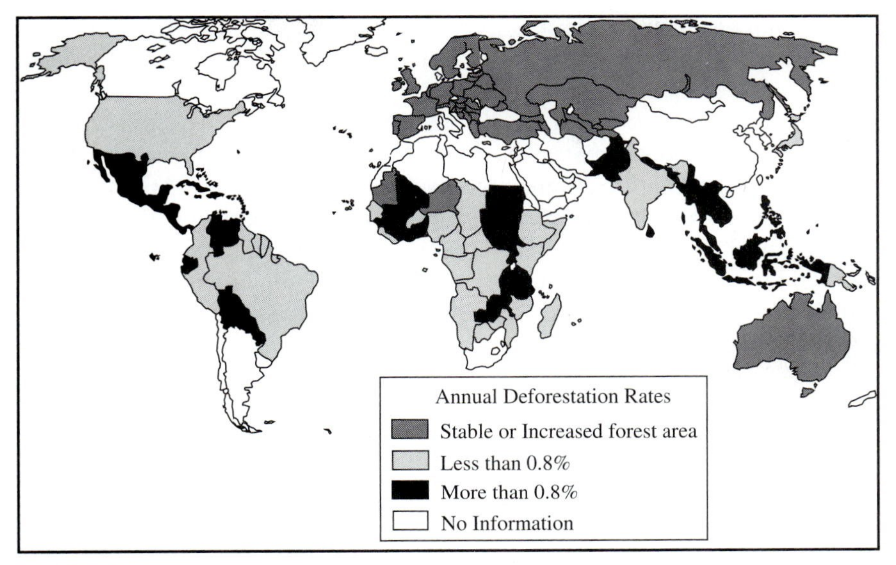

FIGURE 17-3
Estimated annual rates of change in forest area—1981–1990. (From WRI, 1994, Figure 7-1, page 130, by permission. Between 1977 and 1988, the United States forest area has declined by 1.2 percent in total (U.S.F.S., 1978, Waddell et al., 1989.)

Often underestimated are values of *nontimber forest products* in the tropics—gums, resins, nuts, fruits, spices, vegetables, canes like rattan, meat from game animals and fish, animals for pets and tourism, and materials for medical uses. Studies find that in some areas, the net present value of such products might exceed the present value of clearing tropical forests for agriculture (Panayotou and Ashton, 1992).

Tropical forest clearing is an example of concentrated gains and diffuse losses mentioned in Chapter 15. Gains are concentrated in small areas and groups of people, and large scattered populations feel losses in rich biodiversity of irreplaceable ecosystems, increases in greenhouse gases from forest burning, and losses in carbon storage on clearings. Organizing worldwide action to foster sustainable forestry isn't easy, but progress is occurring. In its 1994 statement of objectives, the International Tropical Timber Organization (ITTO), a commodity group of 48 nations, aims "to enhance the capacity of members to implement a strategy for achieving exports of tropical timber and timber products from sustainably managed sources by year 2000" (ITTO, 1994). In 1992, the United Nations Conference on Environment and Development in Brazil—the "Earth Summit"—spawned a number of documents and treaties on protecting biodiversity and the environment. And many nations have pledged funds for environmental conservation programs in developing areas. But the trade-off between economic development and environmental conservation was painfully evident at the Earth Summit. For example, U.S. President Bush did not support portions of treaties he felt would reduce economic growth.

A major force influencing tropical forest policies is the amazing number of nongovernmental organizations, the "NGOs." These include environmental groups, timber trade associations, and professional societies. Roughly 17,000 people representing 7,150 NGOs attended the 1992 Earth Summit: many of these participated in drafting key documents. Activities of NGOs include lobbying the World Bank to reduce lending for commercial logging, urging the Inter-American Development Bank to stop paving roads in Amazonia, pressing the ITTO to support labeling tropical timber from sustainable forests, asking the U.S. Agency for International Development not to fund logging in natural tropical forests, and pushing for local governments in the Netherlands to ban tropical wood use in their projects. The NGOs have been successful in many such efforts (Mohd and Laarman, 1994).

Global Warming and Tree Planting

Increasing levels of atmospheric carbon dioxide are thought to cause global warming. We look to trees to help solve the problem, since they consume CO_2 and store carbon. To stabilize global atmospheric CO_2 would mean reducing current carbon release rates by 3 to 5 billion tons annually. To remove and store just 1 billion more tons of carbon per year would require at least 100 to 200 million hectares of new forests—well more than all the existing plantations in the world (Woodwell, 1992). Also, many fast-growing plantations are on rotations of 10 to 20 years, so the climatic benefits are temporary. Thus, while trees can help reduce CO_2 levels, they can't do the whole job. We need a multipronged approach of reducing fossil fuel use, lowering deforestation rates, and planting trees.

SOCIAL FORESTRY AND AGROFORESTRY[4]

Social forestry has been a popular term in world forestry circles since the 1980s, although it has many meanings. Westoby (1989) gives this definition: "tree planting and management, at the farm, village or community level, by or for small farmers and the landless." Such projects, often supported by groups like the Food and Agriculture Organization (FAO) of the United Nations, the World Bank, or the U.S. Agency for International Development, are usually aimed at the rural poor in developing areas where the major wood use is often for fuel. While many of these projects have been very successful, outside groups introducing them need to be aware of local customs and organizations. For example, in parts of Africa the act of planting and rearing a tree can bring legal title to the underlying land. Sometimes government plantings in such areas have been uprooted by locals who feared they would lose their property rights. International forestry experts also warn agencies to take care that forestry projects actually benefit the poor at whom they are aimed rather than wealthier community leaders.

[4]This section draws from Westoby (1989). See Hyde (1992) for more detail.

Another term with many meanings is **agroforestry**—different combinations of agriculture and trees on the same area. Chapter 15 mentions grazing beneath widely spaced stands of different pines in the United States, New Zealand, and Australia. In Malaysia, some young rubber plantations are intercropped with pineapple and bananas. Later, sheep and poultry may be introduced. In nearly all countries, patches and rows of trees around farms protect lands from soil erosion, wind, and sometimes sun. Shelter from tree plantings in arid regions can reduce soil drying and wind erosion and can increase some agricultural crop yields up to 150 percent (Kanowski et al., 1992). In agroforestry, trees may yield large logs, firewood, poles, orchard crops, or nonmaterial outputs. Nitrogen-fixing trees can improve soil quality for farm crops. In developing countries, some agroforestry is an old and established practice, but many types are experimental. The International Council for Research into Agroforestry (ICRAF) in Kenya sponsors much research in Africa and other tropical regions of the world.

INTERNATIONAL TRADE

The most obvious gains from trade occur when one country has desirable items another country doesn't have, and vice versa. For example, Saudi Arabia produces little timber but has abundant oil. Oil and timber trade readily develops between the Saudis and countries with ample timber but no oil. In this simple example, both partners obtain new goods. But things are usually much more complex. Many countries produce similar products and services but initially at different prices. Importing cheaper goods drives down prices of these goods and some wages in the importing country—a boon to most consumers but a bane to certain industries and workers. And trade can increase prices of goods shipped from the exporting country: there producers benefit and consumers suffer. But worldwide, the gains are positive, assuming no **market failures** (see Chapter 3). Examples are abundant: automobiles and electronic equipment from Asia cause problems for those industries in the United States but benefit our consumers and Asian industries. Our computer industry benefits from exporting. In the forest industry, at times, low-cost wood from Canada has depressed wood prices in the United States, or relatively low-cost wood from the United States depresses European and Japanese timber prices: in the importing country, this benefits consumers and hurts producers. Exporters gain from higher prices received. With such gainers and losers, trade policy creates conflicts. Chapter 3 notes that as long as gainers from a policy can overcompensate losers, welfare increases. But that doesn't satisfy losers if they aren't compensated, which is the usual case.

Some long-run results of world trade are larger worldwide yields of goods and services and greater equality in prices of inputs and outputs. Trade can bring greater specialization, competition, and efficiency. Theoretically, without market failures, net gains are positive, but certain groups lose in the short run. You might speculate that increasing trade, specialization, and productivity will tend to lessen the gap between the haves and the have-nots, but it's a slow and imperfect process.

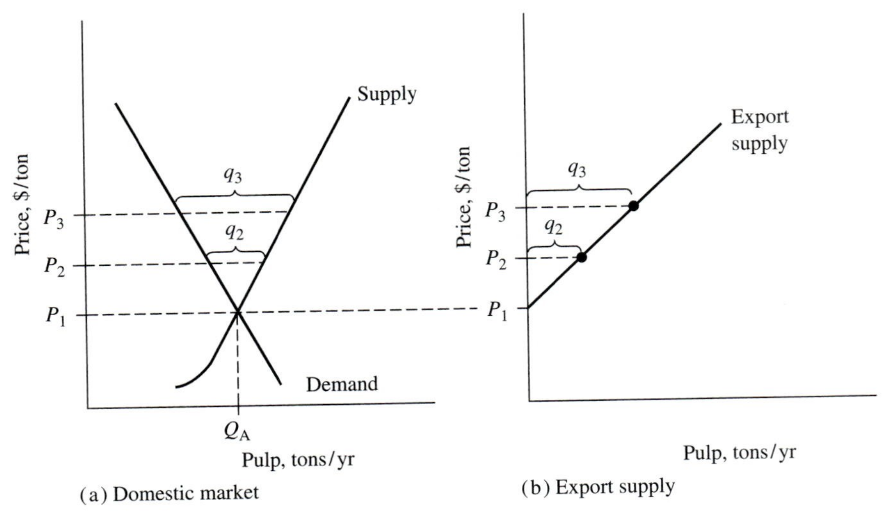

FIGURE 17-4
Country A—potential pulp exporter—no trade.

A Simple Trade Example

Let's see how price differences stimulate trade when two countries both produce the same good. Consider production of one good, pulp, in a two-country world. For simplicity, initially all prices are in dollars and transportation costs are zero. Without trade, Figure 17-4a shows the equilibrium output of Q_A tons of pulp per year at price P_1 in country A. If foreign consumers bid up the price to P_2, domestic consumers would demand less than Q_A in country A, and producers would supply more than Q_A. The excess is shown as q_2 in Figure 17-4a and also in Figure 17-4b as a point on the **export supply curve**, which gives amounts available for export from country A at different prices. Other points on the export supply curve, derived from Figure 17-4a, are no export supply at price P_1, and amount q_3 at price P_3.

Figure 17-5a shows the equilibrium pulp price and quantity, P_5 and Q_B, in country B, without trade. With trade, country B would become the importer, since price there at P_5 is higher than in country A. If imported pulp into country B drove the price down to P_4, domestic firms would supply less than Q_B, and consumers would demand more than Q_B. The domestic shortage is shown as q_4 in Figure 17-5a and also in Figure 17-5b as a point on the **import demand curve** showing pulp imports demanded by country B at different prices. Other points on the import demand curve, derived from Figure 17-5a, are no import demand at price P_5 and amount q_3 at price P_3.

Since the equilibrium pulp price without trade is higher in country B than in country A, there is incentive for A to export pulp to B. Because A can produce pulp more cheaply, it has a *competitive cost advantage* in pulp production. Figure 17-6 includes graphs from Figures 17-4 and 17-5, using different axis scales and showing the combined export supply and demand curves in panel (b). In equilibrium,

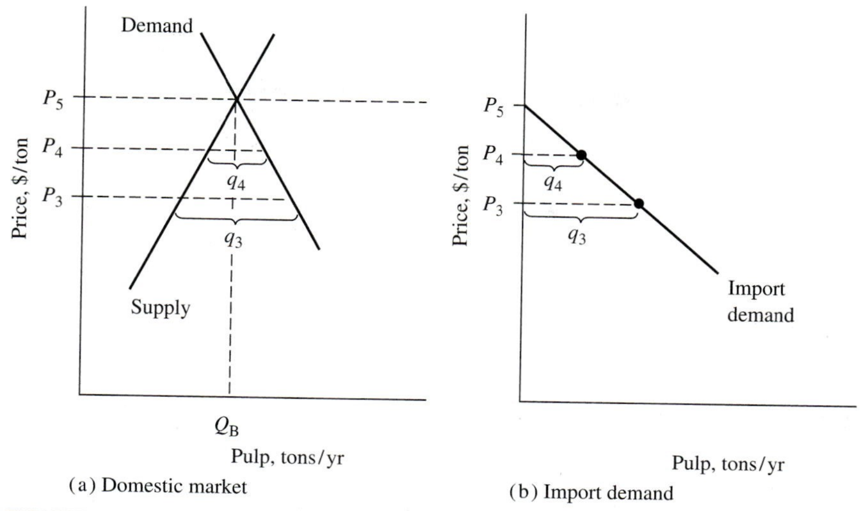

FIGURE 17-5
Country B—potenial pulp importer—no trade.

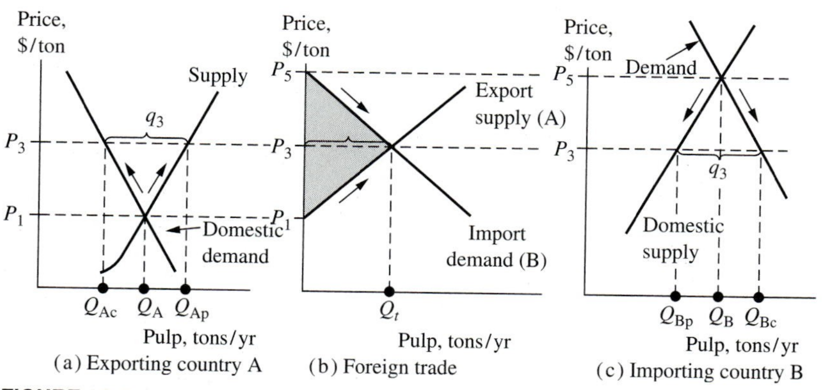

FIGURE 17-6
Equilibrium with pulp trade.

after trade, in panel (b), country A exports Q_t tons of pulp to country B, which is the same as the quantity q_3 in all three panels. In panels (a) and (c), pretrade outputs are Q_A and Q_B. In Figure 17-6a, the equilibrium price after trade is P_3, and domestic *consumption* is Q_{Ac} after trade. Country A's total *production*, including exports of q_3, is Q_{Ap} after trade.

In Figure 17-6c, after trade, country B's pulp price falls to P_3, and q_3 tons of pulp are imported, yielding total *consumption* of Q_{Bc} tons. Country B's domestic pulp *production* falls to Q_{Bp}. So, ignoring transportation costs, the world equilibrium price is P_3 after trade. This tendency toward price equalization is called the **law of one price**. Compared to no trade, country A now produces more pulp and country B less. When only one good is traded, free trade leads to greater specialization in the country that has the lowest production costs, including government subsidies.[5] Complete specialization could occur if price differences were originally great enough and country B's pulp price were driven so low that no firm would produce pulp there.

Relaxing Assumptions Compared to the above simple example, the real world includes all countries and many goods and services, with each country exporting and importing. Also, *transportation costs* drive a wedge between the price of a traded good in two countries. Adding shipping costs in the above trade equilibrium example, pulp prices in the importing country would be higher than in the exporting country by the per ton cost of shipping. And trade volume would be reduced because higher prices would reduce consumption in the importing country. A **tariff** or *duty* on imported goods, by raising their prices, has the same effect as increased shipping costs.

Figure 17-6 shows dollars in country A worth the same as in country B. If one country varies its **exchange rate**, the trade balance can change. Suppose the B$ becomes worth less than the A$; say, it would now take B$2 to buy A$1. That makes country A's pulp more expensive for country B and reduces B's import demand. In Figure 17-6, halving the value of the B$ has the effect of doubling the B$ price scale for country B relative to country A. So in Figure 17-6c, the price P_5, where the demand and supply curves cross, would be half as high on the graph. Thus, demand and supply curves in country B would be lower after devaluation of the B$. That in turn lowers the import demand curve in Figure 17-6b (based on A$), thus reducing the export of pulp from A to B. In fact, if the B$ is devalued enough, B's pulp would become so inexpensive for A that trade would reverse, and A would buy pulp from B. So when countries alter their exchange rates, equilibrium trade levels change. Today the international currency system is a mixture of floating and fixed exchange rates. Most of the major currencies like the United States dollar and the Japanese yen float freely according to supply and demand. In those cases, the exchange rate in the long run is a *result* of changes in trade flows, not a cause, as in the above example.

The degree to which changes in shipping costs, tariffs, and currency values will affect trade flows depends on the **price elasticities of demand and supply** in each country, which determine the elasticities of the export supply and import demand curves.

Gains from Trade Assuming no market failures, no coercion, and no deception, when two people willingly enter into an economic transaction, both are made better off, otherwise no transaction would occur. And so it is with countries. Intuitively, *under the above assumptions,* the exporting and importing shown in Figure 17-6 should benefit both countries, otherwise it wouldn't take place. We'll

---

[5]Trading several goods is more complex: then specialization occurs where there's a relative cost advantage, rather than absolute advantage. For example, country A may produce both pulp and lumber more cheaply than B; say, A's pulp costs are 40 percent below B's, and A's lumber costs are 10 percent below B's. Then it would benefit B to export lumber because it has a *relative* cost advantage in lumber (but not absolute). This is the *principle of comparative advantage*.

see shortly that when the above assumptions aren't met, trade benefits may not necessarily be positive in both countries. But let's first look at the theory of trade gains.

In the middle panel of Figure 17-6, at trade levels from 0 to Q_t, the import demand curve is above the export supply. In that range, importers' **willingness to pay** for more pulp will exceed the price required for exporters to willingly supply additional pulp. Thus, in the shaded area of panel (b), benefits of more trade exceed the costs. So *the shaded area shows the net gain from unrestricted trade*. Actions like quotas or tariffs making trade less than Q_t will clip off parts of the shaded area and reduce gains from trade.

To the right of Q_t in Figure 17-6b, the import demand curve lies below the export supply. That means the price per ton importers would be willing to pay for more pulp will be less than the price exporters would require to supply more. Thus, the costs of expanding trade beyond Q_t will exceed the benefits.

After introducing trade, in the exporting country, pulp prices increase, so producers gain and consumers lose. In the importing country, pulp prices decrease, so consumers gain and producers lose. Although gains will exceed losses in both countries, gainers and losers will be pitted against one another. Appendix 17A gives more detail on the theory and controversy of gains from trade.

Arguments for Restraining Trade

There are parallels between free trade theory and Chapter 3's review of **welfare maximization** in a free market. Both analyses start by assuming no market failures, in which case government intervention—or trade restraints on a global scale—would reduce net benefits. What kinds of actual or perceived market failures might justify trade restraints like import quotas or tariffs?

Usually we refer to market failures, like unpriced environmental damage, in the context of one country. If you go back to Chapter 3's review of market failures, you'll see that they're just as relevant on a global scale. For example, think about these failures (in parentheses are global examples): *property rights not enforced* (global problems with damage to oceans and the atmosphere), *imperfect competition* (monopoly elements of oil cartels, monopsony with large buyers of tropical timber), *immobility of labor* (certainly more relevant globally than nationally), *immobility of capital* (capital doesn't move readily into politically unstable areas), *unpriced negative side effects* (forest burning causing carbon emissions and global warming), *outputs not easily priced* (oxygen output and carbon storage services of the world's forests, option demand for tropical forests and wildlife preserves), *economic instability* (worldwide recessions), *unsatisfactory income distribution* (obviously worse globally than nationally). A few of these are discussed below in an international trade context.

Trade and the Environment One argument for free trade is to promote competition and the resulting efficiency. If one industry can't meet foreign competition, free traders argue that the sector should become more efficient rather than seek protection through trade restraints. But suppose foreign producers' prices are low because they avoid costs of environmental protection and worker safety, hire child labor, use prison labor, and have low health care standards. These are not welfare-increasing improvements in efficiency and are often called unfair trade practices. The arguments for trade restraints are strongest if one country's industries cause direct damage in another country, for example, through acid rain or polluted rivers. The spillover reasoning is becoming more cogent with today's concern over global warming.

But spillover arguments aside, some question one country's right to impose its standards on another in the form of trade sanctions. One way to sort out these questions is to imagine a fictitious case of paper mills in east Texas being able to outcompete those in Oregon because Texas imposed no environmental constraints and Oregon imposed many. Other things being equal, there would no doubt be strong support to level the playing field. Now suppose history had taken another turn, and Texas were part of Mexico. Again, imagine the above hypothetical scenario in paper production. While the United States couldn't rewrite Mexico's environmental laws, one approach would be to impose restraints on paper imports from the "Texas region" of Mexico. Since many trade restraints already exist, a more positive approach is to offer freer trade with those countries willing to improve environmental standards, rather than threatening further sanctions if they don't.

To show the problem of unequal environmental standards, suppose for the pretrade solution in Figure 17-6, country A undersells B only because its pollution control laws are more lax. The free trade equilibrium creates even more pollution by increasing A's pulp production.

The connection between trade and the environment is a controversial issue. Since free trade boosts worldwide economic development and the accompanying environmental degradation, some groups promote trade restraints, hoping to improve the environment.[6] Perhaps the wisest approach is not to take polar views but to aim for freer trade within a framework of international environmental protection agreements. These could include not only standards for emissions and resource management practices but also agreements that firms should be liable for environmental damage, that firms should pay for all pollution control measures without government subsidies ("the polluter pays" principle), and that governments should not subsidize production of outputs that lead to more environmental damage (e.g., agricultural crops, timber harvesting, coal mining, and irrigation water). These approaches would level the playing field in trade, improve environmental quality, and provide gains from trade.

Unfair Competition Sometimes industries seek protection if competing foreign firms have "unfair advantages" like government subsidies or if they practice "dumping" of goods at depressed prices in domestic markets. In 1985 the United States threatened to impose a 15 percent tariff on Canadian lumber imports because it was alleged that the Canadian government subsidized stumpage produc-

[6]See Daly (1993), Bhagwati (1993), and Wisdom (1995) for contrasting views on this issue.

tion. Canada agreed to impose an export duty, but also responded with tariffs on several United States imports, as so often happens in these cases: the result probably reduced gains from trade. The lumber export duty was later dropped when Canadian stumpage pricing was made more competitive. The Canadian lumber case is more complex than it looks on the surface, since builders often prefer Canadian softwoods in some uses because they're easier to work with than, for example, southern pine.

Sometimes "infant industries" are protected until they gain the strength to compete internationally. This is especially important in newly developing countries trying to diversify their economies. But once started, it's hard to wean a sector from protection.

Income Distribution and National Security Most early trade theories assumed immobility of capital, but today capital moves freely throughout the world. Daly (1993) worries about large capital movements from high-wage countries (say, the United States) into manufacturing goods in low-wage countries, which then outcompete United States industries. The result would be decreased United States wages and increased income to owners of capital, which increases income inequality in the high-wage country. But the benefit is higher pay in low-wage countries, presuming their population growth rate isn't excessive (often a shaky assumption).

Employment is always a concern when domestic firms lay off workers due to foreign competition. The flip side is that free trade can increase employment in other industries—again, a distribution issue.

Some industries may be protected for national security reasons because they produce munitions or a basic industrial foundation such as steel or oil. Wood falls less in this category today than it did at one time.

Concentration of Gains and Losses Also important are differing concentrations of gains and losses from policies to increase or decrease trade restraints. Suppose gains from lowering a tariff are diffuse and scattered among millions of consumers at low amounts per person. If losses are smaller but concentrated among a few firms that can effectively fight the policy, it may be hard to implement, even if socially desirable. But if gains from a trade policy are concentrated in a small sector, and larger losses are diffuse, lobbying for this undesirable policy might be effective.

Wood Products Trade[7]

Major wood products trade tends to flow from the largest wood producing regions to the three industrialized population centers shown in Figure 17-7. Central Europe imports from Canada, the eastern United States, South America, Scandinavia, and the former Soviet Union. Canada exports heavily to the eastern United

[7]Unless otherwise noted, this section's data are from FAO (1993).

FIGURE 17-7
Major flows of wood products trade. (From Laarman and Sedjo, 1992, page 167, by permission.)

States. Japan, China, and neighboring countries import from around the Pacific Basin. This only outlines the main arteries: forest products move in hundreds of directions throughout the world.

Compared to goods like jewels and electronic products, wood is bulky and has a low value per pound, and thus its transportation costs are a fairly high percentage of sale value. A study of worldwide exports from Chile found that ocean freight costs as a percentage of delivered price averaged 42 percent for logs, 31 percent for lumber, and 12 percent for bleached pulp from 1981 to 1985. Over the same period, average real ocean freight costs were decreasing (Caamaño, 1988). As you'd expect, the more processed and more valuable a good is per pound, the lower the transport cost as a percentage of value.

Table 17-6 shows roundwood exports of over 126 million cubic meters worldwide for 1991, 25 percent above the 1981 volume. Except for Oceania, most regions export a fairly small percentage of their log production. Worldwide, only 3.7 percent of total roundwood production is exported, reflecting log export bans and the fairly high shipping cost as a percentage of value. These exports, as a percentage of world industrial roundwood, are 8 percent—a more realistic figure, since nearly all log exports are **industrial wood**. For the more valuable processed products, by each product category, worldwide exports as a percentage of world output are much higher—19 percent for lumber, 25 percent for wood-based panels, 17 percent for wood pulp, and 24 percent for paper and paperboard. So most trade is in processed products, not roundwood.

In areas short of timber, roundwood *imports* as a percentage of industrial roundwood production are high—about 30 percent for Asia and 17 percent for Europe, excluding the former U.S.S.R. For all other regions, log imports are only about 1 percent or less of industrial roundwood output (see the last column of Table 17-6).

TABLE 17-6
WORLD EXPORTS AND IMPORTS OF ROUNDWOOD—1991*

Region	Roundwood exports, millions m³	Roundwood exports as a % of roundwood production	Roundwood imports, millions m³	Roundwood imports as a % of industrial roundwood production
Africa	4.7	0.89%	0.8	1.3%
North & Central America	33.6	4.6	7.9	1.3
(U.S.A.; incl. in above)	(30.0)	6.0	(2.2)	(0.53)
South America	6.9	2.0	0.1	0.1
Asia	23.8	2.2	74.7	29.7
Europe (excl. former U.S.S.R.)	31.8	9.5	47.4	16.7
Former U.S.S.R.	14.6	4.1	0.1	0.04
Oceania†	11.2	26.0	0.1	0.25
Total world‡	126.6	3.7	131.0	8.2

*From FAO (1993).
†Primarily Australia and New Zealand.
‡Totals may differ due to rounding. The difference between total world roundwood imports and exports is logs in transit.

The greatest roundwood imbalance is in Japan, where log imports were 178 percent of industrial log production in 1991.

In 1991, global forest products exports were worth over $98 billion, nearly twice the 1981 level. Depending on the inflation measure, the real gain would have been about 35 to 40 percent over that period. The major shippers were North America and Europe, as shown in Table 17-7. Column (4) shows that North America has the largest net exports, although United States imports were 8 percent above exports for 1991. A country's import-export ratio can vary greatly: in 1985, United States forest products imports were nearly twice the exports. United States forest products imports are mainly lumber, pulp, and newsprint from Canada. Asia and Europe are the major net importers [see column (4)]. Japan consumed about 46 percent of Asia's $27 billion wood products imports for 1991. That year, Japan's forest products imports were 5.8 times higher than exports, by value.

Radiata pine is becoming an important species in Asian markets. Plantation areas are increasing dramatically in Chile and New Zealand. One projection shows annual world radiata pine harvest rising from about 25 million cubic meters in 1990 to 60 to 80 million cubic meters in 2020. Much of this will be exported, since a large portion is grown in Chile and New Zealand, where timber harvests far outstrip domestic demand. Australia has about 19 percent of the world's radiata pine plantation area, but due to its strong domestic timber demand, Australia is not expected to be a major exporter (CINTRAFOR, 1993).

Wood Export Bans

To create more timber processing jobs and reduce logging, countries sometimes restrict log exports, thus making logs cheaper for the domestic mill sector, which

TABLE 17-7
GLOBAL FOREST PRODUCTS EXPORT AND IMPORT VALUES—1991*

Region	All forest products exports, $billions	All forest products imports, $billions	Exports minus imports, $billions
Africa	$1.56	$4.33	$−2.77
North & Central America	29.62	16.91	12.71
(U.S.A.; incl. in above)	(12.48)	(13.47)	(−0.99)
South America	2.63	1.11	1.52
Asia	12.33	26.89	−14.56
Europe (excl. former U.S.S.R.)	47.57	55.97	−8.40
Former U.S.S.R.	2.77	0.93	1.84
Oceania†	1.55	1.48	0.07
Total world‡	$98.05	$107.62	§

*From FAO (1993).
†Primarily Australia and New Zealand.
‡Totals may differ due to rounding.
§Export and import totals are not equal, due to goods in transit.

expands. Vincent (1992) argues that, in several tropical countries, this has depressed log values and income by more than the **value added** in processing and later export of, say, lumber or plywood. The issue is further complicated if processed product exports are limited. These arguments may also apply to the current ban on exporting logs from federal lands in the United States Pacific Northwest. The nation forgoes high export log prices. Whether or not that loss is regained in new processing income depends on world prices of processed wood products and the Northwest's processing costs compared with those in other regions. Also, there's the environmental benefit of less logging if export restriction reduces overall harvest levels. Another factor is whether the ban is total or partial. And if an export ban is on, say, logs from federal lands only, private forest owners who export will reap gains of high export prices, but the federal government will not.

To summarize, in addition to environmental benefits if total harvests are reduced, gainers from wood export restrictions are domestic log processors buying cheaper logs on the open market. But under competition it's ultimately domestic consumers who gain from lower wood products prices. Without **full employment**, there will be net gains from new jobs in wood processing. Losers are domestic timber resource owners, who forgo higher wood export prices, and foreign consumers, who pay higher wood prices. The simple theory of gains from trade says that restraining exports will reduce the world's well-being (see Figure 17-6 and Appendix 17A). But that ignores questions about the environment, income distribution, and employment levels. Thus, for export bans on logs or processed wood, it isn't always clear whether net effects will be positive or negative (Wisdom, 1993).

KEY POINTS

- Forestry is becoming increasingly global in terms of investment, trade, and environmental movements.
- Over 50 percent of the world's timber harvest is used for fuel; for many developing countries, the figure is 70 to 90 percent.
- The United States is the world's largest industrial roundwood producer. It has an advantage in wood production, due to efficient transportation and processing plants as well as productive timber sites.
- World industrial wood output has been rising at about 1.3 percent annually between 1969 and 1991. Projections are that supply should be able to meet demand without large price-increases for several decades.
- Forest plantations are producing an increasing portion of the world's wood supply, especially in the southern hemisphere.
- Environmental constraints are reducing timber harvests in many natural forests, but plantations are expected to make up the difference.
- By far, most tropical deforestation is from clearing for agricultural crops, rather than from commercial logging. Tropical forests supply about 10 percent of the world's industrial wood.
- Theoretically, free trade in forest products should bring net gains to the world. But removing trade restraints is controversial because some groups gain while others lose. Some worry about negative impacts on the environment when trade is unrestricted.
- Asia and Europe are net importers of wood; North America is a net exporter.
- When a country bans log exports, revenue losses often exceed gains. But the results become less clear when considering issues of the environment, employment, and income distribution.

QUESTIONS AND PROBLEMS

17-1 Which continent produces the most wood fiber? Which one produces the most industrial wood, excluding firewood?

17-2 What countries are the top three producers of industrial wood? What is the major use for wood in most developing countries?

17-3 What has been the trend of world roundwood production between 1970 and 1990? Roughly what percentage of this is fuelwood?

17-4 For the next few decades, what do projections show for the world's industrial wood supplies and prices?

17-5 Which regions have fairly stable forest areas? Where are forest areas declining? What are arguments for and against decreasing a region's forest area?

17-6 Where are the world's most profitable and productive forest plantations?

17-7 What are the major causes of tropical deforestation? Will more plantations solve such deforestation problems? Why?

17-8 What are social forestry and agroforestry?

17-9 If two countries produce the same product, under what conditions would trade in this product develop between the two countries?

17-10 If country A exports lumber to country B, and both countries have lumber industries, who gains and who loses in country A and in country B?

17-11 In the previous question, where are the gains and losses likely to be the most diffuse and the most concentrated?

17-12 Referring to question 17-10, if country A's currency is devalued relative to country B's, what would be the effect on their lumber trade?

17-13 Outline cases where the gains from trade restraints (tariffs or quotas) might or might not exceed the losses.

17-14 Where are the major market areas to which forest products flow in world trade?

APPENDIX 17A: Gains from Trade

Going back to the simple pulp trade example of Figures 17-4 through 17-6, let's take a more detailed look at the gains from trade. As before, assume a two-country world trading in only one good, pulp. All prices are expressed in dollars, transportation costs are zero, and market failures are ignored.

Figure 17A-1 graphs the earlier pulp trade example. Recall that unrestrained trade led to an equilibrium pulp price of P_3, and q_3 tons of pulp were shipped from country A to country B. In panel (a), country A produces Q_{Ap} tons of pulp and consumes Q_{Ac}, with the difference exported to country B. In panel (c), country B consumes Q_{Bc} tons of pulp and produces Q_{Bp}, with the difference imported from A.

To understand the gains from trade, you'll need to recall the idea of **consumer surplus:** the area under the demand curve and above the price line—the benefit of consumers' extra willingness to pay above what they actually pay (see Chapters 2 and 14). In panel (a) of Figure 17A-1, country A's exporting causes *consumption* to fall from Q_A at price P_1 to Q_{Ac} at price P_3. Thus, consumer surplus declines by area d (a loss in country A). In a supply-demand framework, the mirror image of consumer surplus is **producer surplus:** *the area above the domestic supply curve and below the price line*—the extra income received by some firms above the minimum required. In panel (a), for example, when price increases from P_1 to P_3, the firms that would have been willing to supply Q_A at price P_1 now get price P_3. Thus, the trade-induced producer surplus is area $d + e$. Since the loss is d, the

FIGURE 17A-1
Gains from trade.

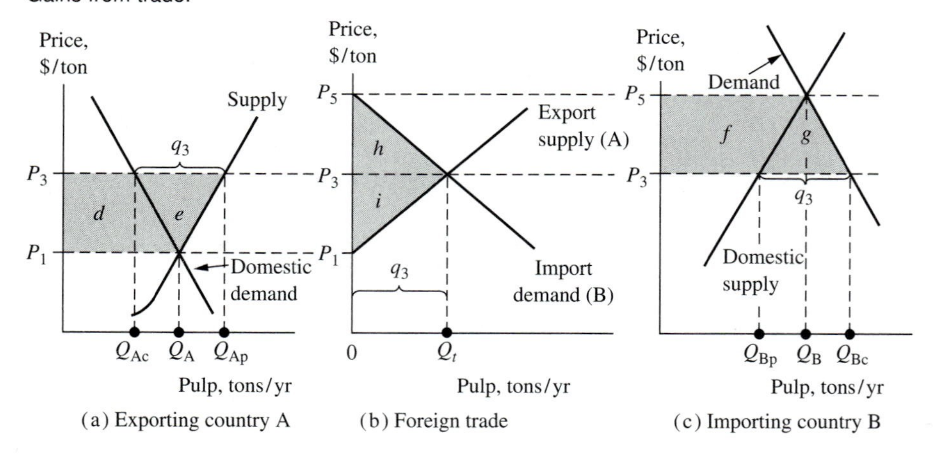

net gain from trade in country A is the area *e*. So in the exporting country, pulp prices rise, and the producers' gain exceeds the consumers' loss.

In Figure 17A-1(c), importing causes country B's domestic pulp *production* to fall from Q_B at price P_5 to Q_{Bp} at price P_3. Thus, producer surplus falls by the area *f* (a loss in country B). But since total pulp *consumption* rises from Q_B to Q_{Bc}, and price falls to P_3, consumer surplus rises by the area $f + g$, a gain. Therefore, country B's net gain from importing is area *g*. So in the importing country, pulp price falls, and the consumers' gain exceeds the producers' loss. Now you can see why trade policy is so controversial: although net gains from free trade are usually positive, the damaged parties want trade restrictions. And so far, I've assumed away market failures: these add more controversy.

Figure 17A-1(b) summarizes the net gains from trade: Since triangles with the same bases and heights have the same areas, area *e* in Figure 17A-1(a) equals area *i* in panel (b): the exporter's gain. And area *g* in panel (c) equals area *h* in panel (b): the importer's gain. Thus the total net gain from trade is the shaded area $h+i$ in panel (b).[8] In the example, this gain (area) is maximized by free trade at level Q_t in panel (b). Therefore, things like import quotas or tariffs, which reduce trade, will shift Q_t to the left, nip off pieces of the area $h + i$ in panel (b), and reduce gains from trade.

Figure 17A-1 shows theoretical gains from unrestricted trade under certain assumptions. The section on "Arguments for Restraining Trade" mentions cases where restricting trade might increase benefits.

REFERENCES

Backman, C. A., and T. R. Waggener. 1990. *Soviet Forests at the Crossroads: Emerging Trends at a Time of Economic and Political Reform.* CINTRAFOR Working Paper 28. University of Washington, Seattle. 382 pp.

Bethel, J. S. 1988. Tropical countries: Suppliers or consumers of forest products? In Johnson and Smith (1988), pp. 193–197.

Bhagwati, J. 1993. Trade and the environment: The false conflict? In Zaelke et al. (1993), pp. 159–190.

Binkley, C. S., and J. R. Vincent. 1988. Timber prices in the U.S. South: Past trends and outlook for the future. *Southern Journal of Applied Forestry.* 12(1):15–18.

Brooks, D. J. 1993. *U.S. Forests in a Global Context.* U.S. Forest Service Gen. Tech. Rept. RM-228. Ft. Collins, CO. 24 pp.

Brown, L., and D. Hammel. 1994. Green certification of wood: Implications for the woodworking industry. *Wood and Wood Products.* September. p. 96.

Brown, L. R., et al. 1991. *State of the World.* Worldwatch Institute. W. W. Norton & Co., New York. 254 pp.

Caamaño, A. J. 1988. An ocean freight cost analysis for Chilean forest product exports. In Johnson and Smith (1988), pp. 44–52.

CINTRAFOR. 1993. Radiata pine: a competitive force in Asian markets. *Fact Sheet.* Center for International Trade in Forest Products. University of Washington, Seattle.

Daley, H. E. 1993. Problems with free trade: neoclassical and steady-state perspectives. In Zaelke et al. (1993), pp. 147–157.

[8]Because supply and demand curves are drawn with the same slopes, net gains from trade are the same in both countries. That wouldn't usually be the case, since differing supply and demand elasticities affect the import demand and export supply curve elasticities in Figure 17-7(b). The more elastic the import demand curve, the smaller the importing country's gain. And more elastic export supply will decrease the exporter's gain.

Food and Agriculture Organization of the United Nations. 1993. *Yearbook of Forest Products—1991.* FAO Forestry Series No. 26. Rome, Italy.

Food and Agriculture Organization of the United Nations. 1982. *Yearbook of Forest Products—1980.* FAO Forestry Series No. 15. Rome, Italy.

Hyde, W. F. 1992. Social forestry: A working definition and 21 testable hypotheses. In Nemetz, P. N. (Ed.). 1992. *Emerging Issues in Forest Policy.* University of British Columbia Press, Vancouver. pp. 430–452.

Ince, P. J. 1994. *Recycling and Long-Range Timber Outlook.* U.S. Forest Service Rocky Mt. Forest and Range Experiment Station. Gen. Tech. Rept. RM-242. Fort Collins, CO. 23 pp.

International Tropical Timber Organization. 1994. Objectives of the new international tropical timber agreement. *Tropical Forest Update.* 4(2):13.

Johnson, J. A., and C. E. Dinsdale. 1993. *The Challenge of Change, the Inertia of Tradition: A View of Opportunities in European Wood Markets.* CINTRAFOR Working Paper 44. University of Washington College of Forest Resources, Seattle. 41 pp.

Johnson, J. A., and W. R. Smith (Eds.). 1988. *Forest Products Trade: Market Trends and Technical Developments.* University of Washington Press, Seattle. 244 pp.

Kanowski, P. J., and P. S. Savill, with P. G. Adlard, J. Burley, J. Evans, J. R. Palmer, and P. J. Wood. 1992. Plantation forestry. In Sharma (1992), pp. 375–401.

Kreinen, M. E. 1991. *International Economics—A Policy Approach.* Harcourt Brace Jovanovich, San Diego. 504 pp.

Laarman, J. G., and R. A. Sedjo. 1992. *Global Forests—Issues for Six Billion People.* McGraw-Hill, New York. 337 pp.

Mohd, R., and J. G. Laarman. 1994. The struggle for influence—US nongovernmental organizations and tropical forests. *Journal of Forestry.* 92(6):32–36.

Myers, N. 1992. *The Primary Source—Tropical Forests and Our Future.* W. W. Norton & Co., New York. 416 pp.

Nautiyal, J. C. 1988. *Forest Economics—Principles and Applications.* Canadian Scholars' Press, Inc., Toronto. 581 pp.

Panayotou, T., and P. S. Ashton. 1992. *Not by Timber Alone.* Island Press, Washington, DC. 280 pp.

Perez-Garcia, J. M. 1992. Global trade model linkages to environmental impacts. In Adams, D., R. Haynes, B. Lippke, and J. Perez-Garcia (Compilers). *Forest Sector, Trade and Environmental Impact Models: Theory and Applications.* Proceedings of a Symposium. CINTRAFOR. University of Washington, College of Forest Resources, Seattle. pp. 15–23.

Perez-Garcia, J. M. 1993. *Global Forestry Impacts of Reducing Softwood Supplies from North America.* CINTRAFOR Working Paper 43. University of Washington College of Forest Resources, Seattle. 35 pp.

Perez-Garcia, J. M. 1995. Market transitions in China and the former Soviet Union: Implications for the global forest sector. *Proceedings of the Southern Forest Economics Workshop.* North Carolina State University College of Forestry. In press.

Repetto, R. 1993. Complementarities between trade and environment policies. In Zaelke et al. (1993), pp. 237–246.

Repetto, R. R., and M. Gillis (Eds.). 1988. *Public Policies and the Misuse of Forest Resources.* Cambridge University Press, Cambridge, UK. 432 pp.

Sedjo, R. A. 1983. *The Comparative Economics of Plantation Forestry—A Global Assessment.* Resources for the Future, Washington, DC. 161 pp.

Sedjo, R. A., and K. S. Lyon. 1990. *The Long-Term Adequacy of World Timber Supply.* Resources for the Future, Washington, DC. 230 pp.

Sedjo, R. A., A. C. Wiseman, D. J. Brooks, and K. S. Lyon. 1994. *Global Forest Products Trade: The Consequences of Domestic Forest Land-Use Policy.* Discussion Paper 94-13. Resources for the Future. 38 pp.

Sharma, N. P. (Ed.). 1992. *Managing the World's Forests.* Kendall/Hunt Publishing Co., Dubuque, IA. 603 pp.

U.S. Forest Service. 1978. *Forest Statistics of the U.S., 1977.* Washington, DC. 133 pp.

Vincent, J. R. 1992. The tropical timber trade and sustainable development. *Science.* 256:1651–1655.

Waddell, K. L., D. D. Oswald, and D. S. Powell. 1989. *Forest Statistics of the United States, 1987.* Resource Bulletin PNW- RB-168. U.S. Forest Service. Pacific Northwest Research Station. Portland, OR. 106 pp.

Westoby, J. 1989. *Introduction to World Forestry.* Basil Blackwell, Ltd., Oxford, UK. 228 pp.

Wisdom, H. W. 1993. An empirical model for estimating the trade benefits and environmental costs of removing a lumber export ban—A case study of the Philippines. *Proceedings of the Southern Forest Economics Workshop.* Duke University. Durham, NC. Pp. 52– 61.

Wisdom, H. W. 1995. NAFTA and GATT—What do they mean for forestry? *Journal of Forestry.* 93(7):11–14.

Woodwell, G. M. 1992. The role of forests in climatic change. In Sharma (1992), pp. 75–91.

World Resources Institute. 1994. *World Resources, 1994–95.* Oxford University Press, New York. 400 pp.

Zaelke, D., P. Orbuch, and R. F. Housman (Eds.). 1993. *Trade and the Environment—Law, Economics, and Policy.* Island Press, Washington, DC.

GLOSSARY

accumulated value See **future value.**
activity day One person visiting a recreation site for any part of a calendar day.
ad valorem tax A tax levied as a percentage of asset value (see **property tax**).
afforestation General term for establishing trees through planting or seeding, whether or not trees were on the site before. *Reforestation* establishes trees on formerly forested areas.
agroforestry Combinations of trees, other crops, and/or livestock on the same area.
alternative rate of return The percent **rate of return** on capital in an investor's best alternative, at a risk similar to new ventures being considered.
amenity rights Society's right to enjoy amenities such as clean air and water, attractive landscapes, and peace and quiet.
amortization The process of gradually reducing some monetary amount over time; can refer to income tax calculations where some cost is gradually deducted over time according to an *amortization schedule.* Loans may be paid off or *amortized* over a given time period.
annualized cost (or **revenue**) An equal annual payment with the same present value as payments that are not annual. May be calculated for a fixed or infinite time horizon [see equations (6-10) and (6-11) and **equivalent annual annuity**].
annuity Equal payments at regular intervals, for example, monthly or yearly.
anthropocentric Human centered.
appraisal The procedure for finding **market value** of an asset.
arc elasticity An **elasticity** of a two-variable function calculated as an average between two points on the curve.
assessed value Taxable value to which the **property tax** rate is applied; may be market value or a certain proportion thereof.
average consumer surplus With reference to a demand curve, the total area under the curve divided by the quantity demanded at a zero price.
average willingness to pay Consumers' total willingness to pay, in dollars, for a given number of units of a good divided by that number of units. With reference to a demand curve, average willingness to pay is the area under the curve up to a given consumption level Q, divided by Q.

basis For income tax calculations, that part of an asset's original purchase cost that is deducted from asset sale price in calculating taxable capital gain. See the section entitled "Depletion" in Chapter 9.

benefit/cost ratio (also called *profitability index*) The present value of a project's benefits (revenues) divided by the present value of costs, using the investor's **minimum acceptable rate of return** (MAR).

bequest value Our willingness to pay for the opportunity to transfer resources to future generations.

biocentric Centered on biology and ecology.

biological rotation A rotation age based on a biological, not economic, criterion—usually based on maximum **mean annual increment.**

biological timber supply shifters Factors like fertilization, planting, precommercial thinning, and genetic improvement, which can change the physical amount of wood in the future.

board foot A measure of timber or lumber volume: a board foot contains the equivalent of an unfinished board measuring 1 foot $\times$ 1 foot $\times$ 1 inch, or $\frac{1}{12}$ of a cubic foot. Actual finished board measures will be less; for example, finished 1-inch lumber is usually $\frac{3}{4}$ inches, and a 2 $\times$ 4 is about 1.5 inches $\times$ 3.5 inches.

bond A certificate indicating a debt issued by a government or private corporation. In return for money paid, the bondholder receives from the issuing agency either an **annuity** and a lump sum at the end of a specified period or a larger lump sum at the end of a period, without an annuity. These returns yield a specified **nominal** interest rate on the original payment.

book value Original purchase cost.

capital As defined here, any store of wealth yielding satisfaction to its owner—includes durable goods, financial assets, land, and natural resources.

capital budgeting Deciding how to invest money, the "capital budget," so that its value to the investor is maximized.

capital gain Difference between the sale price and the purchase cost of an asset.

capital recovery formula The formula giving the equal annual payment with a present value equal to some specified amount [see equation (4-12)].

capitalization rate See **discount rate.**

capitalize To find the present value or to discount. A second meaning is, in income tax calculations, to carry forward a capital expense and deduct it from sale proceeds of an asset to find taxable income.

certainty-equivalent The certainty-equivalent of a risky expected revenue (cost) is the dollar amount that, if received (incurred) with certainty, would give someone the same satisfaction (dissatisfaction) as the risky revenue (cost).

certainty-equivalent present value The **certainty-equivalent** of a risky future cash flow discounted with the risk-free discount rate. This is considered the correct present value of a future risky cash flow.

certainty-equivalent ratio With respect to a risky cash flow of a given **expected value,** the **certainty-equivalent** divided by the expected value.

coefficient of variation For a set of variable values, the **standard deviation** divided by the **expected value.** This is a measure of variability that is useful for comparing variations in populations with different expected values.

commercial timber Standing timber that can be sold at a profit for wood products and is available for harvest, for example, not in National Parks or wilderness areas.

comparable sales appraisal A method of finding an asset's market value by averaging recent sales prices of similar assets, similarly situated.

compound interest Earnings accruing as a percentage of capital value such that earnings occur on the original capital and on all previous earnings, as with interest on a savings account.

compounding Refers to the process whereby a current capital investment (present value) grows over time to a larger **future value.**

concentration ratio The percentage of an industry's total output produced by the largest firms, usually the top four or eight firms.

conservation easement A purchased right to land, less than full ownership, that allows the holder to use or protect the land for purposes like outdoor recreation, education, open space, wildlife habitat, or preserving certain ecosystems.

constant dollars Values expressed in **real** dollars of some base year, excluding inflation.

consumer surplus Consumers' extra **willingness to pay,** in dollars, above what they actually pay for a good. In a demand framework, this surplus is the area above the price line and below the demand curve.

consumer price index (CPI) An index of average prices for a typical market basket of consumer goods. The index is set at 100 for a specified base year. The annual rate of change in the CPI is the **inflation** rate for consumer goods.

contingent valuation method Using either interviews or mail surveys, a way to value **nonmarket** goods and services by asking users the maximum amount they'd be willing to pay for them (**willingness to pay**) or what minimum compensation they'd require to willingly give them up (**willingness to sell**).

conversion return For standing timber, the manufactured end product minus all costs (including profit) of manufacturing, distribution, transportation, and logging. A residual value that is a logger's estimated maximum willingness to pay for stumpage just before harvest.

cost of capital The interest rate firms pay on capital raised for investment. This can be interest on debt (the *cost of debt*) or dividends on stock sold (the *cost of equity*). Also see **weighted average cost of capital.**

cost-push inflation **Inflation** caused by rising prices of inputs such as oil or labor.

cross-price elasticity of demand The percent change in quantity of one good demanded divided by the percent change in the price of another good. When goods are substitutes (complements), cross-price elasticity is positive (negative).

current dollars Values in dollars of the year in which they actually occur, including inflation. Also known as **nominal** values.

cutting cycle In **uneven-aged** management, the number of years between **partial cuts.**

damager liability A bargaining framework allowing no uncompensated environmental damage—where damagers are liable for damage reduction and must compensate damaged parties.

deflate To deflate a **current dollar** value means to express it in **constant dollars** of a base year n years earlier, removing inflation. Divide the current dollar value by $(1+f)^n$, where f is the average inflation rate/100 over the n-year period.

demand In economics, demand refers to the quantities of a good or service that will be consumed per unit of time at different prices, for a given group or individual. A *demand curve* (function) shows this relationship graphically (mathematically) with price on the y axis and quantity on the x axis.

demand-pull inflation **Inflation** caused by too many buyers chasing too few goods; thus, prices are bid higher.

density With reference to timber stands, a measure of stem crowding, for example, number of trees per acre, percent crown closure, or volume per acre.

depletion In income tax calculations, the deductions made for original purchase cost, when assets are sold.

depletion period That period over which a given timber tract is completely harvested.

depreciation An accounting charge for the wearing out of certain assets.

derived capitalization rate The discount rate you estimate the average buyer would have used in computing the price paid for a property. For recent sales of one property type and location, it is the average rate that discounts projected cash flows to exactly equal the sales price, using a consistent cash flow projection approach.

derived demand The demand for an intermediate good that is derived from consumer demand for final products. Thus, sawtimber stumpage demand is derived from demand for housing, furniture, and certain other solid wood products.

development period The number of years it takes after regeneration before partial cutting begins in **uneven-aged** forestry.

diminishing marginal utility In consuming any good or service, the principle that, beyond some point, consuming additional units per unit of time will bring less and less added satisfaction to one individual or one group.

diminishing returns In a production process with certain inputs fixed, the range in which the *added* (marginal) output declines as you add units of one variable input.

direct effects Income and employment resulting directly from constructing and operating a project; does not include **indirect effects** of activities stimulated elsewhere.

discount rate The interest rate at which future values are discounted to present values.

discounted cash flow See **present value.**

discounting The process whereby a **future value** is reduced to arrive at the **present value.**

divisible An investment is divisible if you can invest in part of it, as with adding money to a savings account or acres to a fertilization project (as opposed to **indivisible**).

econometrics A field involving measurement of economic variables; usually applies statistical techniques to estimate demand and supply functions and predict economic outcomes.

economic profit See **excess profit.**

economics The study of how best to allocate or distribute resources to maximize human well-being.

elastic If the absolute value of the **price elasticity** on a demand or supply curve is greater than 1 in some range, the curve is "elastic" in that range.

elasticity In a two-variable function, the percent change in one variable divided by the percent change in another, over some specified range. For example, see **price elasticity of demand.**

equi-marginal principle An axiom saying that you'll maximize the total benefit from using any limited input if you allocate it so that the last unit of input in each activity brings the same added benefit. In any application, the benefit must be some common unit such as income, utility, or sales; the input must be a common unit such as dollars, time, or fertilizer.

equilibrium A state in which some economic variable like price or quantity is at rest with no tendency to change—in a sense a theoretical artifact, since most economic systems are dynamic, with continuing changes in population, tastes, crops, trade, etc. Thus, equilibria can be dynamic processes toward which economic entities are moving.

equity That portion of a firm's assets on which no debt is owed to creditors.

equivalent annual annuity (EAA) An equal annual real income with the same present value, over a project's life, as the project's net present value, all computed at the same **real** discount rate.

even-aged Refers to forests in which trees have been established at about the same time and are thus roughly the same age.

excess profit As defined in this book, the profit or net income *above* the normal or average profit expected from the same amount of capital invested elsewhere with similar risks. This excess profit is usually called **economic profit** in the economics literature.

exchange rate The rate or price at which one country's currency is exchanged for another's. For example, if the United States dollar costs 140 Japanese yen, the exchange rate for the dollar is 140 yen.

existence demand or **existence value** Consumers' **willingness to pay** for the assurance that something remains in existence, even though they'll probably never use or consume it.

expected utility The average utility expected from a risky benefit. For example, if the benefit is an expected revenue, its expected utility is the sum of the possible utilities multiplied by their probabilities of occurrence.

expected value The mean value of some variable in a given population: The sum of the possible values multiplied by their probabilities of occurrence (not always equal to the arithmetic mean). In this sense we can speak of, say, an *expected revenue* or an *expected cost.*

expensing In income tax calculations, the practice of *deducting* or subtracting allowable costs from income to arrive at taxable income.

export supply curve Shows amounts of a good available for export from a country at different prices.

external diseconomies See **unpriced negative side effects.**

external economies See **unpriced positive side effects.**

externalities For negative externalities, see **unpriced negative side effects.** For positive externalities, see **unpriced positive side effects.**

fee timber Standing timber that a firm owns outright on its lands (derived from the legal term "ownership in fee simple").

financial maturity The age beyond which an asset's growth rate is unacceptable or less than the owner's **minimum acceptable rate of return;** can refer to a forest (timber and land combined) or to individual trees. Assets at or beyond this age are *financially mature.*

fixed costs Costs that remain fixed as a firm's output increases. For example, in the short run, if a firm increases plywood output from a given sized mill, property taxes and fire insurance on the mill will be fixed costs.

fixed-income investment An investment from which annual income remains fixed in **nominal** terms (**current dollars**), regardless of the inflation rate—as opposed to a **variable income investment.**

fixed-payment loan A loan on which payments at regular intervals are fixed in **nominal** terms.

forest In financial discussions, forest refers to land and trees combined.

forest value growth percent Annual percent rate of change in the **liquidation value** of trees *and* land.

full employment In reality, 100 percent employment will not occur because some workers will be between jobs. Many economists now feel that at "full employment," we are unlikely to get below a **natural rate of unemployment** of about 6 percent.

future value The value of any income or wealth accumulated with compound interest to a specified future date. A future value is a current value multiplied by $(1 + r)^n$, where r = (interest rate)/100, and n is the number of years in the future.

general equilibrium analysis Analyzing dynamic economic systems in a way that considers interaction between many variables over time.

government failure Government actions that keep an economy from attaining a **welfare maximum.**

gross domestic product (GDP) The market value of all goods and services produced by residents of a nation in a year; excludes income that citizens or firms of that nation earned or received abroad.

gross income Total revenue before subtracting costs.

guiding rate of interest A minimum **rate of return** that projects must earn to be acceptable to an investor.

hedonic travel cost model Based on the theory that demand for any good can be separated into demand for the good's characteristics, this model arrives at values for changes in characteristics of recreation areas, for example, acres of water, hunting success, or average size of fish caught.

Herfindahl index A measure of industrial concentration; the sum of squared **market shares** for all firms in an industry.

holding value The owner's **net present value** of future cash flows from an asset.

hurdle rate A minimum acceptable **rate of return** or "hurdle" that new investments must clear before they are acceptable to an investor.

identification problem In dynamic supply-demand analysis, the difficulty in determining whether a projected price-quantity point is a movement along a supply or demand curve or an intersection of supply and demand curves that have both shifted.

imperfect competition A situation where one or more sellers and/or buyers are not **price takers** and are thus large enough to affect prices of outputs or inputs. Such sellers face downward-sloping demand curves for their outputs and/or such buyers face upward-sloping supply curves for their inputs. **Perfect competition** does not hold.

import demand curve Shows amounts of an imported good demanded by a country at different prices.

income appraisal A method of finding an asset's market value by calculating the net present value of estimated future cash flows therefrom. Also called the *income approach.*

independent Investment options are independent if one option does not preclude the other, for example, fertilization and **precommercial thinning** on the same timber stand.

indirect effects When a new project is started, indirect effects in the region are the resulting new employment and income in industries supplying and servicing anything related to the project.

indivisible An investment is indivisible if it is "all or nothing," as with a vehicle.

industrial wood Sawlogs, veneer logs, poles, pulpwood, and chips—all destined for nonfuel uses.

inelastic If the absolute value of the **price elasticity** on a demand or supply curve is less than 1 in some range, the curve is "inelastic" in that range.

inflation A general increase in prices of all goods and services in an economy, usually expressed as an annual *inflation rate:* a percentage increase per year. This rate is only an average, since not all prices will change at the same rate.

inflation indexing A procedure that adjusts a value or cash flow to increase at the inflation rate.

input-output analysis A technique for measuring interdependencies between different sectors of an economy and making economic forecasts.

input-output table For one region, a table showing the values of sales between different economic sectors or industries.

interest The payment made to lenders of money—often expressed as an *interest rate,* or annual percentage of the outstanding amount loaned. Also refers to earnings on invested capital.

internal rate of return (IRR) For a given project, the interest rate at which the present value of revenues equals the present value of costs (see **rate of return**).

joint cost A cost that produces more than one output is a joint cost if dropping one of the outputs doesn't change the cost. For example, the cost of a forest road can jointly produce access for recreation, fire protection, and timber harvesting.

land expectation value See **willingness to pay for land.**

law of one price Under free trade, the tendency for the price of any traded good to become the same in all areas, except for differences caused by shipping costs.

leakage That part of a region's income that is spent outside the region, for example, on federal taxes, imported goods, or travel.

lessee In a leasing contract, the party making lease payments in return for using the leased property.

lessor In a leasing contract, the party owning the leased property and receiving lease payments.

leveraged Describes a purchase made in part with borrowed money.

liability rule A rule for legally or conceptually considering who is liable for preventing environmental damage: the victim or the damager. See **victim liability** and **damager liability.**

liquidation value For any asset, the value to be received for immediate sale for consumption purposes; for standing timber, the stumpage value for harvesting purposes.

long run An economics term referring to the period of time over which all inputs are variable.

long-run demand curve Shows the amounts of a good that a given group of buyers would demand at different prices in the **long run.**

long-run supply curve Shows the amounts of a good that a given group of producers would supply in the **long run** at different prices.

long-term capital gain Income from selling assets held longer than one year; a definition used in United States income tax laws, as of 1995.

marginal In economics, added or extra, as opposed to total.

marginal cost The *addition* to total cost when you increase *output* by one more unit.

marginal rate of return The rate of return on the last dollar invested in a given enterprise.

marginal resource cost The *addition* to total cost when you add another unit of *input.*

marginal revenue The *addition* to total revenue when you increase *output* by one more unit.

marginal revenue product The *addition* to total revenue from adding another unit of *input.*

market failure A situation preventing a free market system from attaining a **welfare maximum.**

market power A firm has market power if it faces a downward-sloping demand curve for its output and/or an upward-sloping supply curve for an input. Such a firm is not a "price taker" and can influence prices to some degree (see **perfect competition,** where firms have no market power).

market share One firm's total annual sales as a percentage of sales in a given market.

market value An item's most likely selling price; an average expected selling price for similar items, similarly situated.

mean annual increment (MAI) Average annual timber volume growth per unit area, for example, in cubic feet per acre per year. Assuming even-aged management with clear-cutting, MAI is potential **rotation**-age harvest volume per acre divided by harvest age.

merchantable Refers to trees salable for commercial products. Trees to be cut in a **precommercial thinning** are unmerchantable.

minimum acceptable rate of return (MAR) The lowest rate of return that will induce an investor to willingly invest capital rather than spend it.

model A simplified representation of an actual process, situation, or object; also used as a verb (to model).

money illusion The false impression that inflationary increases in values represent real increases in purchasing power.

monopolist The only producer of a good or service in a market. This situation is called *monopoly.* The monopolist faces a downward-sloping demand curve for its output, so the more it sells, the lower the price.

monopsonist The only buyer of an input in a market. This situation is called *monopsony.* The monopsonist faces an upward-sloping input supply curve, so the more it buys, the higher the price.

Monte Carlo simulation A computer-based procedure for simulating uncertain or probabilistic processes.

mortgage A loan, usually for purchasing real estate, where the property becomes collateral to guarantee payment of the loan.

multiple-use forestry Using a forest for many different purposes: can mean many uses on each acre, a mosaic of single uses, one dominant use and other compatible uses, different uses over time, or a combination of these.

multiplier effect The multiplied amounts of income, employment, or sales beyond the initial amounts generated directly from a new project. A *multiplier* of, say, 1.75 for new employment in one industry of a region means that 100 new jobs in that industry will eventually lead to 175 jobs in the region, including the original 100.

mutually exclusive Options are mutually exclusive if choosing one precludes the other, for example, planting a pure hardwood or pure softwood stand on the same area.

natural rate of unemployment The lowest unemployment rate at which inflation is stable at some acceptable level. Many believe that the natural rate of unemployment is currently about 6 percent.

net income Total revenue minus total cost; here, same as **profit.**

net present value (NPV) Present value of future revenues minus present value of future costs (see **present value**).

neutral tax A tax that does not tend to change the way the market allocates resources.

nominal With respect to values or rates of return, in **current dollars,** including inflation.

nonmarket Not traded in the market for a price.

nonmonetary See **nonmarket.**

normal profit The average profit or rate of return on capital that could be expected from an investment under competitive conditions, at a given risk level. Also called *normal return to capital.*

normalize You can normalize investment projects for unequal size by simulating added capital to the smaller project until it equals the larger project's size. You can normalize for unequal lives by simulating repetitions of the shorter project until it equals the length of the longer one. This enables more valid project comparisons.

old growth From an economic view, standing timber so old that net annual volume growth is negligible.

oligopoly A situation with a few large firms, each facing an output demand curve sloping downward but not as steeply as that which would be faced by a **monopolist** for the same market.

oligopsony A situation where a few large resource buyers *(oligopsonists)* face resource supply curves that slope upward, but not as steeply as those of a **monopsonist.** The more of the resource they buy, the higher its price will be.

opportunity cost The cost of an investment opportunity forgone; in financial terms, the best interest earning rate or dollar income that could be earned on a given amount of capital if not in its current use. Can also be in nonmonetary terms; for example, one opportunity cost of an intensively managed timber production area could be certain recreation benefits forgone. The opportunity cost of a wilderness area is the potential net income from marketed products forgone.

optimal rotation age As defined here, the tree harvest age at which net present value of a forest is maximized; may include timber and nontimber values. Refers to mature timber, but not old growth and not thinnings.

optimization The process of finding an optimum.

optimum In a production process, the point at which you have maximized some output such as satisfaction (utility), income, present value, profit, volume, etc. The verb is "to optimize"; the adjective is "optimal." Optimum is used as both a noun and an adjective.

option demand or **option value** Consumers' **willingness to pay** for the option or chance to use some resource in the future, even if they don't use it now.

ordinary income In income tax calculations, any income that is not a **capital gain.**

oriented strand board (OSB) A board made of thin wood strands about ⅝ by 2 ⅛ inches glued together under heat and pressure. Three layers of strands are oriented at right angles to one another.

overrun An extra volume that a sawmill can cut in excess of the originally measured stumpage or log volume, due to measurement inaccuracies.

Pareto optimum A resource allocation where no change could make anyone better off without making someone else worse off; not a true welfare maximum because it ignores changes from which gainers could overcompensate losers (see **welfare maximum**).

partial cutting Any harvest removing part of the timber without clear-cutting.

partial equilibrium analysis Analysis of economic systems that considers the effects of relatively small changes in one variable, holding other variables constant ("other things being equal").

payback period The number of years it takes to recover the initial capital invested in a project.

payoff period The number of years between initial investment and the final yield for investments with primarily one input and one output.

perfect competition A case where each producer, relative to the market, is so small that no matter how much it sells, its output price is the same (or nearly so). Such sellers face horizontal demand curves for outputs. Also, each buyer of an input must be so small that no matter how much it buys, the input price is the same. Such buyers face horizontal supply curves for inputs. Such producers and buyers are **price takers.**

periodic Defined here as occurring at regular intervals of more than one year.

plywood A wood panel made of rotary-peeled veneers glued together under heat and pressure with grains of adjacent sheets perpendicular.

point elasticity An **elasticity** of a two-variable function calculated at one point on the curve.

precommercial thinning A thinning where *stumpage prices* are negative, so that cut trees are left in the woods (logging and hauling costs would exceed the delivered log price).

present value Any future value can be discounted to a present value by dividing it by $(1 + r)^n$, where r = (interest rate)/100, and n is the number of years in the future. Sometimes called *present worth.* This discounting reflects the cost of waiting to receive a future income: the further a given amount of income is in the future, the less it's worth to you now. Discounting is the reverse of **compounding.**

price discrimination A situation where a seller charges different prices to different buyers, or where a resource buyer pays different prices to various suppliers.

price elasticity of demand With respect to a demand curve, the percent change in quantity of a good demanded with a 1 percent change in its price. Or more specifically, the percent change in quantity demanded divided by the percent change in price, over a specified range. If the range is narrowed to one point on the demand curve, elasticity is **point elasticity;** if between two points, it is **arc elasticity.**

price elasticity of supply With respect to a supply curve, the percent change in output of a good supplied divided by the percent change in output price—or the percent change in output supplied with a 1 percent change in output price.

price taker See **perfect competition.**

prime interest rate The interest rate banks charge on loans to businesses with the best credit rating—often called the "prime rate."

probability distribution A graphic or mathematical function showing the probabilities that the value of some variable will fall within given ranges. Also called a *probability density function.*

probability histogram A bar graph showing on the y axis the probabilities of occurrence for given ranges of values for some variable on the x axis. Also called a **probability distribution.**

producer price index (PPI) An index of average prices for a mix of industrial outputs, excluding services, for each year. The index is set at 100 for a specified base year. The annual rate of change in the PPI is the average inflation rate for producers' outputs.

producer surplus The extra income received by some firms above the minimum required to produce current output. For a group of firms producing the same product, producer surplus is the area above the industry supply curve and below the price line.

production function A mathematical or graphic function showing how physical output changes as certain inputs are changed.

productivity tax A forest tax, usually in place of the property tax, levied annually as a percentage of a "productivity value," regardless of timber stocking. The taxable productivity value is based on a formula reflecting the productive potential of the land.

profit In financial operations, total revenue minus total cost (here, costs do not include **opportunity cost** of capital). Thus, profit is a return to capital.

property tax An annual tax levied as a percentage of property value, primarily on real estate and natural resources, although other assets may be taxed, depending on localities—also called an **ad valorem tax.**

public good A good or service not easily parceled out and sold at a price to individuals. You can't exclude those who don't pay for the good from receiving its benefits; more for one person doesn't mean less for another. Examples are national defense and, in forestry, scenic beauty along public roads.

rate of return Earnings on capital, expressed as an annual percent of capital value (see **compound interest** and **internal rate of return**).

rate of time preference The interest rate at which someone discounts future values.

real With respect to monetary values, excluding inflation.

regeneration Process by which trees are reestablished, for example, by planting, natural or artificial seeding, or sprouting.

regression analysis A statistical technique for deriving a *regression equation* that will predict the value of some dependent variable as a function of one or more independent or "predictor" variables.

regulated forest A forest with roughly equal areas of even-aged timber in each age class so that equal harvests could continue in perpetuity every year (or every two years, five years, etc., depending on the spacing of age classes).

reinvestment rate The **rate of return** at which you assume future income from a project could be reinvested.

replacement cost appraisal Finding market value of an asset by estimating its replacement cost minus any value loss due to **depreciation.**

reservation price The minimum stumpage price that will induce a forest owner to sell owned timber for harvest.

risk Refers to the variation in an expected cash flow. The greater the variation, the greater the risk. In a general sense, risk refers to the possibility of loss. Under risk, probabilities of occurrences can be calculated; under **uncertainty,** probabilities are not known.

risk-adjusted discount rate The interest rate for discounting risky cash flows. For *risk-averse* investors, the risk-adjusted discount rate for revenues exceeds the **risk-free** rate.

risk aversion The case where an investor dislikes greater variation or risk in an expected cash flow. A *risk-averse* investor prefers a sure cash flow to a risky one of the same **expected value.**

risk-free Sure or certain; involving no **risk.**

risk-free discount rate The **interest** rate for discounting sure cash flows with no risk. One estimate of a long-term risk-free rate is 3 percent.

risk-neutral Describes an investor who is indifferent between a sure cash flow and a risky one of the same **expected value.**

risk premium The percentage points that a risk-averse investor adds to the risk-free discount rate to account for risk in future revenues.

risk seeker An investor who prefers a risky cash flow over a sure one of the same **expected value.**

rotation Age, in years, at which mature timber is harvested.

roundwood Harvested wood in round or log form.

sawtimber Live trees capable of yielding sawlogs for lumber production.

sensitivity analysis Numerical analysis that tests how sensitive some outcome is to changes in selected variables. For example, you could test how a project's net present value varies as you change the discount rate.

shelterwood cutting Harvesting a mature stand in a series of partial cuts toward the end of the **rotation.** During this time an even-aged stand reproduces in the shelter of the original stand. The final cut occurs when new trees are well established.

shifters Factors that shift demand or supply curves to the left or right. For example, greater per capita income of a group can shift that group's demand curve for a commodity to the right. Also see **tax shifting.**

short run An economics term for the period of time during which some inputs are fixed.

short-run demand curve Shows the amounts of a good that a given group of buyers would demand at different prices in the **short run,** before any demand **shifters** change.

short-run supply curve Shows the amounts of a good that a given group of producers would supply in the **short run** at different prices, before any supply **shifters** like capital and technology can change.

shortage In economics, a shortage exists for any commodity when the amount demanded exceeds the amount supplied at a given price.

simple betterness method A way to rank management options in order of decreasing dollar value accompanied by **nonmarket** benefits ranked by preference. The approach lets you eliminate "dominated" options that have the same or less dollar value accompanied by less nonmarket value than some superior option.

simple interest A percent interest earning rate where interest is earned only on the original capital (the principal) and not on previously accrued interest.

site demand curve A demand curve reflecting some group's **willingness to pay** an entry fee to a recreation site. For example, the *y* axis could be $/visit and the *x* axis visits per unit of time.

site value tax A **property tax** levied on the market value of land only, excluding the value of improvements.

social forestry Planting and managing trees by or for communities for local consumption, generally in developing nations.

soil expectation value See **willingness to pay for land.**

stand A group of standing live trees.

standard deviation A measure of the spread in possible values of a variable in a population. The standard deviation is the square root of the **variance.**

stocking A general term for the quantity of trees on an area, especially with reference to some management objective.

structural panel A wood panel—mainly softwood **plywood, waferboard,** or **oriented strand board**—used in building construction.

stumpage Standing trees in the forest.

stumpage value The estimated or actual amount that buyers would pay for standing timber for immediate harvesting—also called timber **liquidation value.** This amount expressed as a sum per unit volume is called *stumpage price.*

subsidy Favored treatment that a government gives to selected private enterprises. Examples are reduced taxes, direct financial or technical assistance, and guaranteed minimum output price levels.

sunk costs Costs that have been already incurred.

supply In economics, supply refers to the quantities of a good or service that a producer or group of producers will supply per unit of time at different prices. A *supply curve* (function) shows this relationship graphically (mathematically) with price on the *y* axis and quantity on the *x* axis.

surplus In economics, a surplus exists for any commodity when, at a given price, less is demanded than suppliers are willing to supply.

sustained yield A commitment to continued, long-term wood output. A stricter definition requires an even flow of timber over time; also can apply to nontimber outputs.

tariff A tax on goods imported into (or sometimes exported from) a country, either per unit or as a percentage of value; also called a "duty."

tax capitalization Refers to the *shifting* of taxes into lower property values.

tax credit A direct reduction in taxes. For example, a $200 income tax credit will reduce someone's income taxes by $200.

tax shifting The process whereby taxes affect prices: taxes may be *shifted* into lower values of land or other assets, into higher output prices, into lower rates of return, or into a combination of these directions.

technical coefficient In input-output analysis, direct purchases per dollar of sales by one industry from another industry.

technological timber supply shifters Improved processing technology that increases the usable volume of existing stumpage available at any price, for example, utilizing trees to smaller top diameters.

thinning The practice of removing some trees in a crowded stand so that the remaining ones can grow faster. Types of thinning are *thinning from above* (removing mainly the tallest or "dominant" and "codominant" trees), *thinning from below* (removing trees from the understory), *row thinnings* (removing selected rows in plantations), or combinations of these types.

thinning regime A prescribed pattern of thinning over time; for example, cut 30 percent of the volume at ages 15 and 25, and clear-cut at age 35.

timber Refers to trees alone, excluding land, for purposes of producing forest products.

timberland Forest area available for commercial timber production.

timber stand improvement (TSI) An existing stand investment that increases future harvest revenue. Examples are fertilization to yield more timber growth, precommercial thinning and improvement cutting to distribute future growth on the most valuable trees, and pruning to provide more valuable, knot-free wood.

timber supply Stumpage volumes that forest owners will harvest at different stumpage prices for a given region and year.

timber value growth percent The annual percent change in liquidation value of standing timber, excluding land.

time line A diagram of expected revenues and costs for a project, shown on a horizontal scale of time (say, years or months).

total revenue For a firm, total outputs times prices of outputs, *before* subtracting any costs; same as gross revenue.

trade-off In a system of interrelated inputs and outputs, a trade-off refers to the process whereby changing one output can change other outputs. Often, but not always, as you increase one output, another decreases: you trade one for the other.

travel cost method A way of estimating demand curves for nonmarket recreation by using simulated increases in travel costs as a surrogate price per visit.

trend line In a plot of y values over an x axis, a smoothed line drawn through the values—generally done with **regression analysis.** For example, a trend line can be drawn through a series of fluctuating prices over time, in which case you'd have *trend-line prices* and actual prices.

triangular probability distribution A triangular function with relative frequency on the y axis and an uncertain variable on the x axis: the triangle defines the lowest and highest possible values and the most likely value of the variable. The area under the triangle is 1.

trip demand curve In recreation area demand analysis, a curve plotting, for each zone of visitor origin, the travel cost per visit over annual visits per unit of population in the zone.

uncertainty The case where an outcome is not known for sure, and a probability cannot be attached to the outcome.

uneven-aged Refers to stands with trees of many ages. With *uneven-aged management,* cutting is only partial, and regeneration occurs beneath existing trees—sometimes called the "selection system."

unmodified property tax The current property tax rate applied to full market value of property annually.

unpriced negative side effects Uncompensated costs imposed by a producer on other parties ("external diseconomies" in economics). Examples are all forms of pollution.

unpriced positive side effects Free benefits to others resulting from a firm's production ("external economies" in economics). Examples are scenic beauty and climatic benefits to society resulting from a firm's tree planting.

utility In economics, human satisfaction or well-being. In graphs or equations, units of utility ("utils") are not absolute (cardinal) but are only relative (ordinal) in that one amount is greater than or less than another.

valuation The procedure for finding an individual investor's value of an asset; could be either asking price or bid price.

valuation factor A timber tract's **net present value** expressed as a proportion of its current **stumpage value.**

value added The difference between the sale price of goods sold and the cost of materials and supplies used in production.

value growth percent The annual percentage change in **liquidation value.**

variable A measure, the value of which can vary for different members of a population or sample. Examples are per capita income and weight for members of a group, or volume per acre for a group of 40-year-old red pine stands on a given site quality. The opposite of a variable is a constant.

variable costs Costs that change as a firm's output changes.

variable income investment An investment where unit prices of the output and the asset tend to rise at least as fast as the inflation rate (as opposed to a **fixed-income investment**). Examples are forests and farms.

variance A measure of the spread in possible values of a variable in a population: the greater the spread, the higher the variance. The variance of a population is the sum of squared deviations from the mean, each squared deviation weighted by its probability of occurrence. The variance in any variable value is also the square of the **standard deviation.**

veneer Wood that has been peeled, sawed, or sliced into thin sheets.

victim Defined here as the individual harmed by environmental damage.

victim liability A bargaining framework where damaged parties ("victims") are liable for environmental damage reduction, and we determine how much they'd be willing to pay to reduce damages.

volume growth percent The annual percentage change in timber volume for a tree or a stand.

waferboard A board made of randomly oriented wood chips glued together under heat and pressure.

welfare Human well-being; see **utility.**

welfare maximum An allocation of resources (land, people, and capital) such that no reallocation could yield a net gain. At this optimum, the losses from any change would exceed the accompanying gains.

weighted average cost of capital (WACC) A firm's average cost of obtaining capital from debt and sale of stock, expressed as an average **interest** rate weighted by the firm's percentages of debt and **equity.** For example, suppose a firm's assets are 40 percent debt and 60 percent equity, and it pays 10 percent interest on debt and pays a 6 percent dividend on stock. Its WACC is $0.40(10) + 0.6(6) = 7.6$ percent.

willingness to pay A maximum monetary amount an individual or group is willing to pay for a given good or service.

willingness to pay for land (WPL) Starting with bare land, WPL is the **net present value** of all future expected cash flows discounted at some **minimum acceptable rate of return** (MAR). Thus, WPL is the maximum a buyer can pay for land and still earn the MAR. The WPL is usually expressed in dollars per unit area, for example, $/acre. (In forestry, WPL is also known as **land expectation value** or **soil expectation value.**)

willingness to sell A minimum monetary compensation an individual or group would require to willingly give up some good or service and feel no loss of satisfaction.

with and without principle The true benefit from any added investment in a project is the **net present value** (NPV) of the project with the investment minus the NPV without the investment.

yield tax A tax levied as a percentage of harvested stumpage value.

APPENDIX A

ANSWERS TO ODD-NUMBERED NUMERICAL PROBLEMS (Solution procedures and other answers are in the Instructor's Manual.)

CHAPTER 1

1-1. $10.

1-5. 14 units.

CHAPTER 2

2-1. −0.5.

CHAPTER 4

4-1. $21,853.86.

4-3. $5.25.

4-7. $125.

4-9. $135.

4-11. $34.55.

4-13. $239.66.

4-15. 10 percent.

4-17. 9.7 years.

CHAPTER 5

5-1. 5.66 percent.

5-3. 6.7 percent.

5-5. 13.42 percent.

5-7. **(a)** $16,274.50.
(b) $15,353.30, $14,484.20, $13,664.40, $9,087.40.

5-9. $154.85.

5-11. $4,131 in 1993 dollars; $2,848 in 1983 dollars.

5-15. **(a)** Today's income is $19,672; today's forest value is $393,430.
(b) 5 percent.

5-17. Present value of bond income is $100,000.
Present value of forest income is $100,000.

CHAPTER 6

6-1. 4 percent.

6-3. 7 percent.

6-5. $1,096.

6-9. $600.

6-11. **(a)** $759.
(b) 190 nights.

CHAPTER 7

7-1. $320 per acre.

7-3. 8.7 percent.

7-5. 11.32 percent.

7-7. Forest value growth percent drops to 8.5 percent.

7-9. **(a)** 30 years.
(b) 40 years.

7-11. $1,120.

CHAPTER 8

8-5. **(c)** 2 percent per year.

8-7. NPV with uneven-aged management = $1, 514/acre.
NPV with even-aged management = $1, 718/acre.
Annualized cost = $13.35.

CHAPTER 9

9-1. $99.00.

9-3. $119.66.

9-5. Before-tax rate of return = 15 percent; after-tax rate of return = 12.6 percent.

9-9. 5.5 percent.

CHAPTER 10

10-3. $1,638.

CHAPTER 11

11-11. Bid price is reduced by $26,511.

CHAPTER 12

12-7. Rate of return on Baum's outlay is 8 percent.
Social rate of return is 5 percent.

CHAPTER 14

14-7. Total consumer surplus = $51,000.
Average consumer surplus = $6.54 per visit.

CHAPTER 16

16-3. 1.9.

INDEX